VOLUME I

6th INTERNATIONAL CONFERENCE ON CAD/CAM, ROBOTICS AND FACTORIES OF THE FUTURE 1991

Held at South Bank Polytechnic
103 Borough Road LONDON SE1 0AA U.K.

19th – 22nd August 1991

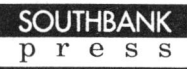

© South Bank Polytechnic, 1992

ISBN 1 874418 00 4 (this volume)
ISBN 1 874418 02 0 (complete set of 2 volumes)

South Bank Press
103 Borough Road
London, UK
SE1 0AA

British Cataloguing-in-Publication Data. A catalogue record for this book is available from the British Library.

All Rights Reserved. No part of this publication may be reproduced, stored in a retrieval system, or transmitted in any form or by any means, electronic, mechanical, photocopying, recording or otherwise, without prior written permission.

Printed and bound by Ashford Press, Southampton.

PREFACE

The unqualified success of the five previous conferences may be judged by the quality of contributions submitted to this, The Sixth International Conference on Computer Aided Design, Computer Aided Manufacture, Robotics and Factories of the Future.

This conference, organised by South Bank Polytechnic on behalf of the International Society for Productivity Enhancement (ISPE) was the first to be held in the United Kingdom. Previous conferences were held in the United States of America (4) and in India (1).

The objectives of the Sixth Conference were to bring together researchers interested in multi-disciplinary and inter-organisational productivity aspects of advanced manufacturing systems utilising Computer Aided Design/Manufacture/Engineering, Parametric Technology, Artificial Intelligence, Robotics, Sensors, Factories of the Future and to address productivity enhancement issues.

Over four hundred contributions from twenty six countries were received by the conference reviewing committee. These concentrated on the state of the art research and development, product development, and novel applications in the area of systems automation.

The nature and breadth of the papers selected for these proceedings demonstrates the vast and varied research that is advancing technology and having an enormous impact on manufacturing. The proceedings contain one hundred and forty-two contributions broadly categorised under sixteen chapters.

The conference organisers hope that you find the papers presented of interest, but perhaps more importantly, encourage researchers to continue their efforts in the knowledge that the future of manufacturing industries depends on their contributions.

HRISHI BERA

RAJ GILL

ACKNOWLEDGEMENT

I would like to express my gratitude to colleagues at the Department of Mechanical Engineering, Design and Manufacture and the staff of the Conference Office for their efforts and continuous assistance.

Much appreciation is owed for the co-operation of all members of the International Scientific Committee for providing constructive support in mounting this conference. I am most grateful.

My sincere thanks to Rt. Hon Baroness Perry for allowing the use of the facilities at the South Bank Polytechnic and for kindly accepting our invitation to welcome the delegates.

I wish to acknowledge with thanks the contribution of all the authors who presented their work at the conference. It is also my pleasure to recognise the role of the Plenary Speakers whose contribution added greatly to the success of the conference.

HRISHI BERA

CHAIR OF CONFERENCE

LETTER FROM THE PUBLICATION CHAIRMAN, I.S.P.E.

The International Society for Productivity Enhancement (ISPE) is entering its eighth year. The Conference you attended was our sixth of the International series on the CAD/CAM, Robotics and Factories of the Future (CARS & FOF). The fifth conference was held at Omni International Hotel, Norfolk, Virginia, Dec. 2-5, 1990, USA. In the last seven years, we have expanded our activities quite significantly. The membership interest and international participation are also growing. During the last couple of years, the society has made tangible progress in the following five frontiers:

JOURNAL The Society is publishing its own journal entitled The International Journal of Systems Automation - Research & Applications (SARA). SARA is an international, multidisciplinary research and applications-oriented journal to promote a better understanding of systems considerations in interdisciplinary automation using computers. The journal is an important reading for design, engineering and manufacturing persons and those with interest in research and development, and applications of productivity tools, concepts and strategies to multidisciplinary systems environments. The journal contains only original archival quality papers. To receive more information about the journal write to: Editor-in-Chief, ISPE - SARA Journal Department, P. O. Box 731, Bloomfield Hills, Michigan 48303-0731, USA.

PROCEEDINGS Starting with CARS & FOF '92 conference, the society is planning to make proceeding and SARA journal available for pick-up at the conference. Selected papers from this conference are also considered for publication in SARA.

CONFERENCES ISPE's annual conferences are now planned until 1995, Seventh International Conference will take place in I.A.E. International, Metz, France, August 17-19 1992. The Eighth, Ninth and Tenth International Conferences will be held in St. Petersburg, RUSSIA, New Jersey, USA and Quebec, CANADA, respectively. ISPE will be sponsoring a new symposium's series on Systems Automation Technology (SAT).

COOPERATIVE PROGRAMMES In 1989, ISPE launched a co-operative programme called the Indo-US Forum for Co-operative Research and Technology Transfer (IFCRTT) in co-operation with West Virginia University and National Science Foundation (NSF). The first joint meeting of IFCRTT was held in December 17-18 1989, at New Delhi, India. ISPE is co-operating with IFAC on the 7th IFAC/IFIP/IFORMS/IMACS/ISPE Symposium on information Control Problems in Manufacturing Technology, May 25-28, 1992. Many similar co-operative programmes are being pursued.

WORKSHOP A second joint IFCRTT and ISPE workshop on CAD, CAM and Robotics took place in New Delhi, India from December 19-21, 1991. A joint ISPE and UUFCRE workshop on "Intelligent Engineering & Manufacturing" is planned at St. Petersburg, Russia, May 18-23, 1992.

As you see, ISPE has made great strides. Significant changes are now taking place in manufacturing sectors due to intense global competitiveness and economic conditions. The productivity enhancement needs are even larger than before. Members need our support. We have to be more dynamic and resourceful. ISPE ought to show some leadership. We need your support in building our strong technical base. If you would like to help us, or if you would like to work on ideas of your own, please write to us. ISPE has needs in the following areas:

- ★ SARA Journal: Readers Committee, Reviewers,
- ★ Productivity Directors and Associates,
- ★ Workshop and Tutorials Organisers, and
- ★ CARS & FOF Conference and SAT Symposia: University, Industry, and International Representatives, Session Organisers, Technical and Programme Chairpersons.

We are still a young organisation and your leadership can play a significant part. Thank your very much for your continued help and participation. With best wishes,

Biren Prasad, Ph.D., Chairman, Publication Committee ISPE, P. O. Box 731, Bloomfield Hills, MI 48303-0731, USA

ORGANISERS

General Conference Chairman
Dr Hrishi Bera (UK)
General Conference Co-Chairman
Mr F. Marcotirchino (France)
Dr Suren Dwivedi (USA)
Dr C B Jennings (UK)
Conference Manager
Mrs Antoinette Dixon (UK)
Conference Chairman
Dr Sumetra Reddy (USA)
Joint Chairmen (Editorial)
Dr Raj Gill (UK)
Dr Homer Rahnejat (UK)
Conference Patrons
Dr Jai N. Gupta (USA)
Dr Don Lyons (USA)
Dr Richard Matthews (UK)
The Rt. Hon Baroness Perry of Southwark, Hon LLD (Bath) MA (CANTAB) UK
International Organising Committee
Dr Bruce Palmer (USA)
Prof P. C. Pandey (India)
Mr Majib Babi (USA)
Ms D. Broni (UK)
Dr Hyung Suck Cho (Korea)
Prof S. R. Dev (India)
Dr F. Robert Dax (USA)
Miss Christine C. Ganpat (Guyana)
Prof V. Giard (France)
Mr Paul Gill (USA)

Dr Jozef Lastowiecke (Poland)
Prof Bruno Maione (Italy)
Dr Sagar Midha (UK)
Dr Yoshikuna Okawa (Japan)
Ms I. P. Podnozova (USSR)
Dr Biren Prasad (USA)
Dr R. Sagar (India)
Dr Sfantsikopoulos (Greece)
Mr Harold Schaal (USA)
Dr S. Srivastava (USA)
Mr Anil P. Thakoor (USA)
Prof G. Ulusoy (Turkey)
Mr V. Vernadat (France)
Prof. Marek B. Zaremba (Canada)
International Programme Committee
Mr J. Adams (UK)
Dr A. Ajmal (UK)
Prof Enzo Gentili (Italy)
Mr Albrecht Von Hagen (USA)
Prof. Mario Hoyos (Colombia)
Dr Alan Long (UK)
Mr L. Rose (UK)
Dr Kenneth H. Means (USA)
Dr J. M. Proth (France)
Dr James E. Smith (USA)
Dr Michael Sobolewski (USA - Program Chair)
Dr Chi Vansteenkiste (Belgium)

The Conference Committee would like to thank the Institution of Manufacturing Engineers for their assistance in publicising this year's conference.

CONTENTS

A: CAD/CAM and Numerical Control

Post-Processor Generators for Numerically Controlled Machine Tools
S. Beuvelet, P. Curnier, E. Debernard, and P. L. Kociemba 1
A CAD Interpreter for Interfacing CAD and CAPP of 2-D Rotational Parts
Y. Zhang, and A. R. Mileham 8
Computer Aided Design in Rehabilitation Robotics
S. D. Prior..14
A Concurrent Engineering System: The Use of Constraint Networks
R. E. Young, P. O'Grady, and A. Greef..20
A PC-Based Gearbox Design System
G. Song, A. Jebb, and S. Sivaloganathan 26
Monocity Preserving Approximation for Curve Definition in Computer Aided Design
J. Wojciechowski 34
Unfolding 3D Developable Geometries
J. M. Sullivan, Jr. and D. H. Fu 40
Processing Isometric Free-Hand Sketches in Computer-Aided Design
S. Sivaloganathan, A. Jebb, and H. P. Wynn..46
Applying Hypermedia Concepts to CAD in an Assemble to Order Environment
M. Cwiakala, and N. Brouer 54
Software CAD Environment for PLC
V. V. Deviatkov 60
Computer-Aided Design of Gating Systems for Cast Iron Castings
R. Skoczylas, A. Gradowski, and J. Zych 66
Computer Aided Design of the Underwater Robotic Vehicle Robust Invariant Control
I. Mandic, J. Marasovic, N. Gacic, A. Batina 72
An Efficient Boundary Data Structure for Integrated CAD/CAM, Robotics and Factories
of the Future
S. R. Ala, and D. A. Chamberlain 78
A Capstone Two-Course Sequence for Design, Build and Test
N. Berzak, and W. W. Walter 84
NC Trajectory Collision Checking
K. Marciniak 90
Maximum Time Control of a DNC Machine Table
F. W. Wright, and F. Azpiazu 98
An Approach to the Real Integration of CAD-CAM Systems
M. Akkurt 104
CAD/CAM Program Package for Cold Metal Forming
M. Tisza, and P. Racz 110

B: Modelling and Simulation

Multi-Arm Robotic Systems: Computer-Oriented Methods for Control
M. M. Svinin, and S. V. Eliseev..118
An Effective Choice of a Time Step for Solution of an Unsteady Temperature Field
J. Hlousek, and F. Kavicka 124
User-Generated Feature Libraries for Maintenance of Recurrent Solutions
H. Grabowski, S. Braun, and A. Suhm 130
Optimal Trajectory Planning for Manipulators with Goal Point Uncertainty
A. Adams, and A. A. Melikyan 138
CAD Simulation of the Factory of the Future
C. Stylianides, and K. Ebeling..145

Cycle Time and Path Optimization of Manipulator Arms In Assembly Operations
    R. S. Redman, Jr., and A. S. El-Gizawy 151
Obstacle Avoidance in a Manufacturing Based Modeller
    I. I. Esat, and R. Lam 158
Flexible Assembly Cell Simulation - A Tool for Layout Planning and Cell Controller Development
    E. Freund, and H.-J. Buxbaum 164
FMS Management Optimization Using Physical Simulation
    G. Cardarelli, A. Dentini, and P. M. Pelagagge 170
A Comparison between Traditional Methods and New Models for Computerized
Production Planning
    A. Sianesi 176
A Simulation Study of Ethernet in Flexible Manufacturing Applications
    J. L. Sevillano, G. Jimenez, A. Civit Balcells, A. Civit Breu and E. Diaz 182
Application of a Learning Technique to Factory System Tuning
    P. Liu, and R. Chou 188
Functional Enterprise Modelling: A Process/Activity/Operation Approach
    F. Vernadat, and M. Zelm 194
Issues in the Formal Definition of Product Model Exchange
    A. Markus 201
Strategic Material Flow Analysis Using Simulation
    V. B. Patange, and U. S. Bititci 207
A Lambda Calculus Model for the Officer-Cell Phenomenon in Manufacturing Systems
    G. R. Liang 214
Simulation Design Techniques in Network Analysis
    C. S. Putcha, and J. H. Kreiner 220
Manufacturing Cell Formation with Identical Machines: A Simulated Annealing Approach
    T. Hamann, J.-M. Proth, and X. Xie 228

C: Robotics

The Study of Micro-Drive Robotic End Effector
    H. G. Cai, L. Hong, and Z. Quan 234
Development of Excavation Robot for Underground Space Construction
    T. Arai, T. Nakamura, K. Homma, R. Stoughton, and H. Adachi 240
On a Robot which Plays the Xylophone
    M. Itoh 249
Teach Control Based Off-Line Programming
    E. C. Morley, C. S. Syan, S. V. Grey-Cobb, and J. R. Wilson 255
Introduction of Robots in Docks
    Z. Domazet, I. Mandic, and T. Pirsic 263
Autonomous Operation of Multi-Robot-Systems in Factory of the Future Scenarios
    E. Freund, and J. RoBmann 269

D: Sensors, Control and Signal Processing

Interactive Motion Specification Using Splines
    M. Caulfield-Browne, B. L. MacCarthy, and C. S. Syan 275
Fast Algorithm for the Calculation of Generalized Torques in Robotics
    B. Levesque, and M. J. Richard 284
Adaptive Control of Machining Using Expert Systems
    P. S. Subramanya, M. 0. M. Osman, and V. N. Latinovic 290
A Reinforcement Connectionist Path-Finder
    J. R. Millan, and M. Becquet 296

'True' Surface Determination Using a Novel Algorithm of Directional Search
through a Point Table Sampled by a Contact Probe
    M. Akbary-Safa, I. I. Esat, and C. B. Besant 302
Robotic Deburring of Parts with Arbitrary Camber Using Hybrid Position/Force Control
    H. G. Cai, Y. Zheng, Q. Tao, and H. Liu 308
Design of Block Networks
    A. A. Rubchinsky..314
An On-Line Control System of Tool Status in Continuing Cutting
    E. Cerreni, G. Maccarini, L. Zavanella, and A. Bugini 320
Temperature Compensation of Sealed Pump-Gauge Zirconia Oxygen Sensors Operated
in the AC Mode
    M. Benammar, and W. C. Maskell 326
A Technique for Coupling Non-Compatible CNC Controllers with Touch Trigger Probe Systems
    E. Zhou, D. K. Harrison, and D. Link...333
Trajectory Planning for Robotic Technology System Control
    V. G. Gradetsky, and A. M. Ermolaev 339
Vision in Pyramids Object Recognition in Real Time
    Z. N. Li 344
Mobile Robot Hydroblast System
    M. K. Babai, R. M. Rice, and S. A. Cosby 350
Generalized Approach to the Control of Assembly Process Using Tactile Sensors and Grippers
with Soft Fingers
    B. Borovac, D. Seslija, and S. Stankovski 356
Monocular Blur Depth Algorithms for an Image-Processing System
    R. P. M. Craven, W. K. Preece, and J. E. Smith 362
Real-Time Sensor-Based Approach to the Improvement of Transitions of Robot
Manipulator Controls
    Y. F. Li, and R. W. Daniel..368
A New Active/Passive Robotic Precision Assembly System with F/T Sensor
    H. G. Cai, and T. Yang..374
Extending the Productivity of Industrial Robots by Efficient Integration of a Novel Force-
Torque Sensor
    J. Wahburg 380

E: Expert Systems and Artificial Intelligence

    EGRD-Expert Gear Reducer Designer
        T. G. Boronkay, M. L. Brown, and C. Vanderhorst 386
    Application of Expert Systems to Process Planning in Manufacturing
        D. Alasya 392
    Intelligence Robotic Assembly Cell in CIM Environment
        D. Seslija, S. Stankovski, and B. Borovac 397
    The Application of AI and RISC Methodologies to Manufacturing Simulation Tools
        I. Astinov, J. F. O'Kane, and D. K. Harrison...403
    Control of an Inherently Unstable System Using the Neural Networks
        I. I. Esat, and T. J. Suh 408
    Object Location Using an Artificial Neural Network
        B. Parsons, M. Stoker, and R. Gill..414
    An Artificial Intelligence Planner for Assembly Process Planning
        Y. P. Cheung, and A. L. Dowd 421
    A Knowledge-Based Approach for Small Parts Manual Assembly in the Space Environment
        R. L. Roman, and S. N. Dwivedi 430
    Building Industrial Expert Systems with Flex
        V. Devedzic, and D. Velasevic 436

A Distributed Knowledge-Based System for Total Manufacturing Support
V. Ram 442
Intelligence Computer-Aided Industrial Ecology System
V. V. Kupriyanov, and N. I. Kupreev 448
An Expert Systems Approach to Setup Reduction
S. L Kim, B. Arinze, and S. Banerjee 453
Application of an Expert System to Robot Gripper Selection
C. T. Ho, M. DeVore, and A. S. El-Gizawy 458
An Object-Oriented Model for Manufacturing Database
P. Ji, and R. S. Ahluwalia 464
Object-Oriented Knowledge Bases in Engineering Applications
M. Sobolewski 470

F: System Architecture: FMS & CIM

MAP Applications in European CIM Pilot Plants
G. L. Kovacs, and G. Haidegger 476
Laboratory Studies on the Integration of a Pilot FMC
G. Cardarelli, C. Giancola, and P. M. Pelagagge 482
Strategic CIM Research Issues for the Planning of Large Scale Production
W. A. Taylor 487
Computer Integration: An Issue of Strategic Direction
M. A. Sanders, W. Hepworth, and R. Leonard 493
Robots and Flexible Pallets: The Solution for Flexibility and JIT
E. Gentili, R. Sala, A. Rovetta, and M. Braglia 499
Design of Future Manufacturing Systems Using ClM-Cybernetic Methodology
A. Zakeri, N. E. Gough, R. Espejo and D. Bowling 505
Configuration Aspects and Control of a Completely Integrated Manufacturing Cell
W. M. Goodman, S. J. Tricamo, and C. Chassapis 511

G: Mechanism, Design and Analysis: Kinematics, Dynamics and Synthesis

Synthesis of the Nominal Dynamics for Manipulation Robots Using Optimal and
Sub-Optimal Procedures
Z. Konjovic, M. Vukobratovic, and D. Surla 518
Quaternion Based Kinematics of Robot Manipulators
A. S. Abdel-Mohsen 524
Task Space Co-ordination of Dual Arm Motion
D. S. Necsulescu, R. Jassemi-Zargani, and W. B. Graham 530
Experimental Analysis of Dynamics of a Flexible Crank
I. I. Esat, and A. I. Ianakiev 536
A Path-Motion Algorithm for Multivariable FMS Structures Based on Graph Networks
G. M. Dimirovski, N. E. Gough, 0. L. Iliev, B. R. Percinkova, R. M. Henry, and A. Zakeri ... 542
Vibration Control in Robots and Other Automation
A. S. White 550
Two-Dimensional Absolute Position Recovery for an Automated Guided Vehicle
E. M. Petriu, F. C. A. Groen, T. Bieseman, and N. Trif 556
Dynamic Impedance Matching of Direct-Drive Actuator
S. Kobayashi, and T. Takamori 562
Design and Analysis of a New Six-Degree-of-Freedom Parallel Minimanipulator
L. W. Tsai, and F. Tahmasebi 568
Fast Learning Control of Industrial Mechanisms and Robots
J. E. Kurek, and M. B. Zaremba 576

H: Integrated Production and Quality Management

Integration of Design and Assembly Planning - Ways to Improve Competitiveness
K. Thaler, and M. Richter 582
Optimal Production Control in a Manufacturing System with Machine Failure Rates of Steady State Probabilities
H. R. Yazgan, and I. I. Esat 588
Role of Electronic Data Interchange in Inventory Management
S. Banderjee, and A. Banerjee 594
Conceptual Cost Estimating Using Complexity Criteria
C. G. Currie, A. R. Mileham, A. W. Miles, and D. T. Bradford 600

I: Automated Process Planning

Process Planning: Survey and Trends
F. Alzate, and H. Bera 606
Intermediate Workpiece Shapes and Specifications in Feature-based Expert Process Planning
M. T. Wang 612
IAGE: A feature Oriented CAPP System
J. T. Hernandez, M. Rodriguez, and E. Gutierrez 618

J: Manufacturing Planning and Control

A Methodology for Concurrent Engineering Process Modelling
L. Korbly, B. Palmer, and G. Trapp 624
The Application of Advanced Factory Control and Management Techniques to Aerospace Composite Manufacture
V. A. Michell 630
Minimizing Total Flow Time with Release Dates
C. Chu 637
A Code System for Job Sequencing
M. Hoyos, and H. Bera 643
Integration of Group Technology and Scheduling Techniques for PCB Assembly Systems: Proposal and Test of a New Framework
M. Perona, and A. Pozzetti 649
Reactive Scheduler for Discrete Manufacturing
J. Lazaro, J. Maseda, G. Escalada, and F. Diaz 655
Economic Evaluation of Schedules in Job Shops
V. Giard 661

K: Measurement and Inspection

Object Segmentation Employing a Monocular Blur Depth Algorithm
R. P. M. Craven, W. K. Preece, and J. E. Smith 667
Toward Fault-Tolerant Manufacturing Systems
M. E. Staknis 673
Computer Aided-Design and Software Prototyping in Manufacture of Precision Engineering Components
V. Gupta, and R. Sagar 679
Integrating CMM and CAD System in Manufacture of Precision Engineering Components
V. Gupta, and R. Sagar 690
Simultaneous Robotic Inspection and Assembly
N. Berzak 700
Measuring Robot of Large-Dimensional Technical Objects
J. Szpytko 706

L: Management Aspects of Manufacturing: Human and Resource Engineering

Education for High Automation Level and Computer Integrated Manufacturing Systems Managing
M. Bonfioli, and C. Noe 712
A C4 Centre - Education and Industry Partnership
C. Stylianides, and J. Stanislao 719
Strategy of Manufacturing Education in the Polytechnic Institute of Bucharest
M. Gheorghe, C. Ispas, A. Szuder, and C. Stancescu 725
Towards a Focused Factory
R. Gill and C. Holly 731
HUMAN ACTIVITY: Friend and Foe of Advanced Manufacturing Systems
B. Trought 738

M: Safety and Reliability

Computer-Aided Design of Safety and Ergonomics at Flexible Manufacturing Systems
M. Mattila, and M. Leppanen 743
SAFE: A Knowledge-Based System for Prevention Incidents
A. M. Yemelyanov 749
Automatic Safety System for FMS-Management Logic
C. F. Marcolli 755

N: Material Handling and Assembly Line Balancing

A Knowledge-Based Approach for Small Parts Robotic Assembly in the Space Environment
R. L. Roman, and S. N. Dwivedi 761
Composite Manufacturing Advisor in the Concurrent Engineering Environment
R. H. Willison, and S. N. Dwivedi 767
Methods and Algorithms for the Generation and Selection of Assembly Sequences
K. Ghosh, and G. B. Reddy 773
Design of a Multi Model Assembly System for Medical Kits
M. Dov, and A. Shtub 779
Automatic Constraint Generation in Discrete Optimal Assembly Schemes
J. Jiang, and K. H Means 785
Low Cost Representation of Parts for the Analysis of Mechanical Assemblies
J. E. Gomez, M. J. Garcia, and J. R. Toro 791
On Generation of Parts Assembly Sequences of Plate Structure
H. Maekawa, H. Horibe, and S. Nakata 800

Miscellaneous

Method for the Evaluation of Manufacturing Systems Test Efficiency
M. Bonfioli, C. Noe, M. Pellegrini, and E. Riva 806
The Stochastic Nature of Tool Life: Experimental Results in Continuous Turning
C. Giadini, G. Pellegrini, A. Bugini, and R. Pacagnella 812
Modelling Method for Low Prandtl Number Fluids Used in the Coatings Industry
G. Thompson, R. Churchill, and J. E. Smith 818
Control of Flexible Process Schemes in Ferrous Metallurgy
A. A. Rubchinsky 824
Skiving, A Part of Manufacturing Deep Holes
H. Hegewald 830

Invited Papers:

Artificial Insects (Silicon Microrobots)
I. Shimoyama, H. Miura, K. Suzuki, and T. Yasuda 836
The CIM Enterprise and Graduate Education
P. Sackett 842
Advanced Research and Development on AGV in Factory and Out-Door
T. Tsumura 849
Walking Robots and 3-D Dynamic Simulator
T. Nakamura, N. Koyachi, and H. Adachi 859
Actuators of the Future for Robotics and Mechatronics
T. Takamori 865
Modelling the Co-ordination of Multi-Robot Equipment
G. M. Acaccia, R. Michelini, R. M. Molfino, and M. A. Recine 871
Productivity Measurement in AMT Systems
E. A. Cahill, and M. E. J. O'Kelly 877
Advanced Studies in Flexible Automation
T. Watanabe 883
Flexible Agricultural Robotics Manipulator System (Farms)
P. S. Gill 889
Toward Next Generation CAD Systems
S. Ohsuga 895

6th International Conference on CAD/CAM, Robotics and Factories of the Future, 19th - 22nd August 1991, South Bank Polytechnic

1. A. S. Abdel-Mohsen (Egypt, G1, G3)
   Department of Mechanics & Elasticity, Military Technical College, Cairo, Egypt

2. G. M. Acaccia (Italy, PS4)
   Industrial Robot Design Research Group, University of Genova, Genova, Italy

3. H. Adachi (Japan, C2)
   Robotics Department, Mechanical Engg. Laboratory, A I S T, MITI, Namiki 1 - 2 Tsukuba, Ibaraki 305 Japan

4. S. Adams (UK, B4, L2)
   Department of Electrical & Electronic Engg., South Bank Polytechnic, Borough Road, London SE1 0AA, U K

5. Prof. A. V. Agashe (UK, L2)
   Middlesex Polytechnic, School of Mechanical & Manufacturing Engg., Advanced Manufacturing Group, Bounds Green Road, London N11 2NQ, U K

6. R. S. Ahluwalia (USA, A4, E1)
   Department of Industrial Engg, West Virginia University, Morgantown, WV 26505, U S A

7. Mustafa Akkurt (Turkey, A1, A5)
   Faculty of Mechanical Engg., Istanbul Technical University, Istanbul, Turkey

8. S. R. Ala (UK, A2)
   Construction Robotics Unit, School of Engg., City University, Northampton Square EC1V 0HB, UK

9. Derya Alasya (USA, E4)
   Engg. Management Department, Old Dominion University, College of Science, Norfolk, VA 23529-00162 U S A

10. Fernando Alzate (Colombia, H1, K2)
    Mechanical Engg. Department, Technical University of Pereira, Colombia, South America

11. M. Amamou (Maroc, J2)
    University Mohamed IER, Faculté des Sciences de Oujda, Department of Physics, Ouja/Maroc

12. T. Arai (Japan, C2)
    Robotics Department, Mechanical Engg. Laboratory, AIST, MITI, Namiki 1- 2 Tsukuba, Ibaraki 305 Japan

13. Bay Arinze (USA, E1)
    Department of Management, Drexel University, Philadelphia, PA 19104, U S A

14. I. Astinov (Bulgaria, E4)
    Department of Machine Tool Techniques, Technical University, Sofia, Bulgaria

15. A. Azmal (UK, J3, F2)
    Department of Mechanical Engg., South Bank Polytechnic, 103 Borough Rd., London SE1 0AA, U K

16. F. Azpiazu (UK)
    Electrical & Electronic Engg., Staffordshire Polytechnic, Beaconside, Staffs., UK

17. Majid K. Babai (USA, D2, L2)
    Press Engg. Division, Material & Process Laboratory, Science & Engg. Directorate, NASA/Marshall Space Flight Centre, AL 35812, U S A

18. A. D. N. Bajpai (India, O1)
    Jabalpur University, Jabalpur, Madhya, Pradesh, India

19. A. Civit Balcells (Spain, B2)
    Department of Electronics & Electromagnetism, University of Sevilla, Avda. Reina Mercedes, s/n 41012 Sevilla, Spain

20. Avijit Banerjee (USA, H1)
    Department of Management, Drexel University, Philadelphia, PA 19104, U S A

21. Snehamay Banerjee (USA, E1, C2, H1, D5)
    Department of Management, Drexel University, Philadelphia, PA 19104, U S A

22. A. Batina (Yugoslavia, A2)
    Faculty of Electronics, Mechanical Engg. & Naval Architecture University of Split, R. Boskovica bb, 58000 Split, Yugoslavia

23. Marc Becquet (Italy, D6)
    Institute for System Engg. & Informatics, Commission of the European Committee, Joint Research Centre, Building A36, 21020 ISPRA (VA)

24. M. Benammar (UK, D3, D6)
    Energy Technology Centre, Middlesex Polytechnic, Bounds Green Road, London N11 2NQ, U K

25. Hrishi Bera (UK, I1, J1, J3)
    Department of Mechanical, Design & Manufacture, South Bank Polytechnic, 103 Borough Road, London SE1 0AA, U K

26. Nir Berzak (USA, A2, K1)
    Mechanical Engg., Rochester Institute of Technology, Rochester, New York 14623, USA

27. Colin B. Besant (UK, D5)
    Department of Mechanical Engg., Imperial College, Exhibition Road, London, U K

28. T. Bieseman (Netherlands, G2)
    University of Amsterdam, The Netherlands

29. S. Beuvelet (France, A5)
    272 Avenue de Grammont, 37000 TOURS, France

30. U. S. Bititci (UK, B1)
    Strathclyde Institute, Glasgow, Scotland

31. Mario Bonfioli (Italy, L2, O1, P2)
    Polytechnic of Milano, Department of Mechanics, Piazza Leonardo da Vinci, 32 - 20133 Milano, Italy

32. Thomas G. Boronkay (USA, E1, E4)
    University of Cincinnati, Cincinnati, Ohio, U S A

33. B. Borovac(Yugoslavia, D2, E4)
    Faculty of Technological Science, University of Novi Sad, 21000 Novi Sad, V. Vlahovica 3, Yugoslavia

34. D. Bowling (UK, F1)
    Syncho Ltd, Aston Science Park, Lovelane, Aston Triangle, Birmingham B7 4BJ, U K

35. Marcello Braglia (Italy, F1)
    Department of Mechanical Engg., University of Brescia, Via Valotti 9, 25060 Brescia, Italy

36. S. Braun (Germany, B3, B5)
    Institut für Rechneranwendung in Planung und Konstruktion (RPK) Universität Karlsruhe, Kaiserstraße 12, 7500 Karlsruhe, Germany

37. D. T. Bradford(UK, H1)
    School of Mechanical Engg., University of Bath, Claverton Down, Bath BA2 7AY, U K

38. A. Civit Breu (Spain, B2)
    Department of Electronics & Electromagnetism, University of Sevilla, Avda. Reina Mercedes, s/n 41012 Sevilla, Spain

39. Nils Brouër (Germany, A3)
    Fulbright Fellow at Rutgers University, RWTH Aachen, 5100 Aachen,

40. Max L. Brown (USA, E4)
    University of Cincinnati, Cincinnati, Ohio, U S A

41. A. Bugini (Italy, D4, P1)
    Department of Mechanical Engg., University of Brescia, Via Valotti 9, 25060 Brescia, Italy

42. H. J. Buxbaum (Germany, B3, B4)
    Institut for Robotics (IRF), Universität Dortmund, Postbox 500500, D-4600 Dortmund 50,

43. E. A. Cahill (Ireland, PS3)
    Irish Productivity Centre, Dublin, Ireland

44. G. Cardarelli (Italy, B3, F1, F2)
    Faculty of Engg., University of L'Aquila, 67040 Roio, L'Aquila, Italy

45. Natele Cattaneo (Italy, L2)
    U. C. I. M. U. Sistemi per Produrre, Italy 7

46. M. Caulfield-Browne (UK, D6)
    Department of Manufacturing Engg., University of Nottingham, Nottingham, NNG7 2RD, U K

47. E. Ceretti (Italy, D1, D4)
    Department of Mechanical Engg. University of Brescia, Via Valotti 9, 25060 Brescia, Italy

48. Denis A. Chamberlain (UK, A2, PS4)
    Construction Robotics Unit, School of Engg. City University Northampton Square, London EC1V 0HB, U K

49. C. Chassapis (USA, F1)
    Design & Manufacture Institute, Mechanical Engg. Department, Stevens Institute of Technology, Hoboken, NJ 07030 U S A

50. R. C. Chaturvedi (India, P1)
    Department of Mechanical Engg., Indian Institute of Technology, Powai, Bombay 400 076, India

51. Atal Chaudhuri (India, A4)
    Computer Science and Engg. Department Jadavpur University, Calcutta 700 032, India

52. J. H. Chen (China, N2)
    Hua Zhong University of Science and Technology, Wu Han, Hu Bei, P. R. China

53. R. Y. Chen (China)
    Hua Zhong University of Science and Technology Wu Han, Hu Bei, P. R. China

54. Y. P. Cheung (UK, E1, E2)
    Department of Engg., University of Warwick, Coventry, CV4 7AL, U K

55. M. R. Chidambaram (India, I1)
    Fenner India Limited, P. O. Box 117, Madurai, India 625 - 001

56. Robert Chou (USA, B2)

57. Chengbin Chu (France, J2)
    Project SAGEP, Inria-Lorraine, 4 rue Marconi, 57070 Metz, France

58. Randy Churchill (USA, P2)
    West Virginia University, Centre for Industrial Research Applications (CIRA), Mechanical & Aerospace Engg. Department, Morgantown, W. Virginia 26506-6101, U S A

59. Steven A. Cosby (USA, D2)
    Press Engg. Division, Material & Process Laboratory, Science and Engg. Directorate, NASA/Marshall Space Flight Centre, AL 35812, U S A

60. Robert P. M. Craven (USA, D1, K1, K2)
    West Virginia University, Centre for Industrial Research Applications (CIRA), Mechanical & Aerospace Engg. Department, Morgantown, W. Virginia 26506-6101, U S A

61. C. G. Currie (UK, H1)
    School of Mechanical Engg. University of Bath, Claverton Down, Bath BA2 7AY, U K

62. P. Curnier (France, A5)

63. Martin Cwiakala (USA, A3)
    GEMCO, 301 Smalley Avenue, Middlesex, NJ 08846, U S A

64. S. Dadunashvili (USSR, G1, G2)
    Institute of Machine Mechanics, Academy of Science of the Republic of Georgia, Tbilisi, U S S R

65. V. Damodaran (Canada, J2)
    Department of Industrial Engg., University of Windsor, Ontario, Canada N9B 3P4

66. R. W. Daniel (UK, D1)
    Robotics Research Group, Department of Engg. & Science, Oxford University, U K

67. S. R. Deb (India, D2)
    Production Engg. Department Jadavpur University, Calcutta 700 032, India

68. E. Debernard (France, A5)
    Department of Industrialisation & Production, 12 rue Pasteur B.P. 76, 92152 Surenes, Cedex, France

69. S. K. Debnath (India, D2)
    Production Engg. Department, Jadavpur University, Calcutta 700 032, India

70. A. Dentini (Italy, B3)
    Faculty of Engg., University of L'Aquila, 67040 Roio, L'Aquila, Italy

71. S. K. Deshmukh (India, L1)
    Department of Mechanical Engg., Indian Institute of Technology New Delhi - 110 016, India

72. V. Devedzic (Yugoslavia, E2)
    Mihailo Pupin Institute, Belgrade, Yugoslavia

73. V. V. Deviatkov (USSR, A1, A2)
    Institute of Control Sciences, Profsoyuznaya 65, Moscow, 117342, U S S R

74. Michael DeVore (USA, E1)
    Industrial & Technology Division Centre, University of Missouri, Columbia, MO65211

75. F. Diaz (Spain, B2, J1)
    Department of Electronics & Electromagnetism, University of Sevilla, Avda. Reina Mercedes, s/n 41012 Sevilla, Spain

76. G. M. Dimirovski (Yugoslavia, G3)
    Division of Automation & System Engg., Cyril & Methodius University, Skopje Macedonia, Yugoslavia

77. Zeljko Domazet (Yugoslavia, C2)
    Faculty of Electrical Engg., Mechanical Engg. and Naval Architecture, University of Split, R. Boskovica bb, 58000 Split, Yugoslavia

78. A. L. Dowd (UK, E2)
    Department of Engg., University of Warwick, Coventry, CV4 7AL, U K

79. Mosche Dov (Israel, N1)
    Department of Industrial Engg., Tel Aviv University, Ramat aviv, Tel aviv 69978, Israel

80. A. K. Dutta (India, D2)
    Production Engg. Department, Jadavpur University, Calcutta 700 032, India

81. S. N. Dwivedi (USA, E2, N2, P1, IPS)
    Department of Mechanical & Aerospace Engg., West Virginia University, Morgantown, WV 26506, U S A

82. Kenneth Ebeling (USA, B4)
    Department of Industrial Engg. & Management., North Dakota State University, Fargo, U S A

83. Sergy V. Elseev (USSR, B5)
    Automation & Technology Physics Department, Science Centre, Lermontov 281, Irkutsk, U S S R, 664033

84. A. Sherif El-Giazawy (USA, B4, E1, E3)
    Industrial & Technology Centre, University of Missouri, Columbia, MO 65211, U S A

85. A. M. Ermolaev (USSR, D3)
    Institute for Problems in Mechanics of Academy of Science 117 526, Moskow, pr. Vernadskogo, 101, U S S R

86. Ibrahim I. Esat (UK, B4, D4, D5, E3, G3, H1)
    Department of Mechanical Engg., Queen Mary College, Mile End Road, London, U K

87. Gonzalo Escalada (Spain, J1)
    IDEIA, Investigación y Desarrollo en Informatica Avanzada, LABEIN, Cuesta de Olabeaga, 16 E - 48013 Bilbao, Spain

88. R. Espejo (UK, F1)
    Aston Business School, Aston University, Aston, Birmingham B4 7ET, & Syncho Ltd., Aston Science Park, Love Lane, Aston Triangle, Birmingham B7 4BJ, U K

89. D. P. Fan (China)
    Hua Zhong University of Science and Technology, Wu Han, Hu Bei, P. R. China

90. Aureo C. Ferreira (Brasil, I1)
    Grucon, Engg. Mechanics, University Federal de Santa Caterina, Caixa Postal 476, 88049 - Florianópolis, SC, Brazil

91. E. Freund (Germany, B3, B5, C1)
    Institut für Roboterforschung (IRF), Universität Dortmund, Otto-Hahn-Strasse 8, W-4600 Dortmund 50, Germany

92. Don H. Fu (USA, A3)
    Department of Mechanical Engg., Worcester Polytechnic Institute, Worcester, Massachusetts

93. N. Gacic (Yugoslavia, A2)
    Faculty of Electrical Engg., Mechanical Engg. & Naval Architecture, University of Split, R. Boskovica bb, 58000 Split, Yugoslavia

94. M. Gannon (UK, J3)
    34 Vane Close, Harrow, Middlesex, London, HA3 9XD, U K

95. Manuel J. Garcian (Colombia, N1)
    DFAC Group, Computer-Aided Design & Manufacturing Group, Engg. School, UNIANDES, Apartado Aereo 4976, Bogota, Colombia

96. Enzo Gentilli (Italy, C2, F1, P1, E2)
    Department of Mechanical Engg., University of Brescia, Via Valotti 9, 25060 Brescia,

97. M. Gheorghe (Romania, L2)
    Machine Construction Technical Department, Polytechnial Institute of Bucharest, Romania

98. Kalyan Ghosh (Canada, N2, P2)
    Industrial Engg. Department École Polytechnic, P O Box 6079, Montréal, Québec, Canada H3C 3A7

99. C. Giancola (Italy, F2)
    Faculty of Engg., University of L'Aquila, 67040 Roio, L'Aquila, Italy

100. Vincent Giard (France, J1, J2)
     Institute National des Science, Appliques de Lyon, 20 Avenue, Albert Einstein, 69621 Villeurbanne, Cedex, France

101. C. Giardini (Italy, P1, O1)
     Department of Mechanical Engg., University of Brescia, Via Valotti 9, 25060 Brescia, Italy

102. Paul S. Gill (USA, M1, PS2)
     Process Engg. Division., Material & Process Laboratory, Science & Engg. Directorate, NASA/Marshall Space Flight Centre, Alabama 35812, USA

103. R. Gill (UK, L1, E3, L2)
     Middlesex Polytechnic, School of Mechanical & Manufacturing Engg., Advanced Manufacturing Group, Bounds Green Road, London N11 2NQ, U K

104. Jaime E. Gomez (Colombia, N1)
     DFAC Group, Computer-Aided Design & Manufacturing Group, Engg. School, UNIANDES, Partado Aero 4976, Bogota, Colombia

105. W. M. Goodman (USA, F1)
     Design & Manufacturing Institute, Mechanical Engg. Deptartment, Stereo Institute of Technology, Hoboken, N. J. 07030, U S A

106. J. Gomez (Colombia, G1)
     Technical University of Pereira, Pereira, Colombia, South America

107. N. E. Gough (UK, F1, G3)
     Manufacturing & Systems Centre, Wolverhampton Polytechnic, Wulfrun Street, Wolverhampton, WV1 1SB, U K

108. Steven A. Gosby (USA, D2)
     Press Engineering Division, Material & Process Laboratory, Science & Engg. Directorate, NASA/Marshall Space Flight Centre, AL 35812, U S A

109. H. Grabowski (Germany, B5)
     Institute fur Rechneranwendung in Planung und Konstrukton (RPK)Univesität Karlsrue, Kaiserstraße 12, 7500 Karlsruhe, Germany

110. V. G. Gradetsky (USSR, D3, D6)
     Institute of Problems in Mechanical & Academy of Science of USSR, 117526 Moscow, pr. Vernadskogo, 101 USSR

111. A. Gradowski (Poland, A3)
     Institute of Foundry Technology and Mechanics Academy of Mining & Metallurgy, Cracow, Poland

112. W. B. Graham (Canada, G3)
     Space Technology, Canadian Space Agency, 3701 Carling Avenue, Ottawa, Ontario K2H 8S2, Canada

113. Arthur Greef (USA, A5)
     Group for Intelligent Systems in Design & Manufacturing, Department of Industrial Engg., Campus Box 7906, North Carolina State University, Raleigh, North Carolina 27695 - 7906, U S A

114. Sue V. Grey-Cobb (UK, C1)
     Department of Manufacturing Engg. & Operations Management, University of Nottingham, University Park, Nottingham NG7 2RD, UK

115. F. C. A. Groen (Netherlands, G2)
University of Amsterdam, The Netherlands

116. Jai Gupta (USA, PS3)
I. S. P. E., P. O. Box 731, Bloomfield Hills, Michigan 48303-0731, U S A

117. V. K. Gupta (India, K2)
V & C India Ltd., University of Engg., F-45, NDSE - I, New Delhi 110049, India

118. Eliécer Gutierrez (Colombia, I1)
DFAC Group, Faculty of Engg., UNIANDES, Apartado Aéreo 4976, Bogotá, Colombia, South America

119. Géza Haidegger (Hungary, F1)
Computer & Automation Institute, Hungarian Academy of Sciences, 1502 Budapest, P O Box 63, H 133 mar @ ella. uucp., Hungary

120. T. Hamann (France, B1)
INRIA Technopôle, Metz 2000, 4 rue Marconi, 57070 Metz, France

121. M. Happiette (France, J1, J2)
E N S A I T, 2 Places des Martyrs, 59070 Roubaix, Cedex 1, France

122. D K Harrison (UK, D1, D3, E4)
Department M C A E, Staffordshire Polytechnic, Stafford, U K

123. Cai Hegao (China, C2, D4, D1)
Robot Research Institute, Harbin Institute of Technology, Harbin 150006, People's Republic of China

124. H. Hegewald (Germany, N1, P1)
Laboratory for Machine Tools, Department of Mechanical Engg., HOCHESCHULE - BREMEN, Germany

125. R. M. Henry (UK, G3)
Faculty of Engg., University of Bradford, Bradford BQ7 1DP, U K

126. W. Hepworth (UK, F2)
U. M. I. S. T., Total Technology Department, P. O. Box 88, Manchester, M60 1QD, U K

127. José T. Hernandez (Colombia, I1)
DFAC Group, Faculty of Engg., UNIANDES, Apartado Aéreo 4976, Bogotá, Colombia

128. J. Hlousek (Czechoslovakia, B5)
Technical University B R N O, Czechoslovakia

129. Chien-Te Ho (USA, E1)
Industrial & Technology Department Centre, University of Missouri - Columbia MO 65211

130. C. Holly (UK, L1)
ITT Jabsco, Bingley Road, Hoddesdon, London EN11 0BU, U K

131. K. Homma (Japan, C2)
Robotics Department, Mechanical Engg. Laboratory, AIST, MITI, Namiki 1- 2 Tsukuba, Ibaraki 305 Japan

132. Liu Hong (China, C2)
Robot Research Institute, Harbin Institute of Technology Harbin 150006, China

133. Hiroshi Horibe (Japan, N1)
Faculty of Engg., Osaka University, Yamada-oka 2-1, Suita, Osaka 565, Japan

134. Mario Hoyos (Colombia, I1, J1)
Department of Mechanical Engg., Technical University of Pereira, Colombia, South America

135. A. I. Ianakiev (UK, G3)
    Queen Mary & Westfield College, University of London, Mile End Rd. U K

136. O. L. Iliev (Yugoslavia, G3)
    Division, of Automation & System Engg., Cyril and Methodius University, Skopje, Macedonia, Yugoslavia

137. C. Ispas (Romania, L2)
    Machine Construction Technology Department, Polytechnic Institute of Bucharest, Romania

138. M. Itoh (Japan, C2, C1)
    Department of Mechanical Engg. , Ashikaga Institute of Technology, Ashikaga, Toshigi, Japan

139. A. Jebb (UK, A3, A4)
    Engg. Design Centre, City University, Northampton Square, London EC1V 0HB, U K

140. P. Ji (USA, E1)
    Department of Industrial Engg., West Virginia University, Morgantwon, WV 26505, USA

141. J. Jiang (USA, N1)
    Department of Industrial Engg., West Virginia University, Morgantwon, WV 26505, USA

142. Zhu Jibei (China, P2, D5)
    Department of Mechanical Engg., Beijing Mechanical College, Beijing, 100026, China

143. G. Jiménez (Spain, B2)
    Department of Electronics & Electromagnetism, University of Sevilla, Avda. Reina Mercedes, s/n 41012 Sevilla, Spain

144. Jong-Yin Jung (USA, A4)
    Department of Industrial Engg. West Virginia University, Morgantwon, WV 26505, USA

145. Kapturkiewicz (Poland)
    Academy of Mining and Metallurgy, 30-059 Cracow, ul. Reymonta 23, Poland

146. S. Karunes (India, L1)
    Centre for Management Studies, I. I. T., New Delhi - 110016, India

147. F. Kavicka (Czechoslovakia, B5)
    Technical University of B R N O, Czechoslovakia

148. Y. B. Kavina (UK, C1, K1)
    School of Mechanical & Manufacturing Engg., Middlesex Polytechnic, Bounds Green Road London N11 2NQ, UK

149. M. Khvingia (USSR, G1, G2)
    Institute of Machine Mechanics, Academy of Science of the Republic of Georgia, Tbilisi, U S S R

150. Seung Lae Kim (USA, B5)
    Department of Management, Drexel University, Philadelphia, P A 19104, U S A

151. Shigeru Kobayashi (Japan, G1)
    Department of Mechanical Engg., Kobe City College of Tech., Kobe, Japan

152. R. Kobiashvih (USSR, G2)
    Institute of Machine Mechanics, Academy of Science of the Republic of Georgia, Tbilisi, U S S R

153. H. Kochekali (UK, C1, K1)
    School of Mechanics & Manufacturing Engg., Middlesex Polytechnic, Bounds Green Road, London N11 2NQ, UK

154. P. L. Kociemda (France, A5)
    Aerospatiale, C C R, 12 rue Pasteur, 92152 Suresnes, Cedex, France

155. Rambabou Kodali (India, L1)
    Centre for Management Studies, I. I. T., New Delhi - 110016, India

156. Z. Konjovic (Yugoslavia, G2)
    C M M Institute, University of Novi Sad, Novi Sad, Yugoslavia

157. Letitia Korbly (USA, J3)
    Department of Industrial Engg., West Virginia University, Morgantwon, WV 26505, USA

158. George L. Kovacs (Hungary, F1, F2)
    Computer & Automation Institute, Hungarian Academy of Sciences, Budapest, Hungary, H - 1111, Kende u. 13 - 17

159. Noriho Koyachi (Japan, C2)
    Robotics Department, Mechanical Engg. Laboratory, AIST, MITI, Namiki 1- 2 Tsukuba, Ibaraki 305 Japan

160. U. A. Kunhumoideen (India, L1)
    Chemical Engg. Department, Government Engg. College, Trichur, Kerala State, India

161. Jese H. Kreiner (USA, B1, B2)
    California State University Fullerton, CA 92634, U S A

162. N. I. Kupreev (USSR, E1, E4)
    Mechanical Engg. Research, USSR Academy of Science, U S S R

163. V. V. Kupriyanov (USSR, E1, E4)
    Mechanical Engg. Research, USSR Academy of Science, U S S R

164. J. E. Kurek (Poland, G1)
    Instytut Automatyki Przemyslowej, Politechnika Warszawska ul. Chodkiewicza 8, 02-525 Warszawa, Poland

165. R. Lam (UK, B4)
    Department of Mechanical Engg., Queen Mary College, Mile End Road, London, U K

166. R. S. Laskari (Canada, J2)
    Department of Industiral Engg., University of Windsor, Windsor, Canada N9B 3P4

167. V. N. Latinovic (Canada, D6)
    Deptartment of Mechanical Engg., Concordia University Montreal, Canada

168. José Lazaro (Spain, J1, J2)
    IDEIA, Investigación y Desarrollo en Informatica Avanzada, LABEIN, Cuesta de Olabeaga, 16 E - 48013 Bilbao, Spain

169. Vladimir Lazarov (Bulgaria, D4)
    Robot Control Systems Research Centre, a. k. G. Boncher Str., B L 2, 1113 Sofia, Bulgaria

170. M. Leppänen (Finland, M1)
    Tampere University of Technology, Occupational Safety Engg., P O Box 52733101 Tampere, Finland

171. B. Levesque (Canada, D6)
    Department of Mechanical Engg., Laval University, Quebec, Canada G1K 7P4

172. R. Leonard (UK, F2)
    U. M. I. S. T., Total Technology Department, P. O. Box 88, Manchester, M60 1QD, U K

173. Y. F. Li (UK, D1, D4)
    Robots Research Group, Department of Engg. Science. Oxford University, U K

174. Ze-Nian Li (Canada, D2, C1)
    School of Computer Science, Simon Frazer University, BC V5A 1S6, Canada

175. G. R. Liang (Taiwan, B1)
Department of Industrial Engg. & Management, National Chiao Ting University Hsinchu, Taiwan, R. O. China

176. Peiya Liu (USA, B2)
Science Corporate Research, 755 College Road East, Princeton, N. J. 08540, U S A

177. David Link (UK, D3)
Department of Mechanical & Computer-Aided Engg., Staffordshire Polytecnic, U K

178. Hong Liu (China, D4)
Robot Research Institute, Harbin Institute of Technology, Harbin 150006, China

179. Alan Long (UK, H1)
School of Engg., City University, Northampton Square, London EC1V 0HB, U K

180. G. Maccarni (Italy, D4)
Department of Mechanical Engg., University of Brescia, Via Valotti 9, 25060 Brescia, Italy

181. B. L. MacCarthy (UK, D6)
Department of Manufacturing Engg., University of Nottingham, NNG7 2RD, U K

182. Hitoshi Maekawa (Japan, N1)
Faculty of Engg. Saitama University, Schimo-okubo, 255 Urawa, Saitama, 338 Japan

183. Shyamal Majumdar (India, A4)
Computer Science and Engg. Department, Jadavpur University, Calcutta 700 032

184. Ivica Mandic (Yugoslavia, A4, C2, A2)
Faculty of Electrical Engg., Mechanical Engg. and Naval Architecure, University of Split, R. Boskovica bb, 58000 Split, Yugoslavia

185. Magdi H. Mansour (Canada, B1)
Dalhousie University, Halifax, N. S., Canada B3H 3J5

186. A. Markus (Hungary, B1, B2)
Computer & Automation Institute, Hungarian Academy of Sciences, 1502 Budapest, P. O. Box 63, Budapest, Hungary

187. Jadranka Marasovic (Yugoslavia, A2)
Faculty of Electrical Engg., Mechanical Engg. & Naval Architecure, University of Split, R. Boskovica bb, 58000 Split, Yugoslavia

188. K. Marciniak (Poland, A1)
Warsaw University of Technology, Precision Mechanics Department, Poland

189. Carlo Felice Marcoli (Italy, M1)
Department of Mechanical Engg., University of Brescia, Via Valotti 9, 25060 Brescia

190. A. Markus (Hungary, B1, B2)
Computer & Automation Institute, Hungarian Academy of Sciences, 1502 Budapest, P O Box 63, H 133 mar @ ella. uucp., Hungary

191. José Maseda (Italy, J1)
IDEIA, Investigación y Desarrollo en Informatica Avanzada, LABEINCuesta de Olabeaga, 16 E - 48013 Bilbao, Italy

192. W. C. Maskell (UK, D3)
Energy Technological Centre, Middlesex Polytechnic, Bounds Green Road, London N11 2NQ, U K

193. M. Mattila (Finland, M1)
Tampere University of Technology, Occupational Safety Engg., P O Box 527, 33101 Tampere, Finland

194. K. H. Means (USA, N1)
Department of Industrial Engg., West Virginia University, Morgantwon, WV 26505, USA

195. A. A. Melikyan (USSR, B4)
Institute of Control Science, USSR Academy of Sciences, Prospect Vernadskogo 101, Moscow 117526, U S S R

196. R. C. Michelini (Italy, B3, PS4, PS1)
Industrial Robot Design Research Group, University of Genova, Italy

197. V. A. Michell (UK, J2, O1)
P. O. Box 273, Kings House, Bond Street, Bristol, BS99 7AL, U K

198. A. R. Mileham (UK, A3, A5, H1)
School of Mechanical Engg., University of Bath, Claverton Down, Bath BA2 7AY, U K

199. A. W. Miles (UK, I1, H1)
School of Mechanical Engg., University of Bath, Claverton Down, Bath BA2 7AY, U K

200. José del R. Millan (Italy, D6)
Institute for System Engg. & Informatics, Commission of the European Committee, Joint Research Centre, Building A36, 21020 ISPRA (VA)

201. S. K. Mishra (India, O1)
C. G. N. College, Gola Gokarannath, Kheri Uttar, Pradesh, India

202. Ahmed S. Mohamed (Canada, C1)
Department of Computing Sciences, University of Alberta, Edmonton, Alberta, Canada T6G 2H1

203. R. M. Molfino (Italy, PS4)
Industrial Robot Design Research Group, University of Genova, Italy

204. Emma C. Morley (UK, C1)
Department of Manufacturing Engg. & Management, University of Nottingham, University Park, Nottingham, NG7 2RD, U K

205. George Moys (UK, L1)
Middlesex Polytechnic School of Mechanical & Manufacturing Engg., Advanced Manufacturing Group, Bounds Green Road, London N11 2NQ, UK

206. George Nacher (Bulgaria, D4)
Robot Control Systems Research Centre, a. k. G. Boncher Str., B L 2, 1113 Sofia, Bulgaria

207. T. Nakamura (Japan, C2, PS4)
Robotics Department, Mechancial Engg. Laboratory, AIST, MITI, Namiki 1- 2 Tsukuba, Ibaraki 305 Japan

208. Shuzi Nakata (Japan, N1)
Faculty of Engg., Osawa University, Yamade-oke 2-1, Suita, Osaka, 565 Japan

209. P. B. Neal (Egypt, P2)
Production Engg. Department, Suez Canal University, Port Said, Egypt

210. D. S. Nesculescu (Canada, G3)
Mechancial Engg. Department, University of Ottawa, Ottawa, Ontario K1N 6N5, Canada

211. Xiong Ning (China, D5)
Department of Electrical & Computer Engg., China Textile University Shanghai 200051, P. R. China

212. Carloe Noe (Spain, L1, L2, O1)
Polytechnic of Milano, Department of Mechanics, Piazza Leonardo da Vinci, 32 - 20133 Milano, Spain

213. A. Nowrouzi (UK, C1, K1)
School of Mechanical Engg., Middlesex Polytechnic, Bounds Green Road, London N11 2NQ, U K

214. Patricia O'Grady (USA, A5)
Group of Intelligent Systems in Design & Manufacturing, Department of Industrial Engg., P O Box 7906, North Carolina State University, Raleigh NC 27695-7906, USA

215. Setsuo Ohsuga (Tokyo, A3, PS1)
Research Centre for Advanced Science & Technology, University of Tokyo, Tokyo

216. J. F. O'Kane (UK, E4)
Department of MCAE, Staffordshire Polytechnic, Stafford, U K

217. M. E. J. O'Kelly (UK, PS3)
Department of Industrial Engineering University of Galway, Galway, Ireland

218. M. O. M. Osman (Canada, D6)
Department of Mechanical Engg., Concordia University, Montreal, Canada

219. R. Pacagnella (Italy, P1)
Department of Mechanical Engg., University of Brescia, Via Valotti 9, 25060 Brescia, Italy

220. S. B. Padwal (India, P1)
Department of Mechanical Engg., I. I. T., Powai, Bombay 400076, India

221. Bruce Palmer (USA, J1, J3)
Digital Equipment Corporation, Maynard, MA 01754, U S A

222. B. Parsons (UK, E3)
School of Mechanical Engg., Advanced Manufacturing Department Middlesex Polytechnic, Bounds Green Road, London N11 2NQ, U K

223. V. B. Patange (UK, B1)
School of Engg., Sheffield City Polytechnic, Sheffield, U K

224. P. M. Pelagagge (Italy, B3, B5, F2)
Faculty of Engg., University of L'Aquila, 67040 Roio, L'Aquila, Italy

225. G. Pellegrini (Italy, P1)
Department of Mechanical Engg., University of Brescia, Via Valotti 9, 25060 Brescia, Italy

226. Marco Pellegrini (Spain, O1)
Polytechnic of Milano, Department of Mechanics, Piazza Leonardo da Vinci, 32 - 20133 Milano, Spain

227. B. R. Percinkova (Yugoslavia, G3)
Division of Automation & System Engg., Cyril & Methodius University Skopje, Macedonia, Yugoslavia

228. Monica Perego (Italy, L2)
L. E. I. Ltd., Milano, Italy

229. M. Perona (Italy, J1, J3)
Consorzio Autofaber, Milano, Italy

230. E. M. Petriu (Canada, G2)
Mechanical Engg. Department, University of Ottawa, Ottawa, Ontario K1N 6N5, Canada

231. Tonci Pirsic (Yugoslavia, C2)
Faculty of Electrical Engg., Mechanical Engg. and Naval Architecure, University of Split, R. Boskovica bb, 58000 Split, Yugoslavia

232. A. Pozzetti (Italy, J1, J2)
Polytechnic of Milano, Department of Economic and Production, Milano, Italy

233. B. Prasad (USA, O1, PS4, PS5, IPS)
West Virginia University, Centre for Industrial Research Applications (CIRA), Mechanical & Aerospace Engg. Department, Morgantown, W. Virginia 26506-6101, USA

234. William K. Preece (USA, D1, K2)
West Virginia University, Centre for Industrial Research Applications (CIRA), Mechanical & Aerospace Engg. Department, Morgantown, W. Virginia 26506-6101, USA

235. S. D. Prior (UK, A5, A4)
Faculty of Engg., Science & Maths, Middlesex Polytechnic, Bounds Green Road, London N11 2NQ, U K

236. J. M. Proth (France, B1, B2)
INRIA Technopôle, Metz 2000, 4 rue Marconi, 57070 Metz, France

237. Chandra S. Putcha (USA, B1)
California State University, Fullerton, CA 92634, U S A

238. Yang Qingdon (China, D5, P2)
Department of Mechanical Engg., Beijing Mechanical College, Beijing, 100026, China

239. Zheng Quan (China, C2)
Robot Research Institute, Harbin Institute of Technology, Harbin 150006, China

240. R. R. Raboy (Israel)
A. D. A., 5 Hashkedim Street, Kiriat, Bialik, 27000 Israel

241. P. Racz (Hungary, A1)
University of Miskolc, Department of Mechanical Engg., Miskolc - Egyetemváros, H - 3515, Hungary

242. Vevek Ram (South Africa, E2, E4, M1)
Department of Computer Science, University of Natal, P O Box 375, Pietermaritzburg, South Africa

243. K. V. Sambasiva Rao(India)
Department of Mechanical Engg., I I T, Delhi - 110016, India

244. D. H. Rahnejat(UK, G2)

245. M. A. Recine (Italy)
Consorzio Genova Ricerche, Genova, Italy

246. G. B. Reddy (Canada, N2)
Industrial Engg. Department, École Polytechnic, P O Box 6079, Montréal, Québec, Canada H3C 3A7

247. R. Reddy (USA, N2, PS1, IPS)
West Virginia University, Centre for Industrial Research Applications (CIRA), Mechancial & Aerospace Engg. Department, Morgantown, W. Virginia 26506-6101, USA

248. S. Reddy (USA, N1, PS5, IPS)
West Virginia University, Centre for Industrial Research Applications (CIRA), Mechanical & Aerospace Engg. Department, Morgantown, W. Virginia 26506-6101, USA

249. Raymond Sid Redman (USA, B4)
Industrial & Technology Centre, University of Missouri, Columbia, MO 65211, USA

250. Robert M Rice (USA, D2)
Press Engg. Division, Material & Process Laboratory, Science & Engg. Directorate, NASA/Marshall Space Flight Centre, AL 35812, U S A

251. M. J. Richard (Canada, D6)
Department of Mechancial Engg., Laval University, Quebec, Canada G1K 7P4

252. Michael Richter (Germany, H1)
Fraunhofer - Institut for Industrial Engg., Noblestr. 12, D-7000, Stuttgart 80, F. R. G.

253. Enrico Riva (Italy, O1)
Polytechnic of Milano, Department of Mechanics, Piazza Leonardo da Vinci, 32 - 20133 Milano, Italy

254. J. Robmann (Germany, C1)
Institut for Robotics (IRF), Universität Dortmund, Postbox 500500, D-4600 Dortmund 50, Germany

255. Aleksander D. Rodic (Yugoslavia, D2, D3)
Department of Robotics & Flexible Automation, Institute "Mihailo Pupin", Belgrade, Yugoslavia

256. Marcela Rodriguez (Colombia, I1)
DFAC Group, Faculty of Engg., UNIANDES, Apartado Aereo 4976, Bogotá, Colombia

257. R. Luis Roman (USA, E2, N2)
Department of Mechancial & Aerospace Engg., West Virginia University, Morgantwon, WV 26505, USA

258. Alberto Rovetta (Italy, F1)
Department of Mechanics, Research Group of Robotics, Polytechnic of Milano, Italy

259. A. A. Rubchinisky (USSR, D4, D5, P1)
Moscow Institute of Steel & Alloys, Moscow, U S S R

260. P. Sackett (UK, PS3)
Cranfield Institute of Technology, Cranfield, Bedford MK43 0AL, UK

261. Mahnaz Akbary Safa (UK, D5, D3)
Department of Mechancial Engg., Imperial College, Exhibition Road, London, UK

262. R. Sagar (India, A1, K2, K2)
Department of Mechanical Engg., I I T, Delhi - 110016, India

263. Remo Sala (Italy, F1)
Department of Mechanicla Engg., University of Brescia, Via Valotti 9, 25060 Brescia, Italy

264. M. A. Sanders (UK, F2)
U. M. I. S. T., Total Technology Department, P. O. Box 88, Manchester, M60 1QD, U K

265. D. Seslija (Yugoslavia, D2, E4)
Faculty of Technological Science, University of Novi Sad, 21000 Novi Sad, V. Vlahovica 3, Yugoslavia

266. J. L. Sevillano (Spain, B2)
Department of Electronics & Electromagnetism, University of Sevilla, Avda. Reina Mercedes, s/n 41012 Sevilla, Spain

267. H. M. Shi (China, N2)
Hua Zhong University of Science and Technology, Wu Han, Hu Bei, P. R. China

268. Isao Shimoyama (Japan, PS5)
Department of Mechano-Informatics, University of Tokyo, 7-3-1 Hongo, Bunkyo-ku, Tokyo 113, Japan

269. Avraham Shtub (Israel, N1)
Department of Industiral Engg., Tel Aviv University, Ramat aviv, Tel aviv 69978, Israel

270. Shao Shihuang (China, D5)
Department of Electrical and Computer Engg., China Textile University, Shanghai 200051, P. R. China

271. Andrea Sianesi (Italy, B3, B4)
Polytechnic of Milano, Department of Economics & Production, Piazza Leonardo da Vinci, 32 - 20133 Milano, Italy

272. N. Singh (Canada)
Department of Industrial Engg., University of Windsor, Ontario, Canada N9B 3P4

273. S. Sivaloganathan (UK, A3, A4)
Engg. Design Centre, School of Engg., City University, Northampton Square, London EC1V OHB, U K

274. R. Skoczylas (Poland, A3)
Institute of Foundary Technology & Mechanics, University of Mining and Metallurgy, al Mickiewicza 30, 30 - 059, Cracow, Poland

275. James E. Smith (USA, D1, K2, P2)
West Virginia University, Centre for Industrial Research Applications (CIRA), Mechanical & Aerospace Engg. Department, Morgantown, W. Virginia 26506-6101, USA

276. M. Sobolewski (USA, E2, E3, PS2, IPS)
Concurrent Engg. Research Centre, West Virginia University, Morgantwon, WV 26505, U S A

277. M. A. M. Soliman (Egypt, P2)
Production Engg. Department, Suez Canal University, Port Said, Egypt

278. G. Song (China, A4)
Department of Mechanical Engg., Shanghai Maritime University, Shanghai, China

279. Mark E. Staknis (USA, K1, K2)
Department of Industrial Engg. & Information Systems, Northeastern University, Boston, Massachusetts 02115, USA

280. C. Stancescu (Romania, L2)
Machine Construction Technology Department, Polytechnic Institute of Bucharest, Romania

281. Joseph Stanislao (USA, L2)
North Dakota State University, Fargo, North Dakota, U S A

282. S. Stankovski (Yugoslavia, D2, E4)
Faculty of Technology Science, University of Novi Sad, 21000 Novi Sad, V. Vlahovica 3, Yugoslavia

283. M. Staroswiecki (France, J2)
Centre d'automatique, University of Science & Technology of Lille 1, 59655 Villeneuvre d'Ascq, France

284. M. Stoker (UK, E3)
School of Mechanical Engg., Advanced Manufacturing Department Middlesex Polytechnic, Bounds Green Road, London N11 2NQ, U K

285. R. Stoughton (Japan, C2)
Robotics Department, Mechanical Engg. Laboratory, AIST, MITI, Namiki 1- 2 Tsukuba, Ibaraki 305 Japan

286. Chris Stylianides (USA, B4. L2)
North Dakota State University, Fargo, North Dakota, U S A

287. P. S. Subramanya (Canada, D2, D6, E3)
Department of Mechanical Engg. Concordia University, Montreal, Canada

288. T. J. Suh (UK, E3)
Department of Mechancial Engg., Queen Mary & Westfield College, Mile End Road, London E1 4NS, U K

289. A. Suhm (Germany, B5)
Institute für Rechneranwendung in Planung und Konstrukton (RPK)Univesität Karlsrue, Kaiserstraße 12, 7500 Karlsruhe, Germany

290. John M. Sullivan (USA, A3)
    Department of Mechanical Engg., Worcester Polytechnic Institue, Worcester, Massachusetts

291. D. Surla (Yugoslavia, G2)
    Institute of Mathematics, University of Novi Sad, 21000 Novi Sad, V. Vlahovica 3, Yugoslavia

292. Mikhail M. Svinin (USSR, B5)
    Osvobozhdenia 133, Flat 40, 664019, Irkutsk, U S S R

293. C. S. Syan (UK, C1, D6)
    Department of Manufacturing Engg., University of Nottingham, NNG7 2RD, U K

294. Janusz Szpytko (Poland, K1, K2, N2)
    Institute of Metallurgical Machines & Automatics, University of Mining and Metallurgy, al Mickiewicza 30, 30 - 059, Cracow, Poland

295. A. Szuder (Romania, L2)
    Machine Construction Technical Department, Polytechnic Institute of Bucharest, Romania

296. Farhad Tahmasebi (USA, G1)
    Robots Branch, NASA/Goddard Space Flight Centre, Greenbelt, MD 20771, U S A

297. Toshi Takamori (Japan, G1, PS2)
    Department of Instrumental Engg., Kobe University, Kobe, Japan

298. T. Tantoush (UK)
    University of St. Andrews, Kinnessburn, Kennedy Gardens, St. Andrews, Fife KY16 9DJ, U K

299. T. Tatishvili (USSR, G1)
    Institute of Machine Mechanics, Academy of Science of the Republic of Georgia, Tbilisi, U S S R

300. Qian Tao (China, D4)
    Robot Research Institute, Harbin Institute of Technology, Harbin 150006, P. R. China

301. W. A. Taylor (UK, F1, F2)
    Department of Information Management, The Queens University of Belfast, Northern Ireland, U K

302. Klaus Thaler (Germany, H1, L1)
    Fraunhofer Institute for Industrial Engg., Noblestr. 12, D-7000 Stuttgart 80,

303. Greg Thompson (USA, P2)
    West Virginia University, Centre for Industrial Research Applications (CIRA), Mechancial & Aerospace Engg. Department, Morgantown, W. Virginia 26506-6101, USA

304. Yang Ting (China, D1)
    Robot Research Institute, Harbin Institute of Technology, Harbin 150006, P. R. China

305. M. Tisza (Hungary, A1, A2)
    University of Miskolc, Department of Mechanical Engg., Miskolc - Egyetemváros, H - 3515, Hungary

306. José R. Toro (Colombia, N1)
    DFAC Group, Computer-Aided Design & Manufacturing Group, Engg. School, UNIANDES, Partado Aero 4976, Bogota, Colombia

307. George Trapp (USA, J3)
    Department of Industrial Engg., West Virginia University, Morgantwon, WV 26505, USA

308. S. Tricamo (USA, E1, F1)
    Design & Manufacture Institute, Mechanical Engg. Department, Stereo Institute of Technology, Hoboken, N. J. 07030, U S A

309. N. Trif (Canada, G2)
Mechanical Engg. Department, University of Ottawa, Ottawa, Ontario K1N 6N5, Canada

310. Brian Trought (UK, L1, L2)
Grantham College, Stonebridge Road, Grantham, Lincolnshire, U K

311. Ling-Wen Tsai (USA, G1, G3)
Mechanical Engg. Deparment & Systems Research Centre, University of Maryland, College Park, MD 20742, USA

312. T. Tsumura (Japan, PS5)
College of Engg., University of Osaka, Prefectate Mozu-Uremachi, Sakai 591, Japan

313. Craig Vanderhorst (USA, E4)
University of Cincinnati, Cincinnati, Ohio, U S A

314. D. Velasevic (Yugoslavia, E2)
Mihailo Pupin Institute, Belgrade, Yugoslavia

315. F. Vernadat (France, B1, B2)
INRIA-Lorraine / CESCOM, Technopole Metz 2000, 4 rue Marconi, 57070 Metz, France

316. A. E. Voevudko (USSR, D5)
Geotechnical Mechs. Institute of the Ukr.SSR Academy of Sciences Dniepropetrovsk, U S S R

317. M. Vukobratovic (Yugoslavia, G2)
"Mihailo Pupin" Institute, Beograd, Yugoslavia

318. Juergen Wahrburg (Germany, D1, D3)
ZESS (Centre for Sensory Systems) and Institute of Control Engg., University of Siegen, Siegen, Germany

315. W. W. Walter (USA, A2)
Mechanical Engg., Rochester Institute of Technology, Rochester, New York 14623, USA

316. Ming Tong Wang (Taiwan, I1)
Department of Industrial Engg & Centre for Automation Technology, Yuan-Ze Institute of Technology, Chingli, Taiwan, R. O. C.

317. Tohru Watanabe (Japan, PS3)
Department of Computer Science & Systems Engg., Ritsumeikan University, Kita-ku, Kyoto 603, Japan

318. R. A. Whitaker (UK, C1, K1)
School of Mechanical Engg., Middlesex Polytechnic, Bounds Green Road, London N11 2NQ, U K

319. A. S. White (UK, G3, G2, L2)
Robotics & Automation Group, School of Mechanical & Manufacturing Engg., Middlesex Polytechnic, Bounds Green Road, London N11 2NQ, U K

320. Robert H. Willison (USA, N2)
Department of Mechanical & Aerospace Engg., West Virginia University, Morgantwon, WV 26505, U S A

321. John R. Wilson (UK, C1)
Department of Manufacturing Engg. & Management, University of Nottingham, University Park, Nottingham, NG7 2RD, U K

322. J. Wojciechowki (Poland, A2, A4)
Institute of Industrial Automatic Control, Warsaw University of Technology, Poland

323. F. W. Wright (UK, A1, A5)
Electrical & Electronic Engg., Staffordshire Polytechnic, Beaconside, Staffs.

324. H. P. Wynn (UK, A3)
     Engg. Design Centre, School of Engg., City University, Northampton Square, London EC1V OHB, U K

325. Xu Xiaoli (China, D5)
     Deparment of Mechanical Engg., Beijing Mechanical College, Beijing, 100026, China

326. Xie Xiaoxia (China, P2)
     Deparment of Mechanical Engg., Beijing Mechanical College, Beijing, 100026, China

327. X. Xie (France, B1)
     INRIA Technopôle, Metz 2000, 4 rue Marconi, 57070 Metz, France

328. F. C. Yang (USA, B5)

329. Wu Yaping (China, P2)
     Department of Mechanical Engg., Beijing Mechanical College, Beijing, 100026, China

330. H. R. Yazgan (UK, H1)
     Department of Mechanical Engg., Queen Mary & Westfield College, Mile End Road, London E1 4NS, U K

331. A. M. Yemelyanov (USSR, M1)
     Department of Economic Cybernetics, Plekhanov Institute of National Economy, Moscow, U S S R

332. Robert E. Young (USA, A5)
     Group for Intelligent System in Design and Manufacture, Department of Industrial Engg., Campus Box 7906, Nth. Carolina State University, Raleigh, North Carolina, 27695 - 7906, U S A

333. A. Zakeri (UK, F1, G3)
     Manufacturing System Centre, Wolverhampton Polytechnic, Wulfrun St., Wolverhampton WV1 1SB, U K

334. M. B. Zaremba (Poland, G2, G1)
     Instytut Automatyki Przemyslowej, Politechnika Warszawska ul. Chodkiewicza 8, 02-525 Warszawa, Poland

335. R. Jassemi-Zargani (Canada, G3)
     Mechanical Engg. Department, University of Ottawa, Ottawa, Ontario K1N 6N5, Canada

336. L. Zavanella (Italy, D4)
     Department of Mechanical Engg., University of Brescia, Via Valotti 9, 25060 Brescia, Italy

337. M. Zelm (Germany, B2)
     IBM Germany, Department 2237, Am Hirnach 2, 7032 Sindelfigen, Germany

338. C. D. Zendron (Brasil, I1)
     Grucon, Engg. Mechanical, University Federal de Santa Caterina, Caixa Postal 476, 88049 - Florianópolis, SC, Brazil

339. H. H. Zhang (China, A5)
     Hua Zhong University of Science and Technology, Wu Han, Hu Bei, P. R. China

340. Yunfeng Zhang (UK, A5)
     School of Mechanical Engg., University of Bath, Claverton Down, Bath BA2 7AY3, U K

341. Young Zheng (China, D4)
     Robot Research Institute, Harbin Institute of Technology, Harbin, 150006, China

342. Wang Zhiqiang (China, D5)
     Dept. of Mech. Engg., Beijing Mech. College, Beijing, 100026, China

343. Erping Zhou (UK, D3)
     Department of Mechanical & Computer-Aided Engg., Staffordshire Polytechnic, Staffs., U K

344. J. Zych (Poland, A3)
     Institute of Foundary Technology & Mechanics, University of Mining and Metallurgy, al Mickiewicza 30, 30 - 059, Cracow, Poland

CAD/CAM and Numerical Control

Post-Processor Generators for Numerically Controlled Machine Tools

by
S. BEUVELET (1)
P. CURNIER (2)
E. DEBERNARD (3)
P.L. KOCIEMBA (4)

(1) Stéphane BEUVELET, engineer from Ecole d'Ingénieurs de TOURS in CIM, doctorate in automated manufacturing from ENS CACHAN, master in industrial computer science, University of TOURS.
(2) Pascal CURNIER, engineer Aérospatiale, DNC expert, research manager in post-processors at the Louis-Blériot Joint Research Center, Suresnes.
(3) Eric DEBERNARD, engineer CESTI, doctorate in automated manufacturing
(4) Pierre-Laurent KOCIEMBA, engineer Aérospatiale, head of the robotics and automation science department at the Louis-Blériot Joint Research Center, Suresnes.

Current State

The ever-increasing automation of French industries has led to an increase of needs in numerically controlled machines. These numerical controls are more and more various (NUM 760, SIEMENS, PLASMA, DIXI,...). For requirements of performances and integration, CAD-CAM softwares are constantly improved and do not always provide data directly exploitable (UNIGRAPHICS II, CATIA, STRIM 100, CVNC ...). Between these two fields in full development, there is an interface between the CAD-CAM software and the machine numerical control : this is the Post-processor.

A post-processor is a software which converts or reformates data produced by the computer-aided design and manufacturing module into a format which can be understood by the machine numerical control. Until today, when a company purchased a machine, it had also to purchase a post-processor which would enable the company to interface the post-processor with the machine. Thus, the post-processor was customized for the machine and the CAD-CAM software.

This is the reason why AEROSPATIALE, a leading company in the aeronautical field, has emphasized the study of a post-processor generator which could be both easy of use and which could generate codes for any type of machine tools.

Although many post-processor generators are commercially available [CAM 91] [Keith 89] [Brandli 80], few of them use parameterization to describe the functional set of the machine. This is the reason why we wish to present you in this article the fundamental principles of this post-processor generator.

This post-processor generator (GPPCAM : Generic Post-Processor for Computer-Aided Manufacturing) is composed of three main parts : a descriptive part of the machine (kinematic modelling), an operational part (code generation) and an optimization part (optimization of the NC file, also called tape).

These different parts use information stored in base files of the generator. Thus, to describe the behaviour of the machine tool, these files need just to be filled and then to be interpreted by the machine tool.

Descriptive part of the machine

In the file (Clfile) provided by the CAD-CAM system, all positions of the tool are given according to a reference called APT reference in which the workpiece to be machined has been designed.

In the file designed for the machine numerical control (tape), the tool positions are given according to a reference called Machine reference of the machine tool.

Moreover, the tool vector has always an orientation defined according to the workpiece area. This condition permits to determine the angular coordinates of each angle link. The second problem is therefore to determine these angles by the resolution of an equation system.

The descriptive part has been designed to give a solution to these problems. The file used by this part is called the MACHINE file. There are two types of information in this file, on the one hand information relating to the operative part of the machine (translation and rotation axes, shifts, tools, spindle rotation , feed rate, tolerances...) and on the other hand information relating to the tape (type, length, machining time ...).

Thanks to these pieces of information it is possible to create transfer matrices which will permit to determine the coordinates of the tool head in the different references, knowing that there are four references :

- the APT or programming reference

- the Tool reference relating to the tool

- the Workpiece reference relating to the workpiece to be machined

- the Machine reference relating to the machine which is fixed.

At the beginning, all the references are mixed up unless there are shifts. From this description on, the programmer establishes the kinematic sequence of the machine, i.e. the of axes permitting to relate the workpiece to the tool. The GPPCAM is then capable to create the transfer matrices (4X4 matrices). These matrices will be automatically updated each time the state of the machine will be modified (rotation of the table, rotation of the tool...).

Let us take the case of a Mandelli 5-axis machining center composed of two rotation axes (rotary plate B and Twist A : rotation of the tool) and three translation axes (X, Y, Z) (see figure 1).

figure 1 : Kinematic modelling of the machine

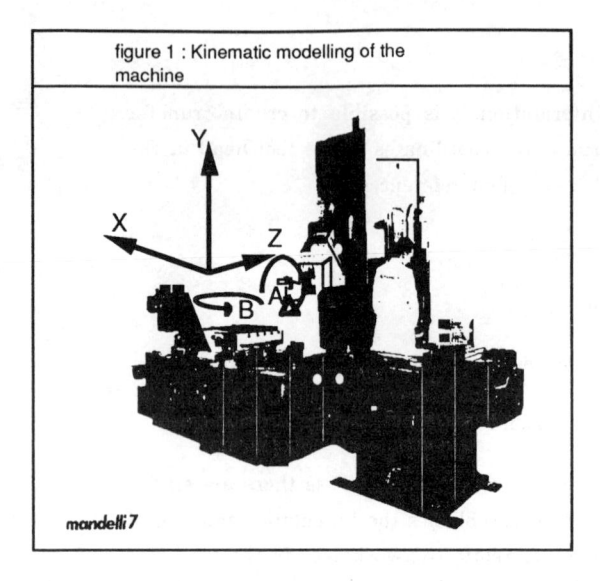

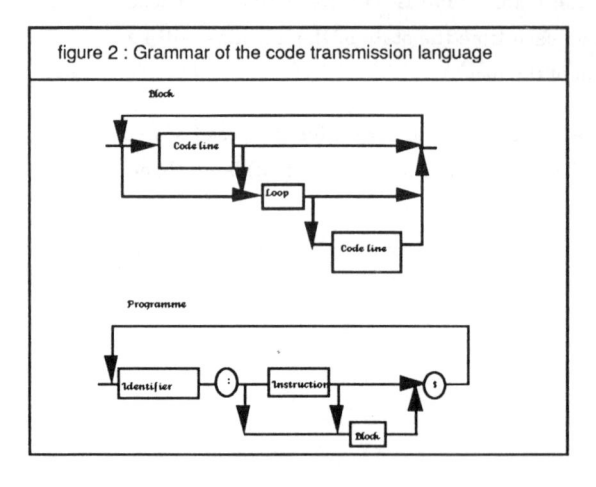

mandelli 7

figure 2 : Grammar of the code transmission language

figure 3 : Example of the TECHNO file

```
ARCSLP :
IF (sens=1.AND.plan=1) THEN
G(2) I(xc) J(yc) K(pas) E(360*z/pas-360*z1/pas)
$
ELSE IF (sens=-1.AND.plan=1) THEN
:
:
$
ENDIF
$
CALSUB/n <,TIMES,t><OPTION,x1,x2,x3,...> :
G(27) NUM(n) DO i=1,nbparam DX=(param[i]) ENDDO
$
CALSUB/UP,n :
JUMPTO/ (n)
$
CALSUB/UP,n1,n2,n3,OPTION :
IF (cond_saut) (n1) (n2) (n3)
$
CHECK/RED,ON :
M(58)
$
CHECK/RED,OFF :
M(59)
$
CHECK/DASH :
G(10)
$
DELAY :
G(4) T(t*3)
$
```

In order to establish the kinematic sequence, we start from the workpiece (put on the table) and we note the axes we meet until we reach the tool. In this case, we have the sequence: B, X, Z, Y, A. This representation is sufficient to describe the behaviour of any machine tool having at most three rotation axes A, B, C and three translation axes X, Y, Z (to which it is possible to add three secondary linear axes u, v, w).

The operational part of GPPCAM

The operational part of GPPCAM can be decomposed into three phases :

- a translation phase of the Clfile

- a processing phase

- a code transmission phase

The objective of the operational part is to be capable of controlling all the codes that are transmitted to the tape. The post-processor sequentially reads the Clfile, transcodes each recording, sets off the necessary procedures and generates a set of codes of numerical control stored in an intermediary file (file of first transfer).

The phase of code transmission to the tape permits on the one hand to program oneself the codes to be generated - i.e. *generation by programmation* -, and on the other hand to read the codes to be generated in a file, i.e.the *automatic generation*.

Generation by programmation

Generation by programmation consists in modifying the set of procedures of the generator to be capable of solving the problems stipulated in the specifications. It is a question of modifications concerning both the kinematic aspect of the machine and the aspect of code transmission.

To be able to carry out these modifications, we can have access to a series of independant procedures. The names of these procedures end up by the identifier "-CODE" and begin by a comprehensible sequence of characters which perfectly reflects the Clfile word which is under processing. For instance, in the ARCSLP_CODE procedure, the whole set of codes to generate on the tape and relating to the word ARCSLP will be found.

Thanks to these procedures, when a new post-processor need to be created, it is possible to carry out top-level programming, i.e. at the level of code transmission. It gives total control over the codes to be transmitted.

Automatic generation

The automatic generation of codes on the tape is obtained thanks to a base file of the GPPCAM called TECHNO file. This file gathers 90% of the codes generated on the tape. The file information are organized and belong to a grammar which constitutes a code transmission language. This language permits among others to carry out tests and loops. This language grammar is built around the data block concept which is assimilated to the way numerical controls represent this information (see figure 2). A block is a sequence of numerical characters and arguments. Characters correspond either to preparatory functions of the machine (ex "G"), technological functions (ex "O") or auxiliary functions (ex "M").

Thus, the TECHNO file enables to gather the code transmission on the tape of each post-processor. An example of the TECHNO file is given in figure 3.

The optimization part of GPPCAM

The objective of this part is to optimize the file (tape) designed for the machine numerical control. The optimization part works on the first transfer file of the operational part. It is decomposed into three parts.
First step (code deletion)
The post-processor ends the processing of recordings, deletes the unnecessary modal codes (i.e. the codes which need not be sent to each tape line) and calculates the machining time. A new intermediary file is created.

Second step (block rise)
The post-processor rises numerical control blocks from a line to the other as long as possible.
Third step (block numerotation and tape cutting)

To automatically carry out steps 1 and 2 , the GPPCAM uses the FICHCODE file which contains the following information :
- the format of the numerical control codes,
- the list of modal codes to be suppressed during the first step if they are transmitted several times without being revoked,
- the list of revocation orders to suppress the modal codes,
- the list of remontée possibilities of codes from onr line to the other to optimize the tape length.

Conclusion

GGPCAM is a very wide generator for access to procedures; moreover it is possible to edit technical data sheets and to calculate machining times. The originality of GGPCAM lies in its code transmission language which differs from usual macro-languages and allows to process both the Clfile and code transmission. This generator offers the possibility to rapidly intervene on the code transmission to the tape and on the coordinates of points to be reached. Control over the generator is reinforced by documentation for each procedure, by a classification of procedures according to the invariancy rate and by a performing software architecture gathering these procedures. Finally, the post-processor generator solution saves much time of development which can be from one to two for 5-axis machines . The post-processor generator is the essential tool to integrate the complete CAD-CAM process which goes from designing to manufacturing.

Stéphane BEUVELET

[Brandli 80] BRANDLI, "Anwendung von Compiler Generatoren zur Generierung von Post-/Preprozessoren im ESPRIT-Projekt CAD*I", VDI BERICHTE, 1980.

[CAM 91] "CAM-POST, le Générateur de post-processeurs", Corporation des Technologies ICAM, 1991.

[Keith 89] KEITH & McKNIGHT, "Generic Post-Processors for Computing-Aided Manufacturing, PRINTED CIRCUIT DESIGN, janvier 1989.

A CAD Interpreter for Interfacing CAD and CAPP of 2-D Rotational Parts

Yunfeng Zhang and A. R. Mileham

School of Mechanical Engineering, University of Bath, UK

Abstract

A new approach for interfacing CAD and CAPP is presented. The CAD interpreter extracts from a CAD DXF (Drawing Interchange File) file all the geometric information and dimensional tolerances and formalises them into a machining feature based product model, which can be used by an automatic process planning system directly. It deals with wire frame models and drawings. The algorithms for machining surface identification and tolerance allocation are discussed. A typical example using this system is given.

1. Introduction

Rapid change and increasing competition has led to the emergence of CAD/CAM systems. The manufacturing activities included in CAD/CAM fall into three categories: (1) Computer-Aided Design (CAD), (2) Computer-Aided Process Planning (CAPP), and (3) Computer-Aided Manufacturing (CAM). To make CAD/CAM integrated, one important task is interfacing the CAD database to the CAPP system, thus enabling automatic product definition input to the planning module. This has led to a considerable number of research approaches on automatic manufacturing feature extraction from CAD models.

Wang [6] developed a machining surface identifier which is able to extract geometric information from a part data file generated using a 2D wire-frame CAD modeller. However, technological information (tolerance, surface roughness, etc) is still inputted interactively. An IGES (Initial Graphics Exchange Specification) post-processor has been reported [5], which can extract the data that defines a 2-1/2D prismatic component, based on a wire-frame IGES file. In [2] a geometric extractor, AUTOGEM, has been developed to interface a specific 3D CAD system named CADAM. All of these research approaches have contributed significantly to this domain. However, research on this topic is far from being conclusive and more effort needs to be concentrated on the following three areas:

(1) The benefit of automatic interfacing is often limited by the fact that only a specific CAD modeller can be used. To overcome this restriction, a CAD neutral format specification should be used. Any CAD modeller can then be used as long as the modeller can convert drawing data to the neutral format.

(2) In order to provide all the necessary data of a component for a CAPP system, geometric and technological information, such as tolerance, should be included in the data interpretation process.

(3) Since there are many ways to model a component using a CAD modeller, the CAPP system has to recognise these possibilities in its data file in order to keep the information requirement to a minimum.

Based on these requirements, a CAD interpreter has been developed to identify the geometric and tolerance data that define a 2D rotational component from a DXF file. The system aims at automating the interface between CAD and CAPP to minimize human intervention.

2. The CAD Product Model

The CAD product models used in today's CAD systems can be broadly classified into three categories: (1) wire frame models, (2) surface models, and (3) solid models. Among these graphic representation schemes, the wire frame model is the simplest one. Despite its limitations, such as non-completeness and ambiguity, the wire frame model is still very widely used in industrial CAD. This is mainly due to its simplicity, and its low requirement for computer storage space and processing time. For 2D and 2-1/2D components, the wire frame modelling system is the best tool to use.

The product model used in this work is generated using a wire frame modeller accompanied by dimensions and tolerances.

3. DXF File

In a CAD system, the drawing data base is usually stored in a very compact format which is system specific. To assist in interchanging component data between different implementations of the same CAD package on different machines, and among different CAD packages, a "Drawing Interchange" file format has been defined [1]. The data file with such a format is called a DXF file. Most of the 2-D wire frame CAD systems on the market have the ability to produce DXF files.

There are four sections of information in the overall organization of a DXF file. Generally, the information contained in a DXF file can be categorized in to two groups as follows:

(1) *General Information of a Drawing*
This category includes information about the font, line type, layer, etc.
(2) *Information of Entities*
This category contains information of all the drawing entities, such as geometric entities and text.

Among these two categories, the entity information contains the data which defines a component, since all the information about geometric elements and tolerance is stored in this section. The entities can be identified from the *Entity Section* of a DXF file. Each entity type is uniquely specified by an *entity type number*. Its dimension is defined by the parameters following its type number.

4. The Manufacturing Product Model

From the manufacturing point of view, a component should be defined by using machining features (e.g. taper, hole) rather than pure geometric entities (e.g. line, arc). In this research, a rotational turned component can be described by the following machining features: (1) face, (2) cylindrical surface, (3) chamfer, (4) taper, (5) fillet, (6) thread, and (7) groove, which can be either internal or external.

This part description method has been used by BEPPS-NC [7, 8], an automated process planning system developed in this research. The objective of the CAD interpreter is to convert the CAD product model to a manufacturing product model, which can drive BEPPS-NC directively.

5. The CAD Interpreter

5.1 Methodology

Chang [3] specifies that there are two major tasks involved in the design/process planning interface: model decomposition and feature recognition. The decomposition task separates the features from a component model, while the recognition task classifies and identifies the semantics of the feature. However, only geometric information can be translated using the above method. In this research, the following procedures are followed in the process of transforming a CAD model into a manufacturing model:

(1) Read and extract only the necessary information from a DXF file. Enhance the parameters of the entities and form a new data file for further processing.
(2) Identify those entities which form the external profile of the component. Remove these entities from the data file and then identify those entities which form the internal profile, if there are any.
(3) Extract any grooves from the external profile, order the external machining surfaces, and recognise any external threads. This procedure is applied to the internal profile if there is one.
(4) Allocate tolerances to their relevant dimensions.
(5) Output this information in a format by which a manufacturing model is defined.

A diagram of these procedures is given in **Figure 1**. Each procedure is discussed in detail below.

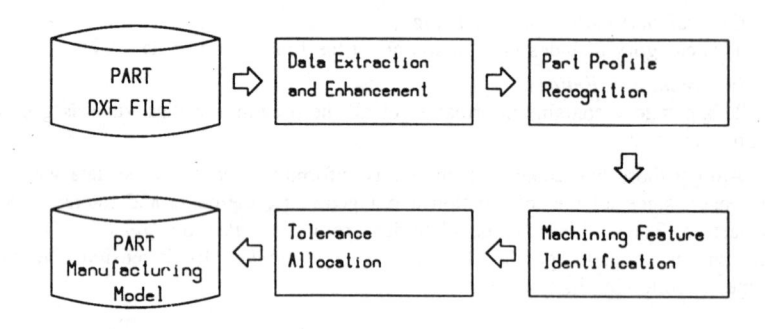

Figure 1. A Flow Diagram of the CAD Interpreter

5.2 Data Extraction and Enhancement

As discussed before, DXF is a text file in which all the geometric and tolerance information is stored in the *ENTITY SECTION*. An entity is either a geometric element (i.e. line, arc) or a text entity. An algorithm has been developed which has the intelligence to extract each geometric element and tolerance information and perform data enhancement as follows:

(1) If the ENTITY is a line, read in its starting point and ending point coordinates.
(2) If the ENTITY is an arc, convert its parameters (center coordinates, starting angle, ending angle) to the form of the coordinates of starting points and ending point and its radius.
(3) If the ENTITY is a text, first check if it is a tolerance. If so, work out its basic size and tolerance values and read in the coordinates of its witness lines.

For a symmetrical rotational part, a one-view drawing in 2-D space provides complete manufacturing information. Further data enhancement is performed to remove those geometric entities which are below the center line from the database. The remaining entities are then used for the identification process. This enhancement procedure leads to a significant decrease of the geometric reasoning time in the feature identification stage.

5.3 Component Profile Recognition

In this stage, both the internal and external profile of the component is identified. The main profile of a component is formed by joining pure geometric machining surfaces (i.e. face, cylindrical surface, arc, etc.). The reasoning process is based on the following characteristics which are unique to the profile of a symmetrical rotational component:

(1) The profile should have continuity at the starting point and ending point of each geometric element.
(2) The profile has its boundary. Externally, it starts from the right most element and finishes at the left most element. Internally, it starts from one ending face of the external profile, and ends at either the other external ending face if it is a through hole or at the intersection point with the center line if it is a blind hole.
(3) If more than one entity is found which start from a connecting point, check the leading direction for each entity. The one that leads to the center line is removed and the other one is treated as the entity of the profile.

Output of the profile recognition stage forms an internal data base which is further processed in the feature identification stage.

5.4 Machining Feature Identification

In manufacturing, a groove and thread are treated as special features and machined using special cutting tools. To construct a manufacturing product model, grooves and threads must be extracted from the profile.

During this stage, the system first checks each recess both external and internal. Based on its dimensional parameters, it is determined whether the recess is a groove or not. If a groove is extracted, the remaining profile is re-arranged (sequence number, feature type, dimensions). This process continues until all the recesses are checked. The next step is to identify the threads with their body numbers and location information from amongst the database. The output of this stage provides a model with all the necessary geometric information that can be displayed on the screen as a 2D model.

5.5 Tolerance Allocation

There are usually two types of tolerance in an engineering drawing: dimensional tolerance and geometrical tolerance. In this research, only dimensional tolerance is considered.

Given a tolerance entity, there are basically two conditions which may help to identify its relevant feature and dimension. The first is called the *size condition* which means that the basic size of the tolerance should be equal to the relevant dimension. The second is called the *position condition* which is based on the information of its witness lines, particularly the tolerance string position and the witness gap. A feature and the relevant dimension can be defined if it satisfies both the *size condition* and the *position condition* of the tolerance.

5.6 Output Format

The final results of the data interpretation process are stored in a data file in which each machining feature occupies a record. The record contains the sequence number, feature type, dimensional parameters and tolerance. The data file can be easily accessed by the other program modules for further processing, or modification.

6. CAD Interpreter Application Example

The program is written in FORTRAN and runs on SUN-3 workstation. DAXCAD [4] is used to design the component and provide the DXF file. **Figure 2** shows the example component drawing and **Figure 3** shows the component images produced based on the results of CAD interpretation.

7. Conclusions

A CAD interpreter has been developed which is able to transform a CAD product model into a manufacturing model. Information extracted in this process includes geometric features and dimensional tolerances. The system has been validated over a wide range of symmetrical rotational components, and represents a major step towards CAD/CAPP integration. It also opens a new research direction in the field of interfacing CAD and CAPP using the DXF file for other groups of component, such as the 2-1/2D prismatic component.

REFERENCES

[1] AUTODESK INC. *AUTOCAD User Reference*, 1986.
[2] Bond, A.H., Melkanoff, M.A., Ahmed, S. Zia, Chang, K.J., Kim, D.H. and Soetarman, B., *Automatic Extraction of Geometric Features from CAD Models*, Intelligent Manufacturing Systems II, pp. 143-160, Elsevier Science Publishers B.V., Amsterdam, 1988.
[3] Chang T C, *Expert Process Planning for Manufacturing*, Addison-Wesley Publishing Company, 1990.
[4] Practical Technology, *DAXCAD (Rev 2.136)*, (c) 1985, 86, 87, 88.
[5] Vosniakos G.-C and Davies B.J., *An IGES Post-Processor for Interfacing CAD and CAPP of 2-1/2D Prismatic Parts*, Int J Adv Manuf Technol 5, pp 135-164, 1990
[6] Wang, Hsu-Pin and Wysk, Richard A., *Intellegent Reasoning for Process planning*, Computer In Industry 8, pp 293-309, 1987.
[7] Zhang Y and Mileham A R, *BEPPS-NC: A Rule Based Expert System for Process Planning*, Proceedings of the Fifth National Conference on Production Research, pp

[8] Zhang Y, *BEPPS-NC: An Process Planning System to Generate NC program from CAD Product Models*, PhD Thesis in Preparation, Bath University, 1991.

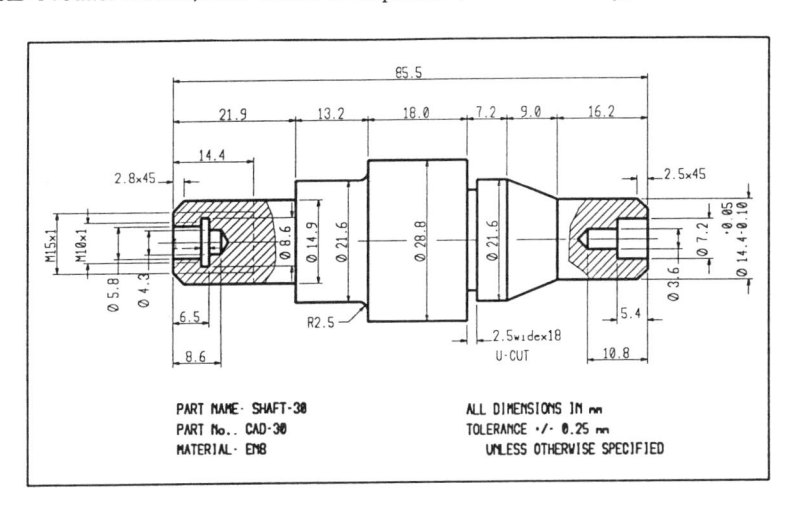

Figure 2. The Example Component Drawing

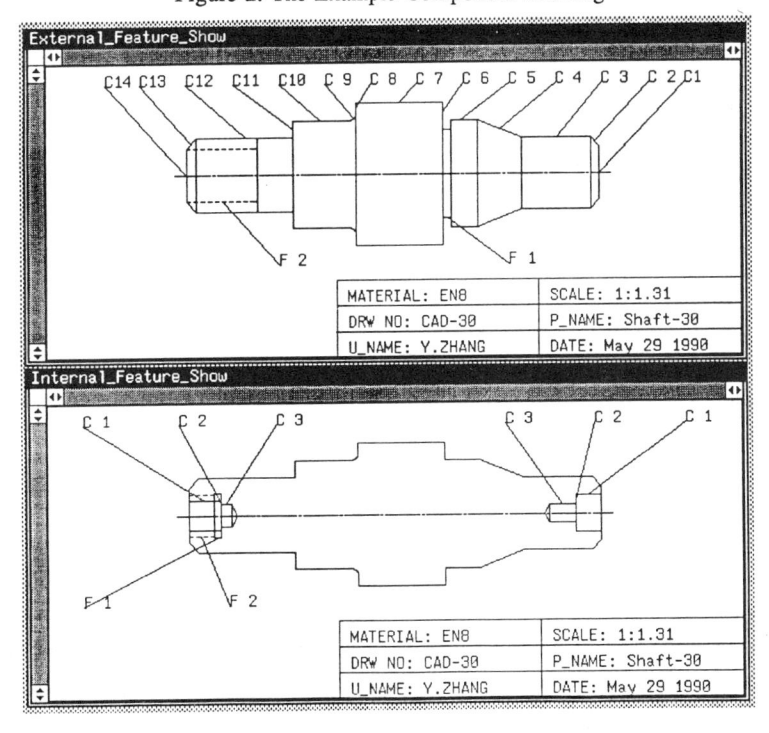

Figure 3. Graphical Disply of the Manufacturing Model

Computer Aided Design in Rehabilitation Robotics

S.D. PRIOR

Middlesex Polytechnic
Faculty of Engineering, Science & Mathematics
Bounds Green Road
London N11 2NQ.

Email: stephen2@uk.ac.mx.cluster

Abstract

This paper describes the area of robotics as applied to assisting disabled people with their everyday needs.

A sophisticated CAD system, namely, an IBM 5080 workstation running CATIA software from Dassault Systemes of France is shown together with the way in which robotic systems can be created, optimised and animated using it to produce realistic tasks. Finally, future trends in computer aided design are discussed.

Introduction

The area of rehabilitation robotics is a combination of industrial robotics and rehabilitation engineering.

The first attempts at producing robotic systems for the disabled began in the late 1960's and early 1970's. Nearly all these systems have failed to reach a production stage due to problems of acceptance by the intended users and high unit costs.

Rehabilitation robotic research can be sub-divided into two main areas, these are:

Robotic Workstations - Usually consist of a table-mounted robotic arm which can manipulate and/or interact with various objects within its workstation environment.

Mobile Robots - Consist of a robotic device mounted on a powered mobile base.

The use of commercial robots in workstation systems tends to increase the cost of the final system beyond that which a disabled person can afford. This limits their use to people in institutional care, and also limits the time that an individual can spend using the system. Since the majority of physically disabled people are living at home with support from family and friends this is an important factor in favour of mobile robotic systems.

Current Research

The current research programme at Middlesex Polytechnic is directed towards the design and construction of an electric wheelchair mounted robotic arm. A robot arm attached to an electric wheelchair and controlled directly by the user, has the advantages that it is always within reach and can be manoeuvred with the wheelchair to perform a variety of tasks inside or outside the home environment.

CAD System Description

The CAD system used in this project is an IBM 5080 workstation system driven by a 4381 mainframe computer and running CATIA (Computer Aided Three-dimensional Interactive Application) software version 2.2.5, developed by Dassault Systemes of France and licensed by IBM.

CATIA Functionality

CATIA is a multi-functional CAD package incorporating 2D draughting, 3D solid and surface modelling, mechanical analysis, colour shading, kinematic design, NC programming and robotic design. The full program consists of over fifty separate functions associated with specific modules, ie, solids, NC, robotics, etc. CATIA is menu driven and when working in the robotics module the two main functions used are Robot and Task.

Robot Function:
This function is used to define and optimise robot cells.

Task Function:
This function is used to define, simulate and record Tasks. Tasks are programmed robot actions defined by the user.

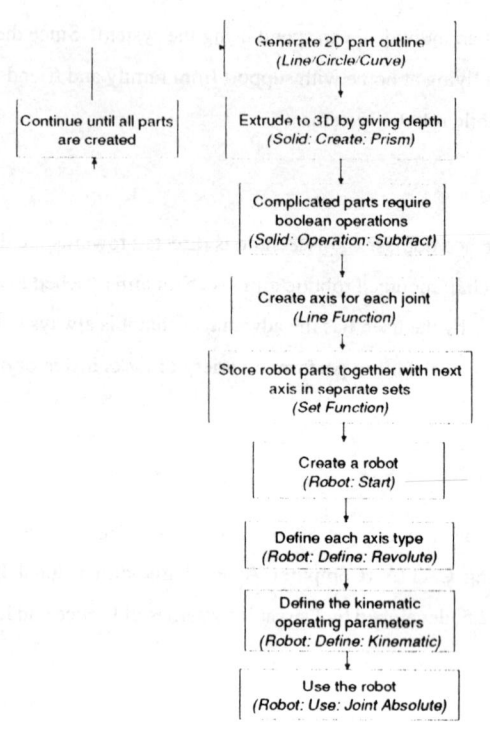

```
                    Generate 2D part outline
                       (Line/Circle/Curve)
                              ↓
Continue until all parts    Extrude to 3D by giving depth
    are created              (Solid: Create: Prism)
         ↑                        ↓
                    Complicated parts require
                       boolean operations
                    (Solid: Operation: Subtract)
                              ↓
                    Create axis for each joint
                         (Line Function)
                              ↓
                Store robot parts together with next
                      axis in separate sets
                          (Set Function)
                              ↓
                         Create a robot
                         (Robot: Start)
                              ↓
                      Define each axis type
                   (Robot: Define: Revolute)
                              ↓
                     Define the kinematic
                     operating parameters
                   (Robot: Define: Kinematic)
                              ↓
                         Use the robot
                  (Robot: Use: Joint Absolute)
```

Fig. 1. Block diagram of robot design using CATIA

Advanced Features

☐ Within the robot function it is possible to specify the robot kinematics, ie, the joint velocities, maximum acceleration, maximum deceleration which are taken into account in the Task function. It is therefore possible to compute the length of time needed to execute a task or series of tasks.

☐ Sensors and events within the robot environment can also be defined which will be used by logical structures for controlling the instruction sequences executed by the task.

☐ Calculating the forces and/or torques to be exerted on the joints by the actuators of the robot in order to preserve static balance with externally applied loads.

☐ The Density Function is used to take into account the effect of gravity on the selected part, based on the definition of gravity and the density of the selected part - this only applies to polyhedral elements.

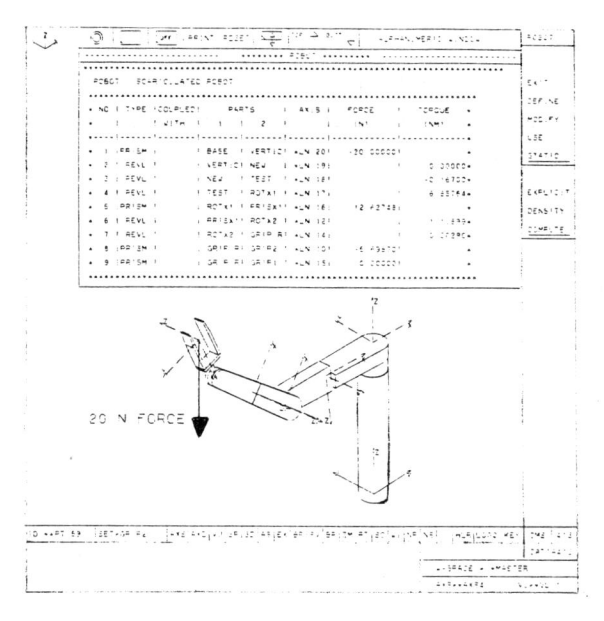

Fig. 2. Forces/torques required to maintain static balance

☐ By placing unfixed axes to the joints of the robot, the Eulerian coordinates at each location can be ascertained.

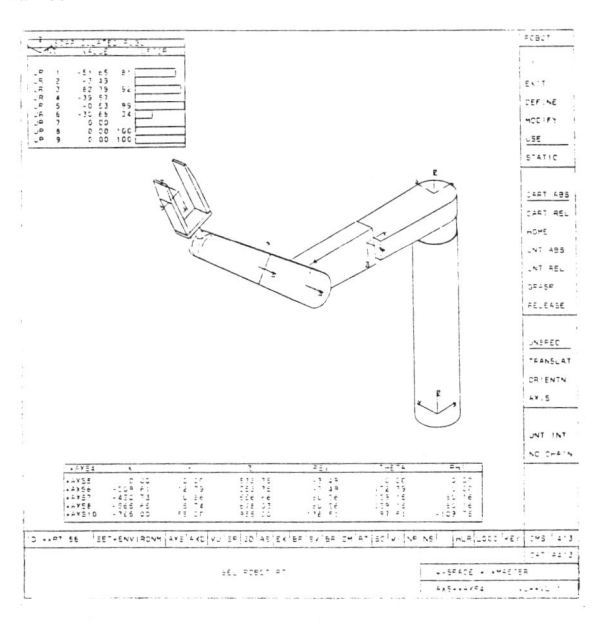

Fig. 3. Euler coordinates obtained from robot joint axes

17

☐ Any part of the robot or associated points can be traced when the robot is simulated in the Task function. This allows visualisation of the motion and accurate tracking of the individual joints, end effector, etc.

Limitations

Robots designed using this version of CATIA are limited to 20 joints, the number of unfixed axes are limited to 5 per robot and not more than one per component part.

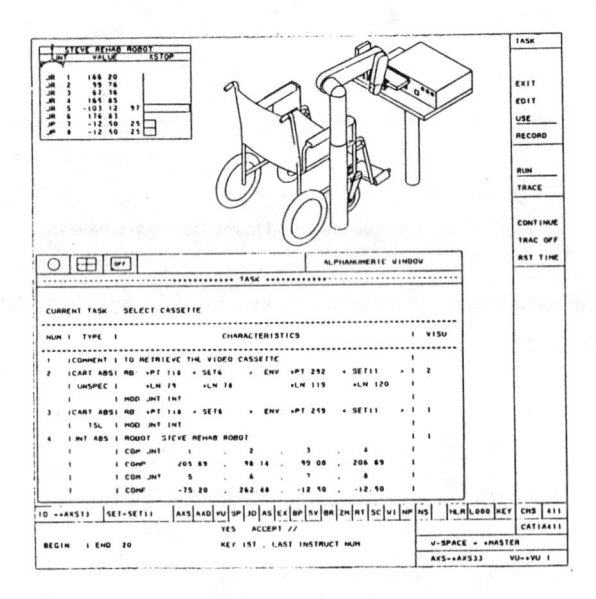

Fig. 4. Conceptual design showing task generation program

Off-Line Programming

Once the task or series of tasks has been defined, simulated and verified as correct, the program can then be recorded before being edited by the post-processor and downloaded to the real robotic cell.

The program can be recorded in Advanced or Basic format, the Advanced format is used with advanced robot controllers which use numerical robotic languages, the Basic format is used for basic controllers which can only handle elementary data such as cartesian or joint positions.

Future Developments

The latest version of CATIA (Version 3.0) has certain enhancements over earlier releases, a selection of these being, robot libraries, greater flexibility in the control of the robot motion, collision monitoring, improved animation algorithms and smooth shading.

The new IBM RISC 6000 stand alone workstation system is already available for use with CATIA allowing greater flexibility and faster operation.

Future trends in simulation of robotics will include more widespread use of transputers to speed up shaded animation, greater interest in virtual reality as a means of training disabled people to use rehabilitation robots and also as a means of teleoperation in dangerous environments, ie, fire fighting, nuclear or military.

Acknowledgement

The author would like to thank Jeff Cooper CAD Demonstrator at Middlesex Polytechnic for his help and encouragement during this project.

References

1. Han, C.; Traver, A.E.; Tesar, D.: Using CAD/CAM in the design of a robotic micromanipulator. *Computer-Aided Engineering Journal.* 4 (1990) 43-48.

2. Aldrich, J.; Schwandt, D.; Sabelman, E.; Van Der Loos, M.: Computer-aided design of rehabilitation devices. *Proceedings of RESNA 9th Annual Conference, Minneapolis, Minnesota.* (1986) 190-192.

3. Kochan, A.: Taking robot programs off-line. *Manufacturing Systems.* 4 (1991) 24-27.

A Concurrent Engineering System: The Use of Constraint Networks

Robert E. Young, Peter O'Grady, and Arthur Greef

Group for Intelligent Systems In Design and Manufacturing, Department of Industrial Engineering, Campus Box 7906, North Carolina State University, Raleigh, North Carolina 27695-7906, USA

Abstract

Concurrent Engineering (sometimes called Simultaneous Engineering or Life Cycle Engineering) consists of bringing a number of factors to bear at the design stage. These factors include product, function, design, materials, manufacturing processes and cost, testability, serviceability, quality, reliability and redesign. In such a manner, designs can be improved from the perspective of the product's life cycle.

This paper describes the use of Artificial Intelligence Constraint Nets to advise the designer on improvements that can be made to the design from the perspective of the product's life cycle. The difficulties associated with performing Concurrent Engineering are reviewed, and the various approaches to Concurrent Engineering are discussed. The requirements for a system to support Concurrent Engineering are indicated. An overview of constraint nets is given and this leads into a description of **SPARK**, an Artificial Intelligence Constraint Nets system for Concurrent Engineering. The operation of **SPARK** is illustrated by considering an example application. The system has been applied thus far to design for assembly, design of cast parts, design of printed wiring boards, financial justification, design for testability, design for maintainability, design of machined parts, and facilities design. The advantages of **SPARK** include being flexible enough to allow the designer to approach a problem from a variety of viewpoints, allowing the designer to design despite having incomplete information, and being able to handle the wide variety of life cycle information requirements. The result is a powerful Concurrent Engineering system that can handle a wide number of application areas.

1. INTRODUCTION

Concurrent Engineering is of considerable importance to manufacturing industry since it is at the design stage that the life cycle requirements are defined. For example, it has been reported that upwards of 70% of a product's manufacturing cost is dictated by design decisions [1,2,3].

Although the evidence is largely anecdotal, it appears that Concurrent Engineering is not generally performed well in western manufacturing industry in that design is carried out without due regard to the various life cycle factors. This can result in designs that are expensive to manufacture, test, service, maintain and redesign.

This paper describes the use of Artificial Intelligence Constraint Nets to advise the designer on improvements that can be made to the design from the perspective of the product's life cycle.

2. CONCURRENT ENGINEERING

The fact that Concurrent Engineering is not performed well can be attributed to three main sources of difficulty [4]: the characteristics of the design process, the volume and variety of life cycle knowledge, and the separation of life cycle functions [5,6,7,8,9,10]. The result is that designers do not adequately consider life cycle factors [5,10,11].

There are a number of techniques and systems that support Concurrent Engineering by advising designers on aspects that reduce life cycle problems [12]. These include the use of design teams, design handbooks, checklists and structured procedures, manufacturing simulation and process planning, and the use of expert systems. There are however drawbacks when these approaches are used for Concurrent Engineering.

3. USE OF CONSTRAINT NETWORKS

As an alternative, we propose the use of a computerized support tool to bring life-cycle information to the designer in a readily usable form. This support tool will, in effect, emulate a good design team by suggesting changes that will improve the design from the life-cycle perspective. The requirements for such a support tool are rigorous and encompass a number of substantial research issues [13]:

- it should be flexible enough to allow the design problem to be approached from a variety of viewpoints;
- it should allow the designer to design despite the absence of complete information;
- it should handle the large volume, variety, and interdependence of life-cycle information;
- it should readily interface to database management and CAD systems;
- it should have a good user interface and be able to explain itself in a manner comprehensible to humans;
- it should support design audits.

4. CONSTRAINT NETWORKS

It is widely accepted that design activities can be regarded as involving search to satisfy constraints [14, 15]. This model of design has been applied to a wide variety of design domains.

A *constraint network* is a collection of constraints which are interconnected by virtue of sharing variables. An example, adapted from Smith [16], and used in, for example, [17], is shown in Figure 1. The variables (ovals in Figure 1) link the constraints (rectangles in Figure 1). The network contains the constraints that must be considered in determining a hole diameter in a printed wiring board. In our work, the selection and testing of variable values is carried out by a combination of humans and computer. This is then a practical proposition for application to Concurrent Engineering. This human/computer combination is termed constraint monitoring.

There are several constraint based languages but these have disadvantages when considered for a design advice system for Concurrent Engineering. The primary disadvantage is that they were mostly developed for fairly narrow application areas and they are therefore not suitable for application to the wide domain of Concurrent Engineering. The development of a viable constraint network language is therefore a crucial step in developing an effective Concurrent Engineering system. The recognition of this has lead to the development of **SPARK** which is a successor to CADEMA [18] and Galileo/Leo [19].

5.   SPARK: A CONSTRAINT NETWORK PROGRAMMING LANGUAGE

SPARK is a constraint programming language based upon an implementation of first order predicate logic combined with frame-based inheritance. This allows users to model concurrent engineering systems as *constraint networks,* e.g., a collection of constraints that are interconnected through shared variables. The constraints, their shared variables, and their interconnections, can be represented graphically as a network. Figure 1 is an example constraint network for hole selection on printed wiring boards. The objective of a **SPARK** program is to find a set of variable values that doesn't violate any of the constraints. Values are propagated bi-directionally among constraints through the shared variables.

A great strength of constraint network-based systems is that logical consistency checking is an inherent property of the underlying technology. Because constraint network technology is based upon first order predicate logic with an underlying constraint propagation mechanism based upon an inference engine, when a constraint is added, deleted or modified, the system immediately tries to find a set of values that will meet all the constraints (i.e., make them logically true). If this is not possible, constraints that are violated (i.e., that are logically false) are identified and the truth maintenance system makes suggestions to the user for correcting the violation. This ensures that a completed constraint network is logically consistent and that when knowledge is updated the resulting system is also logically consistent. As a consequence, an existing problem in creating and maintaining rule-based expert systems is solved automatically by an inherent property of constraint networks.

As an example, the constraint network shown in Fig. 1 describes the constraints inherent in one task in the design of printed wiring boards (PWBs). The task is the selection of hole sizes for component leads on a PWB. A PWB contains many holes, sized for the leads that will attach components to the board and to connect different conducting planes within a multi-layered circuit board. Hole size is determined by the board thickness, the need to accommodate the size and shape of the component lead, whether copper plating is needed to connect different conducting planes within a multi-layered board, and the need to physically and electrically connect the component lead to the PWB with solder. If the hole is too small, then after plating, a lead may not fit into it. If it does fit and there is insufficient oversize, then solder can't wick into the hole. If the hole is too large, then solder won't wick into it. Poor solder wicking results in poor solder joints and is a source of mechanical and electrical failure.

From a manufacturing perspective, hole selection should be constrained by manufacturing capability and cost. Although there is potentially an infinite number of hole sizes, all production facilities have a finite set of drill sizes. The drills are held in the tool inventory and, as noted by our industry partners, adding a tool size to the inventory can cost up to $25,000. To avoid adding tool sizes to the tool inventory, a hole size should be selected that can be drilled by an existing tool. If the hole size variation on a board exceeds the magazine size for the drilling machine, then additional machine setups are required. If we can keep the hole size variation within the maximum drill magazine size, then we can manufacture the PWB with a single setup and avoid increased production time and cost.

6. SUMMARY AND CONCLUSIONS

Concurrent Engineering is important in that it is at the design stage that much of a products costs are specified. This paper has discussed the application of AI-based constraint networks to Concurrent Engineering. The **SPARK** constraint network programming language has been briefly described. An example Concurrent Engineering problem of PWB design has been presented.

7. REFERENCES

1   Andreasen, M.M., Khler, S., Lund, T. (1983). Design for Assembly. U.K.: IFS Publications Ltd.

2   Maddux, K. C., and Jain, S. C. (1986). CAE for the Manufacturing Engineer: The Role of Process Simulation in Concurrent Engineering, in A. A. Tseng, D. R. Durham, and R. Komanduri, eds., Manufacturing Simulation and Processes, ASME: New York, 1986.

3   National Science Foundation (1987). Research Priorities for Proposed NSF Strategic Manufacturing Initiative, Report of a National Science Foundation Workshop, National Science Foundation: Washington, DC.

4   O'Grady P., Bahler D., Bowen J., and Young R.,"Constraint Nets for Life Cycle Engineering", Proposal to National Science Foundation, 1989.

5   Evans, B. (1988). Simultaneous Engineering. Mechanical Engineering, Vol. 110, No. 2. New York, USA: American Society of Mechanical Engineers.

6   Ruiz, C., Koenigsberger,F.(1970). Design for Strength and Production. New York: Gordon Breach Science Publishers Inc.

7   Harfmann, A.C. (1987). The Rationalizing of Design. Computability of Design. USA: John Wiley and Sons.

8   Baxter,R. (1984). Companies must Integrate - or Stagnate!, Production Engineering, April 1984. London: Institute of Production Engineers.

9   Peck, H. (1973). Designing for Manufacturing. London: Pitman Publishing.

10 Gairola, A. (1986). Design for Assembly: A Challenge for Expert Systems. Robotics, Volume 2. Elsevier Science Publishers B.V. (North Holland).

11 Riley, F.J., (1983). Assembly Automation: A Management Handbook. NY, USA: Industrial Press.

12 Jakiela, M., Papalambros, P., Ulsoy, A.G. (1984). Programming Optimal Suggestions in the Design Concept Phase: Application to the Boothroyd Assembly Charts. (ASME Tech. Paper. No. 84-DET-77). New York, NY: American Society of Mechanical Engineers.

13 Bowen, J. and O'Grady P., (1989a), Characteristics of a Support Tool for Life Cycle Engineering, LISDEM Technical Report, North Carolina State University.

14 Gross, M., Ervin, S., Anderson, J., Fleisher, A. (1987). Designing with Constraints. Computability of Design.USA: John Wiley and Sons.

15 Wu, P., 1988, Design for Testability, Proceedings, National Conference of the American Association for Artificial Intelligence, pp. 358-363.

16 Smith L.,"The Design of printed Wiring Boards", M.S. Report, IMSEI, North Carolina State University, 1989.

17 O'Grady P., Young R., Greef A and Smith L.,(1991), "A Design Advice Systems for Concurrent Engineering" Technical Report, Dept. Industrial Engineering, North Carolina State University.

18 O'Grady P. J., Ramers D., and Bowen J., Artificial Intelligence Constraint Nets Applied to Design for Economic Manufacture and Assembly, Computer Integrated Manufacturing Systems, Vol. 1, No. 4, 1988

19 Bowen J. and O'Grady P., (1989b), "A Constraint Programming Language for Life-Cycle Engineering." October 1989, revised March 1990. LISDEM Technical Report, North Carolina State University.

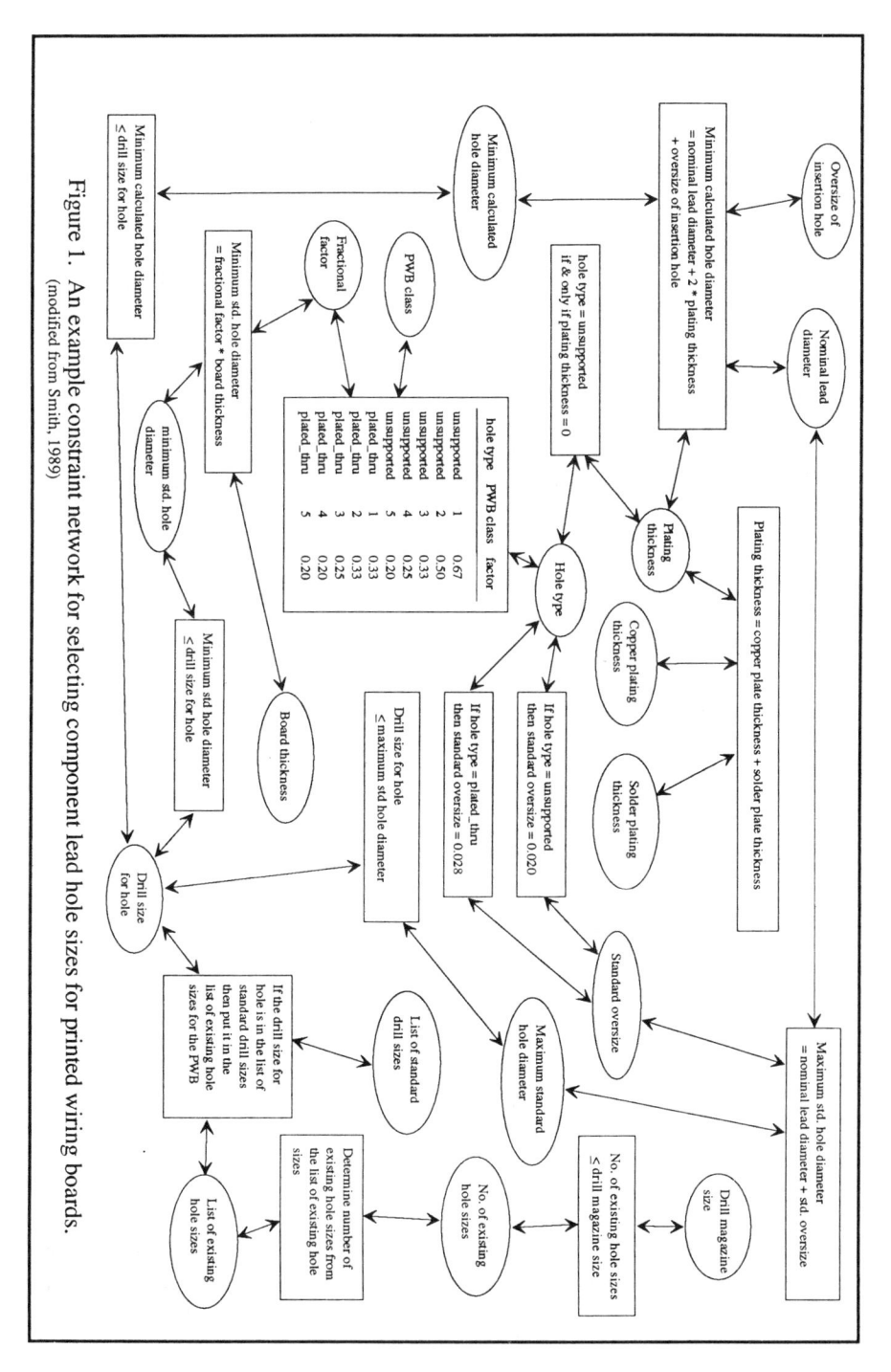

Figure 1. An example constraint network for selecting component lead hole sizes for printed wiring boards.
(modified from Smith, 1989)

A PC-Based Gearbox Design System

G.Song   Lecturer,   Department of Mechanical Engineering,
Shanghai Maritime University, Shanghai, China.

A.Jebb,   Professor,   Engineering Design Centre, School of engineering,
City University, Northampton Square,
London EC1V 0HB

S.Sivaloganathan,   Senoior Research Assistant,
Engineering Design Centre, School of Engineering,
City University, Northampton Square,
London EC1V 0HB

1.0 INTRODUCTION

Industrial gearboxes are commonly used in mechanical systems and represent an important area of mechanical engineering manufacture. The gearbox design is a reasonably difficult problem which involves the satisfaction of many design constraints. It is the job of the designer to specify the major geometrical features of a gearbox in order to ensure consideration of the known requirements. The design process is iterative and time consuming by hand. Currently the general tendency in designing gears is to adopt the ISO standards. It has a comprehensive list of factors which are selected from a wide range, pertaining to the application, and used to alter the stress levels. Inevitably the associated calculations become complex and the volume of data required increased. These make gearbox design more suitable for computers.

The application of computer techniques to gearing design has to date been a well explored field. Various suites of programs have been written, some concentrating only on specialised aspects of gear design, while others claim to have taken a more integrated approach to the subject. Song [1] has produced a survey on these packages. This paper describes a PC based interactive computer system for the design of single or double reduction gearboxes based on the ISO standard and covering a wide range of design functions.

2.0 COMPUTING SYSTEM AND SOFTWARE FEATURES

Figure 1 indicates the outline of the package. It is written in FORTRAN, assembly language and used Graphics Development Toolkit [2] and GINO-F[3]. The program is divided into (a) design programs and (b) draughting programs. The design programs are (i) Gear design (ii) shaft design and (iii) bearing selection. They run on an IBM PC or compatibles with CGA, EGA, VGA or Hercules adaptor and operate in an interactive mode. The computer's role in this application is essentially complementary and at the end of each activity the designer is able to study the results and can decide to proceed to the next stage or to re-run the current stage. In this way, the designer is given considerable freedom to reach an acceptable solution using his ability and experience.

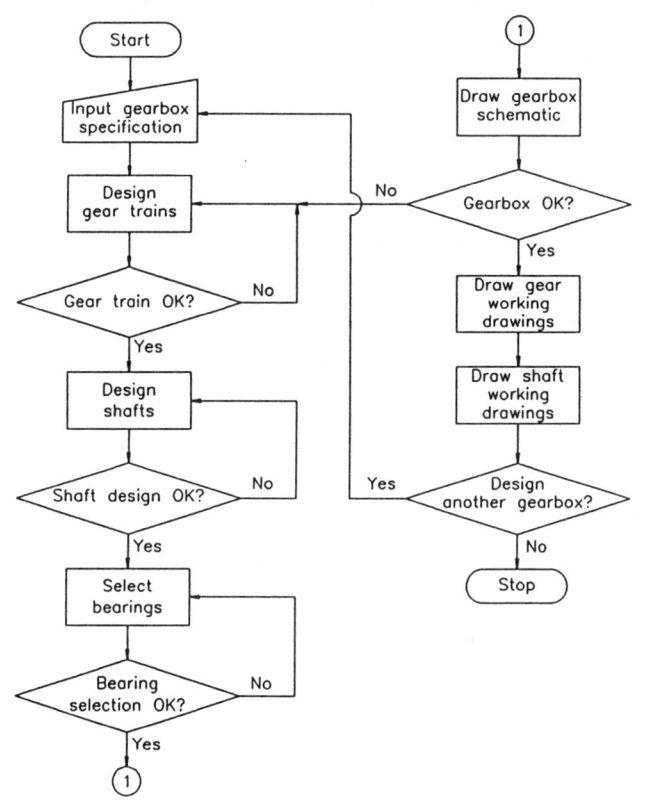

FIGURE 1 GEARBOX DESIGN FLOWCHART

The programs have many built-in features which are designed to make them foolproof and to give clear unambiguous instructions to the user. They may be listed as:

- Inputting data takes the form of pressing a single key wherever possible
- There are safeguards against wrong entries e.g. alphabetic entries will be ignored when numerical entries are expected. This prevents program failure due to wrong entries.
- Detection and warning of inconsistent data which would cause program failure e.g. in bearing selection stage an entry of a non-standard bore diameter will result in a BEEP and the message

No bearing has xxx mm bore diameter. The standard bore diameters near to the required value are xxx and xxx mm. Please input data again.

- Some input data, such as that defining the shaft shape, may be quickly verified through their graphic display.

- The input data can be easily corrected by using the editing facility.
- Help facilities are available to the user by pressing the function key F1 whenever difficulties are encountered in data input
- The output is labelled and self explanatory.
- Whenever the user is not satisfied with the results, he can modify any of the original parameters without having to re-enter all the parameters. This enables him to visualise the effect of the change immediately.
- Both input data and the results are stored in a computer file which could be printed out to study the interim stages of the design.
- Options are provided to visualise the inputs and the results in many instances.
- Optional printing facilities are provided in the appropriate places.

Where technical restrictions are known, either through experience or theory, these have been written into the program. As the software is interactive, the user is only required to enter the answers to the questions asked by the computer. The user need not know the structure or working of the program.

3.0   GEAR DESIGN

This section describes the programs involved in the gear design part of the software, tooth number calculation, tooth profile generation, tooth profile calculation, tooth geometric calculation, gear strength analysis and gear strength design.

3.1   TOOTH NUMBER CALCULATION

Finding the numbers of teeth in a gear train satisfying a specified ratio within a fine tolerance has been a serious problem throughout the history of gearing. A great deal of work has been done in investigating the ways and means of solving problems of this kind. An evaluation of these procedures is given by Song [1]. This program is based on properties of continuous fractions and conjugate fractions. It offers a direct means of finding the required number of teeth in each gear and provides the best possible approximation to the specified gear ratio within the limits of the gear train.

3.2   TOOTH PROFILE GENERATION

In modern manufacturing practice, nearly all involute gear teeth are cut or finished by one of a number of "generating processes". The principle of gear tooth generation is most easily understood by considering the use of a rack form cutter as a means of producing straight spur teeth. This program computes the successive positions of the tooth profile as the rack form cutter, with or without protuberance, is rolled around the gear blank in convenient angular steps. It illustrates the generating process of gear tooth profiles and provides the user with the information on the top land width, the minimum tooth number and modification co-efficient for undercutting to occur. Figure 2 shows this output screen.

It is useful for the understanding of the principles of gear tooth generation, the analysis of the influence of various gear parameters on the tooth forms, and the study of the tooth forms of non-standard and different national gear standards.

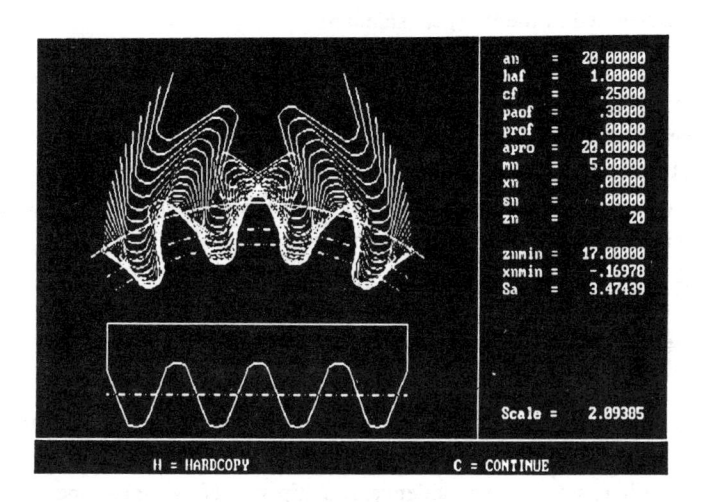

FIGURE 2   GEAR TOOTH GENERATION FROM A RACK FORM CUTTER

3.3   TOOTH PROFILE CALCULATION

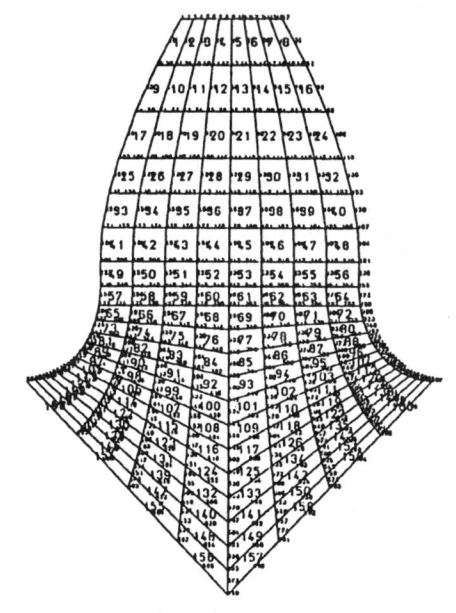

FIGURE 3 PRODUCTION OF GEAR TOOTH GRID

This program is used to (a) provide for visual inspection of the tooth profile prior to manufacture or producing the working drawings of either single tooth or a whole gear (b) estimate the dimensions of the tooth profile and (c) determine the proportions of the critical sections assumed by the ISO gear standard.

The program produces a desirable gear tooth grid, comprised of plane membrane elements with eight nodes. The user is able to see the element connections and the position of each element directly on a display and can change a mesh instantaneously to obtain the best mesh arrangement. Finite element analysis on stresses and strains can now be carried out using commercially available programs such as LUSAS [4]. Figure 3 shows an example of a gear tooth grid produced by the program.

3.4 TOOTH GEOMETRIC CALCULATION

Detailed geometric calculations are necessary for precise strength analysis, manufacture and inspection of a gear set. This program provides the user information on

- Angular dimensions such as transverse pressure angle, operating pressure angle, and base helix angle.
- Linear dimensions such as addenda, root, tip and reference parameters.
- Miscellaneous dimensions such as equivalent number of teeth, contact ratios and pressure angles at the highest points of single tooth pair contact.
- Inspection dimensions such as base tangent lengths and diameters over pins.
- Undercutting, interference and topping if they occur.

3.5 GEAR STRENGTH ANALYSIS

This program can quickly complete the strength analysis of a gear set based on the ISO gear standards. The detailed results presented to the user includes the safety factors against pitting and breakage, all the values of various factors, geometric dimensions and meshing characteristics. From this information the user can judge the design quality of the gear set and make any necessary modifications.

3.6 GEAR STRENGTH DESIGN

This program is used to design spur or helical, external or internal, standard or modified, single or double reduction, reverted or developed gear trains based on the ISO gear standard. The preset face width factors are 0.1, 0.2, 0.3, 0.4, 0.5, and 0.6 and the helix angles $8°$, $12°$, $16°$ and $20°$. The user therefore now has alternative design solutions to choose from. The theoretical number of design solutions depends on the type of gearing and whether the face width factor and helix angle are specified by the user. Some solutions may not be acceptable because of the large dynamic factor or load distribution factor etc. These unacceptable solutions are eliminated to speed up the design process.

When designing double reduction gearing, the user may select the first-reduction gear ratio he prefers, or use one of gear ratio split methods based on the particular requirement of the application. These requirements may be that (i)the first reduction gear set has the same contact strength as the second (ii) the gear box is of the lightest weight (iii) the gearbox is of minimum size (iv) the gearbox employs standard centre distances and (v) employs splash lubrication. However the programmer encourages the user to try different gear ratio split methods to obtain a more optimal design.

When the centre distances of the gearing to be designed are specified, the program will adjust the helix angle, select the suitable modification coefficients or alter the tooth sum to meet the requirements.

The end result will be a detailed document comprising :

- Design specifications and given conditions
- Gear tooth proportion and information on contact ratio, undercutting, interference, etc.;
- Surface in fillet roughness, tolerances on gear tooth, lubricant viscosity, bearing span and pinion offset distances;
- Information on gear strength analysis such as safety factors, load factors, etc.;
- Tangential, radial and axial forces on the gears.

4.0 SHAFT DESIGN ANALYSIS

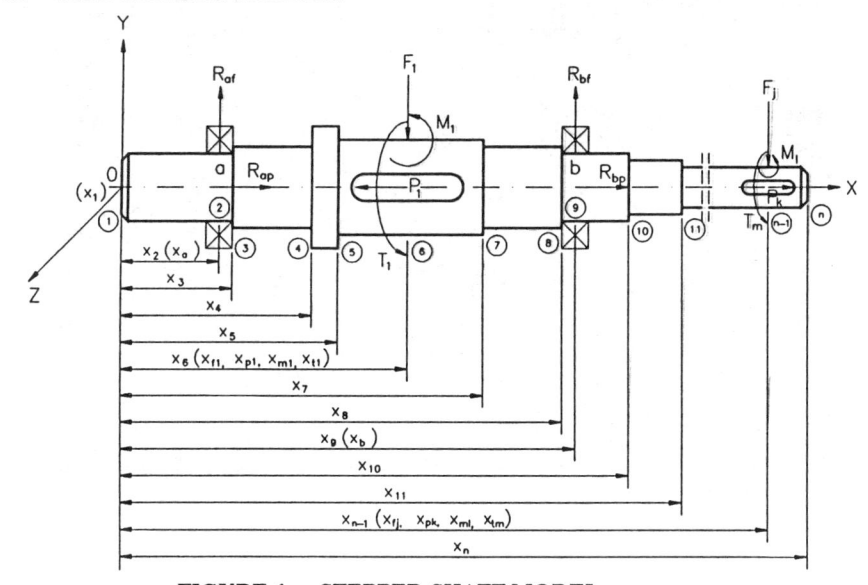

FIGURE 4   STEPPED SHAFT MODEL

Figure 4 shows a stepped shaft model used in the shaft program. It allows for keyways, shoulder fillets or other kinds of stress concentration raisers. In both vertical and horizontal planes, it may be loaded by j concentrated forces along Y axis, k axial forces along X axis, l bending moments and n twisting moments. The directions chosen for X, Y and Z axes are the clues to the sign conventions and for the forces and moments. The program designs shafts according to the maximum shear stress theory and checks their fatigue strength, static strength, rigidity and vibration requirements. The output from the program consistes of

- Shaft diameters
- Combined bending and torsional stresses at critical cross- sections.
- Calculated safety factors at critical cross-sections.
- Forces acting on bearings and bores.
- Deflections and slopes at all nodes along the shaft.
- Angles of twist per unit length.
- Axial load diagram, torsional moment diagram, bending moment diagram, reduced moment diagram, slope diagram and deflection diagram.
- Critical speed of the shaft.

5.0 BEARING SELECTION AND ANALYSIS

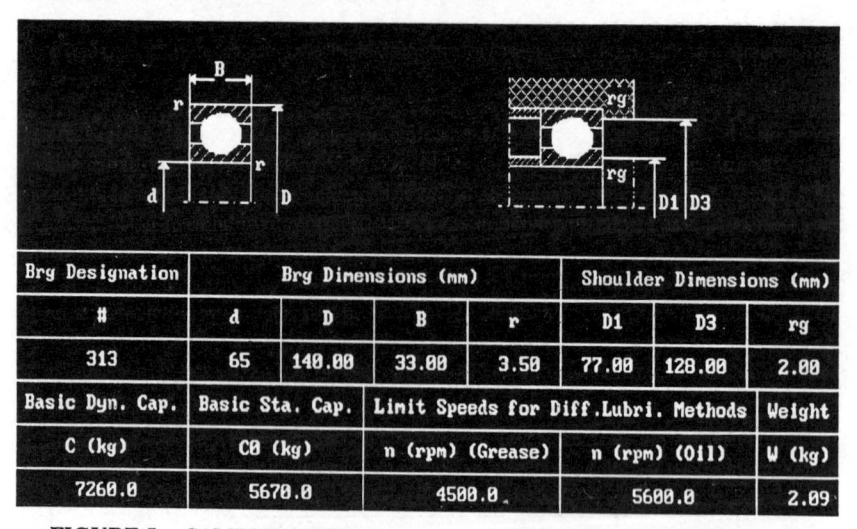

Brg Designation	Brg Dimensions (mm)				Shoulder Dimensions (mm)			
#	d	D	B	r	D1	D3	rg
313	65	140.00	33.00	3.50	77.00	128.00	2.00
Basic Dyn. Cap.	Basic Sta. Cap.	Limit Speeds for Diff.Lubri. Methods			Weight	
C (kg)	C0 (kg)	n (rpm) (Grease)		n (rpm) (Oil)		W (kg)
7260.0	5670.0	4500.0		5600.0		2.09

FIGURE 5   SAMPLE OUTPUT OF INFORMATION ON BEARING 313

This program takes into account the most frequently encountered situation where each end of the shaft is supported by one bearing, and these two bearings are capable of providing sufficient radial support to limit shaft bending and deflection to acceptable values. This is highly desirable and simplifies manufacturing.

The entire information on various types of standardised ball and roller bearings available bearing data covers the bore, outside diameter, width, fillet radius, shoulder diameters, shoulder fillet radius, basic dynamic and static capacities, radial and thrust factors, weight and designation for each bearing.

The program selects the rolling bearings in accordance with the basic dynamic and static capacities and calculates their service lives. If the bearings do not rotate, they will be selected according to the basic static capacity. Otherwise they will be selected according to the basic dynamic capacity and checked for the static strength. When a bearing, selected on the basis of dynamic capacity, fails to meet the requirements of the static strength (which is likely to happen to a bearing rotating at very low speed), it can be selected on the basic static capacity by setting the bearing speed to zero.

This program provides the user with all the suitable bearings in the series of the type of bearing specified, required static and dynamic capacities (if the bearing rotates) and the service lives of the bearings under the given conditions. Information on suitable bearings can be presented on request. Figure 5 shows the sample output of information on bearing 313.

6.0 DRAUGHTING PROBLEMS

Graphical output forms an important and integral part of the whole gearbox design package. The draughting software mainly consists of three programs which automatically make a gearbox schematic drawing and gear and shaft working drawings, based on the data files produced by the design programs. The gearbox schematic drawing provides the user with the principal gearbox characteristics, containing the layout of the gearbox, its major dimensions and specifications. The working drawing program can produce spoked gears, webbed gears, solid gears, and integral gear shafts.

7.0 CONCLUSIONS

The computer package described in this paper is used to design the main components of gearbox for transmitting power between parallel shafts. It not only serves as practical design tool in the industry but also could be used as a teaching aid in the universities. It can form a basis for the input of gearing sytems into solid modelling.

REFERENCES

1   Song G. An aplication of computer aided design and computer aided draughting techniques to Gear design, PhD thesis , City University 1988.
2   Graphics Development Toolkit, IBM, 1984.
3   GINO-F User Manual Issue 2, CAD centre, Cambridge, 1980.
4.   LUSAS Finite element analysis system user's manual, Finite element analysis ltd, London 1986.
5   Handbook of Machine element Machine Design and machine drawing division, The northeast Industry Institute, Peking, 1974

Monotonicity Preserving Approximation for Curve Definition in Computer Aided Design

J. WOJCIECHOWSKI

Institute of Industrial Automatic Control
Warsaw University of Technology, Poland

Summary

A constrained least square approximation has been applied to the definition of the B-spline piecewise cubic curve. The constraints can be imposed on the B-spline coefficients in such a way as to ensure the desired shape of the curve i.e. monotonicity of the function or its derivatives. The problem is one of quadratic programming and can be solved using a gradient projection method. Practical applications have been presented.

Introduction

A sculptured surface is often defined from a set of data points placed in a rectangular mesh. At first a set of curves passing through the given data points is defined. This requires an interpolation scheme. Frequently data come from measurements or a rough estimation of surface points and the position error can even be more than 0.1mm. In such a case curves may have unexpected changes or oscillations in slope and curvature. In order to avoid an improper shape of curves, additional constraints should be used which ensure that the curves will preserve essential information concerning the shape and obey the required constraints.
One of the solutions is to use shape preserving interpolation methods [1,3,5] which result in a curve with the same shape characteristic (e.g. monotonicity) as the given data. If there are more data points than segments of a piecewise function, approximation scheme is needed [2,4]. In many applications, however, the curve description method allows the designer to define an interpolating curve, but the shape obtained is not satisfactory.
In the paper we restrict ourselves to the definition of a planar curve which can be described as a single valued piecewise cubic C^2 function expressed in B-spline basis. The C^2 continuity is essential when NC programming methods for 5C milling machines are to be applied. To achieve a substantial reduction of machining time a surface must have regular changes in curvature. Moreover

C^2 continuity assures that the third derivative (or speed of curvature changes) is limited. Data points are given as n real values y_i for n parameter values u_i. The proposed method is to find the optimal approximating function of the form

$$P(u) = \sum_{j=-1}^{n+1} D_j N_{3,j}(u) \qquad (1)$$

which complies with the prescribed constraints coming from requirements of monotonicity of the function or its derivatives

where D_j are B-spline coefficients (vertices)

$N_{3,j}(u)$ are B-spline basis functions defined by the knots u_i

If the interpolating B-spline curve complies with the constraints, it can be used as the best solution for the approximation.

Constraints

In technical applications curves can always be divided into a few segments with monotonic curvature. This implies that only a few segments with monotonic slope and position exist. Although these segments are not the same (a segment with monotonic curvature can have an extreme in slope) the number of the segments is practically the same. This additional shape information can be provided by the designer as a set of constraints imposed on the B-spline coefficients. The B-spline function is monotonic in the prescribed interval, if the B-spline coefficients (vertices) are monotonic [2] and the monotonicity constraints can be described as linear inequalities of the form of $D_i - D_{i+1} \leq (\geq) 0$. The same rule can be applied for the first and the second derivatives as they can be expressed as B-spline functions of degree two and one respectively. The slope monotonicity constraints can be written as

$$D'_i - D'_{i+1} \leq (\geq) 0$$

where

$$D'_i = \frac{3 \cdot (D_i - D_{i-1})}{d_{i-1} + d_i + d_{i+1}} \quad \text{and} \quad d_i = u_i - u_{i-1} \qquad (2)$$

which are also linear inequalities with respect to D_i coefficients. The constraints, which assure monotonicity in curvature (second derivative) can be written in the form

$$D''_i - D''_{i+1} \leq (\geq) 0$$

where

$$D''_i = \frac{2 \cdot (D'_{i+1} - D'_i)}{d_{i-1} + d_i} \qquad (3)$$

and D_i' are defined by (2). These inequalities are again linear with respect to D_i. Hence all the constraints can be written as $h_k(D) \leq 0$, where $h_k(D)$ are linear form of $D = (D_{-1}, D_0, \ldots, D_{n+1})$. Moreover, the feasible region defined by the constraints is convex. All the constraint inequalities can be generated automatically provided that the information about the monotonicity intervals and a kind of monotonicity (function, slope, curvature, increase, decrease) is supplied by the designer.

The approximation task

For given parameter values u_i such that $u_i < u_{i+1}$ the real function values y_i and non-negative weights w_i are given ($i = 0, 1, \ldots, n$). The function $P(u)$ of the form (1), which minimize the functional

$$F(D) = F(D_{-1}, \ldots, D_{n+1}) = \sum_{i=0}^{n} w_i \left[y_i - P(u_i) \right]^2 \quad (4)$$

and complies with the prescribed set of linear constraints $h_k(D) \leq 0$ is to be found.

This can be seen as the quadratic optimization problem with linear inequality constraints, which describe a convex region of feasible solutions. One of the methods to solve the problem is one of gradient projection.

Solution to the quadratic programming task

The algorithm used for optimization should allow the interpolation curve to be defined quickly, as long as it fulfils the required constraints. Thus the first stage of curve definition is to use an interpolation scheme and to check the constraints. This allows to find the optimal solution without the iterative phase. If the constraints are not satisfied, an arbitrary feasible function must be chosen and the iterative process of finding the minimal value of the functional has to be applied.

The algorithm of gradient projection used for curve definition can be divided into the following phases

1. Determination of the interpolating function.
2. If the constraints are satisfied, the function is the optimal solution to the problem; otherwise
3. Finding any feasible function as a starting solution e.g. by correcting the interpolating function to the required constraints.
4. If the condition for termination is satisfied, the solution is

optimal; otherwise
5. Determination of a direction of improvement by gradient projection.
6. Determination of the minimum of the function in the improvement direction inside the feasible region and passage to point 4.

To terminate the algorithm the Kuhn-Tucker condition of the form (5) is checked.

$$\text{grad}(F) - \sum_{i \in A} \lambda_i \text{grad}(h_i) = 0 \quad \text{and} \quad \lambda_i > 0 \quad \text{for } i \in A \quad (5)$$

where A is a set of active constraints ($h_i(D) = 0$).

The advantage of quadratic programming is that the minimum of the function in any direction can be found in a single step.

Applications

The algorithm has been applied to define several curves on technical surfaces. The approximation of a car bonnet central curve with curvature monotonicity constraints and all weights equal to 1 is presented in fig.1. The curvature is monotonous, but an inflection point occurs near the end of the curve. If such a point is undesired, additional slope monotonicity constraints must be introduced. The approximation error seems to be acceptable (less than 0.35) since the data were taken from manual drawings. It is worth noting that in this case (non-uniform parameterization) the B-spline polygon may have slightly irregular shape despite regular changes of slope and curvature (small drawings in the corners).

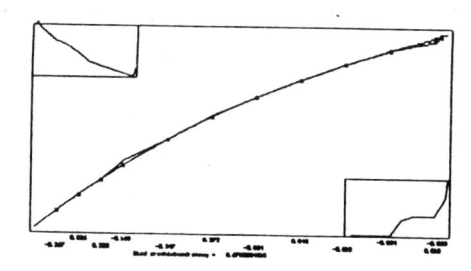

Fig.1. Approximating function with monotonicity of B-spline vertices constraints for the curvature defined for a car bonnet section

The approximation of a marine screw propeller cylindrical section with slope constraints is presented in fig.2a. The shape of the curve is much more regular than that of the interpolating scheme (fig.2b), but the curvature changes are still unacceptable. The points at both ends seems unsuitable for a regular curve. Since their determination is based on other noisy data, their weights

can be substantially reduced. The approximation with additional
curvature monotonicity constraints and weights reduced to 0.4 for
end points is presented in fig.2c. The approximation error for all
but end points is less then 0.1mm, which is equal to the accuracy
of the given data.

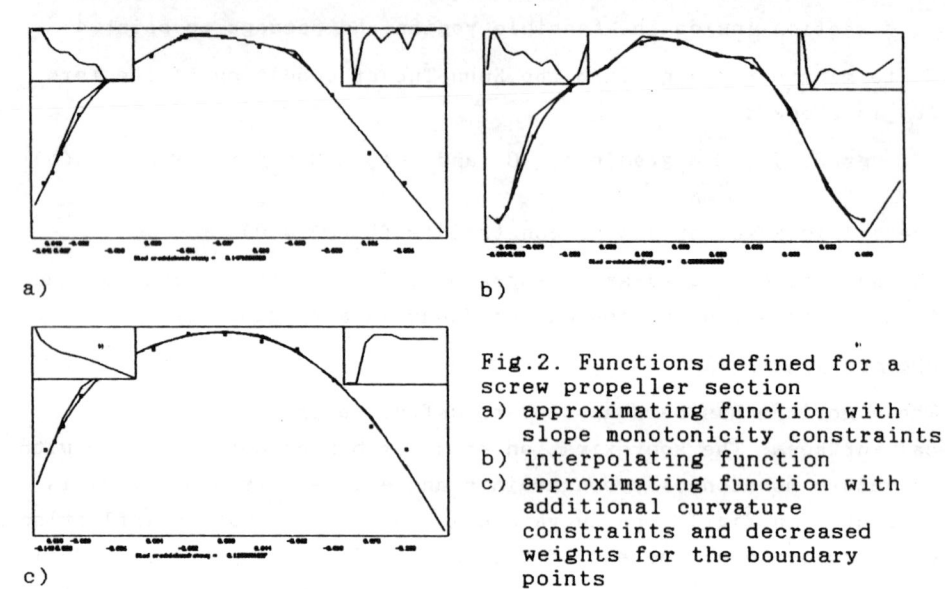

a)

b)

c)

Fig.2. Functions defined for a
screw propeller section
a) approximating function with
   slope monotonicity constraints
b) interpolating function
c) approximating function with
   additional curvature
   constraints and decreased
   weights for the boundary
   points

Conclusion

A technical curve is considered to be smooth if it can be divided
into a few segments with monotonic curvature, which implies the
existence of a few segments (usually divided in a different way)
with monotonic slope and/or position. The proposed approximation
method allows a designer to define such a smooth curve passing
near the given data points. The definition of monotonicity intervals is not difficult for a user and can be automatically transformed into linear constraints for B-spline coefficients. The necessity of interpolation of some data, which may come from matching one part with the other, can be obtained by setting high
weights for these data points.
The interpolation of those points not lying on a smooth curve always leads to undulations and inflections. The application of the
proposed method allows the designer to define roughly the shape of
a curve and not be concerned whether the exact data lie on a
smooth curve or not.

The method presented here is limited to curves described by scalar function of a single variable but can be easily generalized for surfaces described as the scalar function of two variables. Such a generalization leads, however, to an increase in the number of B-spline vertices and thus in the dimension of the approximation problem. This will result in greater requirements for computer resources (processing time, memory). But even in the presented version the algorithm is useful in practical application since at the first stage of surface definition a set of curves must be defined and many such curves are described using functional parameterization.

References

1. Asaturyan, S.; Unsworth.: A C^1 monotonicity preserving surface interpolation scheme. in The Mathematics of Surfaces III ed. by Handscombe, D.C., Clarendon Press 1989. pp. 243-266.
2. Arge, E.; Daelen, M.; Lyche, T; Morken, K: Constrained spline approximation of functions and data based on constrained knot removal. in Algorithms for Approximation II ed. by Mason, J.C and Cox, M.G., Chapman and Hall, London, 1989, pp.4-20.
3. Carlson, R.: Shape preserving interpolation. in Algorithms for Approximation ed. by Mason, J.C and Cox, M.G., Clarendon Press, Oxford, 1987, pp.97-113.
4. Fritsch, F.N.: Monotone piecewise cubic data fitting. in Algorithms for Approximation II ed. by Mason, J.C and Cox, M.G., Chapman and Hall, London, 1989, pp.99-106.
5. Gregory, J.A.: A review of curve interpolation with shape control. in Algorithms for Approximation ed. by Mason, J.C and Cox, M.G., Clarendon Press, Oxford, 1987, pp.131-140.

Unfolding 3D Developable Geometries

John M. Sullivan, Jr. and Don H. Fu

Mechanical Engineering Department
Worcester Polytechnic Institute
Worcester, Massachusetts

Summary
   A large number of products are manufactured by folding thin sheets of material into desired 3D shapes, say the folding of sheet metal to form a heating duct for example. Determining the optional 2D pattern layouts to minimize waste and ensure appropriate folded shapes is an iterative progress involving time and money. We have developed a highly automated and user friendly package that utilizes a reverse engineering strategy for the solution of this problem. Three dimensional geometries of the desired product are created in a PC based CAD system. The geometry surface is faceted automatically and treated as a composite of planar panels whose topological data is provided by preprocessing the given geometries. These panels are represented by an undirected graph which is traversed by the breath-first search technique. This search produces a free tree or acyclic graph that guarantees complete unfolding. With panel connectivity and orientation established each panel is transformed or unfolded. Surface cuts or seams are introduced automatically or via user specified instructions to facilitate the 2D pattern formation. The 2D layout is adjusted for material thickness. Overlaps or fastening flaps can be added to seams. Stress-relief holes can also be created at folded corners. Slots, cutouts, and other geometric entities are preserved in the layout. An animation option enhances the clarity of the unfolding process. Since the 2D layouts were generated directly from the desired 3D object, the ability to reproduce the object within the specified constraints is assured.

Introduction
   A large number of manufactured goods are prepared by folding thin sheets of material into three dimensional shapes. The heating and ventilation industry relies heavily on the ability to fold sheet metal into ducts or conduits for air handling systems. Similarly, file cabnets, component cases and container manufacturers must fold metal or corrogaged sheets into appropriate three dimensional shapes. The process of laying out the two-dimensional patterns to be folded is time consuming, prone to erroneous layouts and wasteful of the raw material.
   With the advent of computer aided design systems (CAD) the designer is free to simulate the desired product in its full three dimensional form electronically. The CAD design can be subjected to numerical analysis for thermal and structural load simulations. The design can be modified to incorporate any geometric change deemed problematic from the numerical analysis or that the designer views as unacceptable aesthetically. The entire iterative design process requires no physical prototypes.

Only after the simulated geometric object passes all aesthetic and in-service load requirements does the design process dictate a prototype model.

Objects made from thin sheets of materials are constructed by attaching multiple 2D layups which have been folded into appropriate shapes. In large sheet metal stamping operations the folded metal is usually riveted or mechanically fastened along overlap seams. The selection of the most appropriate layup patterns accounting the thickness and fastening considerations for these objects requires numerous iterations.

A new solution strategy that we are investigating [1] is to create the desired product in the CAD environment. Consider for example an electrical outlet box design, Fig. 1a. The geometry is discretized into planar polygons. Appropriate transformations are applied to unfold the polygons and geometric shapes (such as the conduit access holes) into the x-y plane yielding the precise 2D patterns to recreate the three dimensional object, Fig. 1b. This unfolding strategy embraces the goals and directions of current CAD systems. That is, start with the 3D CAD object; analyze the object numerically and then use the same numerical mesh to unfold the object.

The primary application of folding is on developable materials, i.e. materials whose Gaussian curvature is zero. The Gaussian curvature, $\mathbb{K}$, is expressed as:

$$\mathbb{K} = \kappa_1 * \kappa_2$$

where κ_1 and κ_2 are the principle normal curvatures at a given point on the surface. In the area of double curvature Calladine has examined "faceted" surfaces composed of triangular elements where the element vertices lie on the original surface.[2] He showed that the solid angle subtended by the vertex common to a set of triangular elements is identical to the angular defect of the faceted surface when unfolded to a common plane. We have used this information to unfold nonzero Gaussian curvature objects.[1,3] Other recent double curvature investigations concentrated on textile

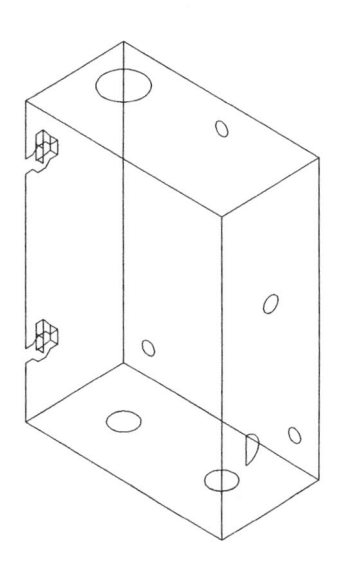

Figure 1a: CAD Drawing of Switch Box

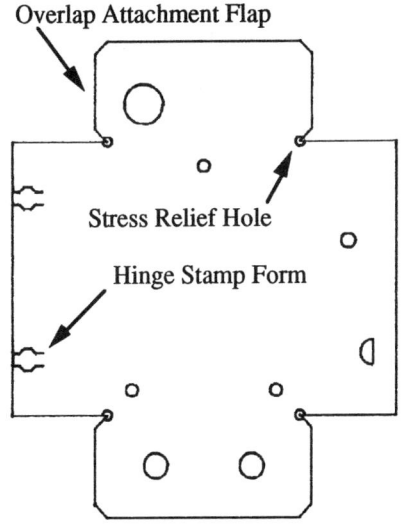

Figure 1b: Unfolded Switch Box with Geometric Entities Preserved. Flap and Relief Hole Options Activated.

fabrics and other sheet materials. [4,5] Therein, they demonstrate that conventional shell theory has little value and one must deal with membrane strains.

We have restricted this work to developable geometries. A large product base is manufactured via developable folding: and user friendly, interactive tools to facilate this task are either lacking or financially unattainable.

Goals of the Method

The major theme of the pattern layout design was to unfold three dimensional geometries as created via a CAD system directly into the two dimensional patterns. The system had to be interactive such that the direction of unfolding could be guided during the process. The magnitude and configuration of overlapping material had to be an option. The user interface must be intuitive, simple to operate and attractive. The system had to be operational within the shell of a CAD system that operates on a PC platform. This latter requirement allows small manufacturers to compete with larger businesses competitively.

Procedure

The user creates a 3D wireframe or surface representation of the object to unfold. The entities can include polygons, lines, arcs, and circles. The entities are processed automatically for adjacency and in-plane connectivities. Closed loops are created from the lines and arcs to form surface polygons. The circles and existing polygons already form closed loops. In essence, a surface is created from wireframe information. This process can be erroneous and must receive corrective information from the user. Consider the transition duct with a baffle plate across the smaller throat, Fig. 2. During the preprocessing of this object it is not known *a priori* that a panel containing lines 1 and 2 is actually a hole, Fig. 2. The situation is corrected with an interactive user option to delete unwanted panels or polygons. This *delete-panel* option is also activated when the system identifies conflicts such as edges with more than two associated panels.

Once the object is represented by valid surface polygons the topology of the geometry is established using single array sturctures for node-to-panel, neighboring panels, and panel-to-edge lists.

Unfolding Tree Traversal

A surface for unfolding is assumed to be simply continuous over the entire domain. Any two points on the surface can be connected by a simple continuous path

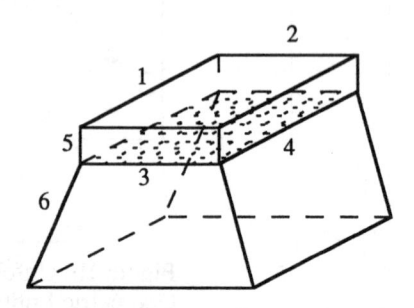

Figure 2: Transition Duct with Baffle Plate

without leaving or crossing the surface. Certainly surfaces exist that have some cross-connections which violate this assumption. However, these infrequent occurances are treated easily with our *delete-panel* option. The faceted surface is represented by a tree graph where all surface polygons are tree leaves. The unfolding procedure identifies a root (user selected base polygon) of the tree and traverses all leaves. Since the order of leaves visited is not important, it is an undirected graph. A valid and effective traversal technique for this application is the breath-first search (*bfs*) which requires a spanning tree (or free tree) from the leaves. The user selected root starts the free tree followed by visits to its neighbors (children of the root) through common edges between the root and neighbors. These neighbors become new parents whose unvisited neighbors are identified in the next loop. The loop process continues until all elements are visited establishing the free tree. The CPU time of the *bfs* is linear with respect to the number of polygons. The traversal on the free tree guarantees a complete unfolding process.[6]

Panel Transformations

The free tree dictates the order of unfolding. The root panel is transformed first to the x,y plane using a 3-point to 3-point transformation. Listed children of the root polygon are unfolded next using points on the common edge with the parent polygon as two of the three points in the transformantion. The third point is any non-colinear vertex of the child polygon. A check of the unfolded panel's normal is made. If the direction of the normal is opposite that of the parent, a 180° transformation is applied to the panel. This procedure is followed throughout the tree.

Adjustments of the 2D Layout

During the unfolding process some edges created as single lines in the 3D object need to be split into two independent lines in the 2D stamp, Fig. 3. These seams or surface cuts are automatically created by the program whenever necessary. The user has the option to *add* or *delete seams* which can change the unfolded layout. This feature is necessary since the location of the automatic seams may not be ideal for a specific application.

Whenever panels are folded fasteners of some form are required along mating seams. An option exists that allows the user to add, remove or modify individual overlap material configurations. As shown in Fig. 1b, overlaps were added to the unfolded switch box. This figure also shows stress relief holes which are commonly

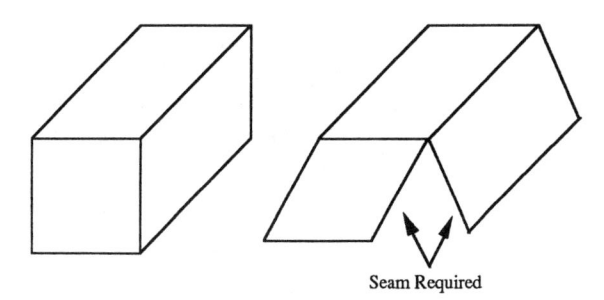

Figure 3: Unfolding Panels Require a Seam

punched in a 2D stamp at corner locations to alleviate stresses. Other options include material thickness, bend allowance, and animation.

The thickness of the material affects the lineal span of the 2D layout. During folding a bend radius R must be established. Usually the radius of the bend is directly related to the material thickness T. The bend allowance $B = \alpha(R+KT)$ is proportional to the bend angle α and the radius to the neutral axis $(R + KT)$. [7] To account for the thinning of sheet metal during folding a K factor is used. Typical K values for sheet metal range from 0.4 to 0.5 with the latter value being the coefficient for a neutral axis without thinning.

The animation feature allows the user to specify the number of frames or snapshots taken during unfolding. Once unfolded the sequence can be viewed in a frame by frame animation. The animation feature may provide insight to the user for a more optimal unfolding sequence.

Example

Consider the manufacture of hard drive covers for PCs, Fig. 4. This 3D geometry has ventilation slots located on the sides of the drive. The back end has openings for the SCSI ports, 8-pin mini plugs and power. The front and bottom of the drive housing are not part of this design. Presumably, they are configured in another design which mates with this example.

To unfold the object the user selects the base panel. One option for this selection is to identify two lines common to a panel as shown in Fig. 4. The system establishes the unfolding sequence and produces the 2D layout orientated relative to the base panel, Fig. 5. The dashed lines indicate mold lines of the part. If the user changes the thickness of the material the program adjusts the 2D layout. Figure 6 overlays the 2D stamps for wireframes of thickness 0.0 and 0.15 in. The blank calculated for the thicker material has reduced lineal spans in the x,y plane. This effect is due to the deviation of the actual part (thickness dependent) from the wireframe or mold line construction of the object.

Conclusions

An unfolding strategy has been presented that allows the user to interactively communicate with the software and guide the unfolding process. The system is ca-

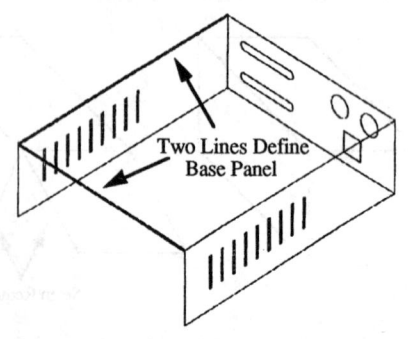

Figure 4: Top Surface is Selected as Base Panel

pable of unfolding 3D developable objects to their 2D pattern layouts while maintaining user specified options. The system operates at the PC level within the shell of CADKEY[8], an inexpensive but powerful CAD package. The menu structure allows expansion of the software system. Since the 2D layouts were generated directly from the desired 3D object, the ability to reproduce the product geometry within the specified constraints and numerical resolution of the original object is assured.

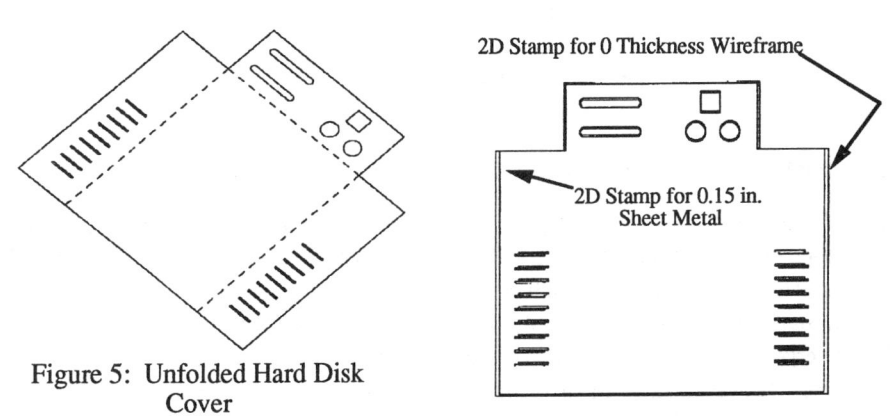

Figure 5: Unfolded Hard Disk Cover

Figure 6: Overlay of 2D Stamps Shows Thick Sheet Metal has Reduced Lineal Span

Acknowledgement

This work was supported, in part, by CADKEY Inc. under WPI grant # 532832.

References
1. Sullivan, J.M.,Jr. and Fu, D.H.: Unfolding three dimensional CAD objects for two dimensional pattern formations. Proc. 3rd Nat. CAD/CAM conf., Mexico, (1989) 219-226.
2. Calladine, C.R.: The static-geometric analogy in the equations of thin shell structures. Thin Shell Theory, ed. W. Olszak, CISM, Udine, (1980).
3. Fu, D.H.: Unfolding 3-D CAD geometries to 2-D pattern layouts. MSME thesis, Worcester Polytechnic Institute, (1990).
4. Amirbayat, J. and Hearle, J.W.S.: The complex buckling of flexible sheet materials - part I. Theory approach. Int. J. Mech. Sci. 28 (1986) 339-358.
5. Amirbayat, J. and Hearle, J.W.S.: The complex buckling of flexible sheet materials - part II. Experimental study of three-fold buckling. Int. J. Mech. Sci. 28 (1986) 359-370.
6. Aho, A.V., Hopcroft, J.E., and Ullman, J.D.: Data Structures and Algorithms, Ch 7., Addison-Wesley, (1987).
7. Sachs, G.: Principles and Methods of Sheet Metal Fabricating. Reinhold Pub. Corp. N.Y., (1966).
8. CADKEY, 440 Oakland St., Manchester, CT, U.S.A. 06040-2100

Processing Isometric Free-Hand Sketches in Computer-Aided Design

S.Sivaloganathan, Professor A.Jebb and Professor H.P.Wynn
Engineering Design Centre
School of Engineering
City University
Northampton Square
London EC1V 0HB

1.0 COMPUTER AIDED DESIGN (CAD) AND SKETCHING

The advent of computers in drafting and design started with the 'SKETCHPAD' program developed by Ivan Sutherland [1] at MIT in the 60's. This made it possible for a man and computer to communicate through the medium of line drawings instead of the traditional written statements. This invention of using the computer as a sophisticated drafting tool has undergone many refinements and has given birth to drafting packages in many application areas. However, none of these packages adequately cover sketching, the physical expression of the thinking process of the design engineer.

The design process often begins with a graphical description of a proposed device or system. This lack of coverage of the sketching stage of the design process is mainly due to (a) the low precision of the sketch and (b) the fact that the sketch is made while the thinking is still in progress. Thus there remains a need for a computer package which will require minimum interaction and accept the sketch as the design engineer makes it. This paper describes the development of a package which accepts free hand isometric sketches made on paper placed on a digitizer. The program then recognises the sketch and converts it into a proper isometric drawing.

2.0 REQUIREMENT ANALYSIS

This section describes the desirable features of the sketching input system from the point of view of the user, the design engineer, to yield the specification of the sketching input system. Since the inputs are isometric sketches it is necessary to look at the features of sketching together with the computer system, in order to draw the requirements. The term 'free-hand sketch' is too often understood as a crude drawing into which no particular effort has gone. On the contrary, it requires considerable effort and care, with the degree of precision depending on the use to which it will be put. Sketches hurriedly made to supplement oral description may be rough and incomplete. On the other hand, if a sketch is the medium for conveying important and precise information it should be created as carefully as possible.

The components of a sketch describing a manufactured component are mainly straight lines, circles, ellipses and circular and elliptical arcs. Isometric sketching relies on the fact that the x and y axes are inclined at 30° to the horizontal while the z axis remains vertical. Straight lines parallel to the axes, called the isometric lines, are easy to sketch. However, lines on objects located by angles, called non-isometric lines, are more difficult. Non-isometric lines are fixed by ordinates at their ends from some isometric line. When objects having cylindrical or conical shapes are placed in the isometric or other oblique positions the circles are viewed at an angle and appear as ellipses.

2.1 REQUIREMENTS OF A SKETCHING INPUT SYSTEM

(i) For the sketching input system described here, the inputs are free-hand isometric sketches made up of four classes of lines (a) centre lines (b) visible lines (c) hidden lines and (d) construction lines. The first requirement is the provision for the entry of these lines. It is not necessary to have all the lines in all the outputs and often the visible lines alone are sufficient to illustrate the sketch. However it is necessary to store all these lines separately in the computer memory, allowing independant access.

(ii) In a traditional sketching situation the solid and other lines are sketched with pencils of differing hardness. Selection from the menu should resemble this.

(iii) As described earlier sketching is the physical expression of the design activity and the designer should be left free to carry out the design process. This means there should be a minimum of interaction or distraction. Thus (a) the menus must be kept to a minimum and (b) no input should be necessary to indicate the start and finish of a line segment.

(iv) One of the main requirements of sketching is the ability to erase whole or part of a line. The erasing process should be simple and easy.

(v) A sketch is made to rough size and hence subsequent extension is an essential feature.

(vi) Free hand sketches would have overstruck lines and provision should be made to accommodate them.

(vii) The user may at times sketch from both ends of a straight line or circle to meet at an intermediate point.

(viii) The program should automatically identify the line segments and vertices.

(ix) The program should automatically identify straight lines, ellipses and arcs and process them accordingly.

2.2 THE SYSTEM

The system in which the sketching input software is developed, comprises (a) an IBM PS/2 model 60 computer and (b) a CALCOMP 2000 series digitizer. The digitizer has a tablet and a stylus. The stylus is similar in appearance to a pen and is used to digitize the point selected. The data transmission parameters are set by three banks of switches at the back of the tablet. They set baud rate, data bits, stop bits, parity and mode of operation. The digitizer is connected to the RS232 communication port in the PS/2 computer.

The digitizer locates and identifies the (x,y) co- ordinates of the point relative to the origin at the lower left corner of the tablet. In the track mode (x,y) co-ordinate pairs are output continuously at the selected sampling rate so long as the stylus is pressed against the tablet. This is the mode suitable for sketching input.

The sketch made on the digitizer is transformed into a series of points over which the stylus has passed. Processing a sketch is the storing of these series of points in four different groups according to the type of the line and subsequent curve fitting in a manner that would realise the three dimensional content of the sketch.

2.3   STRUCTURE OF THE PROGRAM

The program is designed in such a way that it can be ported to any system with a minimum of modifications. To facilitate this, the device and system dependant sketch part is separated from the processing part. In the sketching part the points of the four classes of lines namely (a) centre lines (b) visible lines (c) hidden lines (d) construction lines and a temporary class of line called erased lines, are accepted and written into a file. This file is then read by the processing part of the program. This is divided into two groups (a) two dimensional processing and (b) three dimensional processing. In the two dimensional processing part the file is read, points are broken into groups belonging to individual line segments, terminal points of these line segments or groups are identified and curves are fitted to these line segments. Using these fitted lines and terminal points of the segments, vertices are identified. With these vertices, the edges and their incidence details are established for each of the five classes of lines. An interactive merge facility is then used to accommodate overstriking, subsequent extension, and sketching from both ends to meet at some intermediate point. These visible lines and hidden lines are coupled together to form the final model of the sketch. This model is then accepted by the three dimensional processing part which also uses the information about the origin to transform the connected lines to three dimensions.

For each hole in the solid represented by the sketch, an additional three dimensional point is required for this transformation.

3.0   METHODOLOGY OF THE PROCESSING

Processing starts with the breaking of the conglomerate of points into sub-groups belonging to individual line segments and ends with the production of a list of vertices and edges in three dimensions. The following steps are taken to achieve this.

(i) Break the points into sub-groups belonging to different line segments, with an indication whether it is a straight line or curve. This forms a list of 'Lineseg' nodes (See figure 1 a).

(ii) Analyse the terminal points of the line segments in (i) above and extract the co-ordinates of the points which could be approximated as vertices together with the lines which emanate from them. This forms a list of 'Termpoint' nodes (See figure 1b).

(iii) Fit two dimensional equations for each of the line segments identified in (i) above.

This results in a list of 'Geom_edge2d' nodes (See figure 1c).

(iv) Using the equations of the lines emanating from each terminal point in (ii) above and the co-ordinates of the terminal point as the initial approximation for the numerical solutions, the co-ordinates of the vertices are identified. This forms a list of 'Vertex' nodes (See figure 1d).

(v) Using the equations of the edges, they are classified as one of the twelve classes. The twelve classes of lines include the three isometric lines, non-isometric lines, complete ellipses in the three isometric planes, complete non-isometric ellipses, arcs in the three isometric planes and the non-isometric arcs. The straight lines are classified according to the slope of the lines, while the ellipses and arcs are classified according to the slope of their major axes. The terminal points are used to identify the details of the starting and the ending vertices. This forms a list of 'Edge' nodes (See figure 1e).

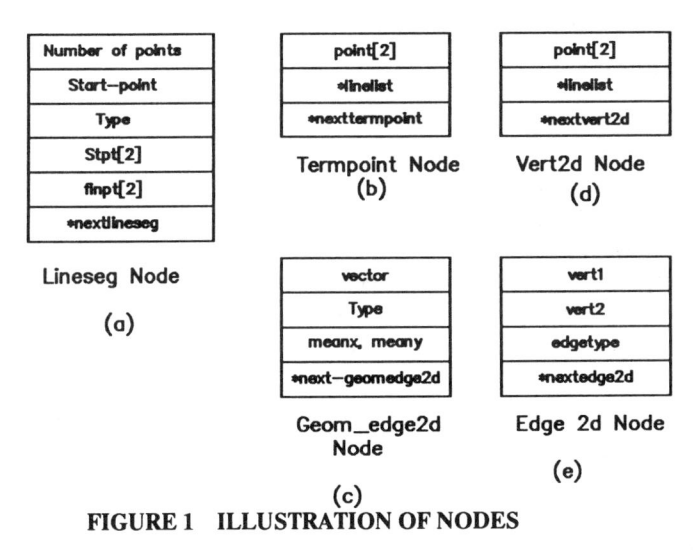

FIGURE 1 ILLUSTRATION OF NODES

(vi) These five lists for the four classes of lines are first merged within themselves to account for overstriking, sketching from both ends to meet in between and subsequent extension before being merged with the erased lines to account for erasing. The hidden lines and the visible lines are then merged to give the final list of lines in the solid sketched.

(vii) To transform the final list of 'Edge' nodes and 'Vertex' nodes into three dimensions, the line type and the two dimensional co-ordinates of the three dimensional origin (in the object) are used. Mathematically speaking a straight line in three dimensions is identified by two end points having six unknowns. To identify them six equations are necessary. Of these six, four are provided by the two dimensional co-ordinates. The line type in (v) above gives one equation such as when the line is vertical in which case it should have the x co-ordinate constant ($x_1 = x_2$). The sixth equation is given by the three dimensional co-ordinate of the origin or any other point in the body. This enables the transformation of the connected body into three dimensions. For each hole in the solid, which is unconnected, an extra point in three dimensions is needed to be known. In the case of non-isometric lines the construction lines or other isometric lines are used to get the extra information.

3.1 THEORETICAL ISSUES

In breaking the conglomerate of points the problems addressed are (a) finding whether the line is a straight line or a curve (b) finding the end point of a straight line and (c) finding the end point of a curve. Fitting straight lines is an easy matter but fitting ellipses is difficult. The method described here is due to Angell and Barber [2]. After finding the line segments and terminal points the co-ordinates of a vertex is calculated by solving more than two equations with two unknowns (an over determined system). The situation is more complex when some curves and straight lines meet.

3.1.1 DETECTING A STRAIGHT LINE

The strategy of detecting a straight line uses one or more of the following methods.
   (i)   perpendicular distance
   (ii)   moving slope
   (iii) generalised conic
   (iv)   dispersion of points

In (i) the perpendicular distance of the points belonging to a sketched straight line is considered to be within a specified small distance from a line connecting any two of these points. In (ii) the slope of the line segment connecting two points is computed from one end to the other and for a straight line it is expected to be varying within a specified limit. Since the points are closely located, three point moving average smoothing is carried out, as the slope computation proceeds from one end to the other. In (iii) an equation of the form

$$x^2 + hxy + by^2 + fx + gy + c = 0$$

is fitted using the points falling within the first centimetre of the line under investigation.

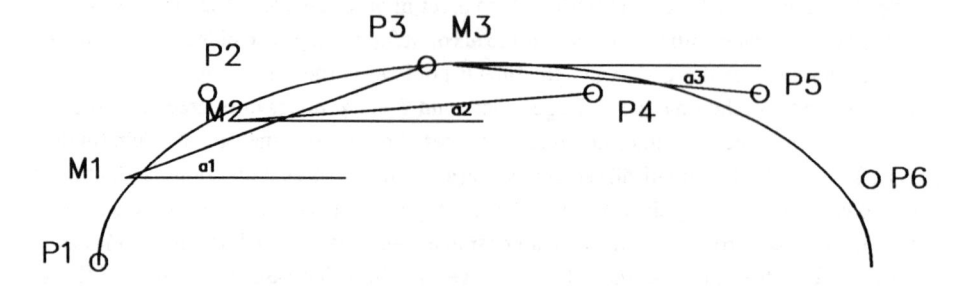

FIGURE 2   DETECTION OF ELLIPSE - MOVING SLOPE METHOD

In (iv) it is expected that the points would fall onto both sides of any straight line connecting any two points under consideration. Since these tests work efficiently only in certain conditions confirmatory tests are often made.

Similar methods are established for the detection of an ellipse. For instance the dispersion test for an arc expects all points to be on one side of straight line connecting the ends of the arc. Similarly the moving slope from one end to the other is expected to vary only in one direction as shown in figure 2.

3.1.2 DETECTING THE END POINT OF A STRAIGHT LINE

In the last section the moving slope method was established as a way of detecting the straight line. The moving slope will continue to vary within the specified limits until it reaches its terminal point. The slope change would then exceed the limit indicating the end point. The next line is expected to start at this end point if the distance between the end point and the next point is less than a specified value. The lines do not meet and are distinct if this distance is more than the specified value. In a similar way the moving slope method identifies the end point of an elliptical arc by the change in the sign of the moving slope. The method succeeds in identifying the end point by the combination of distance test and moving slope test for non continuous lines. In the case of continuous lines the method succeeds in situations as illustrated in figure 3 (a) and fails in situations 3(b) and 3(c). The requirement on the user therefore is not to draw continuous curves with the same sense.

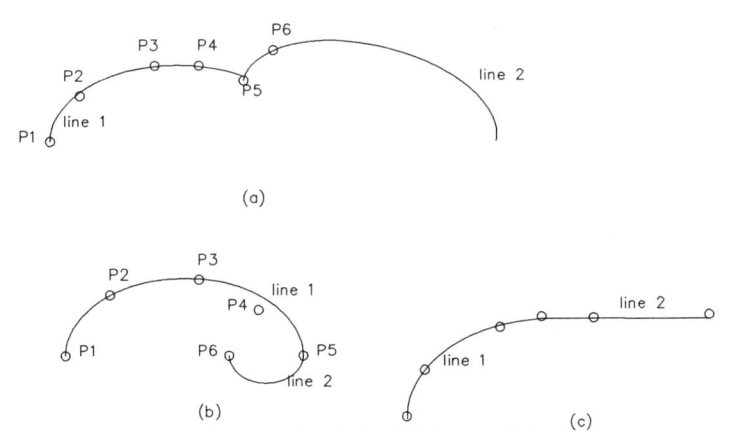

FIGURE 3 DETECTION OF THE END OF AN ELLPSE

3.1.3 FITTING AN ELLIPSE

Fitting the ellipse is achieved by the method after Angell and Barber [2] in which

$$\sum_i (x_i^2 + hx_iy_i + by_i^2 + fx_i + gy_i + c)^2 - \left(\sum_i x_i^2 + hx_iy_i + by_i^2 + fx_i + gy_i + c\right)^2 \text{ is}$$

minimised by equating the partial derivatives with respect to h, g, f, b and c to zero.

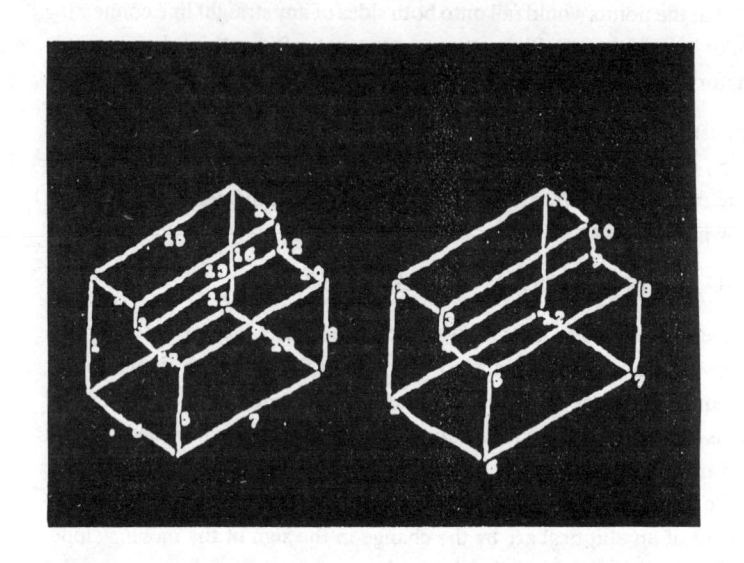

FIGURE 4 PHOTOGRAPH OF THE DISPLAYED SKETCH

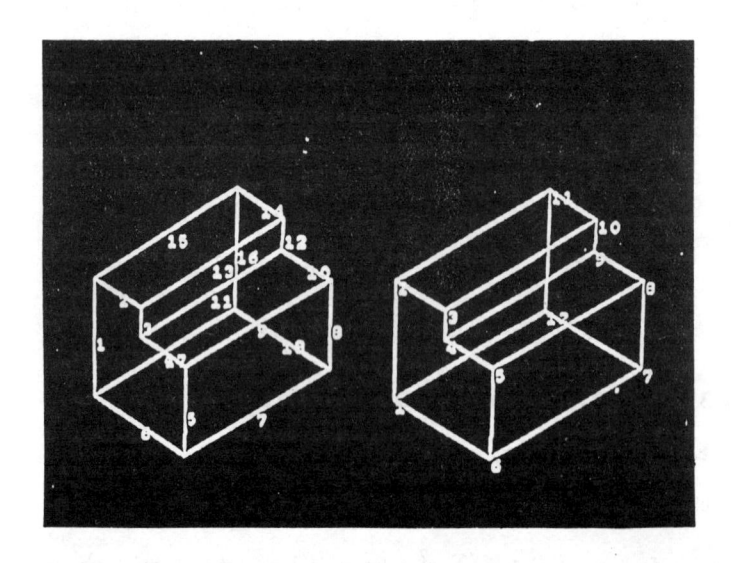

FIGURE 5 PHOTOGRAPH OF THE DISPLAYED FITTED SKETCH

3.1.4 INTERSECTION OF MORE THAN TWO LINES

When two lines intersect, finding the point of intersection is a straight forward solution of two simultaneous equations. But when more than two lines intersect the sytem is overdetermined and finding a solution is difficult. In the case of straight lines a least squares solution gives a point which makes the sum of the squares of the perpendicular distances to all the lines a minimum. When there are ellipses and straight lines the problem becomes more difficult. The equation of the elliptical arcs involved are of the form

$$x^2 + hxy + by^2 + fx + gy + c = 0.$$

Suppose that the solution point (x_1,y_1) gives rise to an error e. In a similar way the equation of the straight line is $px + qy + r = 0$ and suppose the error with (x_1,y_1) is E. Then if

$$S = \sum_i (E^2 + e^2) = \sum_i (p_i x_1 + q_i y_1 + r_i)^2 + \sum_i (x_1^2 + h_i x_1 y_1 + b_i y_1^2 + f_i x_1 + g_i y_1 + c_i)^2$$

the least square solution (x_1,y_1) is given by the solution of the equations $\frac{\partial S}{\partial x_1} = 0$ and $\frac{\partial S}{\partial y_1} = 0$. The co-ordinates of the terminal point are used as an initial approximation for the solution. The solution is obtained by a two dimensional Newton-Raphson procedure in which y_1 is kept constant during an iteration for x_1 and vice versa.

4.0 AN EXAMPLE

Consider the simple example of an 'L' block shown in figures 4 and 5. This block was sketched in the system developed and the points were written into a file. In the processing part these points are read and broken into 18 subgroups or line segments in the order they were drawn. The lines are then fitted and terminal points extracted. The terminal points and the equations of the fitted lines are then used to extract the vertices. Using the vertices the edges are then identified. Figures 4 and 5 show the displays of the labelled sketch before and after processing. The data extracted here are the sketching input used to calculate the three dimensional vertices.

5.0 CONCLUSION

The paper describes a viable approach to sketching input to CAD with some new fitting procedures. The configuration has proved effective for real use in design and portability into multitasking environments and solid modelling suites.

REFERENCES

1 Sutherland I.E.; "SKETCHPAD: A man machine graphical communication system", AFIPS Proceedings, Vol 12, Spring Joint computer conference, 1963.

2 Angell. I. and Barber.J; An algorithm for fitting circles and ellipses to megalithis stone rings, Science and Archaeology, No 20, 1977.

Applying Hypermedia Concepts to CAD in an Assemble to Order Environment

Martin Cwiakala, Ph.D.
Project Engineer
GEMCO
301 Smalley Avenue
Middlesex, New Jersey

Visiting Part-time Lecturer
Department of Mechanical and Aerospace Engineering
Rutgers, the State University of New Jersey
New Brunswick, New Jersey

Nils Brouër
Fulbright Fellow at Rutgers University
RWTH Aachen
5100 Aachen, Germany

Abstract

This paper presents the implementation of hypermedia concepts with CAD to create an interactive main assembly drawing for an "assemble-to-order" environment. An assemble to order environment is one in which the final custom piece of equipment is constructed primarily of standard subcomponents. By utilizing hypermedia concepts with the CAD, an intuitive graphic user interface is created. A general main assembly drawing is used as an interactive menu to select the various sub-components. By "pressing" on a button attached to a part (by using a mouse or other graphic input device) possible choices are displayed. After making the appropriate choice, the selected symbol is loaded. If necessary, a drawing may be "re-sorted" to correct all mating components, thus changing the menu into the final assembly drawing. Two examples are provided. The first is a cabin cruiser assembly. The second example is of an industrial blender with a bill of material information.

Hypermedia and CAD

Hypermedia is a term adapted from Theodor (Ted) H. Nelson's concept of Hypertext. Nelson presents his views of Hypertext, which he defines as "non-sequential writing" in his book "Dream Machines"[1]. The basic concept of Hypertext is that of being able to create links between many small files that make up a Hypertext document. Hypermedia is an extension of Hypertext in the way that it incorporates not only text files but any kind of media including sound recording, graphic or video images. Cwiakala demonstrates how such concepts can enhance CAD systems [2,3]. He presents Command Buttons that can be programmed so that when they are "pressed" using a graphic input device (such as a mouse) they cause the programmed command to be executed. If the executed command is "call up a new drawing", a link between two drawings has been created. A link like this can be established between elements of the same ranking in a hierarchy (i.e. between two drawings) or of a different ranking (i.e. between an element and a drawing). The utility presented here uses command buttons like these to create an interactive assembly environment.

An Assembly-and-Reconfigure Utility

In many industrial applications, custom equipment is created by using standardized parts. One example would be that of industrial blenders. According to the demands of the customer, the manufacturing company assembles a customized product. The customer has the choice of options such as different vessel sizes, optional agitators, different motors. In such an environment, it is especially helpful to be able to quickly provide the prospective customer with a drawing of the customized piece of equipment. Such an assembly drawing can be easily created using the presented utility.

The program has been written for Prime Medusa, a CAD system that runs on Prime minicomputers. Cwiakala lists five general requirements for CAD systems to be able to realize the concept of command buttons [2]. Prime Medusa fulfills all of those (and so should other CAD systems like the PC based platforms of Autocad and VersaCad).

To start the program, the user calls up a drawing named "START-UP". This drawing has a number of buttons on it. These buttons are groupings containing two types of text. Type is a Medusa attribute assigned to text. One piece of text in the grouping is used as a label for the button and is displayed. It has a style which makes it look like a button. The second piece of text is invisible (height set to zero) and stores the command to be executed when pressed. To "press" a button, the graphic cursor is positioned over the grouping and a buffer on the CAD menu is activated. This buffer stores the commands to execute a macro which locates the command text in the button grouping and executes it.

If the button named "A1" is pressed, a menu is loaded to the screen. It asks the user which of the available symbols they want to load at the datum of the button. This is shown in Figure 1. The user types the number of the desired symbol and hits <return>. The menu is deleted and the symbol is loaded at the datum of the button (Figure 2).

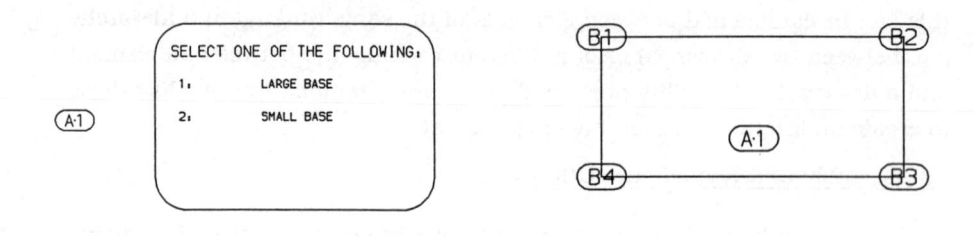

Figure 1

Figure 2

In this case the symbol contains buttons, which in turn may be pressed by the user to load more symbols. Symbols may thus be arranged in a hierarchy. There is no limit to the level symbols may be stacked. Figure 3 shows a first-generation symbol with two second-generation symbols attached to it.

Any button may be pressed at any time. If a button is pressed again, the same menu will appear and the user may make a new choice. The old symbol will then be deleted and the new one loaded in its place. For example, the button named "B2" in figure 3 may be pressed again, and a triangle like the one attached to button B3 may be loaded. The user can also press button A1 and select the small base. This was done in figure 4. Because the datums of the buttons attached to the small base do not match those of the large base, the location of the symbols do not correspond to their respective load buttons anymore. The user now presses a button named "RECONFIGURE" and the drawing is re-sorted (Figure 5).

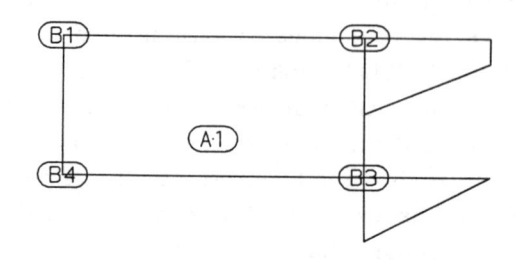

Figure 3

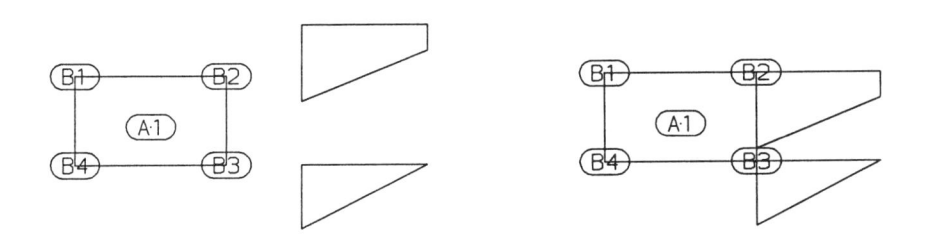

Figure 4   Figure 5

The Loading and Reconfiguration Programs

    Two programs were written to perform the demonstrated operations. The first program performs the loading of a new symbol. It is executed every time a load button is pressed. The program reads all hidden texts as well as the location and orientation of the button. Using that information it first loads the appropriate menu and prompts the user for a selection. Once the selection is made, the menu is deleted again and the new symbol is loaded. If the button has already been pressed the attached symbol is deleted. To create a link between the button and the loaded symbol a unique marker is needed. The system date and time recorded when the button was pressed is used for that purpose. The program writes this marker to a hidden SR1 type text in the button grouping and to a SR2 text located at the datum of the symbol grouping.

    The second program performs the reconfiguration. It is executed when a button named "RECONFIGURE" on the start-up drawing is pressed. It searches the database for load buttons by looking for a string "LOAD-BUTTON" which is included as TSJ text in each load button grouping as an identification label. If it has found a button, it goes to the beginning of the grouping and looks for the SR1 text. If the button has been pressed the SR1 text will be set to a system time. The program then seeks the symbol with a matching system time as its SR2 text and compares the two locations. If they do not match, the symbol will be moved to the location of the button.

The program keeps searching the database for load buttons until all symbols line up with their load buttons.

Two Examples: A Cabin Cruiser and an Industrial Blender:

The utility was applied first to a cabin cruiser assembly. Figure 6 shows an example of a finished assembly drawing. There are two first-generation symbols available (a small and a large hull, button A1) as well as five sets of second-generation symbols (an inboard or in/outboard motor, a small or a large railing, a small or a large cabin, and an option of including a flag and/or a flydeck). The buttons corresponding to the second-generation symbols are those named B1 through F1.

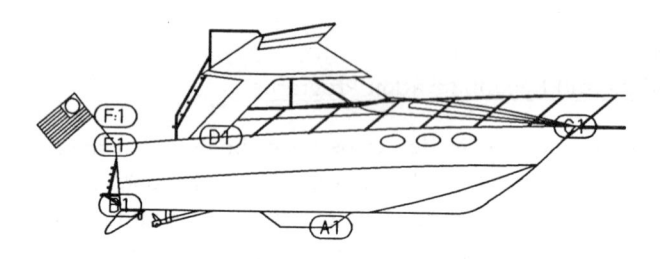

Figure 6: A Cabin Cruiser Assembly

The second example is one of an industrial blender. A blender consists of a vessel, a support structure, a cover, a reducer with attached motor, a valve, an agitator, and an agitator motor (the last two considered optional). Over 1000 different blenders can be assembled using the provided symbols. The finished drawing consists of three views of the blender, it is shown in figure 7. There is one major difference between the previous example and this one: Here it is also possible to create "classes" of symbols. For example, there are two types of covers, solid and vented ones. Each vessel size furthermore requires a cover of a different diameter. Each set of covers of the same diameter constitutes one class of covers. An existing drawing may be changed by changing the vessel to one of a different size. If the drawing is then reconfigured, the old cover will be deleted and one of the new class (i.e. diameter) but of the same type (i.e. solid or vented) loaded instead.

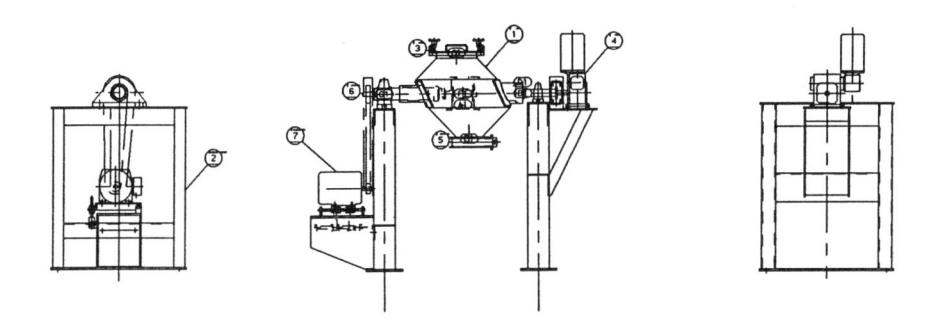

Figure 7: An Industrial Blender Assembly

This example also provides a basic "Help" utility as well as the possibility to create a bill of material. Bill of material information was incorporated with the symbols to generate a list of selected components. This utilized an existing part-list system present in Medusa.

Conclusions

In this paper, a new concept for automatic creation of an assembly drawing using command buttons to load symbols was presented. The presented utility makes it possible to easily change existing drawings by simply "pressing" selected buttons. If necessary, a drawing can be reconfigured to rearrange symbols. Two examples were given: a cabin cruiser and an industrial blender assembly which also included a bill of material.

Acknowledgements

The cabin cruiser symbol drawings were created by John Barry, Dave Sassano, and Paul Boyajian as part of the requirements of Mechanical Engineering Computer Aided Design, a senior level course at Rutgers University.

References

[1] Nelson, Theodor H.: Computer Lib/Dream Machines. Redmond, WA: Microsoft Press, rev. Ed. 1987.
[2] Cwiakala, Martin: Using Hypermedia Concepts to Enhance CAD. Mechanical Engineering, 9 (1990) 44-47.
[3] Cwiakala, Martin: Menu Driven Assembly of a CAD drawing. 1991 Design Productivity International Conference, Vol. 2, 929-936.

Software CAD Environment for PLC

V. V.DEVIATKOV

Institute of Control Sciences,
USSR Academy of Sciences, Moscow

Summary

This paper has its roots in a research project at Moscow Institute of Control Sciences of USSR Academy of Sciences, which developed an software CAD environment called AMPLUA and oriented toward the specification, drawing, optimization, debugging, modeling, generation and documentation for Programmable Logical Controllers (PLC). The target application of this environment is discrete manufacturing systems where the costs involved with the drawing, debugging and maintenance of code for on-line process control can be significantly reduced through the use automated synthesis and analysis techniques given by AMPLUA. The paper features a rule-based high level languages (USLOVIE) which is translated into special model SDPAM (Statement Diagrams of Parallel Algorithms with Memory) composed of interconnected state machines. The concept of automated design approach based on SDPAM is introduced as a means for synthesis, analysis and transformation facilities to change the design from one representation to another and to perform temporal analysis, performance analysis and PLC code generation. CAD AMPLUA features are described involving the structure of CAD, different languages including special language of high level USLOVIE. The approach is illustrated by an example of a simple discrete state system.

Introduction

Discrete control complex including hardware and software for discrete control represent a significant component of computer integrated manufacturing systems. These complexes can range from simple programmable logic controllers that use combinatorial and sequential logic to implement control function, to general purpose computers that use complex scheduling algorithm for process control. The design of

control software and hardware comprises a large part of the lifecycle costs for these complexes and is a non-deterministic prosess requiring a non-trivial decision-making. As a rule in this process one has to:
- describe control algorithms (specification, modeling, synthesis);
- optimize complexity of algorithms (structure, memory occupied by the program and time it requires);
- generate code for discrete controllers (linking, debugging and maintenance);
- produce documentation on designed control complex;
- evaluate complexity and costs of the decisions taken;

In order to reduce these costs, several methodologies have been proposed for the development and maintenance of discrete control systems (complexes) [1]. This paper presents a methodology of an software CAD environment called AMPLUA and oriented towards the specification, drawing, modeling, optimization, debugging, generation and documentation for Programmable Logical Controllers (PLC). Specification involves describing the desired behavior of the controlled system. Modeling involves determining properties of the behavior to synthesize a correct control algorithm and for this purpose uses specific formalism. To be useful the specification and modeling must necessarily have a close correspondence. One way to achieve this correspondence is to use the modeling formalism itself for the system specification. An example of this is the use of Petri nets for the formal specification of the behavior [2]. Another way to relate the specification to a model for analysis is to use separate specification and modeling formalisms, with the translation between the two being accomplished by software. We use this latter approach in a manner that makes it possible for the system designer to work in term of abstract machines with the automated translation of these specifications into the modeling formalism which eliminates the need and difficulties for the designer to think and to develop into a frame of a formal mathematical model. In this environment the structure and functional description of the process control logic are entered into a database on a workstation. The software uses a rule-based high level language USLOVIE [3] which is translated into special formal model SDPAM (Statement Diagrams of Parallel Algorithms with Memory)[4]. The concept of automated design approach based on SDPAM is introduced as a means for synthesis, analysis and transformation facilities to change the design from one representation to another and to perform temporal analysis, performance analysis and PLC code generation. We use SDPAM as our basic modeling formalism. The translation of the state-variable rule-based language USLOVIE into an equivalent SDPAM is performed by software. We have developed a number of tools for analysis SDPAM model. CAD technology based on SDPAM makes it possible for the system designer to work in term of abstract machines while still "thinking USLOVIE" and on approach for the transformation of initial SDPAM discrete system description to another equivalent structured presentation through the use of additional variables and sensing elements.

Rule-based language USLOVIE
_____

USLOVIE has the following structural components: (1)system, (2)block, (3)process, (4)macrosentense, (5)compound sentence, (6)simple sentence, (7)compound condition, (8)compound consequence, (10)simple condition, (11)simple consequence, (12)statement.
The simple sentence (rule) is an elementary component which is composed of a precondition and a postconsequence. When the precondition of a rule is satisfied the postconsequence is computed. The precondition can contain one state variable and the simple condition. The postconcequence can contain one or more statements and one or more state variables. The example of the simple sentence is the following: IF * M * X1 AND NOT X2 THEN VALVE := TRUTH, AN := 5 + E2, EXECUTE PROCESS2, GOTO M1, M2, M3 ;

Compound conditions and compound consequences are constructed as consequences accordingly from simple conditions and simple consequences. Compound sentences are constructed from compound conditions and compound consequences. Macrosentenses are constructed as consequences of compound sentences. We will not consider these kinds of sentences because all they can be easy represented by simple sentences.

Statement Diagrams of Parallel Algorithms with Memory

(SDPAM)

In the following we give the model of concurrent programs named Statement Diagrams of Parallel Algorithms with Memory (SDPAM). SDPAM $G1, G2, \ldots, Gk$ consists of k, $k \geq 1$, blocks $G1, \ldots, Gk$ which are running in parallel. Each block Gi, i = 1,...,k is an independent transition graph with nodes (locations) labeled by $10, \ldots, 1e$. The edges (or transition) in each blocks are labeled by a simple sentence of the form: g:if *l* scond then scons goto l', where g is a number of the transition, 'scond' is a simple condition $sg(X,Y,Z)$ of the transition g, 'scons' is a simple consequence $(Y,Z) := fg(X,Y,Z)$ of the transition g, and fg is the transformation associated with the transition g. X, Y, Z are sets accordingly of input, inner and output variables of the block: $X = (x1,\ldots,xm)$, $Y = (y1,\ldots,yn)$, $Z = (z1,\ldots,zp)$. Different blocks can communicate only through the input and output variables. If location b is activated and $sg(A,B,C)$ is true we say that the transition is enabled for $X = A = (a1,\ldots,am)$, $Y = B = (b1,\ldots,bn)$, $Z = C = (c1,\ldots,cp)$. Sometimes we will wright W instead of Y, Z considering $W = Y \cup Z$, $D = B \cup C$. For a given node l with k outgoing transitions we define $El(X,W) = s1(X,W) \vee \ldots \vee sk(X,W)$ to be the full-exit condition at location l and we do not require that $El(X,W) =$ true for every (A,D). Nor do we require the condition to be exclusive; thus, deadlocks are allowed and each individual process can be nondeterministic. A location whose individual conditions are exclusive is called a deterministic location.

The sets of input and output variables X and Z is accessible and shared by all the blocks. Therefore communication and synchronization between blocks are managed via the shared variables. A full specification of a concurrent programm also includes a specification of the initial values of the variables $x_1,\ldots,x_m, y_1,\ldots,y_n, z_1,\ldots,z_p$. Analysis of SDPAM is based on the use of pair graph [15]. This graph was used for analysis of a finite automation and here we extended and developed the idea of pair graph analysis methodology for more complicated model SDPAM as compared with the finite automation. Some features of this methodology for SDPAM was described in [4].

Example
———

10 BLOCK PUNCH;
20 IMPULSE INPUTS;
SP1=START;
30 BOOLEAN INPUTS
K1=TOP,
K2=BOTTOM;
& TOP and BOTTOM switches, fixing the upper and lower position of a punch &
40 BOOLEAN OUTPUTS
KM1=UPWARDS,
KM2=DOWNWARDS;
& UPWARDS and DOWNWARDS are commands of the control system supplied to the reverse motor &
50 TIMER T1;
60 OPERATION;
70 STEP MO;
80 IF SP1 THEN KM1,KM2 IS OFF GOTO M1;
85 IF NOT SP1 THEN KM1,KM2 IS OFF GOTO MO;
& STEP MO - stand-by for pushing button "START" &
90 STEP MI;
100 IF K1 THEN KM1 IS OFF, KM2 IS ON GOTO M2;

```
110 IF NOT K1 THEN KM1 IS ON, KM2 IS OFF GOTO M1;
```
& STEP M1 - setting the punch in the upper position and issuing the command "DOWN" &
```
120 STEP M2;
130 IF NOT K2 THEN KM1 IS OFF, KM2 IS ON GOTO M2;
140 IF K2 THEN KM1, KM2 IS OFF, WAIT T1=1 SEC THEN KM1 IS ON, KM2 IS OFF GOTO M3;
```
& STEP M2 - a motor is switched off when a punch reaches its upper position then follows one pause, transition to STEP M3 &
```
150 STEP M3;
160 IF NOT K2 THEN KM1 IS ON, KM2 IS OFF GOTO M3;
170 IF K1 THEN KM1, KM2 IS OFF GOTO MO;
```
& STEP M3 - punch returns to the upper position, transition to STEP MO &
```
180 END;
```

References
----

1. Smolianinov, A.V.; Deviatkov, V.V. The system of graphic training of assembly robot modules in the watch industry. An experianse of itroduction. Technical paper, MS91-117, Society of Manufacturing Engineers (1991) 1-17.

2. Gavrilov, M.A.; Deviatkov, V.V.; Pupyrev, Y.I.: Logical dezign of discrete automata (in Russian). Moscow: Nauka, 1977.

3. Deviatkov, V.V. Expert-system aspects of computer-aided design of program-logical control complexes. (in Russian) Computers, Systems, Control, n. 1 (1990).

4. Deviatkov, V.V.: The CAD technology for programmable logical control systems. (in Russian) CAD for programmable logical control systems. Moscow: Industry (1990) 4-28.

Computer-Aided Design of Gating Systems for Cast Iron Castings

R. SKOCZYLAS, A. GRADOWSKI, J. ZYCH

Institute of Foundry Technology and Mechanization
Academy of Mining and Metallurgy, Cracow

Summary

A program for computer-aided design of gating systems for cast iron castings has been developed. The algorithm bases both on principles of fluid flow and heat transfer as well as common foundry practice. In the paper some selected components of the algorithm, like: choke area, pouring time and pouring temperature, are presented. The main features of the program are shortly mentioned.

Introduction

Making products by casting is one of the oldest production technology. The casting technology is in most cases the cheapest and sometimes the unique one. To make a casting, a gating system is necessary that is a network of channels in mould which allow liquid alloy to fill the cavity and retain slag. The flow of metal in a gating system is governed by the principles of fluid flow mechanics. Based on these principles the design of gating systems could be totally reduced to a set of mathematical equations.

There are two general ways of solving such equations using computers. The first one bases on mathematical models consist of partial differential equations usually describing not only fluid flow but also heat transfer phenomena. Numerical methods (like FDM or FEM) allow to obtain excellent examples of simulation of mould filling, cooling and solidification of castings [1, 2]. However, such approach requires powerful computer, sophisticated software, and highly qualified persons to run it. Moreover, it does not ensure success everytime. On the other hand, basing on relatively simple rules and experience of foundrymen, it is still possible to obtain quite good results. This paper concerns computer-aided design of gating of cast iron castings, basing on the second, simplified, approach (work [3] gives similar example for risering system design). Such software saves time, improves casting quality and increases productivity of casting houses.

Gating system

The elements of a basic and very common gating system are [4] the *downsprue*, through which metal enters the *runner*, and from which it in turn passes through the *ingates* into the mould cavity (Fig. 1). At the top of the downsprue is *pouring basin* to minimize splash and turbulence and promote the entry of clean metal only into the downsprue. The gating system must be designated to accomplish such objectives like [5]: 1) Fill the mould rapidly enough. 2) Prevent the formation of dross in the mould. 3) Prevent aspiration of air or mould gases into the metal stream. 4) Avoid erosion of moulds and cores. 5) Obtain a maximum casting yield and minimum grinding costs.

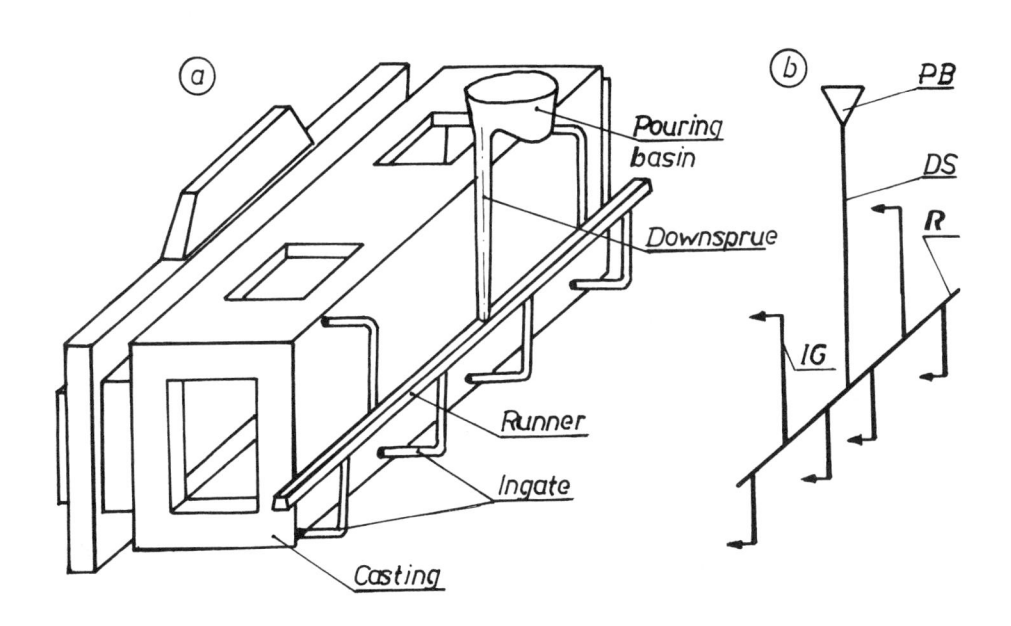

Fig. 1 a) Rough casting with gating system, b) scheme of gating.

Choke area

Choke is the smallest area in the channels which controls the flow rate into the mould cavity and consequently controls the pouring time. The proper area can be calculated by applying a formula based on the application of Bernoulli's theorem [6]:

$$CA = \frac{W}{d\, t\, f\, \sqrt{2gH}} \qquad (1)$$

where CA = choke area, W = casting weight, d = density of molten metal, H = effective height of metal head, f = efficiency factor, g = acceleration of gravity, t = pouring time.

Pouring time

Cast iron is not so sensitive to pouring rate as others alloys. Yet even for cast iron an optimum pouring rate is advocated, which is a function of the casting size and shape:

$$t = F(W, AT) \qquad (2)$$

where AT = average wall thickness.

There are many formulae for calculation of the rough pouring time: $t = \sqrt{W}\,(a + b\, AT)$ [7], $t = a\, \sqrt[3]{W\, AT}$ [8], $t = a\, W^{0.4}\, AT$ [9], $t = \sqrt{W}\,(a - b\, \log W)$ [10], where a, b = constants.

The calculated pouring time must be corrected, if it is outside of valid range of time. The maximum pouring time is limited by a rate of filling of mould cavity (for thin-walled castings) or by mould resistance against heat radiation (for thick-walled castings). The minimum pouring time is limited by mould resistance against erosion and handling a large volume of metal in a short time. If a gating system should also act as a risering system the minimum pouring time is additionally limited.

Pouring temperature

How much temperature loss occurs during pouring is of considerable interest to foundryman. Pouring temperature should be low, but still high enough to ensure casting without misruns:

$$T = TL + \Delta TM + \Delta TG + \Delta TD \qquad (3)$$

where TL = temperature liquidus (start of freezing, close connected with chemical composition of cast iron), ΔTM = temperature loss during filling of mould (depends on minimum wall thickness), ΔTG = temperature loss inside gating system (depends on thermal properties of mould, gates dimensions and rate of filling), ΔTD = alteration of temperature to counteract selected casting defect.

Program "Gating of Grey Iron Castings"
The program is running on IBM PC/XT/AT/386 microcomputers. It has been written in Microsoft QuickBASIC 4.5 and consists of 127 subroutines and about 12 000 lines of source code. Some of the leading ideas when designing the program has been its easy usage and clearness, the ability to keep all the design data in memory and user-friendly interface. From the user's point of view, the program consists of 30 - 60 screens, which one can go through using PgDn and PgUp keys. All choices are made using menus and all numbers are limited to reasonable values. The untouched values remain as they were earlier, if so wanted. This makes fast and easy going through a lot at different designs. Also the possible mistakes in design can be corrected right away. The whole design phase takes a minimal amount of time compared to the traditional computer programs, which proceed step by step and do not allow user to go backwards if he makes a mistake.
This program provides options and features like:
- Entering of chemical composition in four different ways.
- Choosing various gating systems (Fig. 2).
- Estimation of efficiency factor of selected gating system.
- Estimation of an effective height of metal head.
- Determination of an optimum pouring temperature taking into account chemical composition, minimum wall thickness, temperature loss within gates and counteraction casting defects.
- Corrections of pouring time to counteract casting defects.
- Analysis of possibility of feeding through gating system.
- Step by step building gating system using both standardized as well as non-standardized gates.
- Determination of casting yield.
Fig. 3 shows an example of results of calculation. Analyzed gat-

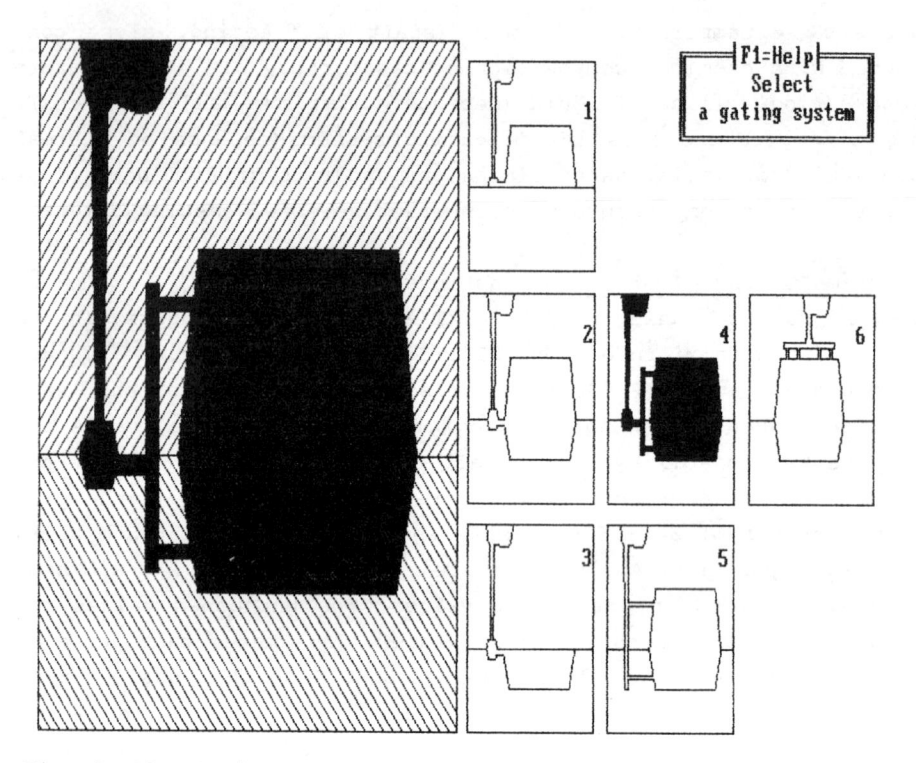

Fig. 2. The *Gating System* menu.

ing system consists of one pouring basin, two downsprues, four slag runners, four whirl-gates, four runners with filters and twenty four ingates (four various types).

Conclusion

The gating system design for cast iron castings can be rationalized and eased significantly with the help of computers. The program "Gating of Grey Iron Castings" is user-oriented. This gives new options to foundrymen who has been hesitating the use of computers in designing the gating system. Similar software seems to be inevitable in casting houses of the future.

References
   1. Kearns, J.K.: Computer Simulation of the Permanent Mold Process. Modern Casting 10 (1986) 29-30.
   2. Corbrett, C.F.: Computer aided thermal analysis and solidifi-

cation simulation. *The Foundryman* 10 (1987) 380-389.

============================= MAIN RESULTS =============================

Casting yield:			72.7%	Pouring time:		25s.
Gating ratio:		2:1.5:1.25:1		Pouring temperature:		1385°C
Total casting weight:			495.1kg	Feeder required		

Element (gate)	Symbol	P.Norm	Area	Element	Symbol	Area
Pouring Basin	PB	ZWcn 21	17.1dm³			
DownSprue	DSx2	WGo 17	16.0cm²			
Slag Runner	SRx4	WRt 12	6.0cm²	Whirl-Gate	WGx4	⌀49x74mm
Runner	Rx4	WRt 11	5.0cm²	Filter	FFx4	37.5x75x22mm
InGate	IGt1x8	WDnt 10	0.5cm²			
	IGt2x4	WDp 185	1.0cm²			
	IGb1x4	WDt 161	1.0cm²			
	IGb2x8	WDo 205	0.5cm²			

Press any key . . . |27.05.1991|09:27:09

Fig. 3. The *Main Results* screen.

3. Louvo, A.; Kalavainen, P.: Method for utilizing solidification simulation in feeding system design of SG iron castings. 56th World Foundry Congress, Düsseldorf, Paper No. 11 (1989).
4. Heine, R.W.; Loper Jr, C.R.; Rosenthal Ph.,C.: Principles of metal castings. New Delhi: Tata McGraw-Hill Co Ltd 1983.
5. Wallace, J.F., Evans, E.B.: Gating of Gray Iron Castings, *Trans. AFS*, **65** (1957) 267-275.
6. Osann, B.: Eingußtechnik und Belastung der Form. *Giesserei* **49** (1928) 1217-1225.
7. Dietert, H.W.: How Fast Should Mold Be Poured? *Foundry* **81** (1953) 205.
8. Sobolev, K.A.: Novyj metod raschota litnikovykh sistem dla otlivok iz serogo chuguna. *Liteishchik*, 10 (1934).
9. Holzmüller, A.; Kucharcik, L: Atlas zur Anschnitt- und Speisertechnik für Gußeisen. Gießerei-Verlag. Düsseldorf 1969.
10. Kotschi, R.M.; Kleist Jr., O.E.: Computerized Gating Of Castings. *AFS Int. Cast Metals J.* **4**, 3 (1979) 29-38.

Acknowledgment

The authors gratefully acknowledge the support of the Research Committee of Poland - *The Project Group RR.I.12*.

Computer Aided Design of the Underwater Robotic Vehicle Robust Invariant Control

Ivica Mandić, Jadranka Marasović, Nebojša Gačić, Anamarija Batina

Faculty of Electrical Engineering, Mechanical Engineering and Naval Architecture, University of Split, R.Boškovića bb, 58000 Split, Croatia, Yugoslavia

Summary

Computer aided design (CAD) of the underwater robotic vehicle (URV) robust invariant decoupling (RID) control have been presented. The RID control of the URV has to be robust on the measurable and unmeasurable external disturbances, robust on the non-infinitensimal variations of the URV parameters, with satisfactory decoupling of the URV subsystems. A new criterion for optimal composite control system structure selection (the minimization of the Euclidean norm of the matrix of invariance) has been proposed. The simulation studies confirmed the rightenss of the suggested approach.

1. Introduction

Modern underwater robotic vehicles are, from the pointview of the modern systems theory, highly nonlinear dynamical multivariable systems, with severe mutual interactions between subsystems, with time variant parameters and with great influences of external disturbances.

The mathematical model of the URV is the conditio sine qua non for the succesfull automatic control and guidance of such the vehicles in hazardous and unpredictable environments |1,2,3,4,5|.

If the correct and exact mathematical model of controlled system in the state space is available and if the disturbances are measurable, there exist basic presumptions for design of the ideal feedforward control unit for perfect system invariance to the disturbances. An appropriate combination of feedforward and feedback control will enable the decoupling of the systems and the invariance to measurable disturbances. Feedback control will also reject the influences of the unmeasurable disturbances and the effect of imperfect feedforward control |6,7,8|.

2. Linearized mathematical model of the underwater robotic vehicle

The underwater robotic vehicle (URV) is six-degrees-of-freedom free body in the water space.

In the references |3,4| the development of the mathematical model of the URV in the state space has been presented. The model is based on the kinematical and dynamical equations of the

URV motion, and on the external disturbances (viscous hydrodynamical forces and moments, hydrostatical forces and moments, forces and moments of the propeller and rudders, the influences of sea streams, the influences of umbilical cable, the influences of collisions with accidentally objects, etc).

The state space mathematical model of the URV is presented on Figure 1. The elements of referent input vector, u, are propeller angular velocity (n) and vertical and horizontal rudders angles (δ_{RV}, δ_{RH}), while elements of disturbance vector, d, are measurable forces and moments acting on UMV via umbilical cable ($F_{MX}, F_{MY}, F_{MZ}, M_{MX}, M_{MY}, M_{MZ}$).

The state variables of this system, x, are the components of the URV velocity vector (V_x, V_y, V_z), URV angular velocity vector (p,q,r), the positions of the URV (ξ, η, ϵ) and the angular positions of the URV (ϕ, θ, ψ).

The elements of output vector, y, are the postitions of the URV center of mass in absolute coordinate system (ξ, η, ϵ).

State space equations are:
$$\dot{x} = Ax + Bu + Ed \tag{1}$$
$$y = Cx$$

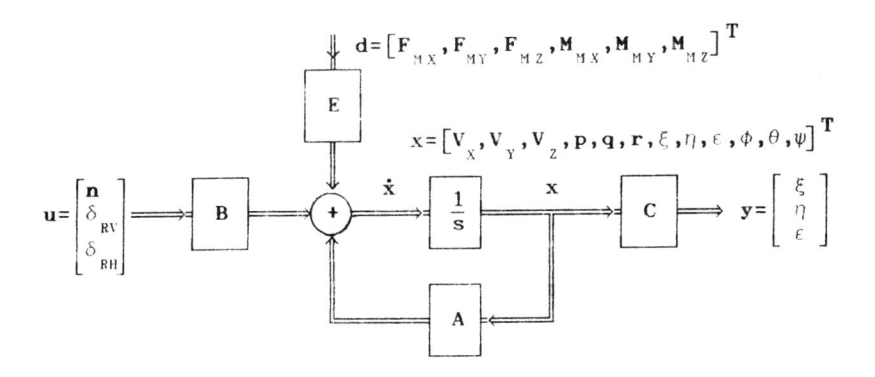

Fig.1. The state space mathematical model of the URV

The elements of the matrices A,B,C and E are presented in |3,4|. The system output vector could be expressed in the form:
$$y = y_u + y_d = Pu + Zd \tag{2}$$
where:
y_u - output vector refered to referent input vector, u,
y_d - output vector refered to disturbance input vector, d,
$P = C(sI-A)^{-1}B$ - system transfer matrix,
$Z = C(sI-A)^{-1}E$ - disturbance transfer matrix.

System transfer matrix, P, is of dimensions 3x3, and disturbance transfer matrix, Z, is of dimensions 3x6. All the elements of these matrices are the rational functions (polinoms) of the sixth order |3,4|.

3. Computer aided design of the multivariable invariant URV control

The objective is to achieve the invariance of the URV to the external disturbances.

Multivariable system is invariant to the external disturbances if their action has no effect on system outputs. Mathematically, it means that transfer matrix which relates disturbance inputs with system outputs is then a zero-matrix.

If the correct and exact mathematical model of controlled system is available and if the disturbances are measurable, we have basic presumptions for design of ideal feedforward control unit for perfect system invariance to disturbances. An appropriate combination of feedforward and feedback control will enable decoupling of subsystems and invariance to measurable disturbances. Feedback control will also reject the influences of unmeasurable disturbances and the effect of imperfect feedforward control |6|.

The design of the composite feedforward-feedback control system is considered as the possible solution of the task. Five different structures of the composite control system will be proposed. All five structures are based on feedforward-feedback configuration with two multivariable controllers: feedforward controller, G, and one of the following controllers: feedback, H, cascade, K, or prefilter, L. (Table 1.) The appropriate matrices of the controllers for decoupling (H, K or L) and for invariance (G) are also presented.

A new criterion of robust invariance for optimal composite control system structure is the minimization of the Euclidean norm of the composite matrix of invariance, M_c |6|:

$$M_c = Y_{di} Y_d^{-1} \qquad (3)$$

where Y_d is output disturbance vector for UR vehicle without control system, and Y_{di} is output disturbance vector for one of the five structures of the control system (i=1,2,3,4,5).

The appropriate matrices of invariance for all the control structures are on Table 1.

The robustness of the proposed criterion is in fact that the non-infinitensimal variations of UR vehicle parameters are allowed, i.e. all of the URV's subsystems parameters can deviate in the range, for instance, ±10% or even more, from their nominal values. The URV transfer matrix, P, in the case of deviation of the parameters from their nominal values, will be labelled with index d, hence P_d (URV matrix, deviated).

In Ref. 8. there are graphical interpretations of the Euclidean norms of the static composite matrices of invariance, M_c, for all the control structures from Table 1. for ±10% deviations of the transfer matrix P parameters from it's nominal values. It is obvious that control structure 1. minimizes the Euclidean norm of the M_c, so the control structure 1. is the

Table 1. Decoupling controller (H,K or L), invariance controllers (G), and composite matrices of invariance (M_c) for five proposed URV - control system structures

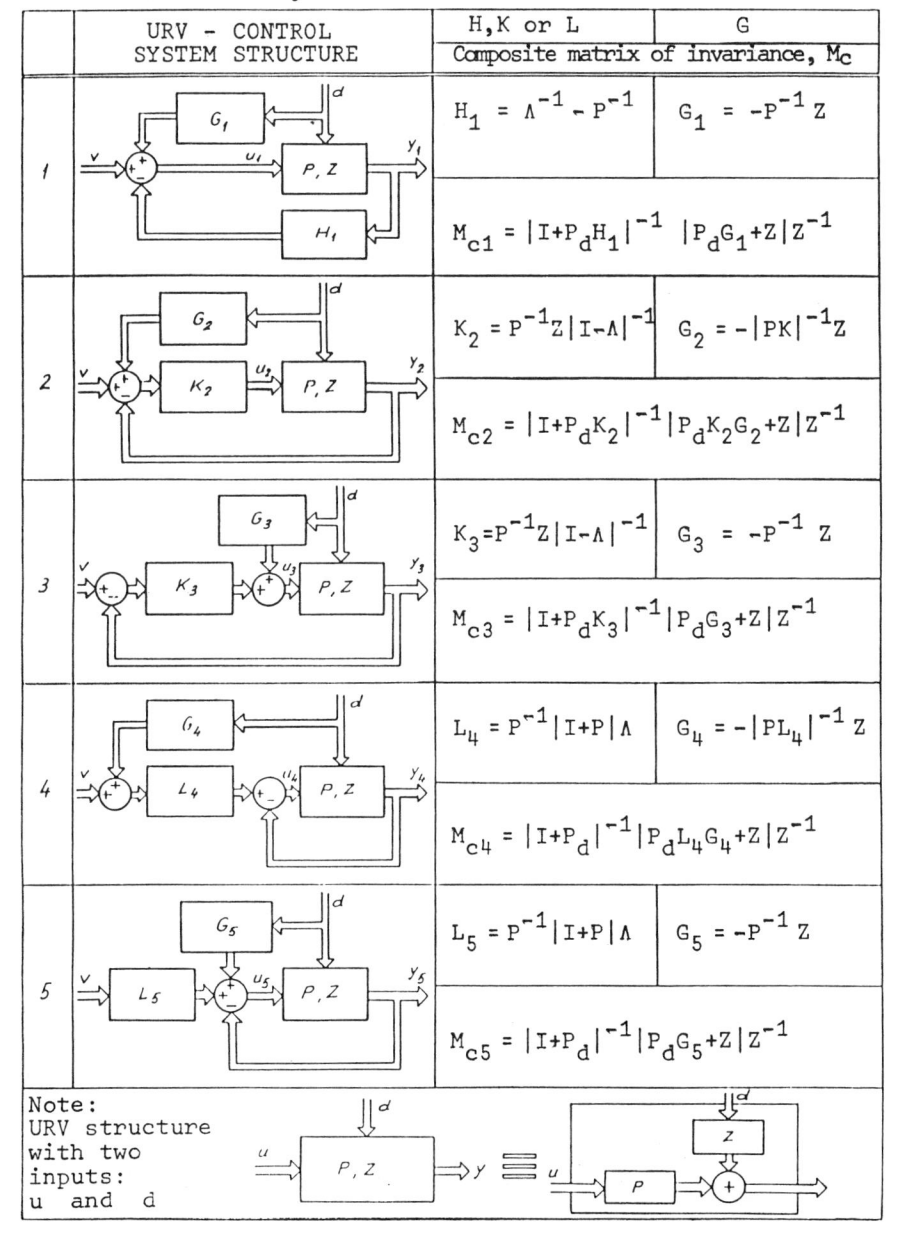

	URV - CONTROL SYSTEM STRUCTURE	H, K or L	G
		\multicolumn{2}{c	}{Composite matrix of invariance, M_c}
1	(block diagram with G_1, P,Z, H_1)	$H_1 = \Lambda^{-1} - P^{-1}$	$G_1 = -P^{-1} Z$
		\multicolumn{2}{c	}{$M_{c1} = |I+P_d H_1|^{-1} |P_d G_1 + Z| Z^{-1}$}
2	(block diagram with G_2, K_2, P,Z)	$K_2 = P^{-1}Z|I-\Lambda|^{-1}$	$G_2 = -|PK|^{-1} Z$
		\multicolumn{2}{c	}{$M_{c2} = |I+P_d K_2|^{-1} |P_d K_2 G_2 + Z| Z^{-1}$}
3	(block diagram with G_3, K_3, P,Z)	$K_3 = P^{-1}Z|I-\Lambda|^{-1}$	$G_3 = -P^{-1} Z$
		\multicolumn{2}{c	}{$M_{c3} = |I+P_d K_3|^{-1} |P_d G_3 + Z| Z^{-1}$}
4	(block diagram with G_4, L_4, P,Z)	$L_4 = P^{-1}|I+P|\Lambda$	$G_4 = -|PL_4|^{-1} Z$
		\multicolumn{2}{c	}{$M_{c4} = |I+P_d|^{-1} |P_d L_4 G_4 + Z| Z^{-1}$}
5	(block diagram with G_5, L_5, P,Z)	$L_5 = P^{-1}|I+P|\Lambda$	$G_5 = -P^{-1} Z$
		\multicolumn{2}{c	}{$M_{c5} = |I+P_d|^{-1} |P_d G_5 + Z| Z^{-1}$}

Note: URV structure with two inputs: u and d

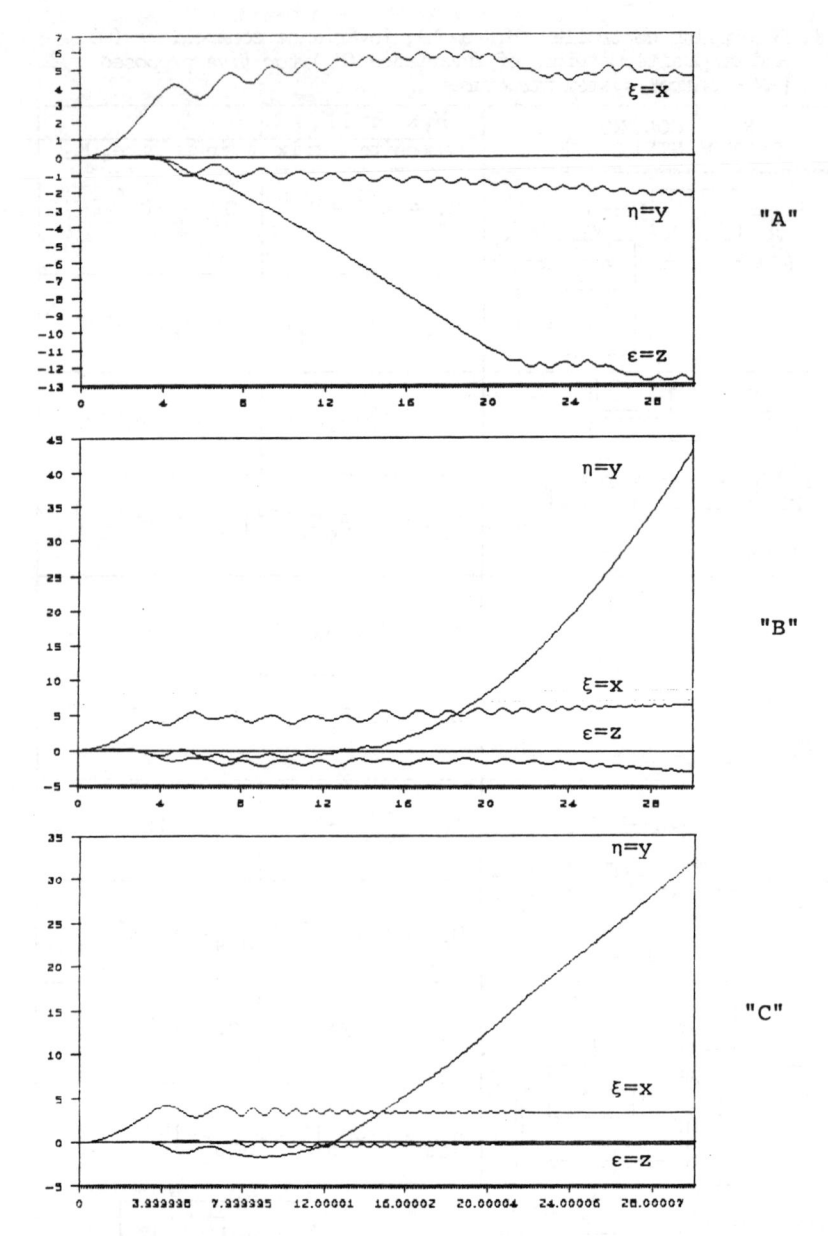

Fig.2. Step responses of the URV (ξ, η and ε) without disturbance and control ("A"), with disturbance and without control ("B"), and with disturbance and control ("C")

optimal one from the invariance point of view.
The simulation studies "A", "B" and "C" on Figure 2. show the behaviour of URV via step responses of the URV positions ($\xi=x$, $\eta=y$ and $\varepsilon=z$) in the absolute coordinate system, for URV system without disturbance and without control ("A"), with disturbance and without control ("B"), and with disturbance and with control system 1. The referent inputs n, δ_{RH} and δ_{RV} are step function of appropriate amplitudes.

4. Conclusions

The mathematical model of the underwater vehicle has been developed. All the influences from the system and from the environment have been taken into account.

The design of the flexible composite feedback-feedforward control has been cosidered as a possible solution of the complex control task. This system is constructed on the basis of four basical control principles: feedforward, output feedback, cascade and prefiltering.

A new criterion for the selection of the optimal control system structure has been suggested, based on the minimization of the Euclidean norm of the comparison matrix of invariance, M_C.

The simulation studies have been done and the results obtained confirmed the rightness of the suggested approach.

The results obtained in this project encourage the people involved to further efforts in the field of RID (robust invariant decoupling) control of the underwater robotic vehicle, not only on the simulation level but also on the level of practical application.

References

|1| Butler,B, and Maryka,S., Evolution of the Dolphin Multi-Vehicle Control System, Proc. of the Int. Conf. Oceans 87. Vol.1, p.596-600, (1987)

|2| Cuomo,C., Pocrio,V., Pelli,E. and Tufano,A., Theoretical and Experimental Aspects of Mathematical Model to Calculate the Submarine Manoeuvring Behaviour, Proc. of the 25 th Anniversary Symposium, Genoa, Italy, p.4.1.-4.40, (1987)

|3| Hrboka,I., The Underwater Vehicle Mathematical Model Development (in Croatian), M.Sc.Thesis, Faculty of Mechanical Engineering and Naval Architecture, University of Zagreb, Yugoslavia (1990)

|4| Hrboka,I., Mandić,I., and Marasović,J., The UMV Mathematical Model Development (in Croatian), Proc. of JUREMA'90. Zagreb, Yugoslavia (1990)

|5| Russel,G.T., The Automatic Guidance and Control of an Unmanned Submersible, Underwater System Design, p.14-18 Aug./Sept. (1981)

|6| Mandić,I., Sensitivity Optimisation of the Complex Systems to the Disturbances (in Croatian), Ph.D.Theses, Faculty of Electrical Engineering, University of Zagreb, Yugoslavia (1987)

|7| Mandić,I., and Vukmirica,M., Robust Decoupling Manipulator Control in the Presence of External Disturbances, Modelling Simulation & Control Journal,B, AMSE Press Vol.30, No 1, p.53-64, (1990)

|8| Mandić,I., Hrboka,I., and Marasović,J., Robust Invariant Decoupling (RID) Control of the Underwater Manipulator Vehicle, Proc. of the 4 th ISME, Kobe, Japan, Vol. II, p. 651-658 (1990)

An Efficient Boundary Data Structure for Integrated CAD/CAM, Robotics and Factories of the Future

S.R.Ala and Denis A Chamberlain

Construction Robotics Unit, School of Engineering,
City University, Northampton Square, London-EC1V 0HB, U.K.

ABSTRACT

Boundary data structures are the most popular data structures in CAD, CAM and Advanced Robot applications. Several data structures have been proposed since the classic Winged edge data structure, these aimed at reducing the storage requirement and increasing information retrieval speeds. The Symmetric data structure has been held to be the most efficient, this having smaller storage and comparable access time to the Winged form. However, more recently a new, compact, fast access time scheme has been proposed which can out-perform these. This improved scheme has been adopted in the implementation of a vision sensing system for two advanced construction robots currently being developed. These elements of the "construction site" factory of the future necessitate extension and integration of the CAD data base.

I. INTEGRATION OF CAD AND ROBOTICS IN THE FACTORY OF THE FUTURE

To achieve the goal of integrated automation it is advantageous to extend and integrate the CAD data model for simulation and real manufacture, including the requirements for vision tasks, like inspection. Such schemes should span the gap between the initial design and planning of the project, and the machine intelligence requirement for the construction robot activity.

Figure 1 outlines an integration strategy currently being pursued in the provision of a data base, the key elements shown in their context. A wall climbing inspection robot and a wall assembling robot [1] are the immediate targets for this work. The provision for finite element analysis arises from the flexible nature of these and the concern for their vibration characteristics. The AutoCad software is being used for design and planning activities, GRASP for kinematic based simulation and in house development of the LEONARDO expert system environment for machine intelligence. Each of these operate on, and extend, the common data base.

Using a common CAD data base for design, planning, manufacture simulation, computer based inspection, robot navigation and manipulation, enhances efficiency and productivity. This single data structure serves all requirements and avoids data integrity problems. In the past, data has been isolated for specific tasks and, whilst this served the short term goal of optimising a specific activity, redundancy was inevitable. This research is aimed at demonstrating the potential of a CAD based approach, working towards the ultimate goal of computer integrated work which embraces both CIM and CICI (Computer Integrated Construction and Inspection).

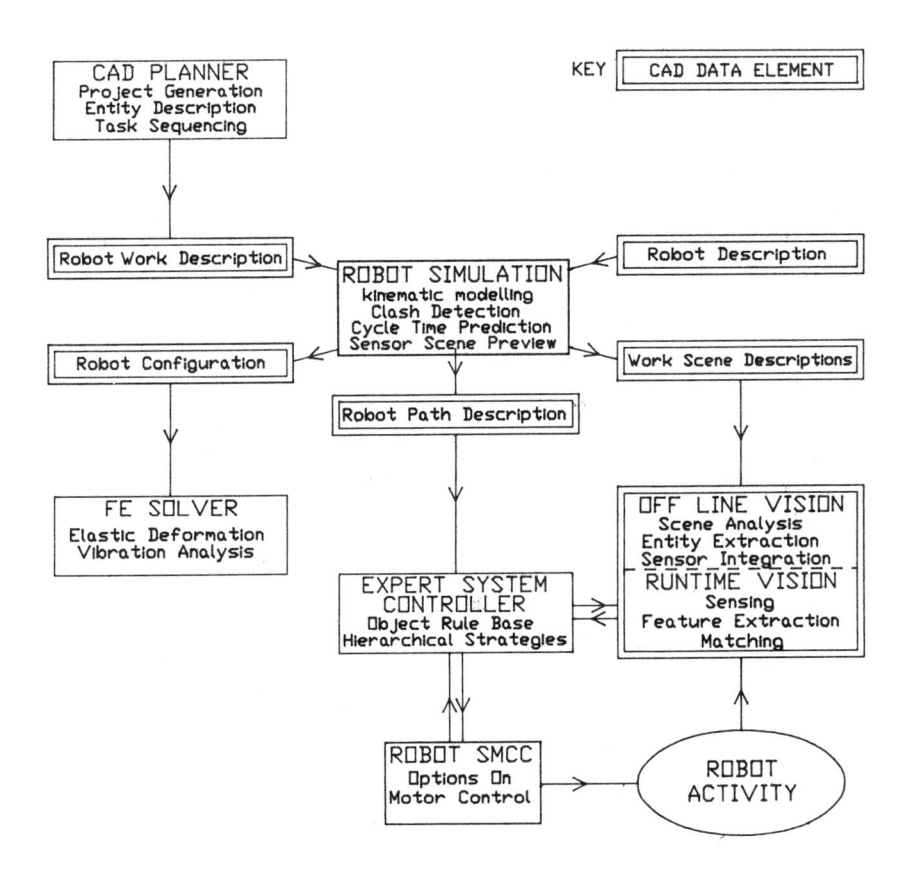

Figure 1.   CAD Data Integration Strategy.

The evolution of the CAD data base and its integration with the other data can be classified into five stages, these being apparent in figure 1. These are considered in the context of the two robots previously mentioned.

(i). Cad planner with 3d draughting

The wall assembly task can be modelled using the "3D block entity" method, this comprising lower order entities such as points and edges. This entity together with other high order entities represents a "parts kit" for the project. A 3D approach is preferred as this overcomes the 2D to 3D interpretation problem. Projects are built up on an interactive basis with setting out aids such as plan grids. Logical task sequencing is build into this by object rule direction. The inspection activity can also be substantially enhanced by the preparation of a terrain and work detail representation. AutoCad is currently being used to develop this facility.

(ii). Robot simulation with runtime attributes

This simulation facility allows the robot and planned task to be integrated and provides assessment of clash detection, production time cycles as well as explicit visualisation of the Camera and eye in hand sensing which can be modelled with key scenes saved. At any stage the robot configuration can be passed for structural performance analysis.

(iii). FE solver for structural assessment

Data for forward kinematic analysis including large deflection static and vibration analysis can be transferred from the simulation facility. As construction robots are likely to be more flexible than their manufacturing industry counterparts, this is an important consideration.

(iv). Offline and runtime vision for inspection and survey

In order to enhance the vision sensing performance, it is necessary to maximise the benefits of the accumulated CAD data. Strategies such as extensive masking off of unwanted detail are extremely important in this. The work scene description derived from the simulation process can be substantially preprocessed for the ensuing runtime vision task. Wall unit alignment and potential unit collision stages are the main targets for this, as well as work location for the inspection robot. Superior data structures are essential for the runtime implementation, further details of which follow.

In the past most of the research in computer recognition involved models built by scanners instead of using a CAD data base. The errors in the data acquisition step due to calibration, noise etc. affect the quality of model and the manual approach limits the number of possible models. Alternatively, most of the parts already have a CAD model which may be used directly or augmented with information necessary for visual tasks.

There are differences in vision and CAD requirements in that recognition is concerned with objects which already exist. On the other hand CAD is concerned with interactive design of new objects. CAD representations also tend to be view independent. Computer graphics is used to render the objects from different view points, for example. In a nutshell, CAD systems stress the interactive design, set operations, rendering, finite element analysis and CNC programming. Vision models need augmentation with imaging and illumination models. Stable orientations, view potential or aspect graph and surface properties (e.g. texture, color, reflectance) information are also needed.

Construction is highly unstructured compared to the manufacturing industry and hence parts and machines can not be tied down or guided by fixtures and other conventional gadgetry of the established manufacturing industry. The machine vision requirement is thus diverse and complex.

(v). Expert system controller for robot activity

This has been approached at a high level by adopting a rule based, object orientated, expert system shell development environment. In this, the completion of task stages are the top level goals, which in turn must satisfy the sub goals for sensed states and sequence forcing. This facility communicates with the robots SMCC in the case of the wall assembly robot, and a development board in the case of the inspection robot.

These collective and integrated facilities represent a blue print for robot cell development in the construction industries factory of the future.

II. APPROACHES FOR MODELLING

Representational Methods in CAD-BASED ROBOTICS

For typical construction project there is an enormous amount of CAD data (e.g. a typical 10 storeyed building will have hundreds of Megabytes of CAD data), thus efficient retrieval is a paramount consideration in the organization of the data. Popular CAD representations for model based robotics are Boundary representation (B-rep), Constructive solid geometry (CSG), and Sweep.

In Boundary representation the object is represented in terms of faces, edges, vertices i.e. the boundary of the object. In the CSG the objects are represented in terms of boolean operations of solid primitives (cuboids, cylinders etc.). In sweep representation, a planar cross section and the axis along which it is rotated (rotational sweep) or translated (translational sweep) represent the objects. Generalised cylinders representation is an example of the sweep representation.

B-Rep requires more storage than the other two, but is the best for access to the shape of the object (e.g. CSG requires cumbersome boundary evaluation). Consequently B-Rep is the most popular form of data representation for robotics algorithms.

Review of popular B-Rep data structures

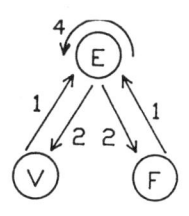

Figure 2
Winged edge
data structure

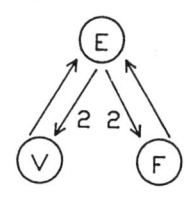

Figure 3
Symmetric data
structure

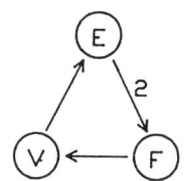

Figure 4
Delta data
structure

Interest grew in B-Reps with the proposition of the Winged Edge data structure [2]. It attempted to access the edge based queries in the best manner by explicitly storing all edge based information viz. the neighboring vertices, faces and edges for each edge. It also stored one neighboring edge for each vertex and face. The other edges being retrievable by examining the explicitly stored data. The schema is shown in Fig.2. V, E and F denote the three basic entities: vertices, edges and faces, of the object's boundary.

Woo [3] propsed the Symmetric data structure (SDS) which requires 8×E a big improvement asymptotically over the 9×E of the WE, E being the number of edges of the object. SDS (see Fig.3) stored for each relation its converse as well and thus possessed symmetry. It stored for each edge its vertices, the converse i.e. for each vertex all the adjacent edges (note: WE stored only one edge). The symmetrical edge-face relations are also stored.

Several other data structures, such as the hybrid edge [4] and Hierarchical SDS [5], which were modifications of the above two basic data structures followed. Also the basic three basic entities grew by two: loop to model objects with holes in their faces and half edge, ostensibly for efficient retrieval. [6] proposed a design methodology for boundary data structures and a benefit of this was the discovery of a new data structure: Δ, the most compact data structure (it requires 6×E storage). The schema for Δ is shown in Fig.4.

III. COMPARISON OF Δ WITH THE OTHER DATA STRUCTURES

Special purpose optimization

There were special purpose data structures which optimised the data manipulation efficiency and the storage. A recent example is the Winged triangle (WT) data structure [7]: it triangulated the faces of the objects and for each of the triangles, WT stored the three vertices and the three bordering triangles. Its storage varied from 4E to 12E. It was optimal only for the face based queries common in the boolean operations of the solids. As noted in Section I, an integrated environment will need access to the other queries (e.g. edge based queries are quite common in computer vision), which will require scanning the whole data base and the access time is clearly unacceptable. Δ performs better in such an integrated environment since it uses its optimization criterion as the sum of the times of all possible queries occurring in integrated robotics environment.

Extension for faces with holes

Δ avoids the usage of the loop entity for modelling faces with holes. This is feasible because of the structure of Δ: as usual the face is represented by an anti-clockwise string of vertices , however holes have clockwise sense of the string and a negative integer in the string denotes the start of the hole boundary. Unlike [8], there is no need for resorting to the extra fictitious edges for bridging the face-hole boundary. Also the scheme possesses the storage savings associated with the avoidance of loop entity.

Compactness has a side benefit

The predecessors to Δ assumed implicitly that the whole of the data resided in main memory. Hence, the only cost was the computation by the CPU and disk access cost does not figure in the optimization criterion.

Modern computers operate in a virtual memory environment which make it unnecessary for the all data to be simultaneously held in the main memory, and makes it possible to query data bases much larger than the main memory. This however comes at a price i.e. it requires frequent transfer of data between the main memory and the disk to get the sought after data from the vast disk store and put in main memory.

Since Δ requires the least storage, a large proportion of its data can be held in the main memory. For example, if the main memory is 2E, a third of Δ's data can be held in the main memory while only a quarter of the SDS's data can be held. Because a large fraction of data is memory resident, fewer disk transfers are required for Δ and consequently savings in the costly disk operations result.

IV. CONCLUSIONS AND FUTURE WORK

The background for the requirement of a fully integrated and enhanced CAD data base for robotic inspection and construction has been presented. For maximum benefit the information technology must cover all stages from the initial planning of the task to the multi-sensor assisted robot activity. Vision in particular is expected to dominate in this due to the diversity and complexity of the tasks. Activity simulation is seen to be usefully linked to the vision facility, providing scene previews for off-line processing. However for practical implementation the nature of the actual data structures merits close attention if fast run time access of vision data is to be achieved. The proposals for this have been detailed including the considerations of optimization, compactness and implementation. A contemporary progress of construction robot development provides the impetus puts for the development and improvement of the run time machine vision implementation.

V. REFERENCES

1. Chamberlain D, Spear P S & Ala S R, Progress in a Masonry Tasking Robot, 8th International Symposium on Automation and Robotics in Construction, June 3-5, 1991, Stuttgart, Germany.
2. Baumgart B G, A polyhedron representation for computer vision, AFIPS National Computer Conference, 589-596, 1975.
3. Woo T C, A Combinatorial Analysis of Boundary Data Structure Schemata, IEEE Computer Graphics and Applications, 5, 3, 1985, pp.19-27.
4. Kalay Y E, The hybrid edge: a topological data structure for vertically integrated geometric modeling, Computer-aided design, 21, 3, 1989, 130-140.
5. Falcidieno B and Giannini F, Automatic Recognition and Representation of Shape-Based Features in a Geometric Modeling System, Computer Vision Grap. and Image Proc., 48, 1989, 93-123.
6. Ala S R, Design Methodology of boundary data structures: ACM/SIGGRAPH Symposium on Solid Modeling Foundations and CAD/CAM Applications, June 5-7, 1991, Texas.
7. Paoluzzi A, Ramella M and Santarelli A, Boolean algebra over linear polyhedra, Computer-aided design, 21, 8, 1989, 474-484.
8. Yamaguchi F and Tokieda T, Bridge edge and triangulation approach in solid modelling in Kunii T L (ed) Frontiers in computer graphics, Springer Verlag , 1985.

A Capstone Two-Course Sequence for Design, Build and Test

Dr. N. Berzak
Dr. W. W. Walter
Department of Mechanical Engineering
Rochester Institute of Technology
Rochester, New York 14623

Abstract

In order to meet the present challenge of introducing more design into the engineering curriculum, a capstone design course has been created. In this non-traditional course, four to five students work as a design team on open-ended problems. To simulate the industrial product development environment, the teams are required to develop an idea into a working prototype. This approach provides an opportunity for close cooperation between academia and industry. Teamwork is carried out in an independent mode which fosters self-discipline, self-reliance, and self-confidence. This paper presents the philosophy and methodology of the course, as well as initial experiences.

Introduction

The goal of engineering education is to prepare future engineers for their professional role in industry. As in any other professional education such as medicine or law, this goal is achieved through two objectives: transmitting information, and training for actual practice.

The major difficulty in meeting the objective of transmitting information came from the explosion of engineering knowledge. As the amount of knowledge increased rapidly, it became impossible in the course of the relatively short period of the educational process to encompass all the knowledge a future engineer would need. The educational system then modified its role based on the assumption that the engineer will continue his education after graduation.

The modified educational objectives then became providing the student with a sufficiently wide basis of knowledge and the tools to carry on his education on his own. Even with this modification, both academia and industry have had to compromise on the desire to provide up-to-date knowledge.

The nature of the development of engineering sciences does aid the objective of providing a wide basis of knowledge. Knowledge is arranged in specific teachable bodies and the analytical approach enhances comprehensive understanding and allows the material to be covered more efficiently. However, by following this trend, two adverse effects were introduced. Engineering knowledge was transmitted to students in small, specialized pieces without combining all these pieces into a harmonious picture of their profession. The extensive application of the analysis approach, left students on their own to develop the skills to apply the synthesis approach, which is essential for engineering practice.

The second effect of the knowledge explosion was that each engineering field was subdivided several times. Each division developed its own jargon and approach to engineering practice to such an extent that one specialized subdivision could hardly communicate with another, let alone understand their major problems. Academia followed the same trend, dividing and subdividing education into specializing fields. This development not only moved students further from actual engineering practice, but also reduced their ability to see the overall picture.

Pressured by time limits and the necessity of making choices, the need for direct experience in actual engineering practice, i.e. design, gave way to more and more engineering sciences and the acquisition of fundamental knowledge. The justification for this change was that the transition from theoretical understanding to actual application is easy and simple. The transmission of the information portion of the engineering education was expanded at the expense of the training portion to a point that the educational system has resigned its original objective of preparing the graduate for assuming an immediate role in industry. This objective is now placed in the hands of the employer.

It is widely accepted that today's engineering curriculum has swung too far in the direction of engineering sciences at the expense of engineering design. The present challenge is to introduce more design into the curriculum in a way which closely simulates "real engineering practice."

In recent years, the topic of engineering design has been given much emphasis in most engineering curriculums. Some universities have revised their programs to include a capstone design course in the senior year which requires students to draw on many of the previous courses in their program. A course of this type provides an opportunity for much industry-university interaction. This has been the case at RIT for the past two years.

Although the need for a change in the general approach to engineering education pertains to all courses, it is especially important in selecting the objectives for a capstone design course. The position of the capstone design course in the last year of the curriculum makes it possible to use the course as an introduction to actual engineering practice.

The two main objectives of a capstone course in engineering design are to bring engineering science knowledge into practice in the solution of a real open-ended engineering problem, and to allow the students to experience, as closely as possible, the actual environment in which they will practice their profession.

To achieve the first objective, it is more important to emphasize the interrelationship between the various engineering aspects of a problem, rather than focus on one aspect. The engineering problem should be sufficiently encompassing to include the application of more than one area of engineering in order to demonstrate the need to see an engineering solution as a compromise of several aspects. A topic restricted to one area, is more suitable for a design project for a course.

In many cases, reaching a solution requires studying topics not covered by standard engineering science courses. Such a situation is a good introduction to the need to integrate continued studies throughout the practice of the engineering. Nonetheless, the additional research required should be limited because of time constraints and the emphasis on integrating the various aspects rather than developing expertise. In most cases, the need for additional knowledge of manufacturing and electrical engineering will be necessary. Whenever possible, interdisciplinary cooperation can be a beneficial experience for the students.

To properly simulate the product development environment, a topic must be transformed by the team into a well-defined engineering problem before proceeding to find a solution. It is important to allow the students to realize that, during the definition of an open-end problem, many solutions are possible and their engineering judgement is needed in order to select the most appropriate one. This unique experience is necessary since in all previous courses, problems were well defined by the instructor to yield only one possible solution.

To achieve the training objective, it is necessary to simulate, as nearly as possible, the actual practice of engineering, i.e., the students should learn to communicate and cooperate, share work and assume responsibility. Students should be required to work as a team with a common goal. The requirement to cooperate with peers is not easy. It is the first time students are presented with the need to organize, by themselves, a team with a very specific goal. This raises the need to develop trusting relationships with minimum supervision. This experience is unique in engineering education. Another aspect introduced in the team environment is the need to be efficient in time management and the ability to share responsibilities.

The Design Capstone Course Outline

The Mechanical Engineering Department at RIT has recently established a required twenty week, two-course capstone design sequence in the fifth year. The objective of this course is to design, build, and test a product by groups of four or five students working in design teams. The same team members work together throughout 20 weeks.

After the teams are formed, each team selects an area of interest from which an idea is chosen for development into a working prototype (See Table II). At this point, the team defines the engineering problem with a set of engineering constraints.

Students brainstorm and evaluate a number of alternative designs and select the best for a detailed design effort. Students then build and test a working prototype. Only prototypes that work are acceptable. Students must do whatever redesign necessary to make their prototypes successful.

Student design teams meet with their faculty advisors each week for "one-on-one" consultations. Faculty advisors help design teams to determine the tasks to complete the project, who will be responsible to complete those tasks, and a timeline or PERT chart to bring the project to a successful, timely completion. Design teams give their faculty advisor a written weekly progress report which details the activities for the week, and the detailed plans for the following week.

Each student is required to keep a bound logbook of their day-to-day activities and the time spent on those activities. All library research, meeting minutes, work on design alternatives, calculations and engineering analyses should be recorded in the logbook. Students find that their logbook becomes a welcome reference when writing their weekly reports.

The required deliverables for the first 10 weeks are a complete set of engineering drawings for the prototype to be built in the second half of the course, and interim and final written reports. It is important that students understand that any successful design depends on a detailed engineering analysis, and this should appear in the final report. Also included should be a PERT schedule of events necessary to complete the fabrication and verification testing of the prototype model. Design teams are required to present their efforts orally.

The required deliverables for the second half of the course are the working prototype and written interim and final reports. The final oral and written presentations will include a performance test/demonstration of the prototype. Students are expected to spend considerable time in the shop on machines during the build phase of their projects, and in the labs during the testing and verification phase. It is expected that test fixtures will need to be designed and built as part of the testing phase, and this should be included in the PERT schedule of events for the project.

Although most of the students involved in these courses are mechanical engineers, some projects have involved mechanical engineering students working together with electrical and industrial engineering students. These interdepartmental projects have been well received by the students involved. These projects allow student teams to work on more problems that were not possible for mechanical engineering students working alone.

In order to maintain a focus on actual experience in engineering product development, lecture and discussion sessions held with the entire class will cover topics that will be directly applicable to the projects. These sessions include information on project scheduling, oral and written communications, videotaping, library research skills, patent issues, ethics, product liability and safety, and entrepreneurship.

Opportunities for Industry-Academia Interaction:

The general approach to this course provides content and an opportunity for close cooperation with industry. The success of such cooperation can be attributed to the format of the course including definition of an open-ended problem, developing a working prototype, and restricted timetable.

Projects done with the cooperation of industry this year are shown in Table I. Projects of this type provide great advantages to students as well as to industry. Some of these advantages will now be discussed briefly.

Students have the opportunity to see the latest in technology, like DFA/DFM software, applied to a real world problem. Their interaction with engineers from the sponsoring companies builds their self-confidence and interpersonal skills prior to going out on permanent employment. Since they do their own fabrication, they develop a better appreciation for the manufacturing aspects of product development. Lastly, the students benefit as a result of the stipends, software, and equipment provided to the university by the sponsoring companies.

For long-term cooperation, the project must also benefit the industry involved. Students offer industry the latest analysis and computer skills. This provides industry with an opportunity to test a fresh approach and new ideas which are not restrained by the way things have been done before with a minimum interference with the day-to-day operation. Students can try out new software and provide an evaluation of it to the sponsor. In general, companies get a very in-depth design effort on their project for very minimal cost.

TABLE I: INDUSTRIAL STUDENT PROJECTS DONE IN 1990/1991

Application of DFA/DFM methodology for improving functional performance, reliability, and manufacturing cost.

1. Ford Motor Company (Taurus)
    A. Door Panel
    B. Dashboard
    C. Rear Window Lift Mechanism
    D. Front Window Lift Mechanism
    E. Bumper and Isolator
    F. Door Latch

2. General Railway
    A. Crossing Gate

3. JI Case
    A. Cable Plow
    B. Wheel and Axle

TABLE II: SELECTED PROJECT AREAS

1. Olympic Sports Equipment Design
2. Design for the Handicapped
3. Thermal Fluids Design
4. Design-for-Manufacturing (DFM)
5. Solar Car Design Projects
6. Vibration Test Equipment Design

Conclusion

The selection of topics for a capstone design course is the most important step of the process. Since the design experience should take the students from an idea to a working prototype, it is necessary to consider time, manufacturing facilities, as well as students' capabilities. To increase the enthusiasm, the students should select topics from areas of interest rather than from a list of projects (See Table II). Such selection forces a preliminary evaluation of the effort involved, provides a sense of strong commitment and self-motivation throughout the execution of the work.

Major attention should be given to the availability of an appropriate workshop and the cost of the projects. It is better to discourage projects which involve high cost or the need for sophisticated manufacturing processes, since the design should stand as the major objective of the course rather than the manufacturing of the prototype.

The project should be carried out in the same manner as a product is developed in industry. The idea should be developed into a conceptual design after all the alternatives have been outlined and analyzed. This stage develops the skill of exercising engineering judgement. The conceptual design should be presented in the form of a design proposal and should be defended orally and in written form. From the conceptual design, a detailed design should be presented in the form of manufacturing drawings for production. After the production of the prototype, a series of tests should be performed to show that the design meets all the functional requirements set forth during the conceptual design. The test results should be analyzed and an engineering evaluation of the design should be performed.

To assist the students during the project, each stage is reviewed technically. Recommendations and suggestions are given. Such supervision is helpful to identify physical phenomena effecting the design, to promote the reliance on engineering calculations, and the need to use engineering and other criteria for making judgement.

From the experience of 75 students working on 20 different projects , it can be concluded that except for minor modifications, the objectives of the course were met to the satisfaction of the faculty and sponsoring industries, as well as the students.

NC Trajectory Collision Checking

K. MARCINIAK
Warsaw University of Technology
Precision Mechanics Dept. POLAND

Summary

Collision checking algorithms for NC packages has been presented. A boundary representation is assumed to describe a part shape. The algorithm consists of three phases. Boundary fragments close to the tool are localized in the first (preprocessing) phase, In the second phase collision is examined under the assumption that the tool is convex. The main algorithm works efficiently with the exception of the vicinity of the tool-surface contact point. To improve it a second order Taylor approximation is used additionally at that point. Extension to non convex tools has been also presented.

State of the art

Several collision checking methods have been proposed.

Duncan and Mair. (1983) applied polygonal representation of the surface for both: tool path generation and interference detection.

Anderson (1978) in case of three axis machining divides the base of the object into squares and keeps track of the cut height above each square. Each tool movement updates the heights if cuts lower than the currently stored height. Similar approach has been introduced by Chappel (1983). The surface is represented by a set of points and vectors normal to the surface in these points. The length of a vector is reduced if it intersects the tool movement envelope. Oliver (1986) and (Oliver and Goodman 1986) used similar approach . but graphical image has been used to choose a limited set of the points. Projection of each pixel on the screen defines one point on the surface. Also (W.P.Wang and K.K.Wang, 1986a,b) had developed an image approach. An extended z-buffer technique has been applied here. The current workpiece z-buffer is obtained by Boolean subtraction of z-buffer of the tool movement swept volume. Van Hook's (1986) method also relies on the extended z-buffer but he subtracted the cutter instead of cutter movement envelope. His method is limited to 3C machining. Atherton et al. (1987) extended van Hook's approach to five-axis machining. The very large ratio between the part dimensions and the machining tolerance cause the visual method to be difficult in use.

In case of local collision checking in the vicinity of the
tool-surface contact point Taylor approximation may prove useful
(Marciniak 1987). Several valuable methods of detecting errors have
been implemented and described by: Drysdale and Jerard (1987),
Jerard et al. (1988), Drysdale et al. (1989), Jerard et al. (1989).

The tool

It has been assumed that the tool is defined as an axis symmetrical
shape defined piecewise by: sphere, cylinder, torus, cone, plane.
The parts of a tool are joined smoothly along circles perpendicular
to the tool rotation axis. Tool collision may be reduced to the
collision of the machined part with some fragments of the mentioned
shapes. In vast majority of cases the tool shape is convex. This
property will be used in algorithms described further.

Boundary parametric representation

Let the B-rep of a solid consist of faces represented by the
mapping $P(u,v)$ over the subset Ω of the rectangle $a<u<b$, $c<v<d$. The
boundary of the region Ω consist of curves $u(t),v(t)$ $e<t<g$ on the
parameter u,v plane. These curves represent also edges of the solid
which are defined as $P(u(t),v(t))$ and vertices defined as
$P(u(e),v(e))$.

Triangular approximation of the boundary

To perform efficiently collision checking an approximation of the
solid boundary is usually applied. Linear interpolation over triangles will be analyzed here. Its approximation error is proportional
to the square of the triangle diameter.

The surface $P(u,v)$ is divided into triangles on the uv (parameters)
plane. This implies curvilinear triangles **ABC** in space. The planar,
interpolating at vertices, triangle **ABC** is used as the approximant.
Let d be the furthest distance between triangle **ABC** vertices, $\varkappa$ be
the upper bound of the normal curvature in the triangle and of its
edges curvatures. The upper bound e of the planar triangle
approximation error is

$$e = \frac{f^2}{R + \sqrt{R^2-f^2}} \cong e1 = \frac{f^2}{8R} = \frac{d^2}{6R} \qquad \text{where} \quad f=d/\sqrt{3}$$

It should be noticed additionally that if $R=1/\varkappa \geq f = d/\sqrt{3}$ then the triangle is entirely inside of the sphere defined by radius f and center **Q**=(**A**+**B**+**C**)/3.

Suspected region localization

The first step to detect collision is to locate the suspected regions. Two solutions: array of voxels and trees of spheres will be described here.

ARRAYS OF VOXELS

If a machined part is complex and much greater than the tool (what is usually the case) the following solution relying on space adjacency may be used to quick localize the suspected (of collision) regions. We start with the boundary divided into fragments. For instance they may be, as described before, triangles. Next, the space is divided into three (two) dimensional array of adjacent voxels (usually rectangular). Each voxel contains pointers to all overlapping it boundary fragments. If the tool overlaps the voxel (the voxel is active) then all its fragments are checked against collision, else no action is taken. The voxel size is a compromise between memory (number of voxels) and time (number and size of boundary elements contained in the voxel). The solution is effective if tool-voxel activation check is cheap.

TREES OF SPHERES

Tree structures may be a better solution in case of solid shapes which are described by voxels in such a way that most of them are empty. As an example, the application of trees of spheres to locate collisions will be described. Spheres have been chosen because:

- it is easy to design the sphere which contains a fragment of a solid boundary,
- the test of sphere collision with typical axis-symmetrical tool shapes: sphere, torus, cylinder, cone is inexpensive and costs from 10 to 50 flops.

In case of a face defined by the surface **P**(u,v) divided into triangles, each triangle is embedded in the smallest possible sphere. Recursive subdivision of the triangle into subtriangles which, in turn, are bounded by spheres is a basis to construct the tree of spheres. The sphere may be constructed in such a way as to have as a diameter a straight line section with the ends in two most distant vertices of the curvilinear triangle. If the sphere curvature is bigger than the biggest principal curvature and bigger than triangles curvilinear edges curvature then the triangle is entirely inside of the sphere. To exclude collision with a triangle it is sufficient to exclude collision with its sphere. After checking spheres down to the radius r the number of remaining suspected spheres is approximately equal to the number of solid boundary fragments of diameter 2r which are not further than 2r from the tool. If r becomes small, than the number of suspected spheres is proportional to $A/(\pi r^2)$ where A is the area of these boundary fragments for which the distance from the tool is smaller than 2r. In the (very unusual) worst case, all spheres of all trees have to be examined. Because of these two factors spheres are usually used only to localize roughly solid fragments which are close to the tool. It should be stressed that the lower parts of the trees of spheres need not be stored. They may be generated while checking, up to the assumed accuracy. It saves the storage but is time consuming.

Collision with triangles

Let the collision with the triangle **ABC** has to be detected. We assume that from the approximation phase the upper bound e of the approximation error by the flat triangle **ABC** is also available.

We start with calculation of distances of vertices **A,B,C** from the tool (the points (**A**+**B**)/2, (**A**+**C**)/2, (**B**+**C**)/2, (**A**+**B**+**C**)/3 may be used additionally here). The calculation of minimal distance of a point from the tool costs approximately 15, 20, 25, 50 flops in case of the tool shape: sphere, cylinder, cone and torus respectively. Then the closest to the tool of these points (let it be **A**) is chosen. Let **D** be the closest to **A** point on the tool surface and **n** be the tool normal versor in **D** sensed outside. *The following procedure works under the assumption that the tool is convex.*
1. If ⟨**A**−**D**,**n**⟩>e and ⟨**B**−**D**,**n**⟩>e and ⟨**C**−**D**,**n**⟩>e then collision has been excluded else 2.
2. If ⟨**A**−**D**,**n**⟩<−e then collision has been detected else 3.
3. If e<ε and |⟨**A**−**D**,**n**⟩|<e and |⟨**B**−**D**,**n**⟩|<e and |⟨**C**−**D**,**n**⟩|<e then a

contact point has been found, else the triangle has to be subdivided. Several subdivision approaches may be applied here. The subdivision into four triangles by bisection of edges (on the parameters plane) is the simplest choice. Such a subdivision costs approximately 300 flop (in case of bicubic patch) and the new upper bound for each of the smaller triangle is approximately equal to e/4. Jerrad et al. (1989) proposed subdivision method presented in fig.1 to improve the shape of the resultant triangles. It should be mentioned here that in rare cases of the triangle edge representing the real edge of the face that subdivision point has to be calculated not by simple subdivision of curvilinear coordinates but additionally adjusted to be the point on the real edge.

If the tool consists of several elementary shapes it may be reasonable to make tests more efficient by preliminary tool approximation with simpler shapes. The tool approximations by spheres and flat ended cylinders has been discussed by MacLellan et al. (1990).

The lower bound of diameter of a triangle which has to be divided may be introduced as presented in fig.2. Let the tool curvature be smaller than $\varkappa$; without loosing generality we may fix our attention on a ball end mill with the radius $r=1/\varkappa$. Let the closest to the tool vertex **A** of triangle **ABC** is at a distance d from the tool and d>e. If e=0 it follows from fig.2 that if a vertex **X** of the triangle is in the hatched region then subdivision is necessary. If **X** is placed as presented in fig.2 then the smallest value of its distance $m=|A-X|$ from **A** is obtained. From elementary calculation one obtains

$$m = \sqrt{2d(r+d)}$$

This result is general if $r=1/\varkappa$ represents upper bound of the tool curvature.

If the triangle is far from the tool (d≥r) then $m \geq d\sqrt{2}$ and m is
    proportional to d.
If r=d then m=2d.
If the triangle is close to the tool (d<<r) then $m \cong \sqrt{2dr}$ and m is
    proportional to the square root of d.
In case of flat ended mill r=0 and $m=d\sqrt{2}$.

Non convex tool

In exceptional situations the tool has to be described by a inside bottom part of torus (Marciniak 1991). This surface is not convex and the algorithm has to be modified.

If the point D closest to any of the vertices is on the inside side of the torus then to exclude collision, the triangle **ABC** is checked against the sphere tangent to the tool at D and with center on the tool rotation axis. Only vertices **ABC** have to be checked because the closest to the sphere point of a triangle must be a vertex. If all vertices are inside of the sphere then collision is excluded else the triangle has to be subdivided.

Contact point vicinity

Let the tool be tangent to the surface at point Q inside of the triangle **ABC**. Undercutting may be detected, as before, via planar triangles but in the vicinity of the contact point the process is time consuming and comparable with two dimensional bisection.
It is natural to use available additional information about the contact point and the common normal in it to improve the algorithm.

Let the upper bound of the surface curvature inside of the triangle **ABC** equals $\varkappa_g$, lower bound equals $\varkappa_d$ and surface approximation error has to be smaller than ε. It has been assumed here that positive curvature value means that the surface is concave looking from outside of the part (from the tool side). Let $\varkappa$ (always nonnegative) be the curvature of the sphere S tangent to the tool in Q with the center on the tool axis.

If $\varkappa_g \leq \varkappa$ then there is no collision between the triangle **ABC** and the tool (at least with the part of the tool which is inside of the sphere S)

If $\varkappa_g > \varkappa$ then the minimal distance r from Q to the point of collision greater than ε is defined by the condition:

$$\varepsilon \geq \frac{1}{2}(\varkappa_g - \varkappa)r^2 \qquad \text{what is equivalent to} \qquad r \leq \sqrt{2\varepsilon/(\varkappa_g - \varkappa)}$$

There is no collision between the tool and the part of the triangle **ABC** which projection on the plane tangent in Q is inside of the circle (**Q**,r). In particular if **ABC** projection is inside of (**Q**,r) then there is no collision with this triangle.

References

Anderson, R.O. (1978). Detecting and eliminating collisions in NC machining. In Computer aided design. vol. **15**, no. 4, pp. 231-7.

Atherton, P., Earl, C, Fred, C. (1987). A graphical simulation system for dynamic five-axis NC verification. Proc. Autofact, SME, Deaborn, Mich. pp. 2-1 to 2-12. 12.

Chappel, I.T. (1983). The use of vectors to simulate material removal by numerically controlled milling. In computer aided design. vol. 15. no. 3. pp. 156-8.

Drysdale. R.L., Jerard, R.B. (1987). Discrete simulation of NC machining. In proc. ACM symposium on computational geometry, ACM, New York. pp. 126-35.

Drysdale. R.L., et al. (1989). Discrete simulation of NC machining. In Algorithmica, special issue on computational geometry. vol. **4**. no. 1. pp. 33-60.

Duncan, J.P., Mair, S.G. (1983). Sculptured surfaces in engineering and medicine. Cambridge university press, New York.

van Hook, T. (1986). Real time shaded NC milling display. In Computer graphics (proc. SIGGRAPH). vol. **20**. no. 4. pp. 15-20.

Jerard, R.B., Drysdale, R.L., Hauck. K. (1988). Geometric simulation of numerical control machining. In proc. ASME int'l computers in engineering conf. ACME, New York. pp. 129-36.

Marciniak, K. (1987). Influence of surface shape on admissible tool positions in 5-axis face milling. In Computer aided design. vol. **19**. no. 5. pp.233-6.

MacLellan, G.M.,Young, G.M., Goult, R.J., Pratt, M.J. (1990) Interference checking in the 5-axis machining of parametric surfaces. Cranfield Institute of Technology. Private communications.

Oliver, J.H. (1986). Graphical verification of numerically controlled milling programs for sculpture surface parts. Doctoral dissertation, Michigan State Univ. E. Lansing, Mich.

Oliver, J.H., Goodman, E.D. (1986). Color graphic verification of N/C milling programs for sculptured surface parts. First symposium on integrated intelligent manufacturing, ASME New York, 1986.

Wang, W.P., Wang, K.K. (1986a). Real time verification of multi axis NC programs with raster graphics. IEEE Proceedings of international conference on robotics and automation. CS press, Los Alamitos, California. 166-71.

Wang, W.P., Wang, K.K. (1986b). Geometric modeling for swept volume of moving solids. IEEE Computer graphics and applications. vol **6**. no. 12. pp.8-17.

FIGURES CAPTION

Fig.1. Subdivision of triangles.

Fig.2. Lower bound of divisible triangle diameter

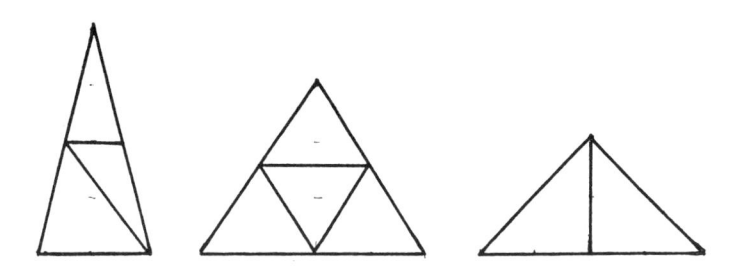

Fig. 1

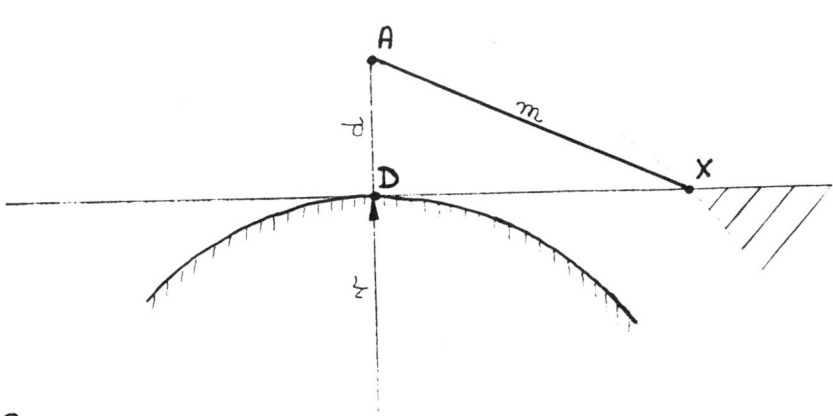

Fig. 2

Maximum Time Control of a DNC Machine Table

F.W.Wright & F.Azpiazu.

Summary.

The minimum time control of a DNC position controlled machine table is rarely used due to the difficulty in the determination of the switching time from maximum acceleration to maximum deceleration and bringing the table accurately to rest over a prescribed distance. The calculation of the switching time involves the use of variational calculus in the form of Pontryagins Maximum Principle. A desired feedback non-linear switching strategy can only be explicitly calculated for the simplest of systems. In this report an alternative approach giving the same structural feedback solution is obtained using modern CAD tools. The strategy has been applied to control the stepper motor drives of a DNC pcb drill table.

1 Introduction.

The operational speed of any DNC machine table is limited by the the maximum acceleration and deceleration available from the amplifier actuator load combination. In positioning applications the solution to motion control in minimum time is simply stated.

Starting at a point A maximum acceleration is applied and at some optimum time during the transient the drive is switched to maximum deceleration to bring the table to rest at the end point B. The difficulty lies in the computation of the optimum switching time.

Traditional DNC machine tables in general use amplifiers with electrical DC machine actuators having integral DC tacho-generator and incremental position encoders on each axis. Analogue tacho-generator feedback may be used to provide a fast critically damped internal velocity closed-loop controller. A step in setpoint control voltage corresponding to a step in velocity from rest to maximum velocity, under normal loading conditions, provides a velocity transient with the maximum possible acceleration. Reversal of the step similarly reveals the maximum deceleration transient. In currently available machine tables the setpoint velocity is derived from the position error via a linear proportional, integral and derivative (PID) series compensator. With this controller the tradeoff between improved accuracy and stability results in a slower speed of positioning response than that obtained using an optimum switching strategy. Similar conclusions can be drawn from position control of DNC drill tables with stepper motor amplifier-drive control in each axis. The stepper drive, under appropriate loading conditions is synchronised to the frequency of the incoming pulse train from the amplifier. Fitting the amplifier with an on board oscillator, generating the driving pulse train through a voltage to frequency converter, provides the means to adjust the maximum acceleration transient. This is arranged to switch and start the motor at a base frequency, within its pull-in performance region and then switch to accelerate so that the frequency rises, normally on a simple exponential, to a maximum that is within the pull out mode of stepper motor operation [1]. The exponential rise is predominantly adjusted for maximum acceleration by a single timing capacitor and resistance. This type of switching arrangement is ideal for the implementation of minimum time digital control of motion.

2 Optimum Switching Controller Theory.

The general solution to switching problems was introduced in 1962 by Pontryagin et al. [2] in the publication of his much celebrated "Pontryagins Maximum Principle". This extended the calculus of variations approach to dynamic optimisation problems to allow for large excursions of control effort to the maximum and minimum limits. To apply this principle the system state equation model is required.

In the context of the present problem each axis is driven from a control voltage u applied to a first order lag to give velocity (v). Integration of the velocity provides the position information (x). The resulting state equations constraining the motion are therfore of the state variable form;

$$\frac{dx_1}{dt} = x_2, \quad \frac{dx_2}{dt} = -ax_2 + bu \tag{1}$$

where $x = x_1$, $v = x_2$, a,b are constant parameters and u is the control, assumed to be linear between the limits $u_{min} \leq u \leq u_{max}$

The object of the exercise is to transfer the system from an initial position to the final position whilst minimising the performance measure,

$$J = \int_0^t dt = t \tag{2}$$

Pontryagin solves this problem by introducing a Hamiltonian function, whose maximum corresponds to minimum J as;

$$H = p_1 \frac{dx_1}{dt} + p_2 \frac{dx_2}{dt} - 1 \tag{3}$$

where the p_i are co-state variables chosen to satisfy;

$$\frac{dp_i}{dt} = -\frac{dH}{dx_i} \tag{4}$$

Substituting equations 1 in 3 provides;

$$H = p_1 x_2 - a p_2 x_2 + b p_2 u - 1 \tag{5}$$

It can be shown that, provided equations (1) represent a stable system, then H is maximised by choosing the switching strategy, $u = u_{max}$ for $p_2 > 0$ and $u = u_{min}$ for $p_2 < 0$, and this can normally be achieved using only one switch when p_2 changes sign only once [2].

3 Hardware Configuration Modelling.

The system to be controlled consists of a three axis "Unimatic" machine table with a high speed PCB drill attached to the z axis. Each axis is driven by a stepper motor supplied by an RS 342-051 Bipolar Chopped Stepper Motor Drive Amplifier fitted with an on board CMOS 4046 voltage controlled oscillator. Each axis is instrumented with a rotary shaft encoder with two quadrature pulse trains, to increase measurement accuracy and provide directional information, and an index pulse for absolute location of an axis at a Home position. The quadrature pulses are conditioned and counted with an HTC 2000 VLSI digital encoder interface. The counters are memory mapped to an Intel 8085 based DDC computer providing a 16 bit count of absolute position from the Home position.

The alternative CAD approach to find the optimum switching time may be used provided the state model and switching is configured to be of the form used in the previous section.

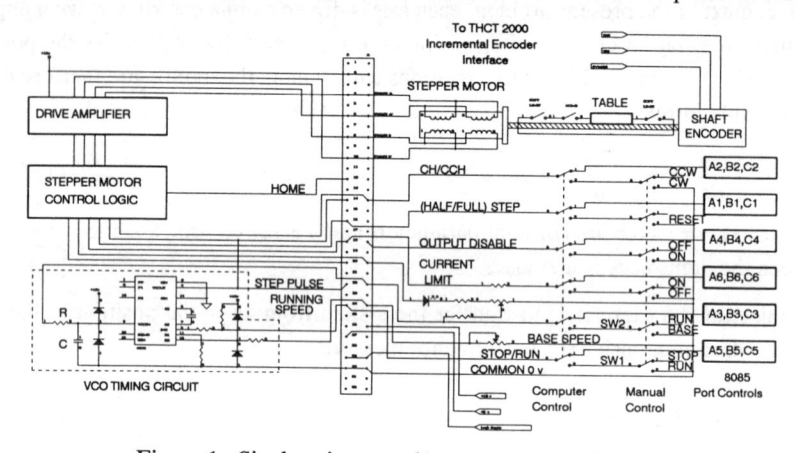

Figure 1.- Single axis manual/computer control system.

To achieve this each stepper motor is set up to give 1 revolution in 400 steps representing 5 millimetres of table travel and motion is essentially controlled by the two switches SW1 and SW2, shown for one axis, in Figure 1 and operating as follows:

SW1 is a run/stop switch to start the motor at a low base speed represented by the base frequency, $f = f_b$ pulses/second, position is then $p = f_b * t$ pulses;

SW2 is a running speed/base speed control that switches a constant voltage onto the VCO via a C/R network, when SW1 is in the run position, to initiate the acceleration to $f = f_{max}$.

When SW2 is switched to the base speed position the system automatically decays to f_b.

A typical transient response is shown in Figure 2. The system exhibits a small dead space and non-linearity in the commercial VCO-amplifier-drive. The dead space represented no more than 15 pulses but the time constant of the retardation is significantly different to the rise time. This is due to the fixed component values used to control the acceleration phase of the transient. This does not however present any problem since a piecewise linear parameter fitting method is used to model the measured data. If in addition the control is implemented in closed-loop accurate positioning can be achieved using proportional

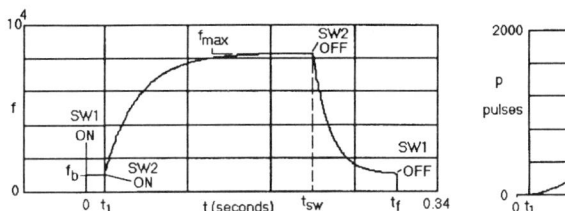

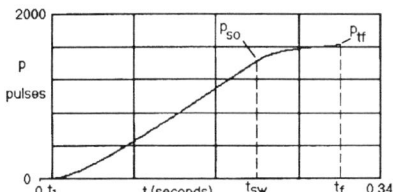

Figure 2.- Minimum time velocity and position transients.

control when the trajectory is close to the desired position. The transient rise in speed may simply be represented by fitting a first order parametric model to give;

$$f = f_b + f_o \text{ (pulses/second)} \quad (6)$$

where f_o is governed from time t_1 to the switching time t_{sw} by;

$$0.032 \frac{df_o}{dt} = f_o + u \quad (7)$$

with $f_o(t=t_1) = 0$, and $u = f_{max}-f_b$ representing the step change.

To stop the machine SW2 is switched to the base speed at time t_{sw} to force the exponential deceleration. The distance travelled in pulses p is simply obtained by integration of the frequency;

$$\frac{dp_b}{dt} = f_b, \ \frac{dp_o}{dt} = f_o, \text{ and } p = p_b + p_o \quad (8)$$

For any given t_{sw} the number of pulses travelled $p(t_{sw}) = p_{sw}$ may be noted. This data is essential to the feedback implementation.

Model behaviour for $t > t_{sw}$ is represented by switching the control u to zero on the retardation model to give the governing transient component f_o where;

$$0.16 \frac{df_o}{dt} = f_o \quad (9)$$

with initial condition $f_o(t=t_{sw})$, f and p are still given by equations (6 & 8).

A terminal time t_f, is defined when the table speed decays to base speed and may be stopped within one pulse time by switching SW1 to the stop position. This may be accurately determined by integrating to give the pulses travelled at f_o as p_o. When p_o is constant to within one pulse the table can be stopped and at this point $p = p_{tf}$ representing the total distance travelled may be noted. The time to reach p_{tf} is the minimum time for the move.

4 Closed-Loop Minimum Time Control.

The model described by equations 6 to 9 was simulated using an Advanced Continuous Simulation Language (ACSL) [3], running on a VAX 780 and operated from a PC using the Procomm terminal emulator package.

Fifty complete solutions were obtained for $0.008 < t_{sw} < 0.208$ with a change in t_{sw}, $dt_{sw} = 0.004$. During each solution $p_{so}(t = t_{sw})$ and $p_{tf}(t = t_f)$ are captured, in a file, and converted to provide a lookup table to be stored in the closed-loop on line controller. The maximum value of t_{sw} is selected by detecting the end of the acceleration phase when $f_o(t_{swmax}) = f_{max} - f_b$. Typical results are shown in graphical form in Figure 3. The horizontal represents the demanded distance for a move and the corresponding vertical value gives the number of pulses travelled when the switch to the retardation phase is to be initiated.

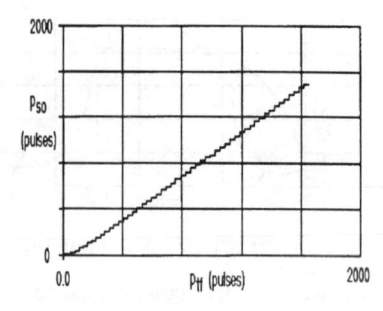

Figure 3.- Feedback switching curve.

The closed-loop system operates from an 8085 based direct digital controller (ddc) initialising to the home position and requesting a demanded position p_d from the host. The demanded position is compared with table values of final position p_{tf} to find that value which is equal to or just less than p_d. The corresponding switching time may then be detected from the table in terms of p_{so} pulses travelled to time t_{sw}. The actual position, p, stored in the HTC 2000 VLSI counter, is compared with p_{so} and when equal the retardation phase is started by switching SW2 to ground. During this terminal phase the error $p_d - p$ is sampled and when zero switch SW1 is operated to stop the motion.

For desired positions greater than the table values the velocity transient moves at a constant velocity f_{max} between the acceleration and deceleration phases. Under this condition the switch to the retardation phase occurs when the error is $p_d - p = p_{tfmax} - p_{somax}$. Once in the terminal retardation phase control reverts to sampling the error $p_d - p$ and when zero SW1 is operated to stop the motion.

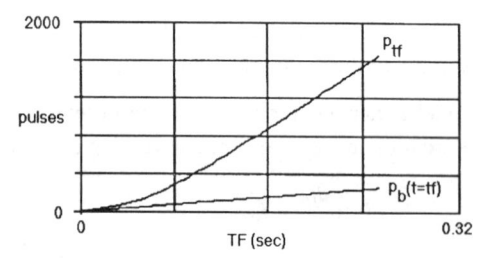

Figure 4.- Comparison of minimum time and base speed positioning algorithms.

The improvement in performance obtained from using this approach as compared to operation at the maximum base speed is shown in Figure 4.

5 Corrections for Error.

Practical tests indicate that the system has a dead-space no greater than 15 pulses when switching from base speed to maximum run speed. Attempts to switch SW1 followed by SW2 with a delay of one pulse (1msec) failed to synchronise the motor. Increasing the delay to 15 msec. solves the problem.

The small errors between the second order piecewise linear model and the actual system can be accommodated by using the switching algorithm conservatively. This is achieved by dropping the value of p_{so} obtained from the table search to the next lowest value in the table. This simply means that the terminal retardation phase will be extended to include a component of constant velocity at base speed f_b.

In the drilling application only the X and Y axis are controlled in minimum time since the Z axis drill feed is only operated with the table stationary and feed rate depends on drill size. The control is operated from a synchronous interrupt routine to implement the switching strategy. The remaining computation time is used to update a display of the table position and request the next movement data from the PC Host computer.

6 Conclusions

It has been shown that CAD tools may be used to construct a non-linear feedback strategy for closed-loop minimum time control of a machine table. This method avoids the use of complex mathematical analysis but provides a controller with the same structure as that obtained from a theoretical solution.

The method is not constrained to the stepper driven table and may be applied to any drive modelled from simple on line test data. Any discrepancies between the identified model and the actual system can be easily compensated by using the switching strategy conservatively to produce a near-optimal working system giving accurate control of position.

References

1. Radio Spares Data Library, "Stepper Motors", No 8199, Nov 1987.

2. Pontryagin et- al, "The Mathematical Theory of Optimal Processes", Wiley, 1962.

3. Mitchel and Gauthier Associates, "Advanced Continuous Simulation Language", Concord, Mass. 01742. USA., 1987.

An Approach to the Real Integration of CAD-CAM Systems

Prof.Dr. Mustafa AKKURT

Faculty of Mechanical Engineering
Istanbul Technical University, Istanbul

Summary

In this paper a method for the real integration of CAD-CAM system is presented. In this method the design and process factors are established in one stage. In this respect the classification of *shapes* (rotational and nonrotational) are made and all the items of this classification are named *variant*. Besides variants two other terms such as *attribute* and *feature* are defined. All the existing variants, attributes and features are stored in the computer so it is sufficient to enter respective parameter in order to produce the required part with processes automatically.

Introduction

The aim of industrial production is the fabrication of workpieces.
A typical manufacturing process consists of three phases, that is, design, production process planning and NC programming. For each of the three phases, there are Computer Aided Design (CAD), Computer Aided Process Planning (CAPP) and Computer Aided Manufacturing (CAM). CAD and CAM are closely related to manipulation of geometric model of part while CAPP deals mainly with the production technology. It is a very serious problem for industry to integrate all these activities [1,2,3,4].

Structure of Integrated system

The main idea of this paper is that the parts are composed from similar geometrical forms which are machined in a similar manner. It is supposed that each part (component drawing) can be broken down into similar geometrical elements. In this respect the classification of shapes (rotational,nonrotational) and surfaces are made and all the items of this classication are named *variant*. Variants are grouped as *base variants* and *features* and besides these other terms such as *attributes* are defined.

Base variants include primitive surfaces such as cylinder,taper,frontal circular,annular etc.;features include factors such as groove, hole, keyway, thread, chamfer, round, flat etc. and attribute include factors such as *material*, *machining* (turning, drilling, milling, boring), *dimensions* and *tolerances*, *surface quality*, *blank* and *tool*. Fig.1 illustrates this classification in detail.

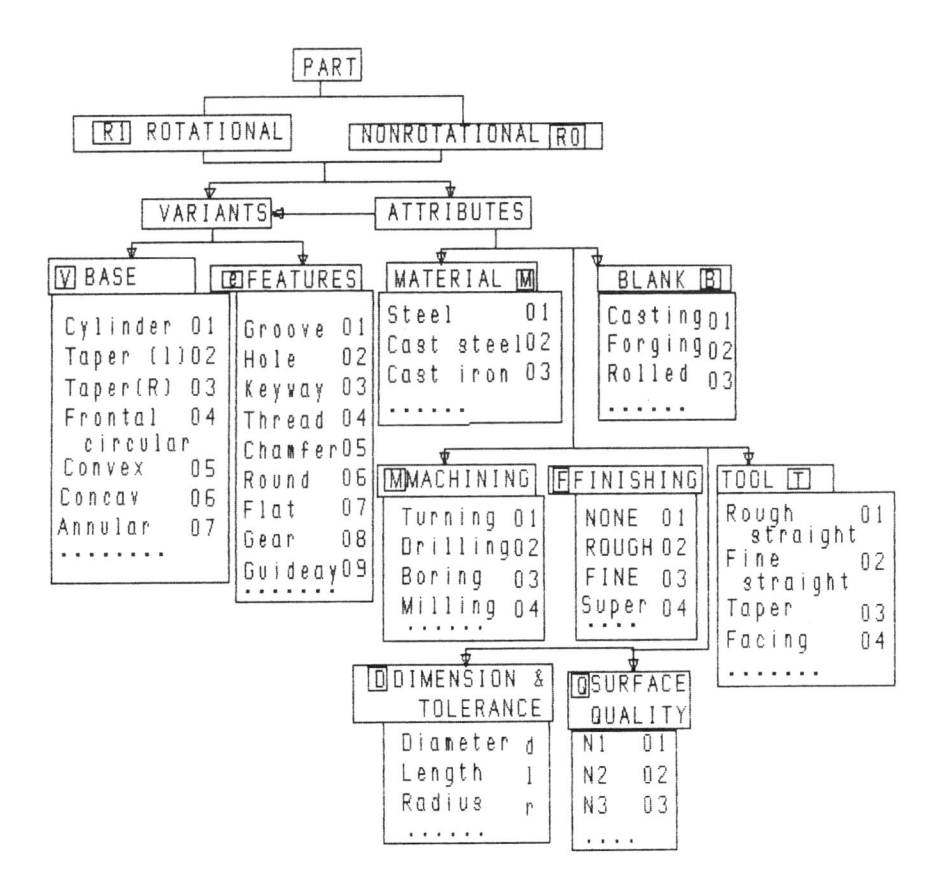

Fig.1. The codification system.

Our codification system which includes geometrical an technological factors are indicated in Fig.1. According to this codification capital letters were adopted for each relevant form element (Fig.2); lower-case letters for features and for items of the sub-groups which need a numerical value such as cutter condition (speed, feed, cutting depth), dimension (diameter, length, radius)   (Fig.2) and two-dijit system for group and sub-group items. For example: A cylinder of 30mm diameter and 45mm length as designed such as V01(d=30,l=45); A turning process with three cutting operations:

$$M01 = \begin{bmatrix} 095 & 1.00 & 2.0 \\ 095 & 1.00 & 2.0 \\ 175 & 0.16 & 0.5 \end{bmatrix}$$

Here the first column reprezent the cutting speed in m/min, the second the feed in mm/rev and the third cutting depth in mm.

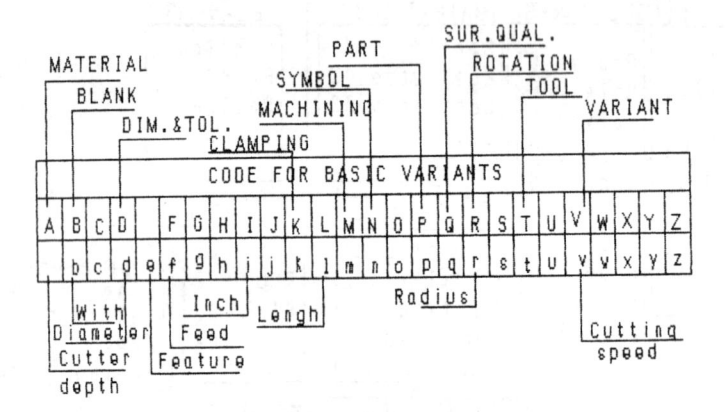

Fig.2. Symbol charracters.

Rules About Part Description

Since usually several of the same form element occur on any one component a count index is added in front of the relevant form element letter. Because the dijit 1 is used for special function it is advantageous to start with 2 as the first count index and then to continue with 3,4 etc. For external form elements the count index is written without parenthesis and for internal form elements with parenthesis, e.g. The first external cylinder according to the direction of counting will be 2V01(d=30,l=45). The first internal cylinder according to the direction of counting will be (2)V01(d=20,l=15).

Form elements with specific functions have a special status; they have a code with two letters. For some cases these are: Variants which limit the part: L0V for left faces and L1V for right faces. Clamping faces: KV; for example for turning: K1V for clamping between center; K2V for clamping in the jaws and K3V for clamping in the jaws and center.

The direction of description is in a clockwise direction or from left to right and top to bottom. For external description always work from left to right, starting at the left limiting face (L0V). For completely hollow components also work from left (L0V) to right. For partially hollow the following rules should be followed. If the hollow begin from left limiting face (L0V), the description works from left to right; if it begins from the right limiting face description works from right to left. for two partially hallow which begin from L0V and L1V, the description works from left to right.

If a variant has several function these should all be indexed. If there should be, for exemple, six similar holes in a pitch circle, the same count index should usually be used for all.

Description of a Part

An example of description of a part with only geometrical variant is illustreated in Fig.3. The full description of surfaces is defined as:

2V04L0:   N01=2V04(d=25)S06M01[90,.50,1.0]T06

2V01:
$$N03=2V01(d=30,l=25)S04M01 \begin{bmatrix} 085 & 1.20 & 3.0 & & 02 \\ 085 & 1.20 & 3.0 & T & 02 \\ 085 & 1.20 & 3.0 & & 02 \\ 150 & 0.50 & 1.0 & & 04 \end{bmatrix}$$

(2)V01:   N21=(2)V01(d=15,l=20)S04M02(d=10)
$$[n=1000,f=0.18]T12M01 \begin{bmatrix} 65 & 0.7 & 2.0 \\ 95 & 0.5 & 0.5 \end{bmatrix}$$

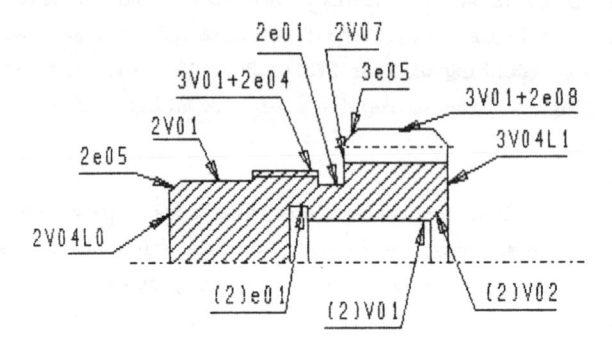

Fig.3. Description of a part.

where N01,N02 etc. are the symbol of surfaces. With the aid of these symbols the full description of the part may be written as:
    P001265=R1A01B03(d=50,l=75){N01N02........N34}

Here P001265 represent the part number or code, R1 rotatinal part, A01 material and B03 blank of material.

All the variants, features, material may be stored in a global geometric file; all the machining processes in a global machining file; all kinds of blanks in a global blank file and all the tools in global tool file. So in practise, the needed factors can be retrieved from these files, and the integrated CAD-CAM part can be generated automatically (Fig.4).

The program that depends on this data handling approach, has two main functions: the first is that the application programmers can compose the product model from separate parts, each of which is conceptually defined with its own world coordinate system and the second is that the application programmers can use it to represent a portion of a workpiece generated by a certain mode of metal cutting.

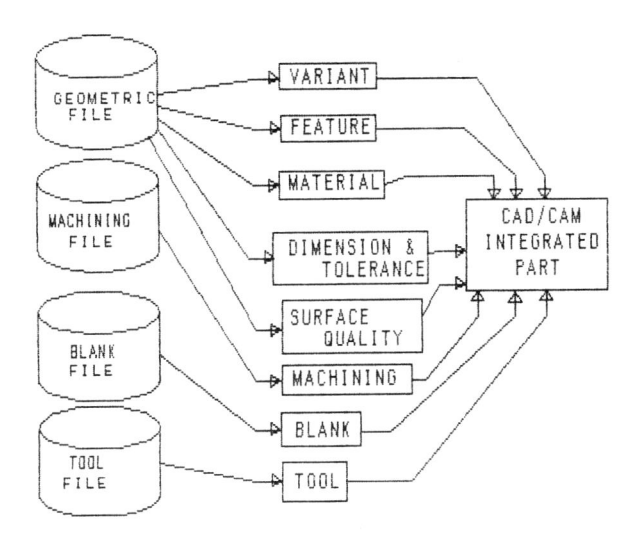

Fig.4. Integration concept of a part.

References

1. Carlier,J.; Peters,J. MOPS-A Machining Centre Operation Planning System. Annals of the CIRP Vol 34/1/1985 409-411.

2. Kishinami, T.; Kanai, S.; Saito, K. MKS:Machining Kernel Software for CAM System. Annals of the CIRP Vol 34/1/1985 419-422.

3. Peklenik, J.;Hlebanja, G. Developement of a CAD-System Based on Part Engineering Model and Binary Coding Matrix. Annals of the CIRP Vol 37/1/1988 135-139.

4. van Houten, M. Strategy in Generative Planning of Turning Processes. Annals of the CIRP Vol 35/1/1986 331-335.

CAD/CAM Program Package for Cold Metal Forming

Dr. M. Tisza
Associate Professor

Dr. P. Racz
Senior Lecturer

University of Miskolc
Department of Mechanical Engineering

Summary

An integrated CAD/CAM system for sheet metal forming processes elaborated at the Department of Mechanical Engineering (University of Miskolc, Hungary) is presented. The paper briefly outlines the general structure of this program package and describes in detail the main modules of it.The efficiency of the elaborated CAD/CAM system has been proved by several industrial applications.

1. Introduction

Recently technological processes in metal forming industry can be characterized by a rapid development and widespread application of CAD/CAM systems. It can be explained by several reasons. The most important among them are as follows:

- to maintain competitiveness, it is necessary to increase the flexibility both in design and manufacturing: the only way to meet these requirements is the implementation of complex CAD/CAM systems to reduce time and labour required for design and manufacturing;
- recent developments in computer hardware and pheripherials make it available even for smaller factories, at a reasonable price.

Most of existing CAD/CAM systems available in the market are supported by mainframes or mega-minicomputers. These systems (like **CATIA**, **CADAM** or **UNI-GRAPHICS**) are very effective with their high processing speed, bulk storage units and sophisticated operation system, but generally they are too expensive for smaller factories to install. Therefore, there has been an ever increasing demand for personal CAD/CAM systems from the very beginning. With the release of IBM PC-s the scene of CAD/CAM systems has radically changed. With their relatively fast processors and acceptable industry standard machines, they became the natural target for a new bread of low cost CAD/CAM systems.

Recognizing these trends an intensive research work has been in progress at the Department of Mechanical Engineering (University of Miskolc) to develope a complete, but as far as possible cost-effective CAD/CAM system for sheet metal forming applying inexpensive IBM and compatible personal computers available even for smaller factories as well. In the following the details of the elaborated CAD/CAM system will be analysed.

2. The philosophy of CAD/CAM system

The elaboration of integrated sheet metal forming technological and tool design, as well as production system is realized continuously following a modular build-up principle. The basic requirements to the system are as follows:

- each module should be applicable independently;
- at the same time they should provide the possibility of their integration into the total system and should contain the necessary input/output connection points;
- the system should be user-friendly;
- all the routine activities of formal and logical character in design tasks should be carried out by the computer;
- the flexibility of system, the intuition and creativity of the engineer are ensured by the dialogue mode of operation;
- the total system should be built on a unified data-base.

The system is running on IBM PC and compatibles. The hardware configuration involves high-resolution color display, a digitizer tablet for simple graphic input and easy menu selection, a plotter and a matrix printer. The graphic functions of the system are based on the AutoCAD® drafting program package.

3. The general concept of CAD/CAM system

According to the original development concept, the system should cover the total technological and tool design process of small and medium lot size and mass sheet metal component production, as well. It should be capable of designing single tool sets for deep-drawing processes and progressive dies for various blanking, piercing, bending and flanging processes as well. The principle of modular built-up of the system assuring the independent applicablity of each module and the hierarchic connections between them is well illustrated in Fig. 1. (It is represented by the computer aided processing of component to be manufactured in a progressive die shown in the right side of the figure.)

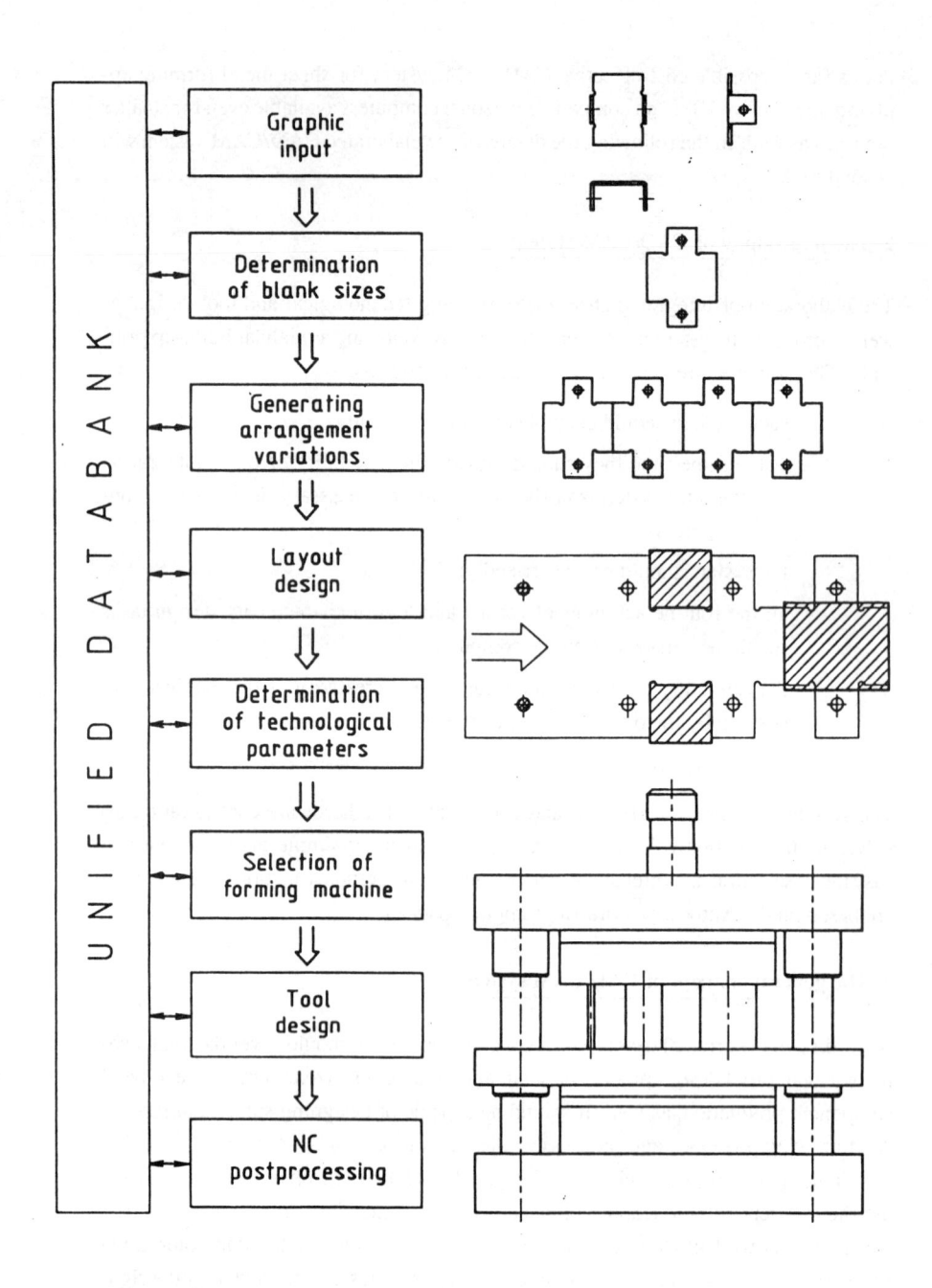

Fig. 1. The hierarchic structure of the elaborated system

4. Modules of the CAD/CAM system

4.1. Geometric description module

The aim of this program module to provide the total and unambiguous geometric input of components as simple as possible. Additional task of it to assure the storage of geometric information both graphically and alphanumerically for further processing.

After the determination of geometrical form and sizes of components, the material input module may be called. The task of it is to provide the material selection for the component to be produced.

4.2. Blank module

This module is required only in that case when the components have bent, drawn or flanged parts. In that case the shape and sizes of blank material should be determined according to analitical expressions. In this module different submodules can be called depending on the component to be manufactured.

The determination of optimimum material utilization is regarded as a key issue in the majority of sheet metal technologies. Therefore in this module there are several submodules for calculation of optimum material utilization offering various blank arrangement alternatives.

In conventional processes mostly strip-coils are used. In this case the optimum blank-layout may be generated by interactive manipulation of blank drawing: displacement, rotation, mirroring can also be handled. The exact blank size determined by the blank-calculation module is equidistantly increased by the value of bridge-width along the blank contour following its configuration. It reduces the control of admissible blank position into the relatively simple interference test of contours in generating blank-layout variations. From these alternative blank-layout variations the optimum solution can be selected taking technological viewpoints into consideration (e.g. single or multi row arrangement, the direction of bending axis, etc.).

4.3. Technological process planning module

The technological process planning module is regarded as one of the most important modules. It is capable of handling single and complex planning of following main forming technologies:

- material processing by cutting
  - blanking, punching,
  - side-cutting,
  - cropping,
- shape-forming technologies
  - bending,
  - deep-drawing,
  - flanging, etc.

The choice of previously listed possibilities can be done on the basis of the technological main menu. Selecting the single technological process planning alternative any of the forementioned technologies may individually be designed. In this module, first the performability test of selected technological operations (from the viewpoint of limit deformability, for example minimum bending radius, minimum diameter to be punched or flanged, etc.) are performed. Then the technological parameters (e.g. forming loads and works, necessary steps of operations, intermediate heat-treatments, etc.) can be determined.

The possibilities offered by this CAD/CAM system can fully be utilized during the so-called complex technological design of forming technologies carried out in progressive dies. The base of it is the blank-layout prepared by the module for optimizing the material utilization. Applying it, the strip-layout (Fig. 2.) is determined interactively, which means the determination of sequence of forming operations, ie. the so-called technological process planning. It simultaneously provides the die-layout design, as well.

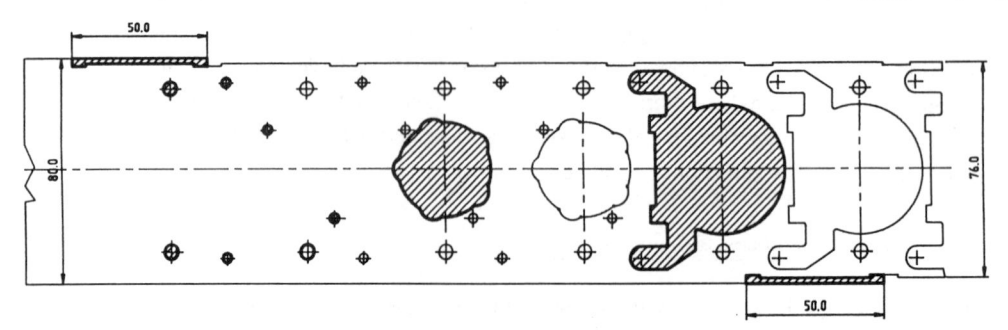

Fig. 2. Strip-layout for progressive dies
elaborated by the CAD/CAM system

These layout variations can be displayed, altered, plotted or stored for further processing, as well. After the strip- and die-layout design, the technological parameters can be determined (in the way described in the single design phase). Finally, on the ground of strip-layout the centre of momentum is determined.

4.4. Tool design module

The tool design module is regarded as another important part of the elaborated CAD/CAM system.

The tool design module similarly to the technological module can be applied for both individual and complex design. By individual design we mean, that the tool design task of a single forming operation is carried out. As an example, let us consider the design of progressive die. In this case the fundamental data of design are complemented by the strip-layout determined in the technological module. Based on it, the die-layout can also be prepared. This is followed by the selection of standard tool set suitable for performing the forming operations determined by the strip-layout.

The part drawings of active tool elements and the assembly drawing can also be documented and stored in the way described formerly. In Fig. 3. an assembly drawing of a progressive die elaborated by the CAD/CAM system is shown.

4.5. The Computer Aided Manufacturing part of the system

Applying the results of geometric, technological and tool design modules, this module aims at working out the control programs for manufacturing the tool elements (eg. punches and dies, etc.) on NC/CNC manufacturing machines, using the data on machines and tools available in the databank.

Since the results of the design process in the formerly described CAD part of the system can easily be converted into standard DXF or IGES files, thus both the numeric and graphic data can be transferred to commonly used CAM systems (like AutoCAM, MASTERCAM, PEPS, etc.). Applying for example the well-known and widely employed CAM system PEPS, the CAD data are transferred through DXF interface. This package includes most of the widely used postprocessors for several manufacturing processes.

The system also offers simulation possibility of elaborated NC-programs which means a tremendous help in analysing undesirable manufacturing problems like collision and it can save significant time and cost at the manufacturing machines.

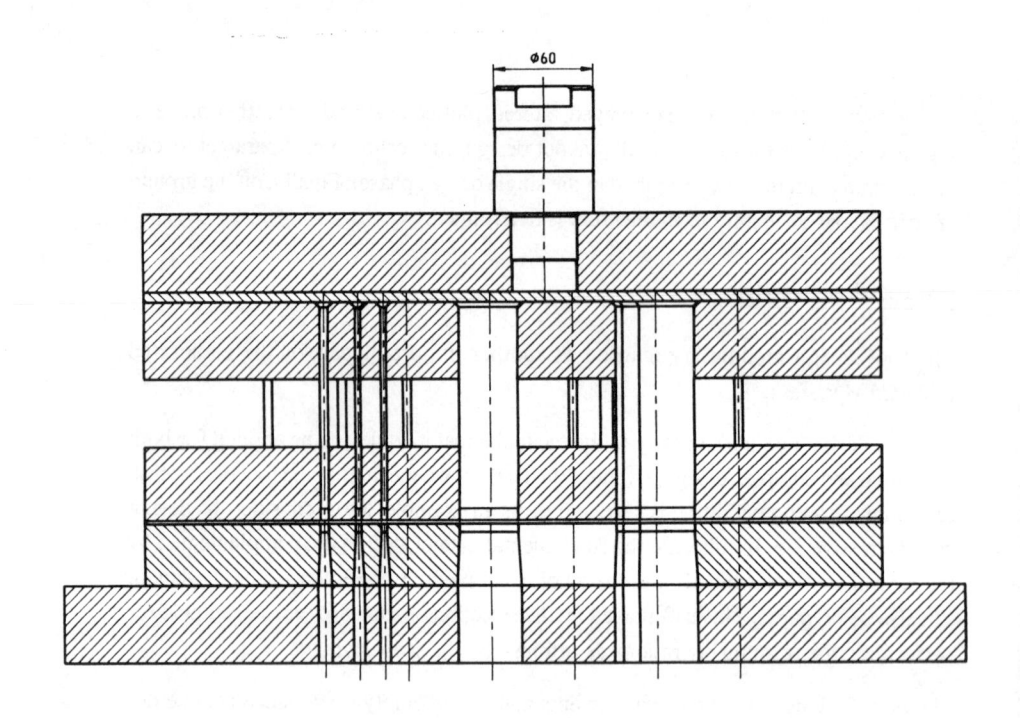

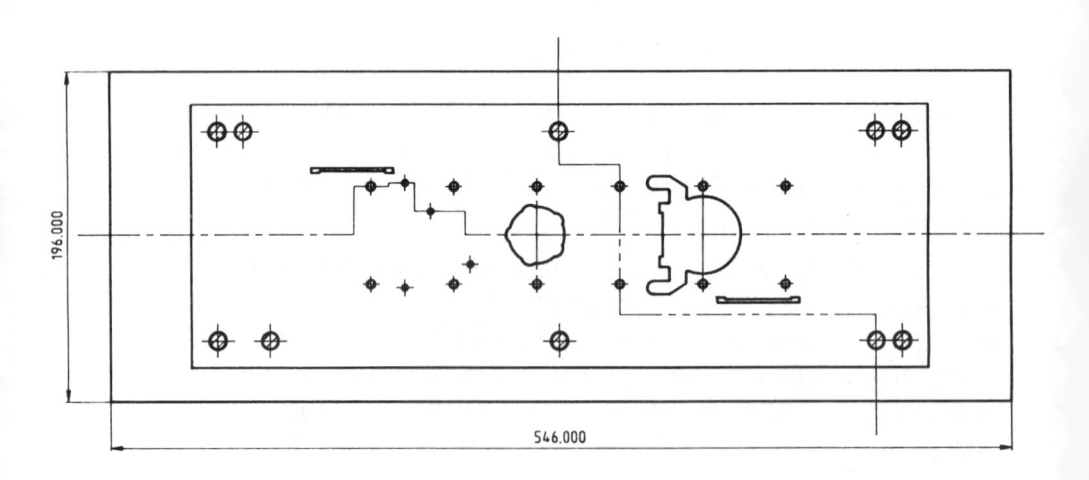

Fig. 3. Assembly drawing of a progressive die
designed by the sheet metal forming CAD/CAM system

4.6. The database handling system and the unified database

Each module of the elaborated CAD/CAM system is in close connection to the unified database containing a great amount of data. It is assured by the database handling system. Moreover this system provides the storage of new data, the modification and cancellation of them, if necessary. The database involves geometric information (eg. geometric primitives, graphic data, etc.), sizes and mechanical properties of raw materials (width, thickness and tolerances of plates and strips, strength and deformation characteristics, etc.), technological parameters of forming processes, fundamental data for tool design, standard tool elements and tool sets, geometric, kinematic and kinetic parameters of forming and manufacturing machines, etc.

5. Conclusions

The paper describes an integrated CAD/CAM system elaborated and continuously further developed at the University of Miskolc. This can be applied for both individual and complex design of sheet metal forming technologies and tools. Applying this system the following main advantages should be noted:

- the time and labour required for design and manufacturing can radically be reduced,
- as a result of it, the system makes possible a more flexible adaptation both in design and production stage,
- comparing to conventional design processes it makes a more optimum design possible also from the viewpoint, that larger possibilities are offered for extremely quick analysis of a great number of variations. From these alternatives the selection of optimum may be realized considering different viewpoints,
- it gets the designer free from the labour and time consuming routine activities (drawings, calculations, etc.) and provides more time and opportunities for creative work.

The suitability and effectiveness of the elaborated CAD/CAM system has been proved by the experiences of several sheet metal forming factories in Hungary where the complete system has succesfully implemented.

Modelling and Simulation

Multi-Arm Robotic Systems: Computer-Oriented Methods for Control

Mikhail M. Svinin *and Sergey V. Eliseev

Automation & Technical Physics Department

Scientific Center, Lermontov 281, Irkutsk, USSR, 664033

Abstract. This paper reports on investigations on coordinate control of systems of interacting manipulators. We have developed a two-stage procedure for organization of cooperative motions and have simulated a control algorithm for coordinated movements of a long-sized object transferred by two manipulators.

1 Introduction

Investigations which are currently being carried out in robotics, are concentrated mainly on one-arm robots and so far multi-arm robots do not find an adequate application in industry. However, with the further development of scientific and technological progress in the fields of mechanics, sensorics, computer art and control, it might be expected that multi-arm robots will be finding ever wider fields of application. This means that already at this stage pertinent theoretical and experimental investigations in this direction of development of robotics are needed.

Most of the control algorithms for multi-arm robots are constructed in the "master-slave" concept [1,2]. However, in this case the degree of cooperation of robots, in particular the influence of each degree of mobility of robots upon the movement of the object is not quite clear as well as the extension of the concept to the case of N robots. In the development of this idea, in [3,4] the T.Mason's theory [5] was extended to multi-arm

---

*Currently visiting scientist at Darmstadt Technical University, Automatic Control Department, System Theory and Robotics Section, Schlossgraben 1, Darmstadt, D-6100. Germany

robots and, under the assumption of a rigid gripping of the object, some variants of position/force regulators were suggested.

We also consider here the regime of equivalent interaction of manipulators; however, unlike [3,4], as the basal system we adopt a scheme for organization of control based on two-stage synthesis of control actions by distinguishing nominal and disturbed regimes.

2 Synthesis of program motions

The simulation employed a system consisting of two identical planar three-link manipulators and the object (Fig.1). The model parameters in the SI system are : $l_1^i = 1, l_2^i = 0.5, l_3^i = 0, l^0 = 1$ are the lenghts of links of the manipulators and of the object; $m_1^i = 5, m_2^i = 2.5, m_3^i = 0$, and $m^0 = 1$ are masses of the links; $J_1^i = 0.125, J_2^i = 0.06, J_3^i = 0.05, J^0 = 1$ are moments of inertia relative to the centers of masses; $k_1^i = 200$, and $k_2^i = k_3^i = 100$ are gear ratios of the reductors; and $j_1^i = 2 \times 10^{-4}$ and $j_2^i = j_3^i = 10^{-3}$ are reduced moments of inertia of the reductors; $i = 1, 2$.

It has been necessary to perform a plane-parallel transfer of the object from the initial lower position $R^0(0) \stackrel{\text{def}}{=} R^{0s} = (1.25; -0.25)^\top$ into the final upper position $R^0(T) \stackrel{\text{def}}{=} R^{0f} = (1.25; 0.25)^\top$, with conservation in the process of motion, of a given orientation $\Psi_{\text{ref}}^0(t) = 0, t \in [0, T]$, and $\dot{R}^0(0) = \dot{R}^0(T) = 0$, where R^0 is a vector of the coordinates of the center of masses of the object in the inertial system, and $q^0 \stackrel{\text{def}}{=} \{(R^0)^\top, (\Psi^0)^\top\}^\top$. Let also q^i, R^i, Ψ^i be vectors of generalized and Cartesian coordinates as well as the orientation angle of gripping of the i-th manipulator.

From geometrical constraint equations for given R^{0s}, R^{0f}, and $\Psi^0(t)$ we determine $R^1(0) \stackrel{\text{def}}{=} R^{1s}, R^1(T) \stackrel{\text{def}}{=} R^{1f}$, and $\Psi^1(t)$, and subsequently, by solving the inverse problem of kinematics $q^1(0) \stackrel{\text{def}}{=} q^{1s}, q^1(T) \stackrel{\text{def}}{=} q^{1f}; \dot{q}^{1s} = \dot{q}^{1f} = 0$. By taking these values, the program trajectory of the 1-st manipulator will be constructed in the class of accelerations - linear functions of time. By specifying in this way the trajectory of the object $R_{\text{ref}}^0(t)$ and determing using the constraint equations $R_{\text{ref}}^2(t)$ and $\Psi_{\text{ref}}^2(t)$, we shall construct, by solving the inverse problems of kinematics of positions, velocities and accelerations we shall construct a program motion $q_{\text{ref}}^2(t), \dot{q}_{\text{ref}}^2(t), \ddot{q}_{\text{ref}}^2(t)$ for 2-nd manipulator.

By knowing program accelerations of the object, we shall find the values of forces $F_{\text{ref}}^0 = m^0(\ddot{R}_{\text{ref}}^0 - g)$ and of the moment $M_{\text{ref}}^0 = J^0 \ddot{\Psi}_{\text{ref}}^0$ such that the object be capable to execute a prescribed motion. By requiring that each of the manipulators receive half the force load of the object, we determine the program values of forces in the grips $F^i_{\text{ref}} = F^0_{\text{ref}}/2$ and $M^i_{\text{ref}} = M^0_{\text{ref}}/2$ and and, subsequently, from the equations of dynamics

$$A^i \ddot{q}^i_{\text{ref}} + B^i = Q^i_{\text{ref}} + G^i + (D^i)^T \Lambda^i_{\text{ref}}, \quad i = 1, 2, \tag{1}$$

we shall find the program moments $Q^i_{\text{ref}}(t)$ in the drives of manipulators. Here : $\Lambda^i \overset{\text{def}}{=} \{(F^i)^T, M^i)^T$, A^i, B^i, G^i, and D^i are, respectively, the inertia matrix (taking into account of the inertia of the drives), a vector of Coriolis and centrifugal forces, a vector of gravitational forces and the Jacobian matrix of the grip of the i-th manipulator calculated on program motions.

3 Coordinated control of manipulators

We consider here the regime of equivalent interaction of manipulators and adopt the scheme for organization of control based on two-stage synthesis of control actions by distinguishing nominal and disturbed regimes .According to this scheme, controls will be specified as $Q^i(t) = Q^i_{\text{ref}}(t) + \Delta Q^i(t)$, where $Q^i_{\text{ref}}(t)$ are program values of motive forces determined at the previous stage, whose introduction takes account of the dynamic interrelation; and $\Delta Q^i(t)$ are stabilizing additions determined at the 2-nd stage, which compensate deviations of the system from nominal. We employed the local control law of the form

$$\Delta Q^i = A(q^i_{\text{ref}})\{\Gamma^i_0 \Delta q^i + \Gamma^i_1 \Delta \dot{q}^i\}; \quad i = 1, 2, \tag{2}$$
$$\Gamma^i_0 = diag\{e^i_j\}, \; \Gamma^i_1 = diag\{v^i_j\}, \; v^i_j \geq 2\sqrt{e^i_j}; \; j = 1, 3.$$

The control law was investigated by the method of mathematical modelling. During the simulations we choose the parameters of the control algorithms $\Gamma^i_0 = 600\, I$, $\Gamma^i_1 = 220\,\sqrt{6}\, I$, (I - the identity matrix).

If the object is assumed to be absolutely hard, then no deformation occurs and additional forces have little effect on the characteristics of motion of the system. In this case one may limit attention to a synthesized local control ΔQ^i, considering the problem to be solved.

In order to verify the validaty of this inference for a nonrigid object as well, we carried out a simulation of a joint motion in which the object was formalized by kinematic chain with 3 rotational elastic degree of mobility q^0_4, q^0_5, and q^0_6 (q^0_5 being at the center of the object, and q^0_4 and q^0_6 to the left and to the right at a distance $l^0/4$ from the center; bending rigidity is

10^3 N/m, which approximately corresponds to a duralumin plate of 0.5 cm thickness). The simulation showed a significant degradation of precision characteristics (max$|\Delta q_3^1| = 0.132 \times 10^{-1}$, and max$|\Delta q_3^2| = 0.195 \times 10^{-1}$) and a nonsteady-state oscillatory character of variation of elastic coordinates of the object (Fig.2).

For purpose of compensating undesirable factors of the force interaction, a study was made of a simple control algorithm for forces in grips, whose ideology is explained thus. Let a deformation axis associated with the object be given, and it is necessary to ensure along this axis an extention force.

We introduce into the program of motion of the object $R_{\text{ref}}^0(t)$ a small correction ΔX such that the position of point C (center of the object) for 1-st manipulator be displaced by $-\Delta X$ along the deformation axis into point C_1 (Fig.3), and the position of this same point for the 2-nd manipulator be determined by a displacement by $+\Delta X$ on the same axis into point C_2. As a result, we obtain two new programs $R^{01}(t) = R_{\text{ref}}^0(t) - \mathcal{M}\Delta X, R^{02}(t) = R_{\text{ref}}^0(t) + \mathcal{M}\Delta X$, where $\Delta X \stackrel{\text{def}}{=} (\Delta\chi, 0, 0)^T; \mathcal{M}(\Psi_{\text{ref}}^0(t))$ is a matrix of orientation of the object.

At a sufficiently small $\Delta\chi$ we obtain also new programs of motion of the manipulators $q^i(t) = q_{\text{ref}}^i(t) + \Delta q^{0i}, \Delta q^{0i} = \pm\{D^i\}^{-1}\mathcal{M}\Delta X$, where D^i is a Jacobian matrix of the grips of the manipulators. The tracking of new programs is virtually equivalent to tracking the old ones, with additional introduction on into the control law of global components $\Delta Q^{0i} = \Gamma_0^i \Delta q^{0i}$.

A simulation of operation on this algorithm, with correction $\Delta\chi = 10^{-3}$, led to an almost total absence of vibrations in elastic coordinates of the object, which decreased 300 times or more. In this case the accuracy of tracking of the trajectories was close to corresponding accuracy of the rigid model, although the power consumption increased slightly.

If, additionally, it is desirable to control not only of the sign but also the amount of forces, then, in the presence of force/torque sensors, instead $\Delta X = const$ one can introduce a feedback in the form

$$\Delta X = k \int_0^t \Delta F(\tau)\,d\tau, \quad \|\Delta X\| \leq \delta, \tag{3}$$

where $\Delta F(\tau)$ are mismatches between the program and actual value of a total force applied to the object. A simulation of operation on this algorithm at $k = \delta = 10^{-3}$ permitted us to still more decrease the value of power consumption, and the accuracy of tracking was as well as for the rigid model, and the reaction forces were close to program ones (Fig.4).

4 Conclusion

The applicability of the two-stage concept with introduction of a program, a local and global control, has been verified by means of computer simulation for the two-arm system. Program control takes account of total dynamics of the system. Local control, that compensates deviations from nominal, has been synthesized on the basis of a model of a "decoupled" system, and global control in the form of force feedback minimizes the destabilizing influence of the mutual connection of the subsystems.

By measuring the forces and ensuring, in the process of motion, a guaranteed extention, it becomes possible to exclude deformations of the object transferred by the manipulators as well as to improve the control quality. This demonstrates the effectiveness of employment of a two-arm robot during coordinated execution of transportation operations of a non-rigid object.

References

[1] Tarn T.J., Bejczy A.K. and Yun X. "Coordinated control of two robot arms" *Proc. IEEE Int. Conf. on Robotics and Automation*, San Francisco, Calif., 1986. - Vol.2. -pp. 1193-1208.

[2] Hayati S. "Hybrid position/force control of multi-arm cooperating robot " *Proc. IEEE Int. Conf. on Robotics and Automation*, San Francisco, Calif., 1986. - Vol.1. - pp. 82-89.

[3] Bruhm H. and Neusser K. "An active compliance scheme for robots with cooperating arms " *Proc. 3rd Int. Conf. on Advanced Robotics*, Versailles , 1987. pp. 469-480.

[4] Uchiyama M. and Dauchez P. "A symmetric hybrid position/force control scheme for the coordination of two robots " *Proc. IEEE Int. Conf. on Robotics and Automation*, Philadelphia, 1988, Vol.1, pp. 350-356.

[5] Mason M.T. "Compliance and force control of computer controlled manipulators." *IEEE Trans.*, 1981. - Vol. SMC-11, No. 6. - pp. 418-432.

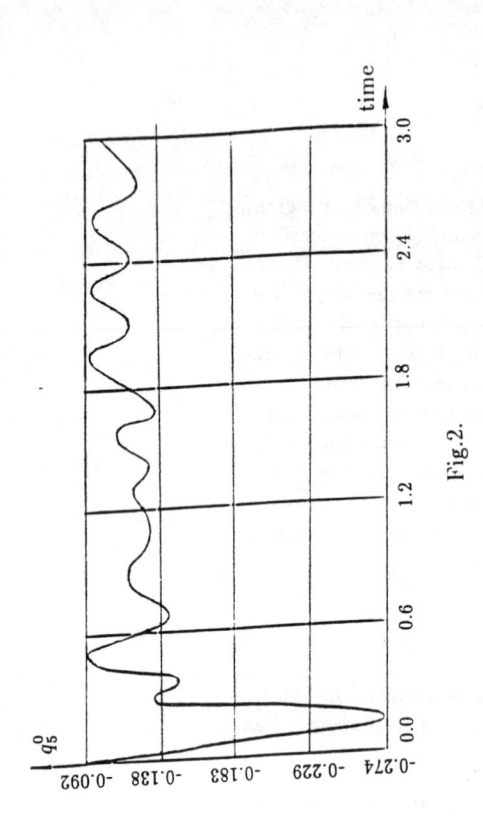

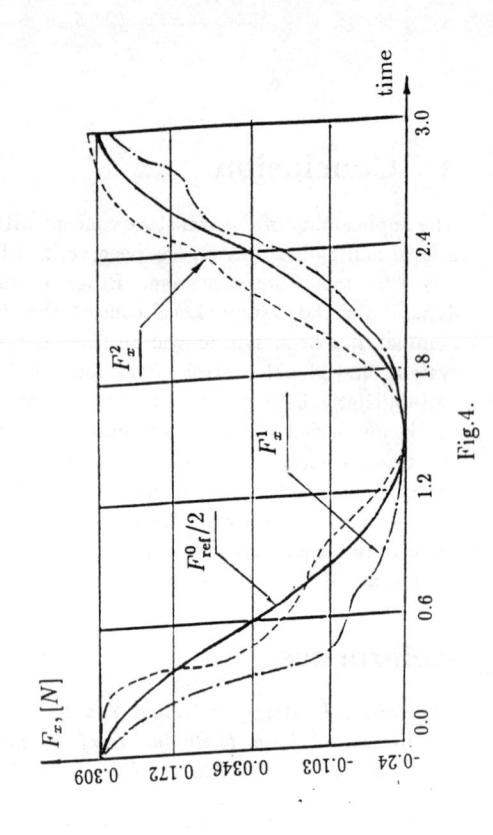

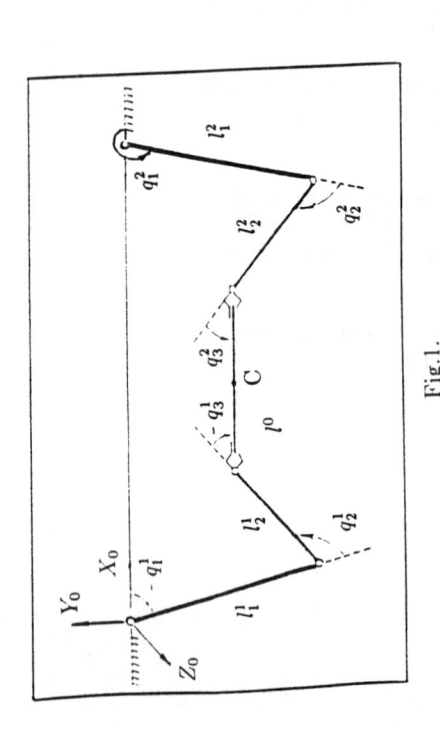

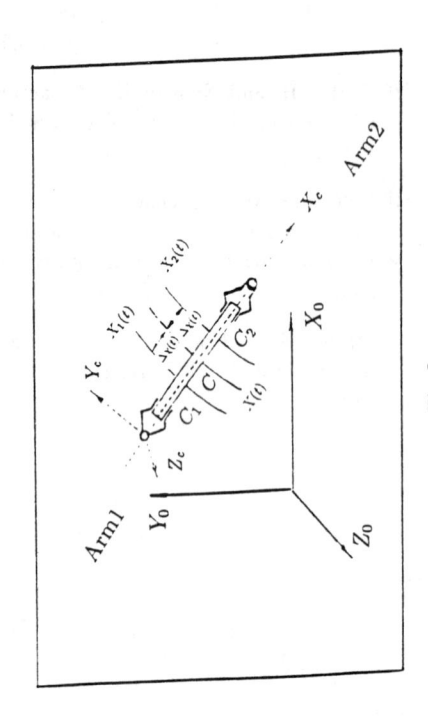

Fig.1.

Fig.2.

Fig.3.

Fig.4.

An Effective Choice of a Time Step for Solution of an Unsteady Temperature Field

J.HLOUSEK & F.KAVICKA

Technical university Brno, Czechoslovakia

Summary

For numerical solution of the temperature field it is often used the Euler's method. With respect to selection very short time step the operational time of calculation is too long. This deficiency can be overcome by using of the method ATSM described in this article.

An explanation of instability causes at numerical solution of a temperature field.

Let us have a problem to solve the temperature field of an infinite plate with a boundary condition of a 1st order. The initial temperature of the plate is constant and the surface temperature suddenly falls down to t_0 at time zero. One half of the plate thickness is L (see Fig. 1.).

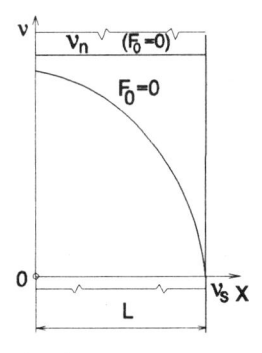

Fig. 1. Temperature course within the plate.

After introduction of dimensionless parameters

$$v = \frac{t-t_0}{t_{in}-t_0} \quad \text{the i.e. the dimensionless temperature}$$

$$x = \frac{x}{L} \quad \text{i.e. a dimensionless distance,}$$

$$F_0 = \frac{a.\tau}{L^2} \quad \text{i.e. the dimensionless time,}$$

(where a denotes a temperature conductivity and τ is time), one can get mathematical formulation of the given problem, as follow

$$\frac{\partial v}{\partial F_0} = \frac{\partial^2 v}{\partial x^2} \quad \text{for } 0 \leq x \geq 1 \quad (1)$$

with the boundary conditions

$$\frac{\partial v}{\partial x} = 0 \quad \text{for } x=0 \quad (2)$$

$$v = v_s = 0 \quad \text{for } x=1 \quad (3)$$

and the initial condition

$$v_x = v_{in} \quad \text{for } F_0 = 0 \quad (4)$$

The system of equations (1 up to 4) is transferred into the system of the ordinary differential equations

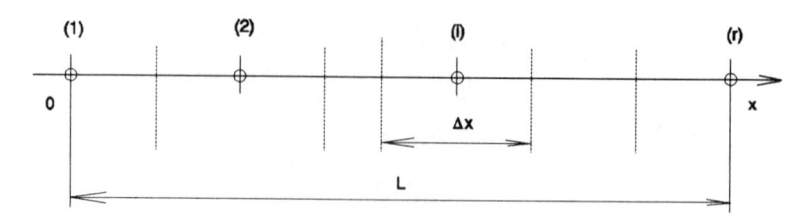

Fig. 2. Distribution of the thickness into layers with nodal points (1) up to (n).

where $\Delta x = \dfrac{1}{(n-1)}$.

$$\frac{\partial v_1}{\partial F_0} = 2\frac{v_2 - v_1}{(\Delta x)^2}$$
$$\frac{\partial v_i}{\partial F_0} = \frac{v_{i-1} - 2v_i + v_{i+1}}{(\Delta x)^2} \quad (5)$$
$$\frac{\partial v_{n-1}}{\partial F_0} = \frac{v_{n-2} - 2v_{n-1}}{(\Delta x)^2}$$
$$v_n = 0$$

By replacing a time derivative by a time difference we get either pure implicite method or Euler's method, so that

$$v_{F_0 + \Delta F_0} = v_{f_0} + \frac{dv}{dF_0} \Delta F_0 \quad (6)$$

The subscripts denote either time F_0 or time advanced at time step ΔF_0 forward. When using the pure implicit method the derivative is in time $F_0 + F_0$ and when using the Euler's method derivative is in time F_0. In practice it is used sometimes one or other method. The pure implicit method is always stable but the Euler's one is either stable or unstable. The stability condition is defined by the relation $\Delta F_0 \leq 0.5$.

Let us elect the most simple case where n=2 and therefore $\Delta x = 1$. The system of equation (5) has been transferred into

$$\frac{dv_1}{dF_0} = -2v_1 \quad (7)$$
$$v_2 = 0$$

Analytical solution has been

$$v_1 = \exp(-2F_0) \quad (8)$$

Numerical solution using Euler,s method has given

$$v_{1, F_0 + \Delta F_0} = v_{1, F_0} - 2v_{1, F_0} \Delta F_0 \quad (9)$$

The attained results have been plotted on fig.3.
It is evident that
a) $0 < \Delta F_0 \leq 0.5$ i.e the results do not oscillate,
b) $0.5 < \Delta F_0 \leq 1$ i.e the results oscillate dumpedly,
c) $1 < \Delta F_0$ i.e the results oscillate undumpedly.

Removing of the instability by means of the method so called the alternating time step method (ATSM)

Let us elect the first time step $\Delta F_0 > 0$. This time step is called the long time step (LTS). Then the short time steps (STS) follow several times according to scheme one LTS and k-times STS (see Fig. 3.).

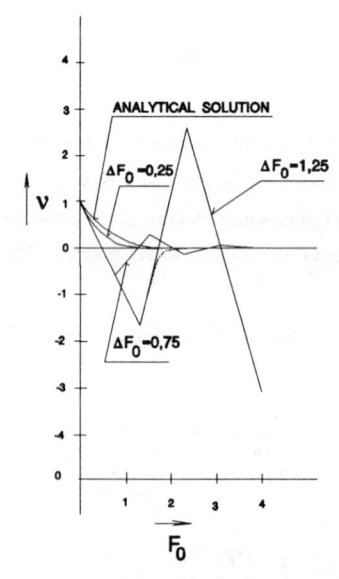

Fig. 3. Dimensionaless temperatures calculated with dimensionaless time steps ΔF_0

Application of the ATSM for solution of the temperature field including a phase change.

The Euler's method is used altogether for calculations of the temperature fields of solidifying casting. Let us solve the temperature field of the solidifying iron casting of the thickness 20 mm. It solidifies within a sand mould which thickness is 3 mm.

The region had been divided into n_c nodal points within the half of the casting ($n_c = 6$) and the whole region (i.e the casting with the mould) had been divided into the total nodal points ($n_m = 17$).

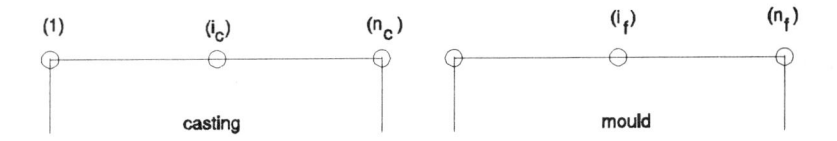

Fig. 4. Distribution of the casting and mould into layers

The physical parameters of the mould and casting had been laid down according to available literature:

a) Parameters of the iron casting are:

Specific heat of the solid phase $(\varrho c) = 4{,}45 \times 10^6$ $J.m^{-3}.K^{-1}$,

Specific heat of the liquid phase $(\varrho c)' = 5{,}35 \times 10^6$ $J.m^{-3}.K^{-1}$,

Latent heat of solidification $l_s = 267900$ $j.kg^{-1}$,

Heat conductivity of the liquid /solid phase/ $k' = \dfrac{16{,}7}{35}$ $W.m^{-1}.K^{-1}$,

a pouring temperature $t_{in} = 1520$ C,

the solidus temperature $t_s = 1439$ C,

the liquidus temperature $t_l = 1486$ C.

b) Parameters of the sand mould are

specific heat $(\varrho c)_m = 1{,}35 \times 10^6$ $J.m^{-3}.K^{-1}$

specific conductivity $k_m = 0{,}7$ $W.m^{-1}.K^{-1}$

the initial temperature $t_{mi} = 20°C$.

Heat transfer had been laid down $h_c = 0{,}1 \times 10^5$ $W.m^{-2}.K^{-1}$ at the contact between the casting and the mould. This value had been calculated according to a granuarity of commonly used moulding sand.

The calculations had been performed on the minicomputer twice.

1) For achievement the stability of calculations had been laid down STS = 0.016 sec. The results had been .

The modal point No	1	3	5	7
had solidified at time (sec)	**209.7**	**202.8**	**180.3**	-
	(209,6)	(202.6)	(181.1)	-
temperatures after solidification	**1438.8**	**1433.2**	**1424.9**	**1412.3**
	(1438.3)	(1433)	(1425)	(1412.4)

In this case there had been calculated 13112 time steps because STS had been constant. The operating time had been 6 min 51.6 sec.

2) Calculations by means of ATSM. In this case LTS=0.64 sec and STS = 0.016 sec (repeated 15 times). The results that had been achieved are in brackets in upper table.

In this case there had been calculated 3818 time steps and operation time had been 1 min. 59.9 sec.

It can be concluded that by using of the ATSM the calculation had been 3.42 times quicker and precision of calculation had been high (0,1 %).

References

1. Myers, G. E.: Analytical methods in conduction heat transfer, McGraw-Hill Book Co., New York, 1971.
2. Hlousek, J.: Příspěvek k řešení přenosových jevů v metalurgických procesech, Technical university Brno, Czechoslovakia, 1981.

User-Generated Feature Libraries for Maintenance of Recurrent Solutions

H. Grabowski, S. Braun, A. Suhm

Institut für Rechneranwendung in Planung und Konstruktion (RPK)
Universität Karlsruhe
Kaiserstraße 12, 7500 Karlsruhe, Germany
Tel.: +49 721 608-2129   Fax: +49 721 661138   Email: suhm@rpksun1.rz.uni-karlsruhe.de

Abstract:

Future CAD/CAM-system technology is based on powerful computer internal product models which provide an integrated support of the product development during the whole product life cycle. These product models are mainly based on feature technology. Features are classified under the aspect of genericity having generic, specific and occurrence features. Generic features are stored in a feature library. The presentation emphasizes the development of a feature library management system which is able to store and manipulate generic feature information in an object-oriented manner. The proposed conceptual model for generic features is based on the language EXPRESS which allow both softcoded and hardcoded features. Hardcoded features are part of the database schema. Softcoded feature are introduced via the proposed feature library management system. Additionally, the feature library management system supports searching and selecting appropriate features in a graphical-interactive manner. At last a system architecture is proposed which illustrates the links in a complex feature modelling environment.

1. Introduction

Future CAD/CAM-system technology is based on powerful computer internal product models which provide an integrated support of the product development during the whole product life cycle. These product models are mainly based on features which provide a mean for storing predefined solutions for recurrent problems.

Features in product models can be classified under the aspect of genericity. Generic features are templates for specific features and feature occurrences. A specific feature is instanced of a generic feature specifying the parameters declared in the generic. Occurrences are created by positioning and orientating the specific feature in an existing part model.

Generic features are stored in a feature library. They are described in an object-oriented manner combining static attribute information and dynamic behaviour. Attributes comprises information about functional intent, geometric parameters with tolerances, technology, etc. The dynamic behaviour is described with procedures for evaluation, validation, recognition of features and their mapping from a design to a production oriented view.

2. Requirements

The intent of product modelling is the modelling of the product life cycle using information technology. All product information is stored in an integrated product model. This computer internal representation of modelling information is generated, accelerated and processed via several modelling methods.

Starting point when building an image of the product development into the computer is the fact, that all modelling methods operate using the same basic data which are stored in the product model. Although the basic data are the same for every method, the different methods need to process the data in different combinations. E.g. in the design stage the designer looks at a product using a functional view, whereas a process planner is looking at it from a production oriented view. Figure 1 impresses the relations between different views.

The regarded views have to be modelled for information processing. Essentially, changing the view to a product includes changing the view from a specific subset of information to another subset. The intersection of the regarded subsets is not empty moreover the intersection has to be restructured for the specific requirements. These views are generated using features. In other words, features offer the possibility to derive different predefined views to a product, based on the same integrated product model.

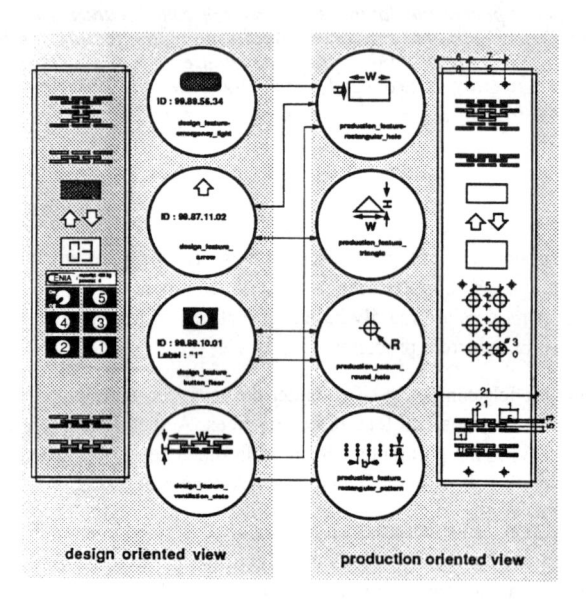

Figure 1:   Different views to a product

At the beginning of the product development the requirements of the development have to be identified and to be modelled via functions which fulfil the requirements. Then the functions are realised with technical elements. Mostly, recurrent solutions are used and a product is not developed from scratch. These circumstances give an impression that basic elements, more or less detailled, will be used in a new product. Detailing the design, the elements will be fully defined, i.e. the recurrent solutions have to be defined as a parameterized object. This recurrent solutions are usable for different product developments and for that reason provided in a feature library. Whereas their specific instances are defined for a certain product and stored in its product model. To realize a function of a product, the objects in the feature library are offered which do not have fixed parameters. Using such objects in a design the connection of the object have to be fixed, i.e. the object is specialised. This specialised object may occur several times in a product. Figure 2 illustrates this fact on the example of a push button.

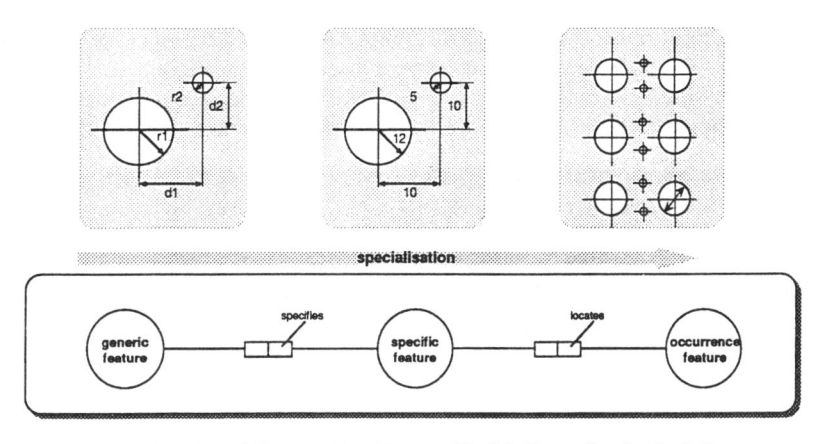

Figure 2:   *Specialisation of Features in a Feature Model; Example of a Push Button*

Looking at the generic features or let´s say the feature library, requirements are definable. If a certain company uses such a modelling system it expects typical features. These features depend on the trade, the material, the machinery of the company and so on. The vendor of such a feature modelling system is interested in offering his modelling system to wide fields of companies. Hence, the system has to have the possibility to process a lot of different features. This conflicting situation can be fixed on the one hand side offering a system having defined as many features as possible or on the other hand side by offering the customer the possibility to define his own features. The first approach was tried with programs for standard parts which have been linked to a modelling system. This effected big object libraries which were difficult to maintain. The second possibility seams to be practical, because only the features of a certain company have to be defined in the system. This set of features is easy to survey.

Supplying a system for user-definable feature the supplier has to offer a management tool for the features. These management tools do not only have the job to provide a possibility to define features, also the feature library has to be maintained by extending, changing or deleting features. Additionally, the features have to be extractable from the library to instance them in a particular product. Even if features of only one company are stored in the feature library the set of features may be numerous. This effects the need of functions for structured feature selection.

3. Concept

For realisation of a system which fulfils the former requirements a global system concept has to be fixed. This concept comprises the development of a product model containing the product, the representation of this product in the database and the development of modelling methods accessing the product data.

As the generation of a model is done manually it may be done in an informal manner or using a formal language for the description. Using a formal language, the database may be generated directly together with its accessing interfaces from the defined model. A usual formal language for data modelling is EXPRESS. EXPRESS is the modelling language used within the STEP development [Sche-90].

EXPRESS is based on an object-oriented approach. All objects are introduced in a describing schema as entities. These entities are described with their attributes and relations between them. Attributes are described either explicit or they are derivable from other attributes. Also inheritance of attributes is possible through the object-oriented representation. This will be effected if an entity is declared as a subtype of another entity.

With this tool the conceptual schema of the feature model is describable and builds the basis for the logical schema of the database. Using processors the logical schema of the database is generated directly from the model description as well as the accessing interfaces.

In this approach the feature model was developed using EXPRESS (Figure 3). Herein the specific and the occurrence features are described, but also the generic features. The advantage of this approach is to declare specific features as subtypes of generic features. With that the attributes described in a generic feature with fixed values are inherited to the specific features. Looking at generic features in more detail, it gets obvious that they are not fully defined with their attributes. Starting from the point of view of a user who wants to define his own features independently of any other feature definition, he needs also to define the geometric representation of the feature. Normally, he wants to get a visual impression of what he has done. To do so, it must be clarified how to evaluate it geometrically. As the feature is defined independent of any other feature, the method for geometrical evaluation is not programable for every feature in a common way. This method depends from the feature and it has to be defined as dynamic behaviour of the feature.

Additionally, features may also interact with each other. In this case a method has to prove the validation of a possibly changed feature. Even here the validation method corresponds directly to the regarded feature and is not predefinable in general.

Also two other methods depending directly to the feature have to be taken into consideration. Through the complexity of today´s products it is possible to have features which are implicitly defined while defining other features. These features are surely introduced in the design but not explicitly known in the feature model. For this reason the feature has to be recognized after the design has been done. This will be realised using feature recognition.

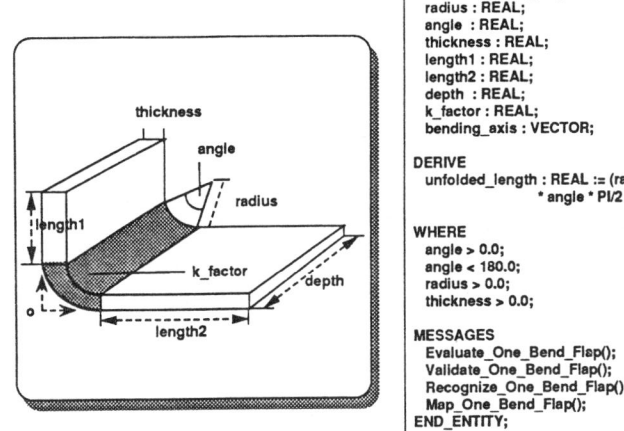

```
Entity One_Bend_Flap SUBTYPE OF (Design_Feature);
   radius : REAL;
   angle  : REAL;
   thickness : REAL;
   length1 : REAL;
   length2 : REAL;
   depth  : REAL;
   k_factor : REAL;
   bending_axis : VECTOR;

DERIVE
   unfolded_length : REAL := (radius+k_factor*thickness/2)
                              * angle * PI/2 - 2 * (radius + thickness);

WHERE
   angle > 0.0;
   angle < 180.0;
   radius > 0.0;
   thickness > 0.0;

MESSAGES
   Evaluate_One_Bend_Flap();
   Validate_One_Bend_Flap();
   Recognize_One_Bend_Flap();
   Map_One_Bend_Flap();
END_ENTITY;
```

Figure 3:   Generic Feature Description; Example of a One Bend Flap

As it has been mentioned in the requirements, features produce special views on a product. These views do not correspond to each other in a one to one relation. For this reason a mapping mechanism is necessary.

Looking at the development of the feature model in EXPRESS all feature information has to be expressed, i.e. not only attributes and relations have to be described, also dynamic behaviour must be considered. This dynamic behaviour is describable with programs. As EXPRESS does not consider programs, it has been extended by introducing MESSAGES [IMP-91].

With this concept softcoded and hardcoded features were mixed together. Describing features in EXPRESS together with their dynamic behaviour allows to transform the model in a database schema which has the programs linked in. So the features are to be seen as hardcoded features. Extending the feature library in this stage causes to have softcoded features which are defined as data. These features can be introduced in a particular design by using an interpreter.

The here described concept lead to the following system architecture (Figure 4).

Basis of the system is the integrated database. Related to the feature representation, the database is structure in two parts. One part comprises the feature description of a concrete product. The other part covers the generic description of features, named the feature library. The data of the model are accessed via a common interface. The management of the generic features is done using the feature library management system. It supports the whole access to the generic features for creation, modification, deletion and usage of the features.

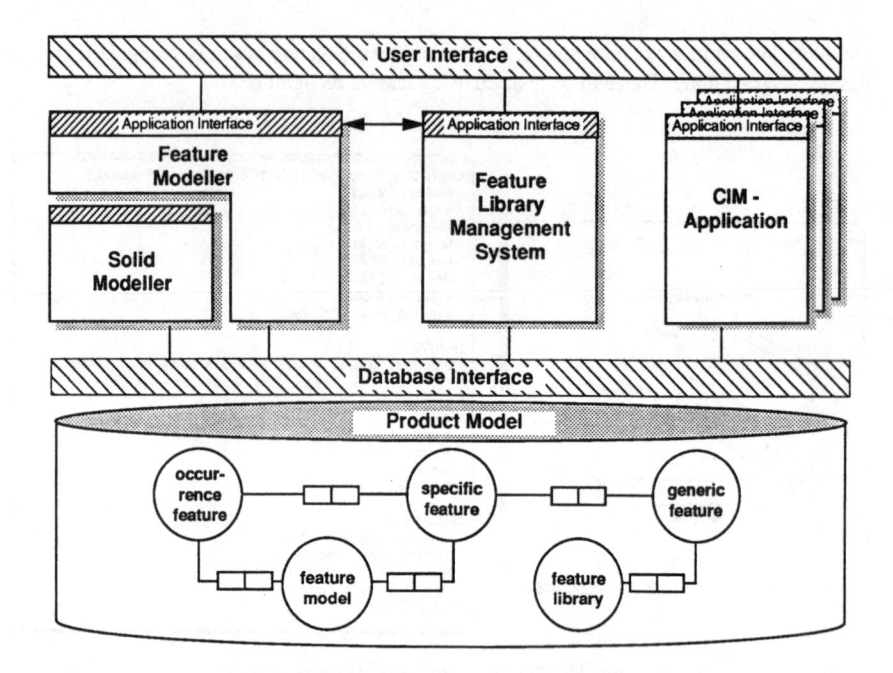

Figure 4: System Architecture of the Feature Modelling Environment

A combined solid/feature modeller supports the processing of features during the product development. Thereby, the modeler accesses features supported by functions of the feature library management system.

4. Application of the feature library management system

On the basis of this concept the feature definition is applicable as follows. The user models the feature geometrically via the graphical-interactive user interface. Additionally, he defines information, like associations, tolerances, etc.. The feature library management system transforms this information to a generic feature description using explicit and derivable attributes as well as validation rules and a sketch of the feature. The information will be stored in the feature library. For structured access the user also has to define the feature as a node in a taxonomy tree.

Now the feature is introduced in a certain design using geometric modelling operations. This procedure will be transformed in an EXPRESS description and stored as a part of the generic feature description.

The feature is now introduced as a recurrent solution in the feature library. The feature is selectable using the user-defined taxonomy tree. Afterwards, the user sets up the values for the attributes and starts the evaluation procedure (Figure 5).

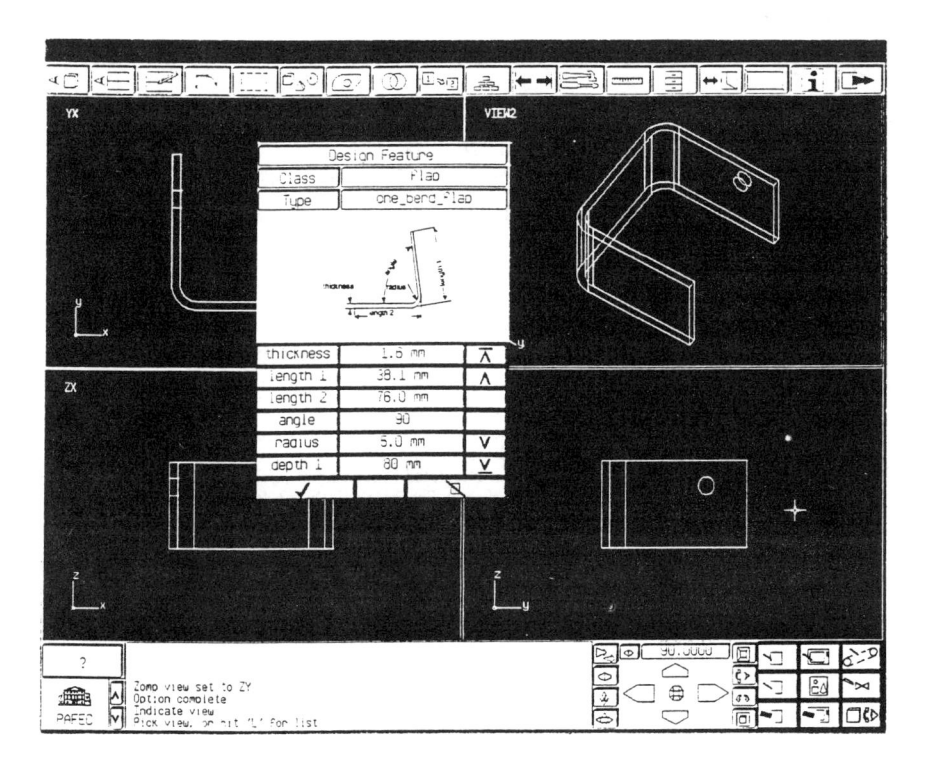

Figure 5:   Example of a User Interface Design for Feature Modelling

5. Conclusion

Feature modelling provides new functionality for the increase of the performance of CAD/CAM-systems. The aim of feature modelling is to develop new CAD/CAM-systems which support the producer during the whole design and planning phase of a product. Also, feature modelling systems will allow the generation of integrated product models which represent not only the product geometry but also additional non-geometric information (such as material, tolerances, technology data, etc.), which is necessary for computer integrated processing.

Acknowledgments

This work was founded by the Commision of the European Communities as part of ESPRIT Project 2165 (IMPPACT). The implementation of the demonstrated system is based on PAFEC products IMAGINER (Solid Modelling System) and HORSES Kernel (User Interface Management Tool).

References

/CAMI-88/	Shah, J., Sreevalsan, P.; Rogers, M.; Billo, R., Mathew, A.: Current Status of Feature Technology, Revised Report, CAM*I No. R-88-GM-04.1, Task 0, 1988
/CAMI-90/	Pratt, M.J.: The CAM*I Application Interface Specification: Formalized final report , CAM*I Report No R-88-GMP-02, May 1990
/Sche-90/	Schenk, D.: EXPRESS Language Reference Manual, DOC No. 466, ISO TC184/SC4/WG1, March 1990
/GAHP-90/	Grabowski, H.; Anderl, R.; Holland-Letz, V.; Pätzold, B.; Suhm, A.: An integrated CAD/CAM-system for product and process modelling; Proceedings of the IFIP WG 5.2/GI International Symposium Berlin, Elsevier Sientific Publisher, Amsterdam 1990
/GAPR-88/	Grabowski, H.; Anderl, R.; Pätzold, B.; Rude, S.: The development of advanced modelling techniques meeting the challenge of CAD/CAM-integration, Proceedings of 4th CIM-Europe Conference Madrid, Springer, 1988
/GARM-88/	N.N.: General AEC Reference Model (GARM), TNO, Institute for Building Materials and Structures, Delft, The Netherlands, 1988
/IMPP-91/	Integrated Modelling of Products and Processes using Advanced Computer Technologies, Proceedings of the IMPPACT-Workshop, Berlin, 1991
/GRBS-90/	Grabowski, H.; Braun, S.; Suhm, A.: Integrated Modelling of Information and Functions, Proceedings of the CIM Europe Workshop, Saarbrücken, Dec. 3-4th, 1990

Optimal Trajectory Planning for Manipulators with Goal Point Uncertainty

S. Adams, A.A. Melikyan

Dept. of Electrical and Electronic Engineering, South Bank Polytechnic, Borough Rd., London SE1 OAA
Institute for Problems in Mechanics, USSR Academy of Sciences, prospect Vernadskogo 101, 117526, Moscow, USSR

Summary

The problem of trajectory planning is considered, while the goal point is known to lie within a previously identified domain of uncertainty. The planned path has two parts; (i) the manipulator is moved to the vicinity of the goal point where the domain of uncertainty is searched, using optical techniques, until the goal point is located; (ii) the end-effector is moved to the goal point. The total time for the complete operation is to be minimised. The corresponding mathematical problem is formulated in the form of a minimax worst case or "game" problem. A solution is developed for the two-dimensional case with a convex uncertainty domain. The search domain produced by the camera mounted on the end-effector is approximated by an infinite semi-plane. The numerical results are presented for both the optimal and the simple trajectories, using an elliptical domain of uncertainty.

1. Goal Point Uncertainty in The Control of Manipulator

All manipulators perform two basic functions; movement and work piece manipulation. The process of movement can be considered as bringing the end-effector of the manipulator to a desired point within the manipulator's work space. The end-effector's trajectory is from a known home position to some desired goal point. It is invariably assumed that the goal point is known with the same accuracy as the manipulator's home position. Unless the manipulator is part of a rigidly defined working environment this is not always a reasonable assumption. In many cases the goal point can be considered as being located within an identifiable work zone (region of uncertainty), as shown in Fig. 1. Being as the location of the goal point, representing the position of the work piece, as unknown prior to the start of the manipulator's work cycle the work piece must be located using sensors mounted on the manipulator. With reference to Fig. 1 it can be seen that a simple ultrasonic range finder mounted at the base of the manipulator is sufficient to obtain coordinates of the work piece. However, due to the nature of the sensor, the spatial resolution is finite and hence, the precise location of the work piece and so, the goal point is unknown.

The goal point can be considered as being located within a region of uncertainty approximated by an ellipse, the dimensions of which are sensor dependent. In order to achieve the goal point when subjected to an element of uncertainty addition sensing coupled with a suitable search strategy is required. In this case a simple dynamic RAM (DRAM) camera is used to search the area of uncertainty and hence, locate the work piece with sufficient accuracy to enable the goal point to be achieved.

Given the inherent uncertainty of the goal point within the work space of the manipulator the time required to reach the goal point may be minimised by employing a time optimal search strategy, developed in the following sections, to obtain the end-effector trajectory. The resulting trajectory consists of three distinct motions; simple motion; time optimal motion and point-to-point motion. The simple motion brings the end-effector to the vicinity of the region of uncertainty, the time optimal motion reduces the worst case search time and the point-to-point motion achieves the goal point following it's location.

2. Modelling of Uncertainty Problem Using Time-Optimal Minmax Approach

Here we give the statement of a mathematical problem, which can be used as a model to describe a situation of previous section. Consider a conventional point-to-point time-optimal control problem [1]:

$$\dot{x} = f(x, u, t), \quad u \in U, \quad t \in [t_0, T]$$

$$x(t_0) = x^0, \quad x(T) = x^1, \quad J = T - t_0 \to \min_u \quad (1)$$

Here $x = (x_1, \ldots, x_n) \in R^n$ is the state vector of the system; $u = (u_1, \ldots, u_m)$ - vector of control parameters; t_0, x^0 - given initial moment and state vector. The moment T isn't fixed and is defined from the condition $x(T) = x^1$. Let the goal point x^1 be not known precisely at the starting moment t_0; it is given only that

$$x^1 \in D, \quad D \subset R^n \tag{2}$$

where D is a given set of uncertainty continuous one or consisting of discrete points (bullets on Fig. 2). The information about point x^1 can be improved during the motion by means of observations, described by moving informational set $G = G(x(t))$, depending upon the current state vector, while

$$G(x) = \{\xi \in R^n : \xi - x \in G_0\} \tag{3}$$

where $G_0 = G(0)$ is given convex set. The point x^1 is considered to be observed (to be precisely known) at the first moment $t_* > t_0$, when the observation condition holds:

$$x^1 \in G(x^*), \quad x^* = x(t_*) \tag{4}$$

From the definition (3) it follows that the set $G(x(t))$ is fixed in moving coordinate system with the origin in $x(t)$ and axes, parallel to those of original system.

According to available information the class of the pairs of open-loop control functions of the type:

$$u = \{U_0(x^0, t_0, t), \quad U_1(x^*, t_*, x^1; t)\} \tag{5}$$

is considered as the set of admissible controls. Here x^0, t_0, x^*, t_*, x^1 are parameters. On the interval $[t_0, t_*]$ the search control U_0 is used, which depends also upon set-parameters D, G_0. After the observation moment t_*, when the vector x^1 becomes known, a point-to-point control U_1 is used on $[t_*, T]$, which brings the state vector from x^* to x^1, Fig. 2. Another requirement to admissible control is to produce finite moments t_*, T.

Let the class of admissible controls be nonempty, and $G(x^0) \cap D = \emptyset$. Then to each couple $u = \{U_0, U_1\}$ and each $x^1 \in D$ unique trajectory $x(t)$, $t_0 \leq t \leq T$, and the time of motion $J(x^0, x^1, u) = T - t_0$ correspond. Now the minmax duration of the process $J_*(x^0)$ and guaranteeing control u^* are defined from:

$$J_*(x^0) = \min_u \max_{x^1 \in D} J(x^0, x^1, u) = \max_{x^1 \in D} J(x^0, x^1, u^*) \qquad (6)$$

It is clear that the second component u_1^* of the optimal control is the solution of a conventional point-to-point problem (1) with complete information and initial point t_*, x^* [1]. So the most important object to be found in (6) is the optimal search control u_0^* and corresponding path.

Relations (1) - (6) describe a point-to-point control problem with incomplete information.

3. Solution of the problem for two-dimensional case

Let $n = 2$ and the system (1) has a simple form

$$\dot{x} = u_1, \quad \dot{y} = u_2, \quad |u|^2 \equiv u_1^2 + u_2^2 \leq 1 \qquad (7)$$

Let D be a convex set with smooth boundary, and G_0 be a half-plane $\{(x, y) : x - a \geq 0\}$. So the moving point $x(t)$, $y(t)$ is at the distance $|a|$ from the boundary ∂G.

Let $a = 0$. According to [2] optimal search paths can be constructed as follows. First the focal point $F \in D$ must be found and two curvilinear singular search paths, starting in F. The focus $F(x_F, y_F)$ is the middlepoint of the vertical segment AB, Fig. 3, such that the sides of direct angle BKA are tangent to the boundary ∂D at the points A, B.

Let the function $y = \varphi_2(x)$, $x_F \leq x \leq x_2$ ($\varphi_1(x)$, $x_F \leq x \leq x_1$) defines the arc AA' (BB') of the boundary ∂D. At the points A', B' the equalities hold $\varphi_i'(x_i) = 0$, which mean that tangents to the ∂D are horizontal lines. Let for convenience the domain

D belong to the semiplane $x \geq 0$, and y - axis be tangent to ∂D, Fig. 3. The following initial value problems

$$y'_i(x) = \frac{\varphi'^2_i(x) - 1}{2\varphi'_i(x)}, \quad y_i(x_F) = y_F, \quad x_F \leq x \leq x_i, \quad i = 1, 2$$

define curves L_1, L_2, may be unbounded, upper and lower singular search paths. The set of tangents to $L = L_1 + L_2$ covers the open domain X_L lying on right hand side of L.

Now for any initial point $P_0 \in X_L$ the optimal search path consists of three parts: of the tangent segment P_0P', the singular arc $P'F$ and the part of segment FK, $x \geq 0$, which is tangent to both curves L_1, L_2. The worst locations for goal point are the arcs AA', BB'. If initial point lies between y-axis and X_L, then optimal path is non-unique [3].

This optimal strategy is generalized for the polygonal [3] and discrete D, where L_i are broken lines, algorithms and programs are developed for numerical solution.

4. Some numerical results

Here we give a comparison of the times of process for optimally and non-optimally planned search paths for the system (7). The domain D is bounded by the ellipse with half-axes $p = 3$, $q = 5$, rotated on the angle α, lying in the half-plane $x \geq 0$ and touching to the y-axis in the origin; G is a half-plane with ∂G parallel to y-axis. Non-optimal (simple) search path is the straight line, connecting home point and the center of ellipse. The discrete approximation of the ellipse and numerical algorithm of [3] were used. The optimal (T_0), simple (T_S) times, the difference $\delta = T_S - T_0$ and corresponding goal point coordinates are given in the tables below for several values of the parameters of the problem, while $V = 80$. Here V is the velocity of the end-effector; time is measured in sec, distance in cm. One can see from these data that the optimal (worst case) time isn't always the shortest one.

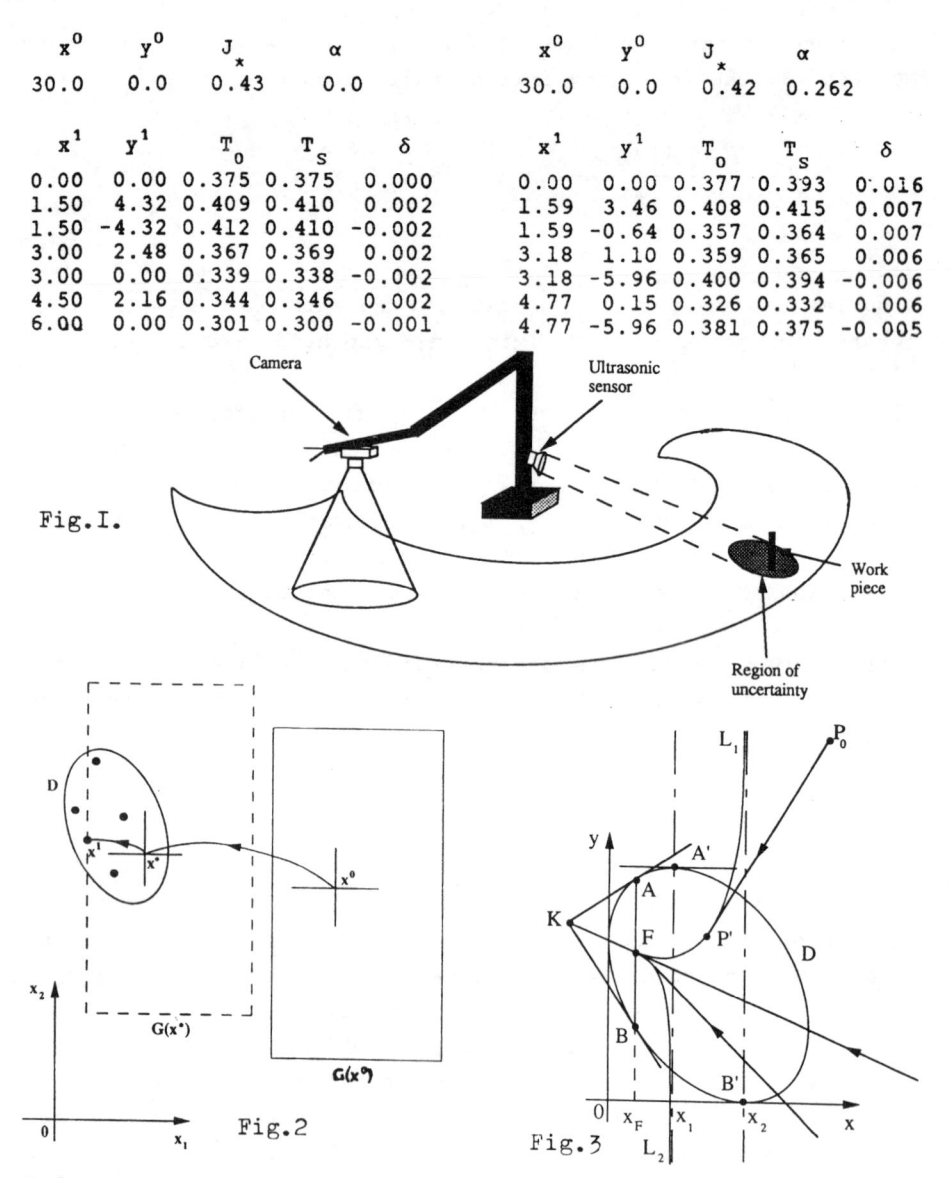

x^0	y^0	J_*	α		x^0	y^0	J_*	α
30.0	0.0	0.43	0.0		30.0	0.0	0.42	0.262

x^1	y^1	T_0	T_S	δ		x^1	y^1	T_0	T_S	δ
0.00	0.00	0.375	0.375	0.000		0.00	0.00	0.377	0.393	0.016
1.50	4.32	0.409	0.410	0.002		1.59	3.46	0.408	0.415	0.007
1.50	-4.32	0.412	0.410	-0.002		1.59	-0.64	0.357	0.364	0.007
3.00	2.48	0.367	0.369	0.002		3.18	1.10	0.359	0.365	0.006
3.00	0.00	0.339	0.338	-0.002		3.18	-5.96	0.400	0.394	-0.006
4.50	2.16	0.344	0.346	0.002		4.77	0.15	0.326	0.332	0.006
6.00	0.00	0.301	0.300	-0.001		4.77	-5.96	0.381	0.375	-0.005

Fig.I.

Fig.2

Fig.3

References

1. Bryson, A.E.; Ho, Y.C. Applied optimal control. Blaisdell Waltham, Mass., 1969.
2. Melikyan, A.A.: The problem of time-optimal control with the search of the goal point. Prikladnaya matematika i mechanika (PMM), vol. 54, No 1 (1990) 3-11 (in Russian).
3. Bocharov, V.Yu.; Melikyan, A.A: The problem of time-optimal guaranteeing control with the search of goal point within the polygonal set of uncertainty. Izvestiya AN SSSR. Tekhnicheskaya kibernetika, No 1 (1991) 82-92 (in Russian).

x^0	y^0	J_*	α		x^0	y^0	J_*	α
30.0	0.0	0.43	0.0		30.0	0.0	0.42	0.262

x^1	y^1	T_0	T_s	δ	x^1	y^1	T_0	T_s	δ
0.00	0.00	0.375	0.375	0.000	0.00	0.00	0.377	0.393	0.016
1.50	4.32	0.409	0.410	0.002	1.59	3.46	0.408	0.415	0.007
1.50	-4.32	0.412	0.410	-0.002	1.59	-0.64	0.357	0.364	0.007
3.00	2.48	0.367	0.369	0.002	3.18	1.10	0.359	0.365	0.006
3.00	0.00	0.339	0.338	-0.002	3.18	-5.96	0.400	0.394	-0.006
4.50	2.16	0.344	0.346	0.002	4.77	0.15	0.326	0.332	0.006
6.00	0.00	0.301	0.300	-0.001	4.77	-5.96	0.381	0.375	-0.005

References

1. Bryson, A.E.; Ho, Y.C. Applied optimal control. Blaisdell Waltham. Mass., 1969.
2. Melikyan, A.A.: The problem of time-optimal control with the search of the goal point. Prikladnaya matematika i mechanika (PMM), vol. 54, No 1 (1990) 3-11 (in Russian).
3. Bocharov. V.Yu.; Melikyan, A.A: The problem of time-optimal guaranteeing control with the search of goal point within the polygonal set of uncertainty. Izvestiya AN SSSR. Tekhnicheskaya kibernetika, No 1 (1991) 82-92 (in Russian).

CAD Simulation of the Factory of the Future

Chris Stylianides, Kenneth Ebeling
Department of Industrial Engineering and Management
North Dakota State University
Fargo, North Dakota

Abstract

The Factory Of the Future (FOF) includes converting conventional machining stations into integrated Flexible Manufacturing Systems (FMS). This paper discusses how simulation can be incorporated within Computer Aided Design (CAD) systems to simulate and animate the evolution to such factories. It draws on industrial studies and experiences that involved animation of modern manufacturing operations.

A FORTRAN based simulation program has been imbedded within a mainframe CAD system and uses a Graphics Programming Language (GPL), to provide real time animation in 3-D wire frame graphics. A batch oriented Material Requirements Planning (MRP) system is simulated to provide daily updates of outstanding production center loadings over a monthly planning horizon. The importance of meeting due dates is preserved in the simulation while accommodating unforeseen scheduling changes.

The application of the program to managing a manufacturing facility provides a means of:

a) Assessing of the economic impact an FMS system has on production operations,
b) identifying operational bottlenecks and assists in identifying the changes needed to smooth production operations,
c) evaluating the impact of replacing, retiring, selling, or reassigning existing equipment based on facility average utilization, and
d) uses real time animation of manufacturing operations to graphically depict the state of the system.

Introduction

CAD/CAM/CAE/CIM, (C4), has received a lot of emphasis over the last decade. Many USA companies are looking at this technology as a key to making the transition from the twentieth into the twenty first century. They are realizing that the advancements achieved by implementing C4 technology can benefit their operations and provide them with an edge over companies from other countries that are slow in utilizing these manufacturing systems.

Many factors are affecting the swift transition from conventional factories to Factories Of the Future (FOF) [1]. These include:

a) The development of new, powerful micro-computers and workstations based on 80386 and 80486 microprocessors, RISC architecture, parallel processing, etc.
b) The development of easy to use simulation software that can run on these machines and can tackle almost any engineering project.
c) The advances in computer graphics and solid modeling which provide animation, so that engineers and designers can visualize systems in operation in real time.
d) The general effort to link industrial islands of automation into Flexible Manufacturing Systems through computer networks.
e) The development of techniques to transfer simulation results into off-line programming for industrial robots.
f) The decrease in fixed and operating costs of many machining tools and flexible manufacturing cells.
g) The training and specialization of people in operating and maintaining complex systems.
h) The advances and implementation of artificial intelligence and expert systems in many areas of manufacturing.

The Integrated Model

The factory with a future includes Conventional Numerical Control (NC) with Computer Numerical Control (CNC) machines, grouped together to form work centers, intermixed with Flexible Manufacturing Systems (FMS) and Cells (FMC). Group Technology principles are also applicable and plant layouts are rearranged to accommodate them. Cost estimating systems can play an important role in estimating the machining and tooling costs for each part and generating summary reports for all direct labor/machine operations, classifying the parts to be produced, generating the work center routing of parts, establishing tooling requirements, and producing other related information.

Parts being processed in the manufacturing facility will be routed to conventional work centers, FMSs and FMCs. The simulation software developed can be used to demonstrate the routing of the parts through such an integrated manufacturing facility. An automated manufacturing cost estimation system has been developed and is capable of being integrated with the simulation software.

The manufacturing facility used as an example to test the developed simulation software consisted of [2]:
a) Eleven conventional work centers made up of one or more vertical and horizontal end mills, and
b) an FMS consisting of four machining cells.

There are five distinct events occurring during the simulation run:
a) The arriving of a job at the manufacturing facility,
b) The end of an operation in the FMS, e.g. the end of a machine operation,
c) The end of a transporter operation,
d) The departure of a job from a work center in the job shop or FMS, and
e) The end of the simulation run.

The simulated manufacturing facility was used to process the monthly production of 106 different parts. Some parts were machined in the conventional centers only, some in the FMS only, and some in both. The routing of the parts was read by the simulation software at the start of the simulation. This file can be the output of the cost estimation software. The parts arrived at the manufacturing facility location based on their due dates and enter a work center queue on a first-in, first-out basis. The researchers assumed that there was no loss of continuity between successive days of operation of the manufacturing facility. The manufacturing facility was simulated for an 8-hour day and 21 days per month.

The Animation Model

Real time animation of a simulation model was accomplished by integrating the simulator with the graphics of a Computer Aided Design (CAD) system. This integration can be done in two ways:
a) Extend the CAD system to include the simulator and integrate it with its other modules, and
b) extend the simulator to access a CAD system or another library of graphic routines, such as Graphical Kernel System (GKS), or Programmer's Hierarchical Interactive Graphics Standard (PHIGS), to support its graphics and animation requirements.

Both approaches to animate various physical conditions have merits and limitations. Existing CAD systems provide an extensive library of graphic generation capabilities. The CAD systems provide graphic manipulation and visualization features that are used in automatic finite element mesh generation, simulation of NC tool paths, and off-line robot programming. These systems could extend their standard software application packages to include discrete and continuous event simulation capabilities as well. This would provide manufacturing and design engineers with an additional tool to address the problems of designing optimal manufacturing systems. Miller [1] stated that engineers in the future will be looking for "complete" systems that could provide the required tools to assist in the design process from conceptualization to implementation of producing the product in the most efficient way.

The animation system developed was used to [2]:

a) Represent in 3-D wire frame the manufacturing facility's physical layout including the conventional work centers and FMS components.
b) Represent in 3-D wire frame the parts that are being processed in the plant and show their movement in real time
c) Represent in 3-D wire frame the transporter, with or without parts, in the FMS system in real time.
d) Represent any and all queues built up at work center loading areas.

Figure 1 shows a "snapshot" taken from the computer terminal display of an animated manufacturing facility. The animation of the integrated model was developed within the CAD system to include next event scheduling logic [3] and integrate it with other CAD modules. A mainframe CAD system, namely Control Data's Integrated Computer-Aided Engineering and Manufacturing Design, Drafting and Numerical Control (ICEMDDN), running on a Control Data CYBER-180/830 computer was used to produce animated graphics on a Tektronix terminal. The interface of the CAD system with the FORTRAN simulator was accomplished using the Graphics Programming Language (GPL) [4], which also resides on the CYBER-180. Figure 2 shows a macro flow chart for the integrated animation model. It also illustrates the relationship between ICEMDDN, GPL, and FORTRAN.

Example Results

When simulation is done from within a CAD system, user benefits can be found in three basic areas:

a) Model development time can be reduced considerably.
b) Accuracy of design is improved.
c) Presentation and communication of the design and design output are improved.

Major disadvantages of CAD systems are their high cost and the limited portability that exists in the CAD world. In the near future the cost of CAD systems will drop and with the implementation of international standards for graphics, the portability from one CAD system to another will increase.

The generated report of results includes:

a) The average delay of individual parts in a conventional work center queue.
b) The average overall part delay in a conventional work center queue.
c) The average number of parts in a conventional work center and an FMS station queue.
d) The average utilization of conventional work centers and FMS stations.
e) The average delay of individual parts in a conventional work center and an FMS station queue.

Conclusions

The software can be used to:

a) Evaluate proposed Flexible Manufacturing System@(FMS),
b) assess the impact of FMS on existing manufacturing facilities,
c) give answers to various "what if" questions, and
d) graphically animate in real time on the screen the operations of the integrated system.

References

1) Miller, R. R.: Manufacturing Simulation: A New Tool for Robotics, FMS, and Industrial Process Design. Madisson, GA: SEAI Technical Publications 1987.
2) Stylianides, C.; Radi, G.; and Ebeling, K. A. Use of Simulation in The Analysis of Shop Floor Operation. In H. Eldin (Ed.), Proceedings: 9th Annual Conference on Computers and Industrial Engineering, New York: Pergamon Press (1987) 144-148.
3) Law, A. M., and Kelton, D. W.: Simulation Modeling and Analysis. New York: McGraw-Hill 1982.
4) Control Data Corporation: ICEM GPL for NOS, Reference Manual Revision C. Minneapolis, MN 1987.

Fig. 1. A "snapshot" of an animation screen

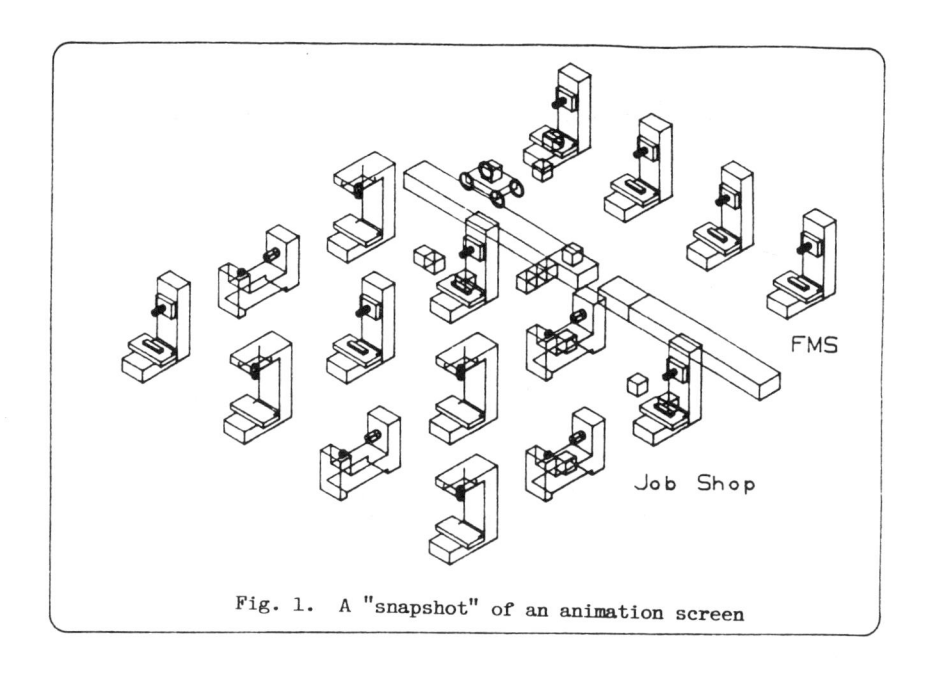

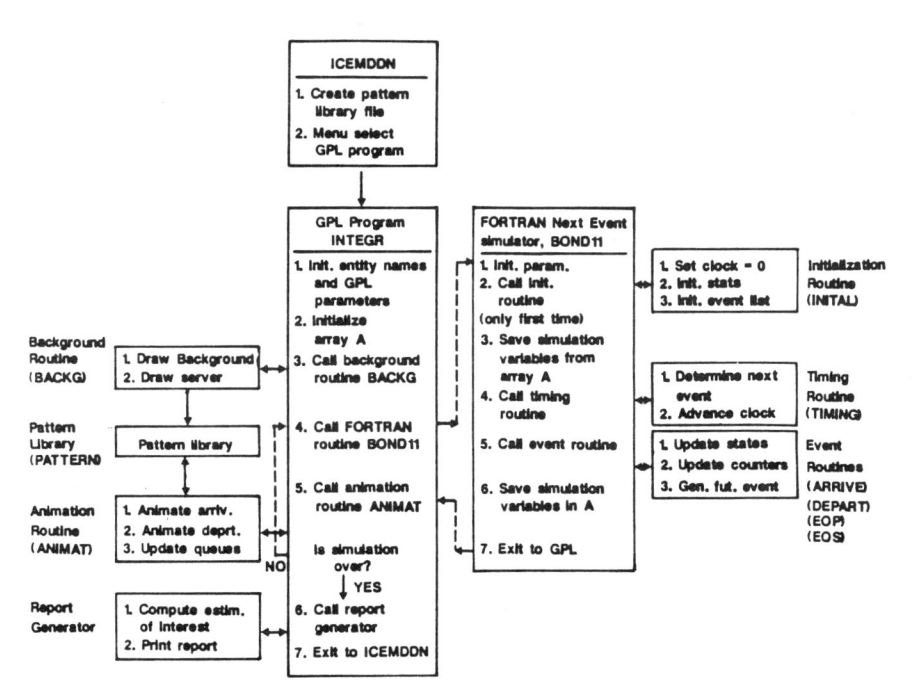

Fig. 2. Macro flow chart for the integrated animation model

Cycle Time and Path Optimization of Manipulator Arms In Assembly Operations

Raymond Sid Redman Jr. and A. Sherif El-Gizawy
Industrial and Technological Development Center
University of Missouri-Columbia
Columbia, MO 65211

Summary

Optimum path planning of manipulator arms in assembly applications involves the selection of the optimum combination of the robot control variables under the constraints imposed by the robot's physical capabilities and the condition of the working area. The present paper describes an approach based on numerical optimization techniques to plan collision-free paths and on Taguchi parameter design methodology to optimize the control parameters of the pick-and-place operation that would yield minimum cycle time.

1.0 Introduction

The ability to plan and optimize a collision-free path in pick-and-place operation is essential in order to improve both productivity and cost effectiveness of robotic assembly operations. Optimum path planning of manipulator arms in assembly applications involves the selection of the optimum combination of the robot control variables under the constraints imposed by the robot's physical capabilities and the condition of the working area (presence of obstacles). The path planning optimization in the assembly applications has two objectives. The first one involves the determination of the shortest path of the manipulator's end effector from a pick position (feeders and magazines carrying individual components) to the place position (assembly fixtures), such that it avoids collisions with obstacles in the assembly area. The second objective is concerned with the determination of the levels of the robot control variables that will yield the minimum cycle time in order to increase productivity.

The problems associated with path planning and optimization in robotic applications have been emphasized in the past [1 - 4]. Most of the reported solutions to these problems involve complex, nonlinear iterating procedures [5 - 8]. Furthermore, these solutions were reported to lack the accuracy required in many applications such as assembly and thus they are limited to applications where precise path planning is not critical. Different methods for off-line cycle time estimation and optimization have been developed. Almost all methods reported use empirical formula or approximation techniques. Tanner [9] provided a "rule of thumb" method for estimating the time required by each element of the motion such as one second per move, 0.5 second to open or close the gripper, etc. This method was found to give a margin of error of about ±30% from the actual time. Owen [10] presented a method based on distance, velocity, and acceleration. This method is simple and provides reasonable accuracy for straight discontinuous paths, but it is rather complex for continuous path motions. A model called time-distance functions (TDF) was derived by Rabie and Zhuhu [11] for determining the cycle time for simple sequence robots. The model is a function of the working conditions, including workplace layout and weight of the end effector. Generation of the TDF requires the conduction of a large number of experiments using different payloads and by moving each joint of the manipulator at constant velocity through an array of displacements. From this method the cycle time is bounded and depending on a particular payload a cycle time can be interpolated. More accurate techniques using computer-aided engineering were used in the last five years to determine optimum velocities, accelerations, and torques to minimize the time a robot spends moving along a path. For example, Wu has developed an efficient algorithm to determine the optimal feed rate that would maximally utilize the capacities of the robot [3]. This method assumes the path to be defined by the operator.

This paper describes an approach based on numerical optimization techniques to plan collision-free paths and on Taguchi robust design methodology to optimize the control parameters of the pick-and-place operation that would yield the minimum cycle time.

2.0 Algorithm for Two-Dimensional Collision-Free Path

2.1 Object and Gripper Representation

For simplicity, the object and gripper are represented by circles of radius r_c and r_g respectively and the object was assumed by be able to interfere only with the gross motion segment of the pick-and-place operation. The object center is at coordinates (x_c, y_c). The gripper center will move along the path (see Figure 1). Different shaped objects can be handled in this algorithm by circumscribing a circle around them and using its radius (r_c) and center coordinates (x_c and y_c) for the analysis. The equation for the circle representing the object is as follows:

$$y = y_c \pm \sqrt{r_c^2 - (x - x_c)^2} \qquad (1)$$

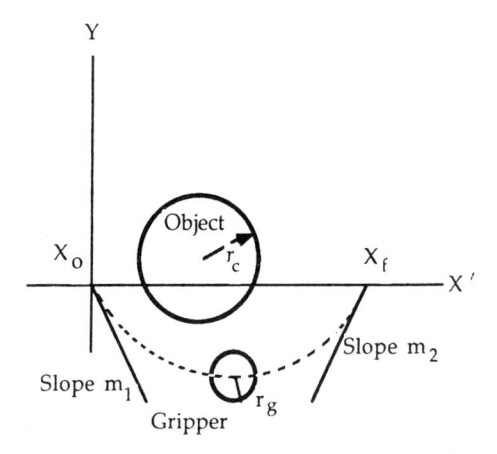

Figure 1: Geometric Features of the Collision-Free Path

2.2 Path Representation

The path of the manipulator end effector during the gross motion is assumed to be continuous and is represented by a third order polynomial of the following form:

$$y = a_0 + a_1 x + a_2 x^2 + a_3 x^3 \qquad (2)$$

Four conditions must be known in order to determine the values of a_0, a_1, a_2, and a_3.

The initial and final positions (the pick and place positions) are know. Thus the initial and final position vectors are set as follows:

$$X_0 = \begin{Bmatrix} 0 \\ 0 \end{Bmatrix}, \qquad X_f = \begin{Bmatrix} x_f \\ 0 \end{Bmatrix}$$

The other two conditions required to solve equation (2) are obtained from the initial and final directions (slopes) m_1 and m_2 of the trajectory (see figure 1). These slopes (m_1 and m_2) represent the independent variables of the problem.

Using the above four conditions with equation (2), the polynomial constants will take the following values: $a_0=0$, $a_1=m_1$, $a_2=-(2m_1+m_2)/x_f$, $a_3=(m_1+m_2)/x_f^2$.

2.3 Objective Function

Once the path is mathematically defined, the objective function (path length) can be formulated as follows:

$$\text{min: } S = \int_0^{x_f} \sqrt{\left(\frac{dy}{dx}\right)^2 + 1} \; dx \qquad (3)$$

2.4 Constraint Representation

There is one major constraint: the path must be object avoiding. This means that at the path's closest point to the object, there must be clearance between the gripper and the object. In addition, limits are assigned to x ($0 \leq x \leq x_f$) and y ($-x_f \leq y \leq 0$) so that the surrounding environment will not be endangered (see figure 1). These constraints are formulated as follows:

$$g_1 = \min \sqrt{(x - x_c)^2 + (y - y_c)^2} - r_g - r_c \geq 0 \qquad (4)$$

$$g_2 = x \geq 0 \qquad (5)$$

$$g_3 = x_f - x \geq 0 \qquad (6)$$

$$g_4 = -y \geq 0 \qquad (7)$$

$$g_5 = y + x_f \geq 0 \qquad (8)$$

The Solution

The constrained minimization problem of the above section was solved using a numerical optimization routine called PEN developed by Gabriele and Ragsdell [12]. PEN is a FORTRAN subroutine that solves the constrained nonlinear programming problem by the penalty function approach. This method consists of converting the constrained problem into unconstrained problems, whose solutions converge at the limit to the solution of the constrained problem. The value of the objective function was obtained by numerical integration using the trapezoid rule method.

2.5 Graphical Simulation

The developed algorithm was tested using three case studies. Table 1 shows the set of objects for which the object-avoidance problems were solved. The solutions for the three cases are graphically simulated in Figures 2, 3, and 4.

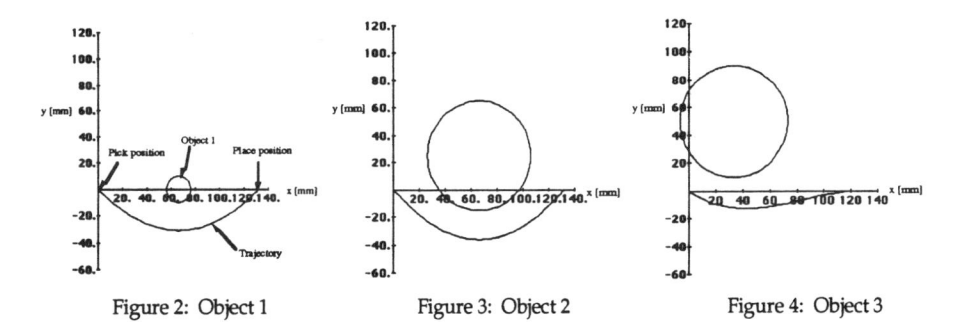

Figure 2: Object 1   Figure 3: Object 2   Figure 4: Object 3

Table 1: Object avoidance path solutions

Object	x_f	x_c	y_c	r_c	r_g	m_1	m_2	g_1
					[mm]			
1	132	66	0	10	21	-0.93	0.94	0.008
2	132	66	25	40	21	-1.09	1.09	-0.026
3	132	33	50	40	21	-0.65	-0.06	-1.097

3.0 Cycle Time Optimization Using Parameter Design

3.1 Taguchi Robust Design Methodology

The parameter design technique developed by Taguchi and described in reference [13], was used to optimize the control parameters of the manipulator motion during assembly operations. The factors that can affect the process performance are classified as control parameters and noise factors as indicated in Figure 5. The quality characteristic or the process response to be optimized in the present case is the cycle time to complete an assembly step. The noise factors include the resolution of the internal system clock used to measure the cycle time of the pick-and-place experiments, and the robot's previous motion since it has a direct effect on the present motion. The control parameters and their levels considered in this study are summarized in Table 2. It should be mentioned that the length of the path in this part of the study is assumed to be an uncontrolled parameter since its level is predetermined in phase one of the present approach. Also the payload, although it has an effect on cycle time, is considered to be an uncontrolled parameter.

A standard L_{27} orthogonal array was used to conduct the experiments. This array consists of only 27 experiments as compared with 3^{13} (1.6 million) experiments if we were to use a full-factorial design.

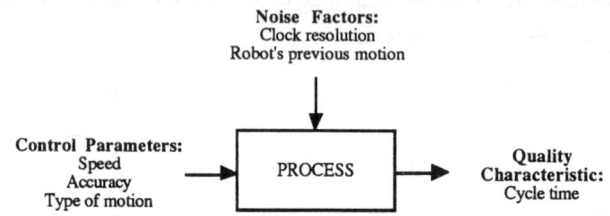

Figure 5: Block Diagram of the Pick-and-Place Process

Table 2: Control Factors and corresponding level values.

Factor	Controls			Level 1	Level 2	Level 3
	Speed:					
A		Pick Motion:	Velocity	50	75	100
B			Acceleration	35	50	65
C			Deceleration	50	75	100
D		Gross Motion:	Velocity	50	75	100
E			Acceleration	60	80	100
F			Deceleration	50	75	100
G		Place Motion:	Velocity	50	75	100
H			Acceleration	35	50	65
I			Deceleration	35	50	65
J	Accuracy			NONULL	COARSE	FINE
K	Motion:		Move(s)	MOVE	MOVES	
L			Depart(s)	DEPART	DEPARTS	
M			Appro(s)	APPRO	APPROS	

3.2 Analysis and Verification

The regular analysis method was used to predict the combination of parameter levels to reduce the cycle time average. The mean cycle time for each factor at each level was calculated and plotted in the graph of Figure 6. From this figure, it is seen that factors A and D have the strongest effect on cycle time. Factors B, C, and M have a small effect on cycle time, while factors E, F, G, H, I, J, K, and L affect cycle time the least.

The following superposition model was used to predict the optimum response under optimum conditions of the significant process parameters:

$$\zeta_{opt} = m+(A_3-m)+(B_3-m)+(C_1-m)+(D_3-m)+(E_3-m)+(M_2-m)$$

where: ζ_{opt} = optimum response time and m = overall mean value. Using the optimum value of each parameter gives ζ_{opt} = 2.032 seconds.

In order to verify this prediction, a pick-and-place experiment was conducted using the AdeptOne robot at the predicted optimum levels. The actual cycle time measured was found to be 2.000 seconds. We therefore conclude that the model prediction have been sufficiently verified (within ±2.0%) and can be used with confidence for similar applications.

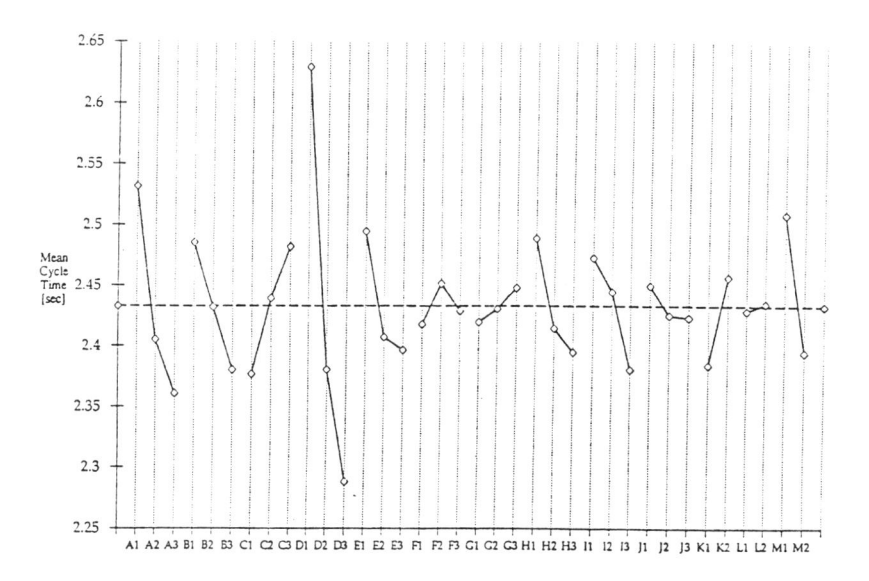

Figure 6: Regular Analysis Response Graph

References

1. Reif, John H., "Complexity of the Mover's Problem and Generalizations", 20th IEEE Symposium on the Foundations of Computer Science, 1979, pp. 421-427.

2. Lin, C. S., Chang, P. R., and Luh, J. Y. S., "Formulation and Optimization of Cubic Polynomial Joint Trajectories of Mechanical Manipulators" IEEE Transactions on Automatic Control, Vol. AC-28, No. 12, December 1983, pp. 1066-1074.

3. Wu, W. D., "Optimization of Feedrate in Planar Translational Motions of Robot Manipulators," Journal of Manufacturing Systems, Vol. 8, No. 4, 1989, pp. 306-316.

4. Kant, Kamal, and Zucker, S., "Planning Collision-Free Trajectories in Time-Varying Environments," Proceedings IEEE International Conference on Robotics and Automation, 1988, pp. 1644-1649.

5. Luh, J. Y. S., Lin, C. S., "Optimum Path Planning for Mechanical Manipulators," Transactions of ASME Journal of Dynamic Systems, Measurement and Control, Vol. 102, June 1981, pp. 142-151.

6. Kim, B. K., Shin, K. G., "Minimum Time Path Planning for Robot Arms and Their Dynamics," IEEE Transactions on Systems, Man and Cybernetics, Vol. SMC-15, No. 2, March/April 1985.

7. Freund, E., and Hoyer, H., "Collision Avoidance for Industrial Robots with Arbitrary Motion," Journal of Robotics Systems, Vol. 1, No. 4, 1984, pp. 317-329.

8. Trabia, M. B., "Automatic Generation of Near-Minimum Length Collision-Free Paths for Robots", Robotics Research, Vol. 14, 1989, pp. 93-98.

9. Tanner, W. R., "Can I Use A Robot?", Robotics Today, Spring 1980, pp. 43-44.

10. Owen, T., Assembly with Robots, 1985, pp. 43-53.

11. Abdelrahman R., Zhuhu, Wu, "Model for the Determination of Cycle Time for Robots" Autofact, Conference Proceedings, 1986, section 13, pp. 45-52.

12. Gabriele, G.A., Ragsdell, K. M., OPTLIB An Optimization Program Library, University of Missouri-Columbia, 1984, pp.. 102-115.

13. Madhav S. Phadke, Quality Engineering Using Robust Design, 1989.

Obstacle Avoidance in a Manufacturing Based Modeller

I I Esat and R Lam

Department of Mechanical Engineering,
Queen Mary and Westfield College
Mile End Road, London, UK

ABSTRACT

Manufacturing based modellers are designed to bridge the gap between design and manufacture. The gap usually associated with the fact that geometrical information alone is not sufficient to describe all the necessary manufacturing operations. CAD/CAM integration is also difficult due to a number of other reasons such as the varying degree of 'informationally completeness' of geometrical modellers and complexity of boundary file evaluation. Constructive capabilities of solid modellers form the basis of many CAD/CAM integration techniques. Deforming Solid Geometry (DSG) employs half space primitives representing tools acting on the solid representing a workpiece. Since the half space represents both material removal and the space where tool motion is not restricted, this ensures machinability. Although DSG is a powerful method of CAD/CAM integration, it suffers from a number of weaknesses associated with the use of half planes as solid primitives. The paper outlines the problems and presents an 'obstacle growth' algorithm capable of operating with respect to a workpiece. The algorithm is implemented to improve the tool definition in an inhouse manufacturing based solid modeller to detect undercut or prevent overcut and tool-fixture clash. The paper also presents an extension of this algorithm applied to CSG models.

1. INTRODUCTION

Design and Manufacturing historically developed as two separate discipline. The isolated evolution of these disciplines is the main cause of the problems associated with CAD/CAM integration. This trend continued after the introduction of computers in each discipline. The early graphics packages were developed to generate and display drawings. Soon, the value of computer graphics for engineering drafting was recognised which led to evolution of graphics standards and latter the early CAD systems. The capabilities of these systems were relatively modest, generating/editing, storing and reproducing drawings. The data structures of such models represented the line drawings with no reference to the object the drawing meant to represent and little concern about the suitability of these systems to aid manufacturing. Similarly, CAM software developed mainly to aid NC code generation and punch tape preparation. These systems were usually text input and command driven. The early graphics systems gradually evolved into geometrical modellers gaining formalism and scientific basis for representing solids. The solid modelling theory based on regular set operations has evolved. A number of schemes addressing 'informationally completeness' was proposed. CAM also continued to evolve, mainly as NC code generation languages and often with a specific machine type in mind.

The importance of integration of geometrical modellers with design and manufacture is recognised partly due to efficiency requirements of modern industry. It is possible classify the research on the integration under two main headings, integration based on application (particularly manufacturing) oriented modellers and intergration of general geometrical modellers with manufacturing systems. In general, the integration usually refers to the second type. This may be expressed as a problem of translating geometrical information into manufacturing information. The degree of integration depends on the type of modeller and application area, but there are some common problems. Geometrical modellers contains primerly geometrical information. Manufacturing, beside geometry data, requires non geometrical information. The boundedness of geometry does not equate the realisability of the corresponding physical object. But more important of all, that there exist no analytical frame for the integration problem.

Over the last decade researchers investigated suitability of various methods for CAD/CAM integration. *Expert systems and AI* tools were used as front end to solid modellers providing manufacturing expertise and advising on manufacturing implications of the solid operation., in boundary file evaluation and recognition. Geometrical algorithms such as *Volume Decomposition* has been developed for decomposing the objects into simpler elements for engineering analysis. This method can also be used for material removal operation by associating decomposed volumes with available cutting tools. In *'Guiding tool'*, a common method in industry, the machine operator interactively picks the line segments representing the required boundary cutting from the display device of a cad system. *Manufacturing based modellers* attempt to generate manufacturing information as an integral part of the geometry representation. The modellers based on the notion that material removal may be represented by the set difference operation between the half space primitives and solid representing tools and workpiece respectively. The analytical basis of this approach, Deforming Solid Geometry (DSG) concept has been proposed by Erbab (1), (2).

2. BASIC FORMALISM of DSG

A cutting tool is defined by Erbab as a triple $\langle N, F, P \rangle$, where N, F and P are sets of points constituting a tri-partition of E^3. Tool is defined as $t = \langle N, F, P \rangle$ and void region of a tool, cutter side is given by $V_t = N \cup F$. The set oparations between tools $t_1 = \langle N_1, F_1, P_1 \rangle$ and $t_2 = \langle N_2, F_2, P_2 \rangle$ are given as:

$$P = P_1 \cap P_2 \qquad F = (F_1 \cup F_2) \cap \overline{P} \qquad N = (F \cup P)'$$

The tool $t = \langle N, F, P \rangle$ is said to be valid if P is an open set and F is a patchwise smooth surface. An example of tool manipulation is shown in Fig 1.

In DSG formalism, W represents the set of all workpieces and T represents as set of all tools. Application of a tool t ∈ T to a workpiece u ∈ W, which implies a material removal operation may be defined as,

$Apply(t,u) = u - Vt$, Where Vt is material removed by tool t.

The DSG, as proposed above, ensures that modelled objects are manufacturable. This is a statement of the method's potential. There are several problems concerning its implementation. In fact the practical value of DSG in its original form is questionable. It is possible to make following observations about DSG.

a) The main contribution of DSG over CSG is associated with the tool definition and analysis of validity of tools.
b) The validity of tools described in terms of tool geometry alone.
c) The contribution of this concept over CSG is somwhat limited since many solid modellers offer half space primitives as the standard.
d) Although tool definition is an integral part of DSG modellers, this conceptual tool may not have its corresponding real equivalent.
e) The DSG concept pays very little attention to the fact that the consequences of set operations between tools may be meaningless when translated to physical tools.
f) Half plane operations increases the danger of the tool clash.

3. A PRACTICAL DSG

A modified version of DSG concept was introduced by Esat and Lam (3). This modeller was designed to eliminate some of the problems described above. The problems associated with tool operations are partly eliminated and partly avoided by adopting a sequential tool application method. The modeller, instead of using general half space primitives as tools, employs predetermined parametised tools described in terms of bounded tools, Fig 2. The advantage of using half space was to describe a space where tool movement was not contrained. Replacing this with ordinary bounded tool definitions would eliminate the fundamental premise behind machinability. Instead the modified DSG uses bounded tools with an 'entry face'. The bounded tools, ensures that the outcome (workpiece) is also bounded. The entry face concept investigated by many researches in the area of boundary file recognition. The entry face in the improved DSG, is the face through which tool is free to enter and operate. Fig 3 shows a turning operation, where relevant section of a workpiece is operated by a facing tool. The parametised tools in the DSG modellers developed by authors, incorporate parametised tool paths which automate the complete process.

4. OBSTACLE GROWTH ALGORITHM

The growth algorithm proposed by Lozano-Perez (4), to aid robot path planning. This is a technique of growing the obstacles, and reducing the robot to a point. The method is useful in path planning due to availability of algorithm to find the shortest path of a point movement. The Fig 4, demonstrates the basic principle of the obstacle growth. R represents a 2 D robot, triangular in shape. To simplify the demonstration a rectangular obstacle is chosen. The growth procedure is simply the trajectory of a point P on robot R due to the translational motion of R in contact with the rectangle. The original growth algorithm is proposed for convex robots and obstacles.

5. INCORPORATING GROWTH ALGORITHM WITH DSG

According to the DSG tool definition, the cutting face or tip of the tool should be able to follow the half space boundary. The extended DSG ensures this, the parametric tools are defined in terms of tool angles therefore the cutting face of the tool coincide with the half space boundaries. Early experiments carried out using the simple DSG modeller developed previously, showed that this does not always avoid the tool holder, or even the non-cutting part of the tool hitting the workpiece. The problem was associated with the selection of the tool angles. The tool angles which describe a half space could not represent the geometry of tool body or the tool holder since this would require a concave tool definition. This problem can partially be solved if the convex hull of the tool and its holder set is employed as the half space tool, but this would exclude a large class of solutions. Any other proposal to be effective, should be capable of distinguishing the tool holder as a moving obstacle.

Two approaches proposed in this paper to solve tool clash problem at two levels. The first proposal designed specially to operate with the DSG or extended DSG. The growth is performed relative to the tool tip with respect to tool holder and non-cutting edges of tool body. The growth is carried out with respect to the workpiece after each tool application. The growth algorithm applied to DSG situation, involves the relationship between, volume, representing material to be removed, the final shape to be produced and obstacles represented by fixtures and clamps.

The first approach require solid definition of work clamp or fixture Fx, workpiece Ut at each stage of operation and solid definition of the physical tool together its half space counterpart. The algorithm does not require to distinguish between the tool holder or tool body or the cutting face of tool. The growth algorithm together with the following constraints ensures machinability.

$(\forall i)$,

$$U_{i+1} = U_i - {}^*t_i \quad (1)$$
$$(U_{i+1} \cup Fx) \cap {}^*t_i = 0 \quad (2)$$
$$G((U_{i+1} \cup Fx), T_i, p) = 0 \quad (3)$$
$$t_i \cap {}^*U_n = 0 \quad (4)$$

The first statement defines the workpiece U at stage at $i+1$, the second equation states that the halfspace tool does not intersect with combined workpiece at $i+1$ and the fixture. Another way expressing this is to state that the fixture has no intersection with tool t_i. The next statement is the growth function G. $G(S_1, S_2, p)$ is an algorithm for growth of S_1 with respect to p a point on S_2. This constraint equation states that the result of the growth of the combined workpiece at $i+1$ and the fixture with respect to the physical tool T_i should be null. It is important to note that the growth algorithm should be able to operate on concave geometries. Finally, the last equation states that at no stage the tool i should have any intersection with workpiece n, the final machined component. Fig. 5 demonstrates a situation where $G((U_{i+1} \cup Fx), T_i, p) \neq 0$.

In the manufacturing based modellers machinability is ensured, because the set difference operation correspond to material removal and the operator is aware of this fact. That is, the manufacturing skill of the designer is crucial to the success of the modelling. The second method represents an attempt to use the properties of the half spaces in describing material removal action of tools in non-manufacturing based modellers. This should eliminate one of the main drawbacks of DSG, their narrow and restrictive definition of constructive capabilities. This approach will create the basic foundation for a general integration of solid modellers with manufacturing systems. The basis of this approach is an algorithm for growth of the final product with respect to half space tools. The relevant definitions and algorithm are as follows,

Volume to be removed at stage i = $(U_i -^* U_n)$

t_i applied on U_n gives growth region(s), these regions are not disjointed.

The effective growth due to the action of i the tool is given by,

$$G_i(U_n, t_i, p) -^* \bigcap_i G_i(U_n, t_i, p)$$

Therefore the workpiece at stage $i+1$ is given by,

$$U_{i+1} = U_n \cup \{G_i(U_n, t_i, p) -^* \bigcap_i G_i(U_n, t_i, p)\}$$

This representation is necessary, since material removal by t_i can be measured only through the growth operation. Finally the following algorithm should generate all the possible tool operations,

While $U_n \cap^* U_i \neq 0$ **do**
$$U_{i+1} = U_n \cup \{G_i(U_n, t_i, p) -^* \bigcap_i G_i(U_n, t_i, p)\}$$

6. CONCLUSION

The paper introduces the basic principles behind DSG formalism. An extended version of DSG is also outlined. The advantages and disadvantages of DSG is discussed. The paper describes, the way, these disadvantages were eliminated by linking the set theoretic tool definition with the physical tools. The major improvement offered in this paper over the extended DSG is an improved tool validity check through the use of the growth algorithm. The paper also introduces a new algorithm based on an extension of DSG tool definition. This algorithm is capable of operating processing CSG modellers.

REFERENCES

1. Arbab F., "Requirements and Architecture of a CAM-Oriented CAD system for design and manufacture of mechanical parts", PhD dissertation, University of California, Los Angeles, 1982.

2. Arbab F., Cantor D., Lichten L. and Melkanoff M.A., "The MARS CAM-Oriented modelling system", Proc. Conf. CAD/CAM Technology in Mechanical Engineering, 1982.

3. Esat I. and Lam R., "A manufacturing based modeller in 2D", Sixth International Conference on Computer Aided Production Engineering, London, Nov. 1990.

4. Lozano-Perez T. and Wesly M., "An Algorithm for Planning Collision Free Paths Among Poyhedral Obstacles", CAMC Oct., 1979, Volume 22, No.10.

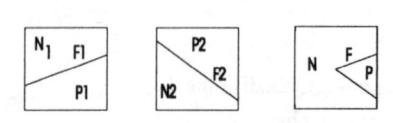

Fig. 1. Tripartition of E^3 space

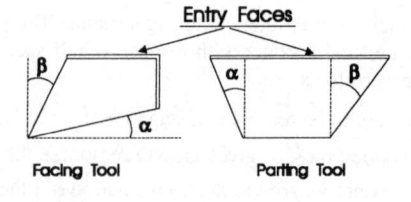

Fig. 2 Facing and parting tools in turning operation

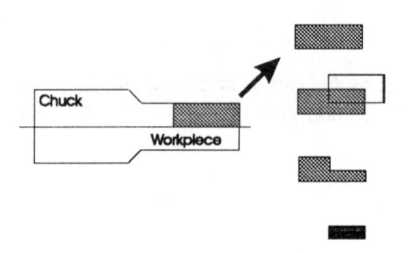

Fig. 3 Workpiece and a facing tool operation

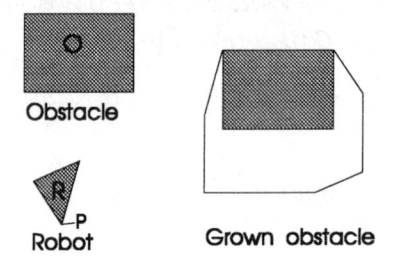

Fig.4 Rectangular Obstacle, triangular robot and obstable growth with respect to a point P on the robot

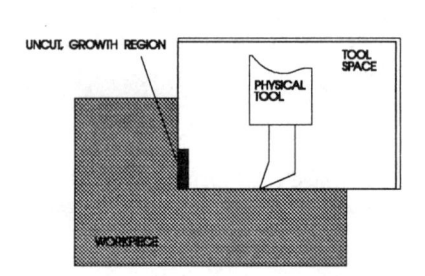

Fig. 5 The growth algorithm applied to workpiece, tool space and real tool set

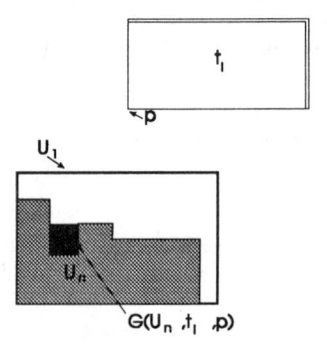

Fig. 6 Thegrowth algorithm applied to Tool space and CSG object set

Flexible Assembly Cell Simulation - A Tool for Layout Planning and Cell Controller Development

E. Freund, H.-J. Buxbaum

Institut für Roboterforschung (IRF), Universität Dortmund
Otto-Hahn-Strasse 8, W-4600 Dortmund 50, Germany

Summary
Robots in the industrial factory plant build the backbone for CIM, which is outlined to be the factory automation strategy of the future. This paper deals with the planning of robotic work cells for flexible assembly applications. The tasks of detailed cell design and cell controller development can be supported significantly by the use of today's simulation tools. This is examined on examplary applications for flexible assembly. Both the presentation of the cell components arrangement and the animation of the production process are realized. The control message flow is simulated too for the purpose of implementing a modular cell controller system. The cell controller coordinates the components and additionally offers an open interface to control systems beyond cell level. Full production flexibility by individual product identification as well as universal cell configurability are characteristic for the cell controller system.

Introduction
In the past the major applications of robots were production plants for large batches with very small product variations. The actual trend to small batches, large product variaty and far-reaching considerations to individual customer requests shows the necessity to develop new factory automation strategies with the aim to enlarge the production flexibility. A migration from normal machine control technology to a comprehensive factory wide planning, coordination and control structure is indicated. The robot itself is a suitable device for the purpose of flexible automation, if only because of its kinematic structure.
The term of flexibility denotes, in this context, to use the production facilities for families of products. The variants of the product families can be manufactured by means of programming or data transfer. It is necessary to have the opportunity to reequip the machinery, so that product changes do not require a lot of time. The overall operational live time of flexible production facilities should exceed the life cycle of the actual product.
The use of computer based simulation systems allows offline investigation of the production facilities and the automation process. This gives the chance to check the real-

izability and to optimize the technology in an early planning state. Also the development of application specific control software as well as interfaces to other CIM systems can be started, proved and valued.

Flexible Assembly Work Cell
The flexible assembly work cell is the smallest cooperating machinery unit in the factory floor which carries out entire assembly operations on unique products autonomously. The objective of those operations is to bring a product from one defined state of manufacturing to the next. A work cell usually consists of a few automation devices such as robots, machines and transport systems. The device controllers are coordinated by a cell controller, which is an order driven system based on state table process plans [1]. The assembly orders are emitted by a hierarchical higher authority, which may be a factory control computer system or a human worker as well. The job of the cell controller is the supervising of the assembly process by coordinating the device controllers by task commands.

The complete production process is described by the process plan. This plan arranges the sequence of the concerned automation devices and includes corresponding parameters for the device controllers. For a batch of unique products - as a single variation of the well known product family - a particular process plan is needed. The demand for flexibility makes an assignment of an arriving product to the corresponding process plan inevitable. This assignment is done by product identification at the very beginning of the manufacturing process for each unique product.

The process activities within the flexible assembly work cell can be devided to three fundamental functions
- product identification,
- internal logistics,
- assembly process.

The product identification must be done everytime a new product arrives in the work cell for assembly. The tasks of internal logistics are the transport and the storage of products and parts within the work cell. The assembly process itself is delegated to assembly devices - notably robots for reasons of flexibility. Fig.1 shows the characteristic control hierarchy in a flexible assembly work cell.

Layout Planning
The computer based layout planning will be demonstrated on a flexible work cell in a printed circuit board (PCB) production line. A typical application for assembly robots in the PCB production is the odd component placement, components which are hard to install because of their large complexity. This placement task requires, next to high

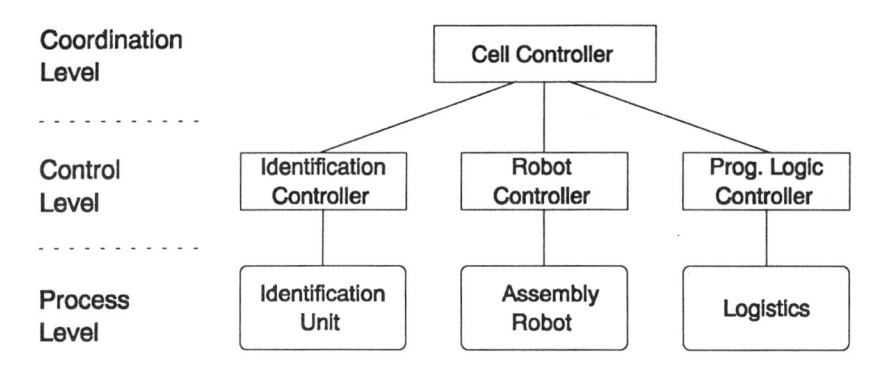

Fig.1: Control hierarchy in a flexible assembly work cell

precision robotics, sensors and tools a far-reaching quantity of flexibility. It is to be automated by a SCARA type assembly robot, integrated in a line-structured product transport system. The components will be supplied for assembly by a parts magazin. Fig. 2 shows the CAD model of the flexible assembly cell with the elements identification system, transport system, assembly robot and parts magazin [2,3].

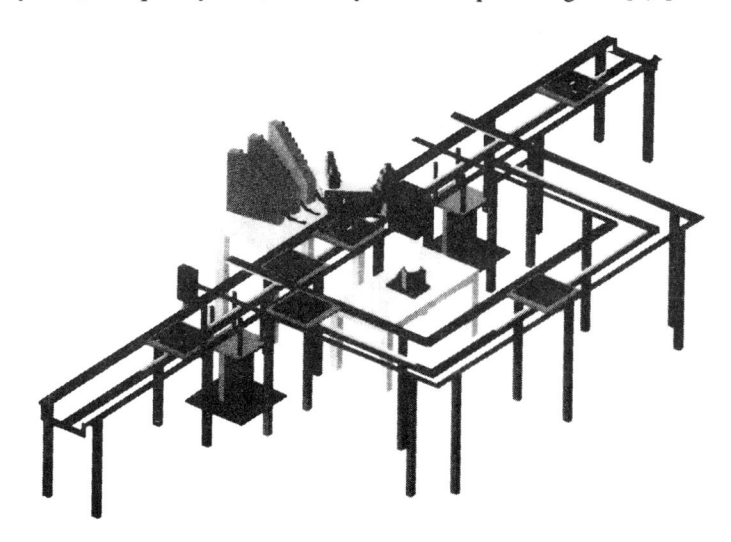

Fig. 2: CAD model of the flexible assembly cell in PCB production

The cell layout takes into consideration, that some PCBs must not be handled by the robot. For those PCBs a bypass structure is added to the transport system, so that the assembly process can be continued without interruption. The work cell has defined import and export points for the PCBs. The product flow direction is displayed from the left side to the right. Incoming PCBs will be sensed and identified automatically by

a barcode reader system. This enables the individual processing for each product (local flexibility). The bypass structure allows to remove certain products from the sequence which do not need to be handled by this work cell (global flexibility). The assembly process within the work cell is as follows.
- A PCB reaches the import point of the work cell and will be identified.
- A process plan is assigned to the PCB. The cell controller knows now how to handle the PCB.
- If no component placement is required for processing, the PCB leaves the work cell via the bypass structure.
- If components have to be placed, the PCB is transported into the robots workspace.
- The robot performs the job, which is described by the individual process plan.
- The PCB is transported to the export point and leaves the work cell.

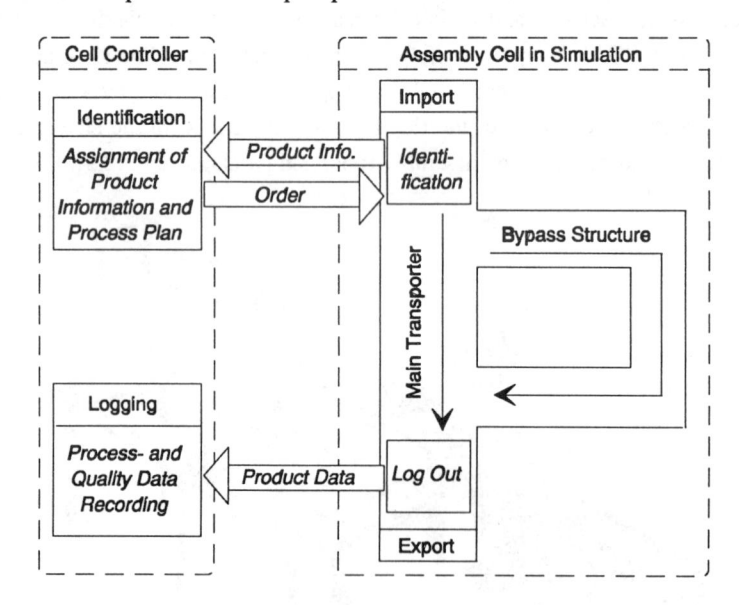

Fig. 3: Coordination of the assembly cell by the cell controller

Fig. 3 shows the coordination of the assembly process by the cell controller. After the PCB arrives, the identification data is reported to the cell controller. The cell controller assignes a process plan to the arriving PCB and coordinates the succeeding assembly process according to this plan. Before the product leaves the work cell the cell controller will be informed. Statistical product data which has been gathered during the assembly process will be stored for purposes of offline analysis.

Another example of a computer based layout planning is given in Fig. 4. Here a multi-robot system based flexible assembly cell in a robots-build-robots factory project is exposed. The robots perform a coordinated operation in the assemblage of robot control systems. After the cabinet identification the first robot in the flow line prepares the computer control system rack and the servo box, the second one configures the robot controller by attaching those components and further elements to the cabinet. In an additional manual workstation the wiring and testing of the system is done.

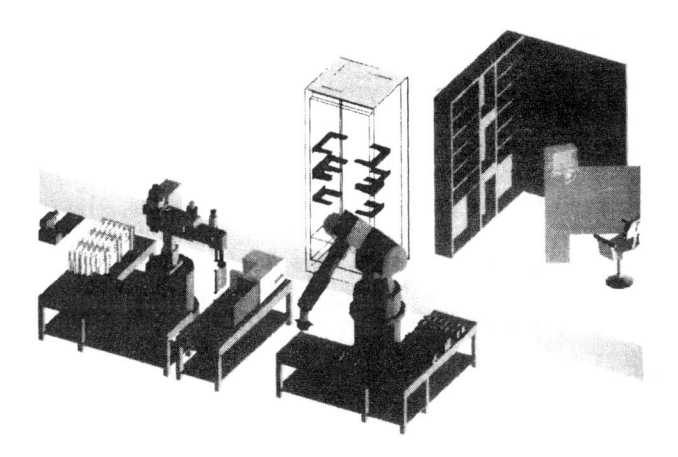

Fig. 4: CAD model of a multi robot assembly work cell

Cell Controller

The cell controller coordinates the assembly process in the work cell by instructing the device controllers. The significant variance of the diverse automation devices and the demand to universal applicability leads to a control concept based on a library of specific device drivers. For each device in the work cell configuration a specific driver has to be installed. A defined communication protocol between the device drivers and the controller kernel gives the opportunity to standardize already on the level of the cell controller's application software and control interfaces. This protocol allows to initialize the device, to start any program on the device controller and to read status variables on the device controller. Full production flexibility via free programmability of the process sequence in conjunction with individual product identification on the one hand and universal cell configurability via integration of device specific driver software on the other hand are characteristic for the cell controller system.

The cell controller works on assembly orders which are given by an operator or a factory control system. An assembly order specifies the identification range of products which belong to this order and includes a reference to a process plan. The process plan points out all process steps to be done on a product which fits in the accompanying assembly order's identification range. An examplary process plan for the flexible odd component placement cell in the PCB production is shown in Fig. 5. Related to the process plan in each step a program is started on the associated device controller. After the ready message is received, the execution process continues with the next step. A conditional branch depending on the execution result of the previous step is possible to handle error states as well as to allow intelligent processing.

Process Step Number	On Result of Previous Step	Start Device Name	Start Program Name	Next Process Step Number
1		Transporter	ImportToNode	2
2		Transporter	AssemblyPosition	3
3	NotOK	OperatorConsole	PositionError	3
3	OK	Robot	Program2705	4
4	NotOK	OperatorConsole	RobotError	4
4	OK	Transporter	Export	End

Fig. 5: Process Plan

Conclusion

The design of robotic work cells in flexible assembly applications as well as controller development and application programming can be done in an early planning state by the use of today's simulation tools. Next to an increase of planning security by various checking and optimizing possibilities also a notable reduction of development time can be achieved. The cell controller concept is independent of the process or the product and therefore convertable to other fields of application. The open interfaces permit easy link up to the logistic and production control systems which are already employed in a factory.

References

1. Buxbaum, H.-J. and Hidde, A. R.: "Flexible Zellensteuerung - Bestandteil eines produktunabhängigen Fabrikautomatisierungskonzepts". Werkstattstechnik 80 (1990), 133-136 and 262-264.

2. Müller, F.: "Erstellung und Integration eines CAD-Modells zur Simulation von Produkt- und Informationsfluß in Arbeitszellen flexibler Fertigungssysteme". Institut für Roboterforschung, Universität Dortmund (1990).

3. Freund, E. and Buxbaum, H.-J.: "Robotergestützte flexible Montage graphisch simulieren". Arbeitsvorbereitung 28 (1991), 29-32.

4. Buxbaum, H.-J.: "Fortschrittliche Robotersysteme für Montage und Handhabung". Proc. 3. Industrierobotertagung - Handhabungstechnik, Dortmund (1991).

FMS Management Optimization Using Physical Simulation

G. Cardarelli, A. Dentini, P.M. Pelagagge
Faculty of Engineering - University of L'Aquila - Italy

SUMMARY

The work examines the causes that can reduce the productivity of a Flexible Manufacturing System with the aim of defining the management criteria capable of limiting their effects. The study has been carried out by means of experimental analysis on a physical model that can reproduce in scale or in functional analogy the different FMS configurations.

INTRODUCTION

In this work the first results of a study on the management logics of FMS are presented. The study is based on the analysis of the causes that can determine non-productive times on the workstations. There have been considered productive scenarios that underlined the effects due to a specific category of causes and implicitly eliminated all of the others. This way of proceeding has allowed for the individualization of some control criterion, the validity of which is not closely linked to the particular production situation. The study has been carried out by means of physical simulation on an experimental equipment capable of representing several FMS configurations [1]. The simulation of complex manufacturing systems by means of a laboratory equipment that reproduce them in scale or in functional analogy, presents advantages with respect to other methods of analysis [2-18]. In particular, the high level of visualization of the experimental system allows an immediate evaluation of the effects of the management choices, and a rapid improvement of the criteria that has determined them.

ANALYSIS OF THE CAUSES OF NON-PRODUCTIVE TIMES

The minimization of non-productive times represents one of the simplest objective functions for much of the scenarios that with major frequency are found in the FMS production [19]. To define the management criteria that follows such function, it seemed useful to carry out an analysis of the links between the non-productive times of the single workstations and the causes that determine these non productive times. A distinction can be made between:

      primary non-productive times
      induced non-productive times

The first have been directly produced on a generic workstation by:

- unbalanced work loads,

- unavailability of FMS subsystems for breakdown or preventive maintenance,

- unavailability in the FMS of ready jobs to be loaded on the workstation,

- inpossibility to unload the workstation for insufficient capacity of the interoperational storage,

- delay in the loading or unloading of the workstation due to interference with the transport system.

The primary non-productive times can then produce induced non-productive times on other workstations, due essentially to the unavailability of ready jobs or to the insufficient capacity of the interoperational storage.

In table I the links between the causes of non productive times and the management factors have been schematically shown.

The choices of the part-mix and the maintenance policies help to define the scenario in which the FMS operates, but they are not usually part of the aims of the real time control system. To point out the phases where the control software has to operate some choices, it is appropriate to distinguish the scheduling from the material handling system management. The first phase is represented by the jobs scheduling, consisting of two fundamental moments [20-22]: loading and dispatching. Loading involves making new jobs accessible for the production system, dispatching involves the choice of the job to be loaded on a workstation or, eventually the workstation to be assigned to a job[1].

MANAGEMENT FACTOR	CAUSES OF NON-PRODUCTIVE TIME	
	PRIMARY	INDUCED
part-mix	unbalanced work load	
maintenance policy	subsystem unavailability	
scheduling	unavailable ready job	unavailable ready job
	unavailable interoperational storage	unavailable interoperational storage
material handling	load-unload delay	

Table I - Links between non-productive time and management

From every dispatching choice there is evidently the necessity to carry out a transport operation. Naturally queues can be created for the material handling system. The second phase, subsequently indicated as *handling*, will involve the choice of the transport operation that has to be carried out first. To point out the influence of these control phases on non-productive times, particular scenarios have been examined that would render pratically negligible the effects of all the non-productive times causes, except those exclusively influenced by the loading or the dispatching or the handling.

PHYSICAL MODELLING OF THE PRODUCTION SCENARIOS

For the study a physical simulator has been used, capable of representing in functional analogy FMS configurations operating in diversified productive scenarios. In particular the experimental system allows to reproduce workstations with or without pallet- change devices, robot or shuttle based material handling systems, carousel interoperational storage, sequential input-output warehouses. The control software is structured in a way that allows for experimentation of several control criterion. In all cases, the phases that feature the production progress control are kept distinct. The loading is run directly by the operator who, according to defined logics, loads the new jobs on the input warehouse. Messages on the PC video allow to adopt logics also depending on the real time production situation of the system. For the dispatching, the control software manages a queue with the transfers, of all the jobs present in the system, from the actual position[2] to the next in the working cycle.

If present, all the alternative transfers at alternative machines or at interoperational storage will be put in queue. With the occurring of events and according to criteria depending on the logic adopted, the dispatching chooses the transfers to carry out and

---

(1) In the case where the next operation can be done on alternative machines.
(2) The positions are identified with the stations, the input-output warehouse and the interoperational storage.

puts them in the queue of the transport system. The handling chooses from this last queue the sequence of transfers.

TESTS AND RESULTS

Handling. The first aspect analyzed regarded the effects of the handling on the non-productive times caused by interferences with the transport system. Bearing this objective in mind, a scenario was represented that eliminated not only all of the other causes of primary non-productive times but also the induced non-productive times. In particular, stops for breakdown are not considered and the interoperational storage is of highly elevated capacity that always permits the workstation unloading. Furthermore, there is, in every moment, at least one ready job for the workstation loading. This is obtained using working cycles where every workstation can receive at least one job directly from the input warehouse, which is always kept full. As foresaid the dispatching queue contains the transfers of all the jobs present in the system. Just for the jobs in the input warehouse, the dispatching queue contains only one transfer for every type of job; at the moment in which a job is taken from the input warehouse, the transfer of a new job of the same type is put in queue. In the hypothesized scenario the non-productive times do not depend on the dispatching. The transfer choices are therefore carried out by the handling with the aim of minimizing the interference with the transport system. A handling rule based on transfer priority has been experimented. The priority has been calculated as the sum of contributions linked to the situation of the loading and unloading workstations. The first priority scale has been based on the transfer ability to reduce non-productive times or even their probability. There have been considered for both the loading and unloading workstations, 5 classes of situations with a decreasing priority (figure 1):

1- a transfer is necessary and sufficient to immediately start the machining,

2- a transfer is necessary but not sufficient to immediately start the machining,

3- a transfer is necessary and sufficient to allow a new machining immediately after the one being carried out,

4- a transfer is necessary but not sufficient to allow a new machining immediately after the one being carried out,

5- a transfer does not give immediate effects.

The first tests have been carried out using *immediate transfers*: at the moment in which the transport

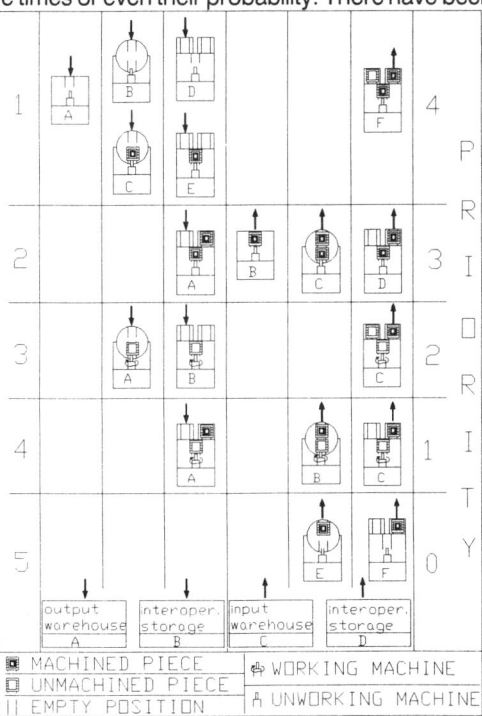

Figure 1

system is available the priority of all of the immediately feasible transfers[3] is calculated and that of the highest priority is carried out. Obviously if the transport system is unoccupied and a transfer becomes feasible, this is immediately carried out.

Being immediately highlighted by the physical simulation, with these criterion some transfers choices can be less effective:
- the immediate transfers toward the interoperational storage could heavily increase the transport system workload,
- the low priority assigned to the departure from the interoperational storage favoured the overtaking of jobs of the same type,
- the low priority assigned to the transfers toward the output warehouse favoured the increase of the flowtime and of the WIP.

The behaviours observed from time to time have allowed a rapid refinement of the transfer priority contributions, the resulting scale is schematically represented in figure 2. It appeared convenient furthemore to delay the transfers toward the interoperational storage until the moment in which they became strictly necessary for the start of a new machining on the departing station.

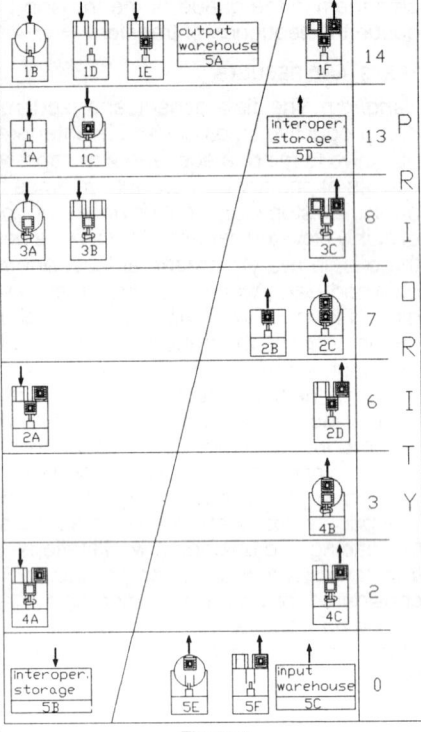

Figure 2

Dispatching. Likewise for the handling, a scenario that automatically eliminated all of the non-productive times not caused by the dispatching was represented. With respect to the previous scenario there has been considered:

a - assigned part-mix,
b - any working cycles (not all stations necessarily receive jobs from input warehouse),
c - stations with same total workload[4],
d - in the input warehouse (and in the dispatching queue) a job of any type is always present until the assigned quantity of the part-mix is reached,
e - the transport system has a very low utilization ratios.

---

(3) The job is in the output buffer of the departing station and the input buffer of the destination is free.

(4) For the workload balancing there has also been considered, besides the machining time, the time strictly necessary for the job change: pallet-change time for machines provided with sequential I/O buffer and the sum of the average times of unloading and loading transfers for machines without pallet-change device.

The last hypothesis pratically eliminates the effects of the handling; the hypothesis -d- defines a loading rule that does not limit the choices of the dispatching even if it does not consent any control of the *mix-deviation*. The mix-deviation is defined as the deviation of the part-mix percentage of each producing moment with respect to the assigned part-mix percentage. The first tests have been carried out using a dispatching rule of the FIFO type. As aspected, two effects occur in these conditions:
- increase of non-productive times due to the unavailability of ready jobs for the stations that do not directly receive jobs from the input warehouse,
- incontrolled increase of the WIP.

The tests have shown that generally, the major non-productive times emerge when production progresses with a strong mix-deviation that does not evidently guarantee the workload balancing within a generic time interval. In the subsequent tests great care was taken to maintain the production progress of any job types closest to the assigned part-mix percentage, by introducing a job priority contribution proportional to the mix-deviation. As a result a modest reduction of non-productive times compared to FIFO dispatching rule has been obtained. The description capacity of the physical simulation has allowed to observe that the variation of the job priorities occurs with delay respect to the dispatching choices that it should have influenced. In the next tests the job priority contribution has been based on the deviation of the percentages of jobs already taken from the input warehouse with respect to the assigned part-mix percentage. The observation of the dispatching choices effects has shown that it was preferable to always assign a higher priority to the transfers from the stations and from the interoperational storage than to the transfers from the input warehouse. This dispatching rule allows for a good control of the production progress together with a remarkable reduction of the non productive-times. Nevertheless, the physical simulation has allowed to underline how this dispatching rule also makes choices that subsequently produce non-productive times, which can occur when:

- a job is transferred from the input warehouse to the input buffer of a working machine, and
- another job becomes available for this machine before the end of the machining.

However, in some cases it has been observed:
- after the initial transient the system operates with a periodic stationary state until the final transient,
- non-productive times occur only during initial and final transient,
- the taking out sequence of the jobs from the input warehouse periodically repeats itself,
- the job group taken from the input warehouse, at every period, respects the percentage of the assigned mix.

This last observation suggested the verification of the system's behaviour when the loading of the job group respecting the assigned mix is imposed.

Loading. With reference to the previous scenario, a job group respecting the percentage of the assigned mix is loaded into the input warehouse. Only when the last job of this group is taken from the input warehouse, an identical group is loaded again. Several tests with different part-mix, working cycles and dispatching rules, have been carried out. In all cases the system reaches a periodic stationary state without non-productive times. The part-mix percentages, the working cycles, the dispatching rules and the numerosity of the job group loaded, influence only the initial and final transient duration and the non-productive times emerging during these transients. The maximum WIP and therefore the suitable capacity of interoperational storage also depend upon the numerosity of the job group loaded.

CONCLUSION

In the study it has been observed that a loading rule based on a job group respecting the assigned part-mix allows to reach a periodic stationary state without non-productive times. For this to happen it is sufficient that the FMS fulfills the following conditions:
- no stops for maintenance,
- balanced total workload,
- low utilization ratio of material handling system,
- suitable capacity of interoperational storage.

It is evident that the hypothesized scenario does not occur often in an actual production situation, but it can be used as a ideal reference system. The performance of FMS in other scenarios can approach the reference system performance by means of appropriate control rules. With this aim, control logics are being studied that consider the system inefficiencies as effects of perturbations on a periodic stationary state.

REFERENCES

[1] G.CARDARELLI, P.M.PELAGAGGE, *Physical Modeling of Flexible Manufacturing Systems*, 22th International Symposium on Automotive Technology and Automation, Florence, may 1990.

[2] Z.S.STERN, D.G.GINZBURG,*Physical Simulation of a Two Stage Control Algorithm for an FMS*, Comput. Ind. Engng., vol.18, no.4, pp.535-545,1990.

[3] J.JUNG, S. CHANG, *An Interactive Animated Simulation Model for FMS*, 1th International Conference on Automation Technology, Taipei, July 4-6, pp. 45-54, 1990.

[4] G.CARDARELLI, P.M.PELAGAGGE, *Robotics Laboratory Experience*, 1th International Conference on Automation Technology, Taipei, July 4-6, pp. 349-354, 1990.

[5] S.S.MARTIN, R.H.CHOI, *Evaluation of Assembly Routines With Multitasking Execution in a Physical Robotic Cell*, Comput. Ind. Engng., vol.17, nos.1-4, pp.221-226, 1989.

[6] R.M.SUNDARAM, L.BLAIR, *Enhancement of Robot Work Envelope in a Flexible Manufacturing Cell*, Comput. Ind. Engng., vol.17, nos.1-4, pp.209-214, 1989.

[7] C.S.CHEN et al., *Design and Development of a Physical Simulator for Robotic Palletization*, Comput. Ind. Engng., vol.17, nos.1-4, pp.202-208, 1989.

[8] B.E.WHITE, *Using Model Manufacturing Systems as an Aid to Understanding Computer Integrated Manufacturing Systems*, Comput. Ind. Engng., vol.17, nos.1-4, pp.196-201, 1989.

[9] J.R.DROLET, C.L.MOODIE,*A State Table Innovation For Cell Controllers*, Comput. Ind. Engng., vol.16, no.2, pp.235-243, 1989.

[10] S.PAIDY, *Realtime Simulation of Manufacturing Systems*, 3rd International Conference on CAD/CAM Robotics and Factories of the Future, vol.1, pp.102-106, 1988.

[11] R.G.HURLEY et al., *The Use of Physical Model Simulation to Emulate an AGV Material Handling System*, IEEE Int. Conf. on Robotics and Automation, vol.2, pp.1040-1045, 1987.

[12] B.E.WHITE, *Using Computer Controlled Physical Models to Analyze Computer Integrated Manufacturing Systems*, Proc. 8th An. Conf. on Comput. and Ind. Engng., march 1986.

[13] R.H.CHOI, E.M.MALSTROM,*Physical Simulation of Work Scheduling Rules in a Flexible Manufacturing System*, Proc. 8th An. Conf. on Comput. and Ind. Engng., march 1986.

[14] K.H.DIESCH, E.M.MALSTROM, *Physical Simulator Analyzes Performance of Flexible Manufacturing System*, Industrial Engineering, vol. 17, no. 6, pp. 66-77, 1985.

[15] R.E.YOUNG et al., *Physical Simulation:The Use of Scaled-Down Fully Functional Components to Analyze and Design Automated Production Systems*, Comput. Ind. Engng., vol.8, no.1, pp.73-85, 1984.

[16] S.B.O'REILLY et al., *On the Use of Physical Models to Simulate Assembly Plant Operations*, Proc. Winter Simulation Conf., Dallas, dec. 1984.

[17] T.H.AISHTON, R.A.MILLER, *Evaluation of Physical Modeling as a Tool for the Design and Implementation of Computerized Manufacturing Systems*, Instit. of Ind. Eng., 1982 An. Conf., New Orleans.

[18] S.Y.NOF et al., *Using Physical Simulators to Study Manufacturing Systems Design and Control*, AIIE Spring Annual Conference, pp. 219-228, 1979.

[19] G.CARDARELLI et al., *Computer Simulation and Management of FMC with a Materials Handling Robot*, 19th International Symposium on Automotive Technology and Automation, Montecarlo, oct. 1988.

[20] M.GARETTI et al., *On-line Loading and Dispatching in Flexible Manufacturing Systems*, Int. J. Prod. Res., vol.28, no.7, pp.1271-92, 1990.

[21] T.WATANABE et al., *The Automatic Improvement of Job-Shop Schedules and New Dispatching Rules*, Sym. on Flexible Automation, pp.1133-8, Kyoto, Japan, 1990.

[22] M.J.HENNEKE, R.H.CHOI, *Evaluation of FMS Parameters on Overall System Performance*, Comput. Ind. Engng., vol.18, no.1, pp.105-110, 1990.

A Comparison between Traditional Methods and New Models for Computerized Production Planning

Andrea Sianesi
Politecnico di Milano, Dipartimento di Economia e Produzione

Introduction

The economic scenario during the Eighties has been characterised by a progressive increase in market instability and this has meant that competition between companies has become more aggressive. The demand has become extremely variable through time, both quantitatively and qualitatively (from the point of view of mixes of products). Thus, a "competitive" company must have the capability to follow this variability, offering the market products with the specifications requested and in the quantities required. To be in a position to provide this flexibility, the production system must possess certain "hardware" type requisites, and, above all, "software" type requisites. In the first case, this involves recourse to computer aided technological solutions such as FMS, FAS, robots etc. which, however, must also be supported in terms of software. Production planning and control systems must be organised, capable of rapidly generating production schedules which can contain costs, maintain competitive margins and respond immediately to the quantitative and qualitative variations in the demand. This production scheduling function is particularly important in process industries which are highly capital intensive since, in these cases, production capacity is precisely assigned and defined and generally cannot be expanded in the short term, except in a very marginal way by costly recourse to overtime working and sub-suppliers.
Given the existence of constraints on production capacity, it is extremely important to exploit available resources to the full in order to guarantee that the system has the necessary flexibility. To achieve this aim, the first necessity is to ascertain the production capacity which really is available on the production line, that is to say production capacity, *net of the time consumed by setup*, in order to be able to choose the size of the lots to be sent into production correctly and accurately. It is well-know that this calculation is very difficult.
It must be remembered that, in the process industries, several families of different products are normally produced in the same plant and this causes problems in terms of setup. If a "family" is defined as a group of finished product codes which are similar from the point of view of production characteristics then the effects of setups of products belonging to the same family ("internal setups)

are negligable, at least at the aggregate level, in which the reference time-frame is the medium period. On the other hand, setups of different families ("external" setups) have an enormous effect. The time required to setup a production line when passing from one family to another is often quite high and consequently there is a big reduction in the production capacity which is effectively utilisable for manufacturing the products. In addition, *the time consumption due to external setups cannot be estimated a priori, before knowing the sequence in which the families will be produced*, because line setup is an operation which depends on both the family which was previously in production and the family which will subsequently be produced. As a result of these factors, we can affirm that *the production capacity which is effectively available on a plant is not an input data, but it is something which must be assumed, together with the Master Production Schedule (MPS) as an output of the production planning process*. This makes Aggregated Planning an even more difficult activity since, the complications deriving from the capacity constraints inherent in the formulation of an MPS in a "single-line multi-product" situation are aggravated by the fact that it is not possible to know the extent of these constraints in advance. Thus, aggregated production planning is a critically important activity in process industries since costly production resources must be fully exploited and it is difficult to estimate with accuracy the production capacity which is effectively available because setup times are very high and also highly variable. For these reasons, it is extremely important to have models available which can accurately estimate line time consumption due to setup, so that an MPS can be formulated which guarantees the best possible compromise in terms of production costs.

The traditional approach in aggregated planning

The Master Production Schedule establishes *which* families of products are to be manufactured and *when* and *how much* of each family is to be manufactuted. It also indicates the *sequence* in which each family is manufactured on the line. In single-line multi-product situations, such as the situation in the process industries, the formulation of the MPS is a complex problem for which no optimising solution exists because of the presence of numerous objectives and constraints. As we know, the primary objective is the creation of an MPS which minimises the total production cost, the sum of inventory costs, setup, overtime and sub-supplier costs. In its turn, this objective is linked to the optimisation of the existing trade-off between setup costs and inventory costs and the limitation of recourse to overtime working and sub-suppliers. In addition, the schedule which is created must also be possible, in other words capable of being implemented with the resources available, and thus can conflict with the optimisation objectives already mentioned.

Many models have been proposed and implemented to solve this problem but, up to now, none of these models has succeeded in taking into account all the variables and all the constraints which exist in a single line-multi product situation subject to production capacity constraints. The methods which attempt to offer optimising or "lower bound" solutions (Linear programming, Wagner and Whitin [2]) are those which have to ignore the largest number of parameters while the most complete solutions are the heuristic ones, such as the Karni and Roll [3] algorithm. The most complete method implemented to date is the Aucamp model and, even in this case, the estimate of setup costs and times is not precise but approximated on average values. This restriction is due to an intrinsic limitation of the approach which has so far been adopted in the formulation of the MPS. This approach can be defined as SIS or "Sequence Independent Setup" since the setups are considered independent of the sequence in the stage of sizing the aggregated production lots. With this method, the *MPS is determined by means of two separate sequential stages in which the size of the production lots is decided in the first stage and the timing of the lots in the second*. As a result, methods of the SIS type have two major limitations which relate to an *incorrect evaluation of setup costs and times*. With the SIS approach, the first component of the setup costs (cost of lost production cost) is ignored since it is linked to the production sequence, while the second component (imputed costs) can only be evaluated as an average. Similarly, during the first stage (lot sizing) even the setup times are estimated only as averages (see, for example [1]), when they are not completely ignored ([2] [3]).

Negative effects of the traditional approach

The consequences of simplifications in the modelling of setup times and costs referred to in the previous paragraph, have negative repercussions on the optimisation of the total costs of the final schedule and its feasibility, since the planning decisions taken by means of an SIS approach are not guaranteed to be really "economical" for the production system and they generate an MPS which, in certain cases, may not be feasible in relation to the real time consumption of the line due to the setups. The higher the cost variances and the setup times of the group of products of the plant in question, the more marked are the negative effects and the penalisation of the SIS approach. As regards the minimisation of the total costs of the MPS in particular, two types of consideration apply.

1- The first relates to the opportunity costs of setup which depend on the level of saturation of the plant, in the sense that these are different from zero each time the necessary local production capacity exceeds the available local production capacity. When, in a certain period, the market requires a quantity of product higher than that which the production system is capable of supplying then a decisison must be made on how to meet excess orders. Let us assume that the set objective is to succeed in satisfying the demand which emerges in every period, then the options available are the following:

a- to bring forward the production of the output requested in period t to period $t-k$ in which there is unutilised capacity (where k is the number of periods in which production is brought forward). In this case, inventory costs are introduced and the setup costs can vary, depending on whether the product in question is already been manufactured in the period $t-k$ or all production is eliminated in the period t.
b- recourse to overtime working is made if this is possible
c- recourse to sub-suppliers is made if this is possible.

The choice of one of these options is taken on the basis of cost considerations which, in the case of SIS - type planning models, ignore the existence of setup opportunity costs. As a result, a decision which seems to be economically advantageous according to an approach which considers setup to be dependent on sequence. This concept can be clarified by an example. Let us support that we have a demand matrix with a profile of the type previously described, in which the production required exceeds plant capacity in certain periods. Let t be one of these periods and let us suppose that the Karni and Roll model [3] is used as the planning tool. The Karni and Roll algorithm proposes a heuristic planning technique based on the shifting of product quantities from one period to another. According to Karni and Roll, the unfeasibility of period t can be reduced by shifting lots to preceding periods (or subsequent periods if this does not lead to stockouts). Let us now suppose that there are several different products which can be shifted; Karni and Roll will choose the shift of the product i to the period $t-k$ which presents the best compromise in terms of additional setup and inventory costs. However, this decision does not take into account the opportunity setup cost savings which are created by reducing the unfeasibility of a period. A model with a concept similar to that of Karni and Roll, but which was also capable of knowing the production sequence of the aggregates while they were being lot-sized, would be capable of estimating those savings and could make a different and more economically advantegeous choice.

2- The second consideration with regard to the minimisation of total costs relates to the imputed costs. If the choice of the lot sizes is based on their average values then we are forced to make choices which are only an average reflection of the real structure of the setup costs and, therefore, this can lead to decisions which are not very advantageous in certain cases. Finally with regard to the feasibility of solutions generated by SIS-type models, it must be emphasised that, where provided for, these models make an estimate of "time consumption" (i.e. time subtracted from plant availability because of setups) during the sizing of the lots based on, at best, average values. In this case, there is the possibility that the production matrices which are generated create an unfeasibility after the timing stage, with consequent repercussions on the final costs.

The new method of aggregated planning

On the basis of an examination of the principal limitations of SIS approach to aggregated planning in multi product single line situations, a new methodology is proposed which eliminates the "Sequence Independent Setup" restriction and which can evaluate setup times and costs (including the cost of lost production) as a function of the current schedule sequence.
This approach has been called SDS or "Sequence Dependent Setup". *This methodology is based on the principle of carrying out sizing, sequencing and timing of aggregated production lots in a single logical stage.* This new methodology permits:
- *the formulation of a schedule which is always feasible* because it is prepared on the bais of the plant capacity which is effectively available, net of time consumption due to setup
- *better minimisation of the total costs of the timed MPS* since all the setup cost components are evaluated in a complete way and the line production sequence is known.
The new SDS methodology has been implemented by means of an expert system (named COPASES) described in detail in [4].
MELPAP [*] has been chosen as the reference model. This is an SIS-type model derived from the Aucamp model [1] and partially based on Linear Programming. This choice was made because of the high level of performance offered by MELPAP (as confirmed in [5]) and because of its intrinsic characteristics. It is, in fact, the only model implemented to date which attempts to estimate the time consumption and costs of lost production due to setup (even though based on close averages) coming close to the philosophy of the new SDS approach in this. The results of a wide series of test demonstrate that the performance of COPASES inmproves as the values of the Shift Factor increase and, in particular, that with low, medium-low, and average S.F. the total cost of the MELPAP plan is lower than that of the COPASES plan while, when the S.F. is medium-high, high and very high, the MPS's formulated by COPASES are those with the lowest total cost. However, it must also be noted that even in the S.F. classes in which MELPAP offers the better performance, the costs of the plans generated by COPASES are not very much higher

apart from the average values. Thus, the series of tests carried has validated the effectiveness of the COPASES system and permitted the identification of the optimal field of application of the model as being in the higher ranges of Shift Factor (medium-high, high and very high) that is to say when the plant is operating in conditions of stress (the level of saturation of production capacity has been assumed as being equal to 0.9).

Note

This research has been carried out in the context of a research project with 60% funding from the Ministry of Education.

References

1. **Aucamp D.C.**, 1987. A Lot-Sizing Policy for Production Planning whith Application in MRP. *Int. J. Prod. Res.*, Vol. 25, No 8.

2. **Wagner H., Whitin T.**, 1958. Dynamic version of the Economic Lot-Sizing Model. *Management Science*, Vol. 5, No 5.

3. **Karni R., Roll Y.**, 1982. A Heuristic Algorithm for the Multi-Item Lot-Sizing with Capacity Constraints. *IIE Transaction*, Dec 1982.

4. **A.Sianesi, S.Ronchi**, 1991. A new model for production planning under capacity constraints. *APICS World Conference*, Dublin, Jun 1991.

5. **A.Sianesi, M.Croci, A.Meroni**, 1991. An Improvement of the LP - based Aucamp Model for the Solution of the Multi-item Lot-sizing Capacity-constrained Problem. *Production Planning & Control (PPC)* - it's being printed.

A Simulation Study of Ethernet in Flexible Manufacturing Applications

J.L. Sevillano, G. Jiménez, A. Civit Balcells, A. Civit Breu, E. Díaz.
Dep. Electrónica y Electromagnetismo. Universidad de Sevilla.
Avda. Reina Mercedes, s/n. 41012 Sevilla. Spain.

Summary: In this paper, we study the performance of an *Ethernet*-like network in manufacturing applications. We develop an event-driven simulator and, as an example, we analyze a message based priority scheme, a modification of that provided by the i82596 LAN controller, considering a typical industrial environment.

Introduction.

One of the most important aspects of CIM (Computer Integrated Manufacturing) is the efficient networking between computers, machine controllers, robots, etc. and between departments.

For this manufacturing applications, specific protocol standards for the OSI Reference Model are specified, known as MAP specification (Manufacturing Automation Protocol) [1]. In the lower layers, MAP uses the IEEE 802.4 standard, known as *token bus*, because it provides prioritized access to the channel and an upper bounded message transmission time. So MAP developers chose token bus instead of other simpler and more widely used protocols like CSMA/CD (eg, the IEEE 802.3-10Mbps *Ethernet*). However, an efficient utilization of Ethernet-like networks would permit us to implement the communications in a Flexible Manufacturing Cell (FMC) in a cheaper and simpler way. This cell could easily be connected to a MAP network using a *bridge*. The existence of a set of protocols called TOP (Technical and Office Protocols), based in Ethernet, in such a way that MAP and TOP are fully compatible in the middle and upper layers, makes this interconnection easier.

CSMA/CD models are usually used to predict Ethernet performance. However, most authors agree that these analytical models are not good enough, especially if you require accurate descriptions or a study over a wide range of parameters values [4]. Due to the inflexibility of experimentation, we have to use a simulation approach. In our case, the simulator is simplified because we only have to model the system at the level of the Medium Access sublayer, and some aspects of the physical layer (MAP and TOP use common protocols in the upper layers).

The Simulation Study.

Simulation studies reported in the literature are of the *discrete-event* type. Usually, each node is described by means of a finite-state machine representing the functions of stations in a real network. Some times, a general module handles some continuous time processes such as carrier sense. In our simulation study only one state machine is executed, modeling the whole network. This approach gives a computationally simpler model.

The system may be in one of the following states:

    0 *Idle*: One user senses the channel idle.

    1 *Single user*: One user is transmitting on the channel.

    2 *Safe transmission*: The packet has been heard by all of other users, so they will not interfere with it. The system remains in this state until the transmission is completed.

    3 *Collision*: Two or more users are transmitting at the same time.

We do not consider the possibility of a packet whose transmission has ended, but it has not been detected by other distant users, since the Ethernet have a minimum packet length of 72 bytes, while the bus is not longer than 2500 m.

Each station is modeled by means of a list of structures which represent the packets. The elements of this structure include a pointer to the next packet, and some information like arrival time, packet size or the number of successive retransmission attempts. The rate of arrivals can be different from one station to another (thus allowing non homogeneous traffic), following a Poisson distribution.

A variable called *delay* is assigned to each node, handling the delay impossed by the collision resolution algorithm and the propagation delay. Stations must wait for a null *delay* when they want to transmit.

In a *Star* topology, the time the system remains in state 1 before passing into state 2 is τ, with τ representing the end-to-end propagation delay. The time a colliding station must wait before it senses the channel idle again is $2\tau+Jam$, plus the estimated collision detection time [2].

However, when in a bus topology, it is very difficult to determine *a priori* whether a station would participate in a collision or not (because it detects the transmission of a closer station). This is due to the utilization of a general model, which can not handle certain station functions. We overcome this difficulty assuming that the network behave as in a star topology when it is in the collision state. This approximation was also made in [3]. This is a pessimistic assumption but, when validating the simulator by comparison with experimental and simulation studies reported in the literature [4] we obtain a good approximation to the behaviour of a real net [10].

We consider that there are no packets lost because of line errors or electromagnetic noise. The buffer size of stations may be variable. Packets generated when the buffers are full are discarded.

The input of the simulator specify the number of stations and their positions along the bus, the network configuration parameters and the load characteristics. The configuration parameters were chosen following the IEEE 802.3 standard specification [5]. However, they can be altered to optimize the communication system, as if you were using an Ethernet controller such as the i82596 LAN VLSI chip [6]. The priority mechanisms of the i82596 can also be included. The simulator outputs include both queueing and system delay (mean and variance), throughput, percentage of collisions and packets lost because of an excessive number of collisions.

The program is run for a transient period before it begins to collect statistics, thus enabling the system to reach a steady state. The simulator was written in C under the UNIX O.S.

Application:

The need for prioritized access to the channel is a crucial aspect when supporting real-time applications. Standard Ethernet does not support priority functions, but the i82596 LAN controller can provide station priorities. There are two different approaches:

- Linear priority: Each station is assigned a value between 0 and 7 which specifies the number of slot times that a station ready to transmit waits after Interframe-spacing or after backoff. Therefore, a station with a high priority level can acquire the channel while the lower priority stations wait.

- Exponential priority: A value between 0-7 alters the random backoff delay, so a station involved in n collisions computes a number between $[0, 2^{\min(n+\text{Exp\_pri}, 10)})$. A low priority level is achieved using a high Exp_pri value.

These mechanisms provide an improved service for some stations at the expense of others. The linear scheme offers almost constant response time for the highest priority stations, although the performance of the low priority ones suffers greatly [7]. The exponential scheme introduces additional delay only in colliding stations, so we will choose it for our study.

However, when establishing the network traffic characteristics, we note that usually a station do not have constant timing requirements, but it depends on the message being sent[1]. Therefore, a message-based priority scheme would be more accurate.

Then, we use the simulator described above to study the performance of an alternative mechanism: a Exponential message-based priority scheme, a slight modification of the i82596 scheme. We consider a load description commonly found in industrial environments, described in [8], originated by three possible stations: workshops, telephones and factories. Each station can generate five classes of messages: files, transactions, telephones, sensors/actuators and alarms, each one with different frame length, priority (added as an extra overhead of the packet) and timing requirements. This is depicted in figure 1. To study the performance of the system when the traffic changes, these arrival rates are altered proportionally. The simulator provides statistics for each message class. In these cases, the load is defined as $G=\sum_i \sum_j T_{ij}/\vartheta_{ij}$ with $i=1,\ldots,N$ (number of stations) and $j=1,\ldots,K$ (number of messages classes), T_{ij} = message transmission time and ϑ_{ij} = time between arrivals.

TRAFFIC	LENGTH Bytes	EXP_PRI	DEADLINE (ms)	AVERAGE MESSAG/SEC/STATION		
				Workshop	Telephone	Factory
Files	2000	7	50	0.5	0	0.1
Transactions	200	6	20	10	0	2
Telephones	250	5	15	0	32	14.6
Sensors/Actuators	30	3	5	80	0	20
Alarms	30	0	0.9	1	0	0.25

Figure 1. Workload Description.

Note that standard Ethernet would not admit a packet longer than 1500 bytes. However, for our study, we will assume that it is not necessary to fragment the 2000 bytes packets. Similarly, the shorter packets would have to be filled with useless data up to the minimum packet length (72 bytes). We could avoid this last point by using the programmable controller i82596,

---

[1] A message is considered lost if it does not meet its timing requirements.

which permits us to change the minimum packet lenght. It must be taken into account that the resulting packet transmission time must exceed the *slot-time*. This parameter can be programmed as well. Note that a small *slot-time* produces shorter delays in the collision resolution periods.

In our simulation experiment, the network was comprised of 18 stations (3 clusters -workshops, telephones and factories- of 6 stations each), uniformly distributed on a 1000 m bus. We chose the standard IEEE 802.3 for the rest of network parameters. Although the real traffic would be small (Figure 1) we consider the possibility of a sudden increase of the load. Each station has only one packet buffer, in order to separate queueing delay from the delay properly imposed by the protocols.

The results for *alarm* messages are shown in figure 2, in which we also represent the deadline of these messages. At low loads, the differences between the two protocols are small, because the number of collisions is not significant, and in both cases the delay is minimal. However, for higher loads the mean packet delay for the standard method become higher than the deadline, so the system fail. For the prioritized network, however, the mean delay of alarms is always below the deadline. *Sensors* improve their performance as well but, as we said before, lower priority messages delays become longer, although for this range of loads they never reach their deadlines (unfortunately, there is not enough space for all the figures). Anyway, this is usually acceptable since the loss of lower priority classes is less crucial. Due to the imposed delays for the lower priority classes, the total throughput for the new method is lower than for the standard scheme, as shown in figure 3.

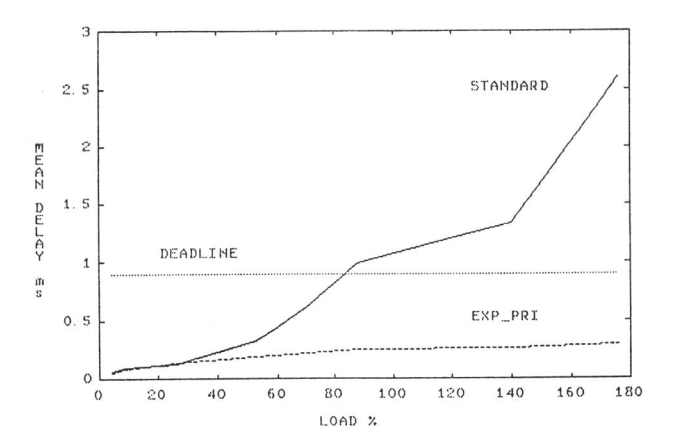

Figure 2. Load-Delay Characteristic for Alarms.

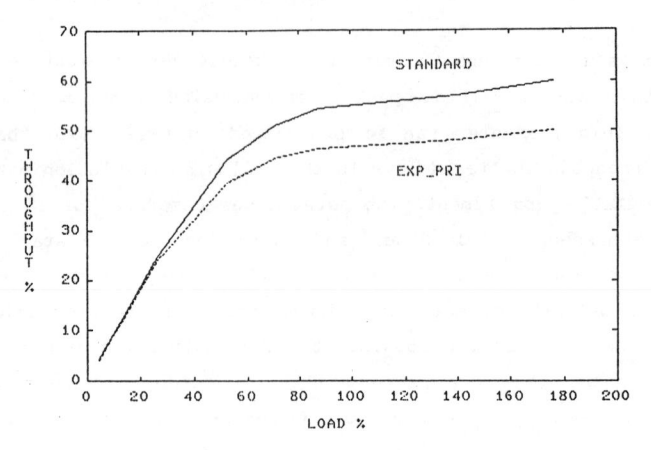

Figure 3. Throughput vs. Load.

The total percentage of lost messages decreases with the new method. However, it is well known that standard deviation of packet delays rises rapidly when throughput tends to its saturation value [9]. Therefore, there are always a small percentage of packets lost even for the highest priority classes, due to the randomness of the protocol. When this situation cannot be acceptable, deterministic networks would have to be used, for example CSMA/DCR [6] or *token bus*.

References:

[1] A.S. Tanenbaum, "Computer Networks". Prentice-Hall, 1988.
[2] F.A. Tobagi, V.B. Hunt, "Performance Analysis of Carrier Sense Multiple Access with Collision Detection". Computer Networks, vol. 4, no. 5, pp. 245-259, Oct-Nov. 1980.
[3] G.J. Nutt, D.L. Bayer, "Performance of CSMA/CD Networks Under Combined Voice and Data Loads". IEEE Trans. on Commun., vol. COM-30, no. 1, pp. 6-11, Jan. 1982.
[4] T.A. Gonsalves, "Measured Performance of the Ethernet" in Advances in Local Area Networks, Ed. K. Kümmerle, F.A. Tobagi, J.O. Limb. IEEE Press, 1987.
[5] IEEE Standard 802.3-1985. *Carrier Sense Multiple Access with Collision Detection CSMA/CD*. 1985.
[6] Intel Corporation, "82596 User's Manual", 1989.
[7] I. Chlamtac, "A Programmable VLSI Controller for Standard and Prioritized Ethernet Local Networks", in Local Area & Multiple Access Networks, Ed. R.L. Pickholtz. Computer Science Press, 1986.
[8] T.F. Znati, "Adaptive Priority Schemes for CSMA/CD LANs Supporting Real-Time Applications". Proc. of the ISMM Int. Sym. Computer Applications in Design, Simulation and Analysis, New Orleans, U.S.A., pp.128-131, March 1990.
[9] J.D. DeTreville, "A Simulation-Based Comparison of Voice Transmission on CSMA/CD Networks and on Token Buses" in Advances in Local Area Networks, Ed. K. Kümmerle, F.A. Tobagi, J.O. Limb. IEEE Press, 1987.
[10] J.L. Sevillano, "Estudio y Evaluación de redes CSMA/CD en Fabricación Flexible". Trabajo de Investigación. Universidad de Sevilla, Nov. 1990.

Application of a Learning Technique to Factory System Tuning

Peiya Liu and Robert Chou
Siemens Corporate Research,
755 College Road East,
Princeton, NJ 08540, USA

Summary
The focus of this paper is the application of a learning technique for automatic tuning of factory equipments. In the factory, it requires a multistep tuning procedure and manipulation of a large number of interactive controls to optimally tune the equipments. The tuning is often a tedious, time-consuming, temperamental, and knob-twiddling process. It is a process of adjustment of system parameters so that the best system performance can be reached through their appropriate cooperation.

The learning technique we proposed here was originally for system coordination and it was inspired by a recent learning technique[7] developed in the MIT Robotics Lab to learn to coordinate the behaviors of artificial insects. However, the system tuning and coordination share similar problem characteristics and the new learning scheme we designed has more robust and more predictable characteristics. Thus, it can be applied to factory system tuning.

1. Introduction
A learning architecture is introduced based on percepts and actions from environments. It is suitable for an intelligent agent to learn skills for achieving tasks. The learning problem is inspired by Maes[7]. We formalize this problem and develop a different robust learning algorithm based on the incremental and distributed characteristics. The architecture and its application to system tuning is discussed.

2. The Factory Tuning Problem
The **tuning process** is a process to adjust the equipments to achieve a better operating conditions. Any equipment is really a function which transforms an input in some domain, such as frequency or time, into an output in a transformed domain for analysis. The equipment tuning is with respect to that transformation.

In a factory, each operation of the factory equipment requires a multistep tuning procedure and manipulation of a large number of interactive controls to optimally tune the equipments. From the operator's point of view, the tuning process is often a tedious, time-consuming, temperamental, and knob-twiddling process. The manual tuning of one set of controls or adjustments to optimize certain characteristics is likely to influence and upset the other characteristics of the equipments optimized during a prior tuning stage. How to automate this process is then a challenging task.

3. Mapping Behavior-Based Learning to the Tuning Problem
In this section, we first introduce the model of a behavior-based learning agent. It is general enough for behavior-based autonomous agents which are equipped with learning capability. Then we discuss how the model fits the tuning problem and, finally, we describe the learning algorithm used. However, the task-dependent design issues are not addressed in this paper.

3.1. A Behavior-Based Autonomous Agent

An autonomous agent in our model consists of three major components: sensors, the central unit, and the electro-mechanical body as shown in FIG-1.

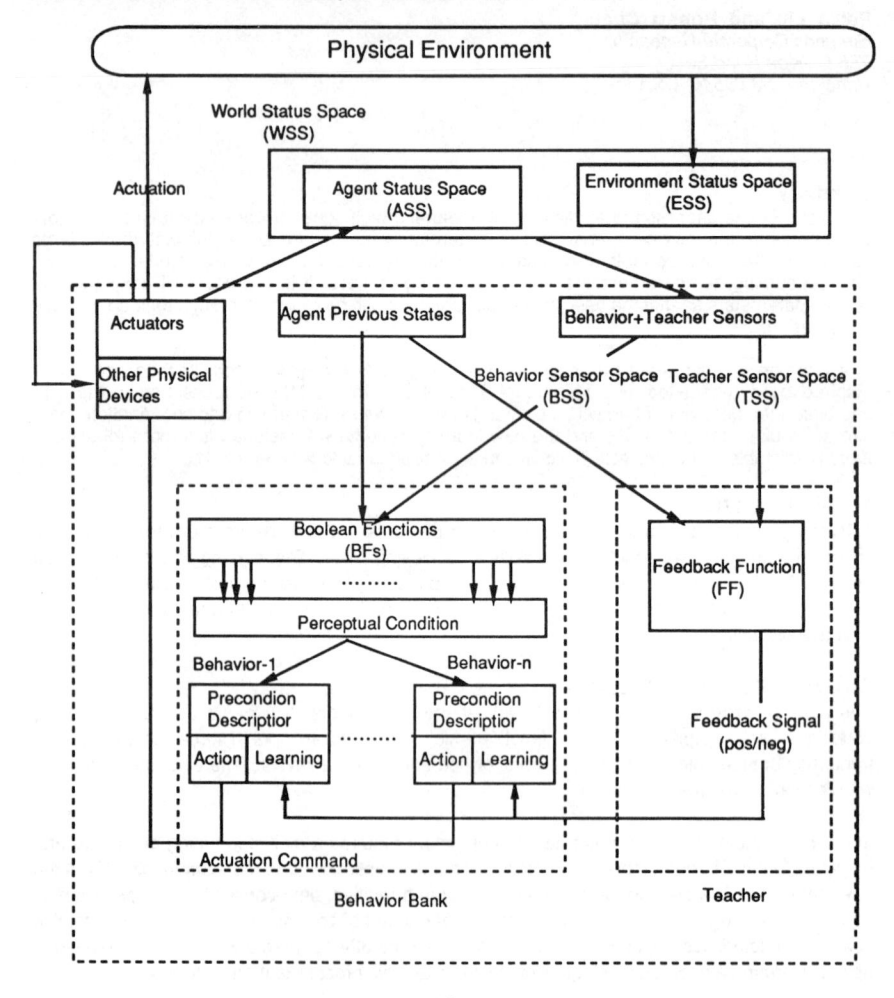

FIG-1

Feel The World Through Sensors

The agent uses its sensors to collect the current status of outside environment as well as the current status of its own electro-mechanical physical body. For a given task, only certain such statuses are used. The all possible agent status used with respect to the given task forms Agent Status Space (ASS); the all possible environment status used with respect to the given task forms Environment Status Space (ESS). World Status Space (WSS) consists of these two spaces, ASS and ESS. Sensors are designed/selected for an agent to perceive WSS instances.

The Central Unit

The central unit of the agent provides the intelligence. In the behavior-based agent, it consists of three subparts: a behavior bank, a teacher, and a vector of the agent's previous state. The behavior bank contains all of the agent's distributed task-achieving primitives, behaviors. Through the interaction between the behaviors and the environment as well as the feedback provided from the teacher, the agent learns the pre-conditions of its behaviors under which it can activate the behaviors in order to achieve the goal of the task.

In the behavior bank, the sensory data of a subset of sensors, Behavior Sensors(BSs), are used in the determination process from which certain behaviors are to be activated. The space they form is called Behavior Sensor Space (BSS). Depending upon the task, a set of Boolean functions (BFs) may be needed to transform the content of the previous state and the raw sensory data gotten from the BSs into a vector of predicate values. The output vector of this transformation is used as the the perceptual condition for behaviors.

Besides being used for behaviors, some sensory data are used for the teacher to produce reinforcement feedback to the behaviors activated. The space they form is called Teacher Sensor Space (TSS). A feedback function, FF, takes sensory data and current state as input arguments and sends an output signal, either positive or negative, to the behaviors.

Each behavior consists of a pre-condition descriptor, an action, and a learning component. When the current perceptual condition satisfies its precondition descriptor, a behavior activates its action. Based upon the signal received from the teacher, the learning component of a behavior decides how to modify its pre-condition descriptor incrementally.

Affect the World Through Actuators

The electro-mechanical physical body of an autonomous agent contains actuators and other necessary devices, such as power supplies and on-board computers. When an action command is received from the Behavior Bank, certain actuators start to execute the command. The actuation of an actuator may change the current physical environment and/or change the agent's own setup.

3.2. Factory Equipment Automatic Tuner

There is similarity between the generic tuning problem and the behavior-based learning problem as shown in Fig-2.

A complex equipment system has to adjust its parameters so that their cooperated performance can be improved. That performance is measurable in terms of some features. To tune such a system, a tuner must have sensors to fetch the features of the equipment and have actuators to adjust the parameters of the equipment. The teacher of the tuner evaluates the effect of its previous adjustment and provides its feedback. The evaluation can be based on the improvement measure or the current value of features. Tuner Previous States provides the history of the tuning process. All of the possible adjustments for parameters are in the tuner's Adjustment Bank. There may exist more than one way of adjustment for one parameter. Each adjustment action of a parameter has its pre-condition to verify and it learns to modify its pre-condition descriptor incrementally when negative feedback arrives.

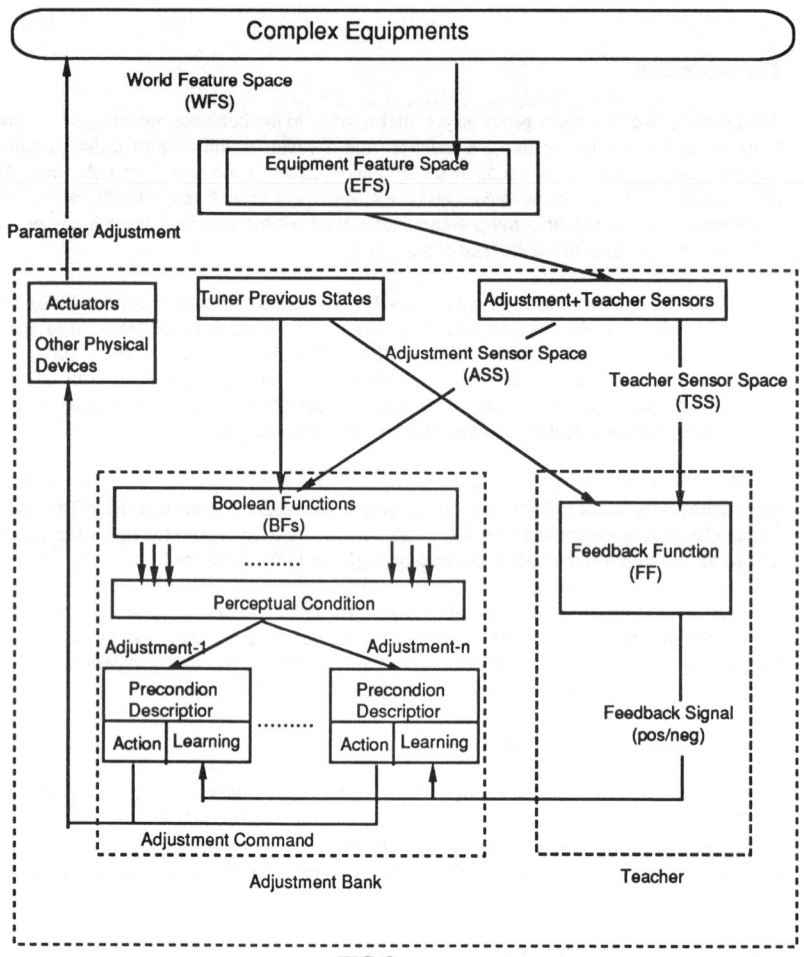

FIG-2

3.3. A Perceptual-Condition-Based Learning Algorithm for Tuning

As previously mentioned, the tuner's tuning capability is expected to be incrementally learned. The learning samples are instances input from WSS during the operation. The generation of such instances are totally dependent on the current capability of adjustments.

In our model, the basic discipline of the learning process is "for any adjustment, never make the same stupid action decision again". In other words, an adjustment should never activate its action under the same perceptual condition which has induced a negative feedback before. At the same time, it will never prevent the behavior from making action decision under any untried perceptual condition. This principle make the proposed learning algorithm distinct from others.

Data Structure

The data structure this learning algorithm working on is the pre-condition descriptor of an adjustment. It is an one-dimensional array and its size equal to the number of all possible perceptual conditions. Each perceptual condition uniquely identifies an entry of that array. Each entry of the array is a bit with the value ON/OFF. The ON value of an entry tells the corresponding perceptual condition that it is allowed to activate the action; The OFF value of an entry tells the corresponding perceptual condition that it is not allowed to activate the action. The initial values of all descriptor entries of all behaviors are ON. This implies that initially all actions can be activated at any perceptual condition. An adjustment uses the value of the current perceptual condition as the index to look up its pre-condition descriptor to decide whether to activate its tuning action.

The Learning Algorithm

1. Look up the pre-condition descriptors of all adjustments with the current perceptual condition value as the index.

2. Select an adjustment which have an ON bit in the corresponding descriptor entry.

3. Activate its action parts.

4. If the feedback is positive, go to step 1,
   else
       for the adjustment last activated,
           turn off its descriptor bits which were previously referenced in step 2.
       go to step 1.

The system starts with the freedom of activating any adjustment at any circumstance. It learns when a negative feedback occurs. The pre-condition descriptor of the adjustment that activated an action that resulted in the current negative feedback should be modified. The entry indexed by the perceptual condition of that action is turned off. Consequently, other later WSS instances producing this particular perceptual condition will not activate this adjustment again. Along the learning process, the number of OFF bit in the pre-condition descriptor of an adjustment is monotonically increasing and it converges to a stable stage of tuning capability.

4. Related Works

There are four major technologies to approach the tuning problem. Each technology has its assumptions and constraints for being applied to the tuning problem.

Adaptive Pattern Recognition(Neural Nets) approach- This approach is based on numerical models and implicit assumptions about relationships between features and adjustments[5] The tuning process adaptively adjusts the parameters from new patterns. The patterns are certain measurements of equipments and can be obtained at real-time during the tuning process.

Expert Systems Based on Heuristic Classification- The expert systems approach[9] is based on empirical rules with incomplete information about relationships between features and adjustments and domain experts available. Preconditions for adjustments are too many to be enumerated and they can be Incomplete or unknown.

Model-Based Diagnosis- View tuning problems as diagnosis problems and is based on functional and behavioral information. The tuning process can be viewed as a process of diagnosis. The parameters to be tuned can be considered as the faults to be adjusted in troubleshooting. Thus, we can use the techniques of diagnosis for tuning. An example of our robotic workcell diagnosis is shown detailed in [1][2][3] and [4]. The diagnosis system uses the model of the workcell to determine from symptoms where adjustments are needed.

Learning approach- Empirical learning algorithms exist[6] for discovering Boolean features from examples (given instances). In this learning model, the target concept is a subset of a given instance space and the teacher presents to the learner a set of instances from the instance space. An instance is labeled positive if it is an element of the target concept and, it is labeled negative otherwise. However, what concerns us here are: (1) efficient measurable preconditions, rather than certain types of DNF or CNF forms. (2) the agent must get instances from environments based on percepts rather than being freely given by the teacher. The teacher and learner are built-in to the same agent. The teacher also gives feedbacks to the learner based on percepts from environment. The teacher may not be able to give all instances to the learner since teaching examples are obtained restrictively from past environment and (3) pre-conditions are based on *transformed* instance space (perceptual conditions) rather than original instance space.

Our work is also related to learning to coordinate behaviors[7]. The learning algorithm works on a real six-leg machine, called Genghis, for walking. The original algorithm is based on statistical sampling.

5. Conclusions
We formalize a behavior-based learning architecture for autonomous agent and for equipment tuning. The algorithm we developed was shown to be very robust and was tested on simulation of different intelligent machines including Genghis. The distinguished feature of this learning algorithm, which is crucial for tuning towards achieving the final goals, is that the convergent property can be preserved.

References
1. Liu, P., M.Y. Chiu, S. J. Clark, and C. N. Lee, Knowledge-based diagnosis for Robotic Workcells. *The Proceedings of Robots 11/17th ISIR, Chicago, Illinois, 1987.*

2. Liu, P., M.Y. Chiu, S. J. Clark, and C. N. Lee, Diagnosis of Workcells by Behavioral Models, *CAR&FOF'87, San Diego, California, 1987.*

3. Lee, C. N., P. Liu, M.Y. Chiu, and S. J. Clark, Model-Based Hierarchical Diagnosis of Robotic Assembly Cells, *SPIE Vol 848 Intelligent Robots and Computer Vision: Sixth in a Series, 1987.*

4. Lee, C. N., P. Liu, M.Y. Chiu, and S. J. Clark, A Hierarchical Symptom Classification for Model Based Causal Reasoning, *The Proceedings of The First International Conference on Industrial And Engineering Applications of AI and Expert Systems,* Tullahoma, Tennessee, 1988 ACM.

5. Grant, P. M., A. R. Mirzai, K. E. Brown and T. M. Crawford, Intelligent techniques for Electronic Component and System Alignment, *Electronics&Communication Engineering Journal*, January/February 1989.

6. Pagallo, G and D. Haussler, Boolean Feature Discovery in Empirical Learning, *Machine Learning*, Vol 5, No.1, 1989.

7. Maes, P. and R. Brooks, Learning to Coordinate Behaviors, *The Proceedings of 1990 International Joint Conference on AI ,* Boston.

8. Valiant, L. G, A Theory of the Learnable, *Communication of ACM*, 27:1134-1142, 1984.

9. Rauch-Hindin, W. B, AI in Business, Science, and Industry, Prentice-Hall, 1985.

Functional Enterprise Modelling: A Process/Activity/Operation Approach

F. Vernadat

INRIA-Lorraine/CESCOM
Technopole Metz 2000,
4 Rue Marconi, 57070 Metz, France

M. Zelm

IBM Germany, Dept. 2237
Am Hirnach 2
7032 Sindelfingen, Germany

Summary
Enterprise-wide modelling is a prerequisite to CIM system design. Functional modelling is at the heart of CIM enterprise modelling and is tightly coupled with information and resource modelling. While previous approaches, and especially SADT-based approaches, rely on one central construct for functional modelling, the activity (also called function), this paper presents a modelling approach centered around three basic constructs to model functionality: process, activity, and operation. Advantages of this paradigm include: more accurate modelling in terms of system semantics, clear separation of enterprise functionality from enterprise behaviour allowing to modify one aspect without changing the other, support for formal verifications, support for performance analyses, and possibility of producing an executable model at all modelling levels. This approach serves as the basis for the function view modelling paradigm used in CIM-OSA, the European initiative for CIM Open Systems Architecture. The process/activity/operation approach is event-driven.

1. Introduction
Enterprise modelling (total or partial) is a prerequisite activity to any CIM system design process. From a CIM point of view and according to CIM-OSA, the European proposal for a CIM Open Systems Architecture [1], an enterprise has to be modelled in terms of:
   - its function view to describe enterprise functionality and behaviour
   - its information view to describe the enterprise information system
   - its resource view to describe the resources used and their management
   - its organization view to describe responsibilities and authorities on functions, information and resources

Other views such as a computational view and a technological view could be considered such as in ATCP as suggested by CAM-I [2] or in EIF. Anyway, it has long been recognized that central to any enterprise modelling paradigm are the function view and the information view [3]. Other modelling approaches have already been proposed and used to model manufacturing enterprises [4,5,6]. They all emphasize the function view which is essentially SADT-based and thus they suffer from the following drawbacks:
   - their function view is based on only one concept: the activity (also called function)
   - the behaviour of the enterprise is not correctly modelled (no clear separation between functionality and behaviour)

- the function view is not well integrated with the information view, the resource view or the organization view, if they have any
- they usually result in limited models, i.e. static descriptive models with limited capabilities for formal verifications, performance analyses or computer execution.

In this paper, we present a process/activity/operation approach as a basis for the function view of a CIM system modelling paradigm which overcomes the previous disadvantages and offers the following advantages:
- true integration of the function view, the information view, the resource view and the organization view of an enterprise
- well-defined and non-overlapping relationships of constructs of the various modelling views
- the modelling approach results in a *computer processable model* which can represent the flow of control, the flow of information, the flow of materials, and the resource control of the enterprise in an integrated way, i.e. as part of the same model
- non-ambiguous and precise definition of the modelling constructs which makes possible formal verification of well-formedness of the model as well as performance analysis using state-transition techniques such as Petri nets [7].

This process/activity/operation approach is based on four basic constructs (event, process, activity and operation), but it also has tight links with constructs of the information view and the resource view as briefly introduced in the next section.

2. Overall Modelling Paradigm
The overall enterprise modelling paradigm spans over the four modelling views and is made of a finite set of modelling constructs, each construct belonging to exactly one view.

From a CIM point of view, an enterprise is essentially made of functions, information and components (including resources and humans) which are placed under some responsibilities and which need to be integrated [1]. Among the components, we distinguish a special class which is a subset of the set of resources, and which contains so-called *functional entities*. Functional entities are active components of the enterprise capable of receiving/storing/processing/sending information and performing basic actions, or units of work, which we call *functional operations*. We distinguish five classes of functional entities: humans, machines (including information technology machines such as computers and manufacturing technology machines such as machining centres or robots), applications (including commercial software packages), data stores (such as database management systems) and communication devices. Any function which can be performed by a functional entity (such as for instance get data, send message, load NC program, etc.) is a functional operation. A passive component such as a tool of a machine-tool is not a functional entity. It is a passive resource object.

Functional operations, as atoms of modelling in the function view, can be procedurally employed to form an elementary task or a piece of functionality. Elementary tasks are called *activities* and they must have well-defined inputs, outputs, and transfer function (i.e. the task). Then, activities can be chained to form a *process*. Processes can be organized into larger processes and so on.

They represent pieces of behaviour of the enterprise. Both activities and processes can be re-used in larger processes. Processes are triggered under some triggering conditions which may involve *events*. Events are real-world happenings occurring within or outside the enterprise.

During their execution, activities need inputs and produce outputs according to their transfer function. They also need time and resources. Most inputs and outputs are object views or information elements of enterprise objects. Both enterprise objects and object views and their information elements are described in the information view. Resource objects are described in the resource view as well as their associated capabilities.

Processes and activities, events, object views and enterprise objects as well as resources need to be placed under some responsibilities and authorities. These are defined in the organization view.

In the next section, we concentrate on the function view of the modelling paradigm.

3. Process/Activity/Operation Approach

This approach is event-driven. It is based on four modelling constructs for functional modelling of manufacturing enterprises: event, process, activity, and operation. However, these constructs are closely related to constructs of the information view named information element, enterprise object and object view as well as to a construct of the resource view named resource object.

3.1. Constructs of the Information View: In this paper, we only discuss information element, enterprise object and object view as constructs of the information view. Other constructs could be defined for this view.

Information element: An information element is any piece of information or data which can be named and, for the purpose it is being used, is indivisible. Each information element is defined by its name and its data type (integer, real, boolean, string, array, record, date, ...). At a given instant, it must hold a value. An information element v_i of type t_i is denoted by $v_i: t_i$.

Enterprise Object: The enterprise object is a construct used to represent objects of the enterprise. Any enterprise object is defined by its name, its abstraction hierarchies and its list of properties. Each property p_k is either an information element, an enterprise object, a set of information elements or a set of like enterprise objects. Three types of abstraction hierarchies ah_l, $l = 1$, 2 or 3, are used to respectively model the "is-a" hierarchy allowing object generalisation and property inheritance, the "part-of" link for object aggregation, and the "member-of" link for grouping together heterogeneous objects. An enterprise object can be defined as a 3-tuple:

EO: <EO-name, {ah_l}, {p_k}>, {p_k} $\neq \emptyset$, $l \in [1,3]$, $k \in [1,p]$, $p \in N+$

Object view: Users and applications of the enterprise handle/use/process objects of the enterprise (indeed they are themselves objects). However, they never directly manipulate the objects but views of them at a given instant. An object view is defined by a name, a list of properties {p_i} and a set of enterprise objects {EO_j} it refers to. Each property p_i is either an information element,

another object view, a set of information elements or a set of like object views. An object view can be defined as a 3-tuple:
OV: <OV-name, {p$_i$}, {EO$_j$}>, {p$_i$} ≠ ∅, i ∈ [1,m], j ∈ [1,n], m, n ∈ N+

3.2. Constructs of the Resource View: In this paper, we only consider the resource object construct. A *resource object* is an enterprise object which is used in support to the execution of one or more activities and has special properties for resource management described in the resource view. Resource objects are sub-divided into two major classes: passive resources and active resources. Passive resources are enterprise objects which are not capable of performing any action and are just employed (such as a tool, a probe, a cart, etc.). Active resources are functional entities capable of performing functional operations (such as a robot, a CNC machining centre, an AGV, an operator, etc.). Resources also need to be categorized according to four other criteria: the possibility of being moved, the possibility of being scheduled, the possibility of being replicated and the possibility to be shared.

3.3. Constructs of the Function View: Essential constructs of the function view are event, process, activity and operation. Additional constructs could be defined to model the domain of analysis, or universe of discourse, function objectives and constraints or declarative rules.

Enterprise *events* describe real-world happenings of the enterprise, which require certain action. Examples of events are the arrival of a customer order, raising a signal indicating a machine failure, sending a management order. In many cases, events carry information (e.g., the customer order, machine indications, the management order). The event information is described in the form of an object view. Thus, an event can be defined as a 5-tuple:
EE: <EE-name, source-list, process-list, object-view, timestamp>
where source-list is a list of names of constructs which can generate the event (resource objects, activities or external world only), process-list is the list of processes which can be triggered by means of this event, object-view is the name of the object view containing the information attached to the event and timestamp is the instant of origin of the event.

Functional operations, or simply operations, are basic units of work, i.e. atoms of work or atoms of functionality, which can be performed by functional entities. From a functional viewpoint they are like elementary action requests sent to functional entities and which will be executed with success or failure. Thus, a functional operation can be defined as a 4-tuple:
OP: <OP-name, FE-name, success, parameter-list>
where OP-name is the operation name, FE-name is the name of the functional entity capable of executing the operation, success is an optional boolean variable indicating the ending status of the operation (success or failure) and parameter-list is a list of parameters (made of information elements) required by the functional entity to execute the operation. For instance, Rotate ROBOT-12 30 will ask robot ROBOT-12 to rotate of a 30 degree angle from its current position in the positive direction.

Activities are elementary tasks, i.e. pieces of functionality of an enterprise, to be performed to achieve one of the basic objectives of the enterprise (usually under some constraints). Activities require allocation of time and resources for their execution. They use function input to produce function output according to their transfer function and using their resource input. They operate under the influence of their control input and additionally they produce control output and resource output. Graphically, they can be represented as a squared box with six legs (three inputs and three outputs). Thus, an activity can be defined as a 10-tuple:

EA: <EA-name, Aobj, Aconst, FI, FO, CI, CO, RI, RO, δ>,

Aobj $\neq \emptyset$, FI $\cup$ FO $\neq \emptyset$, CO $\neq \emptyset$, RI $\neq \emptyset$

where EA-name is the activity name, Aobj is a non-empty list of activity objectives, Aconst is a list of activity constraints, FI is the function input defined as a non-empty list of object views, FO is the function output defined as a list of object views, CO is the control input defined as a list of object views which can provide run-time information to the activity (for instance, object view associated to the event triggering a process for which this activity is the first activity) or which can constrain the execution of the activity (information used but not modified such as, for instance, a work schedule), CI is the control output indicating the ending status after the execution of the activity and eventually the events which can be generated by the activity during its execution, RI is the resource input defined as a non-empty list of resource objects (including the functional entities performing the functional operations of the activity) and RO is the resource output defined as an object view on the resource objects and indicating their status after the execution of the activity. δ is the transfer function modelling the task of the activity. It is a procedural algorithm employing functional operations to perform the task. Thus, δ (FI, CI, RI) = (FO, CO, RO) subject to Aconst. In this modelling approach, an activity can be suspended and resumed or cancelled while a functional operation cannot. Activities can be categorized and are subject to standardization. For instance, manufacturing activities can be classified according to four generic classes: move, make, verify, rest.

Processes are recursive constructs used to model the behaviour, i.e. the logic of control, of the enterprise. Processes are used to chain activities and/or sub-processes to model large business functions achieving major objectives of the enterprise under management, administrative or operational constraints and rules. Processes are triggered under some triggering conditions (involving events) and operate according to their set of procedural rules. Procedural rules are control structures relevant to CIM activities and covering sequential control, conditional control, parallelism, rendez-vous and iterative control. Control structures operate according to values of ending statuses of the processes and activities they govern. For instance, assembly of a car is a manufacturing process. Thus, any process can be defined as a 7-tuple:

BP: <P-name, Pobj, Pconst, {DR$_\alpha$}, {PR$_\beta$}, {EE$_\gamma$}, {ES$_\delta$}>,

Pobj $\neq \emptyset$, {PR$_\beta$} $\neq \emptyset$, {ES$_\delta$} $\neq \emptyset$, $\alpha \in [1,u]$, $\beta \in [1,v]$, $\gamma \in [1,r]$, $\delta \in [1,t]$

where P-name is the process name, Pobj is a non-empty list of objectives, Pconst is a list of constraints, {DR$_\alpha$} is a set of declarative rules, i.e. management, administrative or operational rules conditioning the execution of the process and acting as pre-conditions, {PR$_\beta$} is the ordered set of procedural rules, {EE$_\gamma$} is the set of events involved in the triggering condition of the process and {ES$_\delta$} is the set of ending statuses of the process (these ending statuses are in fact ending statuses of the last activities of the process). Each

procedural rule PR_β is defined by a unique sequence number in the set, a triggering condition and the next action(s) to be executed as follows:
  PR: <Rule-no, triggering-condition, trigger>
where triggering-condition is made of either names of event connected by logical operators (AND, OR), or the name of a process or of an activity along with one of the values of its ending statuses, and trigger contains the name(s) of the action(s) to be started next. The first rule in the set must begin with START and the last one must have FINISH in its trigger clause.

Example: Let us consider a process to be triggered when events EE-1 and EE-2 occur and made of sub-processes BP-1, BP-2, BP-3 and BP-4 and activity EA-1, with BP-3 and BP-4 being executed in parallel and EA-1 being the last action of this process. This can be conveniently tabulated as follows (& means AND):

Rule no.	Wait For	Ending Status	Trigger
1.	START EE-1 & EE-2		BP-1
2.	BP-1	OK	BP-2
		NOT-OK	BP-3 & BP-4
3.	BP-2	Done	EA-1
4.	BP-3 & BP-4	Done	EA-1
5.	EA-1	Done	FINISH

A full example involving processes, activities, events as well as information view constructs appears in [8]. It is not given in this paper for lack of space.

4. The CIM-OSA Approach

Definitions presented in this paper are adapted from an advanced subset of the CIM-OSA constructs [1], but presented with a more mathematical formulation. CIM-OSA operations are called Functional Operations and activities are called Enterprise Activities. CIM-OSA distinguishes two kinds of processes: Business Processes and Domain Processes. Domain Processes are self-contained processes which cannot be re-employed. They are unique to an enterprise and are the only processes which can be directly triggered by events. Business Processes are sub-processes of Domain Processes. They are triggered by parent processes only and can be re-employed. Both Domain Processes and Business Processes are allowed to have function input and function output.

In addition to the function view, CIM-OSA provides constructs to represent Objectives and Constraints, Declarative Rules as well as Domains and Domain Relationships. Domains are non-overlapping sub-sets of the enterprise achieving major business objectives of the enterprise under some constraints. (Two Domains D1 and D2 are said to be non-overlapping if there does not exist a Domain Process dp belonging either to D1 and D2). In the resource view, a construct called Capability is defined to describe abilities required and capabilities installed for Resources.

5. Behaviour Analysis and Performance Evaluation

Behaviour analysis and performance evaluation is an important aspect of the CIM model evaluation. It must deal with execution times and system states defined by occurrences of events, processes, activities, operations, resources and object views.

Once the CIM system has been modelled using the process/activity/operation approach, average durations can be assigned to operations and activities. A probability law must also be provided for each conditional control structures in the sets of procedural rules. Finally, a probability law modelling the appearance of occurrences must be assigned to each instance of event defined and a failure rate must be defined for each resource.

Then, the control part of the process/activity/operation graph defined can be converted into a state-transition diagram such a Petri net [7] in which tokens in places represent occurrences of events, resources and object views and transitions represent activities or operations with their duration time. Aggregated Petri nets can also be analyzed at the process level.

6. Conclusion

The process/activity/operation approach has been presented to model the function view of a CIM enterprise. It is event-driven. Since it has clear relationships with constructs of the information view and resource view of the full CIM system model, it can be conveniently used to describe the flow of control, the flow of information, the flow of materials or the use of resources. This approach has three major advantages:
1. thanks to the concepts of process and activity, it provides a clear separation between functionality and control, allowing to modify one aspect without changing the other;
2. thanks to its formal definition, it lends itself to formal verification of well-formedness of the model, consistency checking and can result in a computer executable model; and
3. it provides a bridge to more abstract analysis formalisms such as Petri nets for behaviour and performance analyses.

References
[1] CIM-OSA. Special Issue. Int. J. Computer-Integrated Manufacturing 3:3 (1990) 144-180.
[2] CAM-I. An Architecture of CIM. Techn. Report R-88-ATCP-01. 1988 (Also ISO paper ISO TC184/SC5/WG1-N90).
[3] ICAM. Integrated Computer-Aided Manufacturing Architecture ICAM Project Part II Vol. IV - Function Modelling Manual (IDEF0), US Air Forces.
[4] Doumeingts, G. Méthode GRAI: Méthode de Conception des Systèmes de Productique. Thèse d'Etat, Bordeaux, France. 1984.
[5] M. Roboam, M Zanettin, L. Pun. GRAI-IDEF0-Merise (GIM): Integrated methodology to analyse and design manufacturing systems. Computer-Integrated Manufacturing Systems 2:2 (1989) 82-98.
[6] CIM Reference Model Committee, Purdue University. A reference model for computer-integrated manufacturing from the viewpoint of industrial automation. Int. J. of Computer-Integrated Manufacturing 2:2 (1989) 114-127.
[7] Peterson, J.L. Petri Net Theory and the Modeling of Systems. Englewood Cliffs, N.J.: Prentice-Hall 1982.
[8] Vernadat, F. Modelling CIM enterprises with CIM-OSA. 2nd Int. Conf. on Computer-Integrated Manufacturing, Troy, NY (May 1990) 236-243.

Issues in the Formal Definition of Product Model Exchange

A. Márkus
Computer and Automation Institute, Hungarian Academy of Sciences
1502 Budapest POB 63
h133mar@ella.uucp

Abstract

The paper reports on-going work on the formal definition of product model exchange. Although the discussion is based on the STEP standard proposal of ISO, major problems outlined here (around subtyping and transformations of schemas and product models) are to be faced by other kinds of product modelling systems as well.

1. Introduction

Among people working with CAD/CAM systems, there is an unanimous agreement on the need of a standard that captures the information necessary to compose a product model in a neutral form so that the model can be exchanged among CAD/CAM systems of different vendors [3,6]. Research and development towards this aim have been done in the ISO / TC184 / SC4 since l983, and have produced a working draft of a standard, inofficially abbreviated as STEP, in l988. The discussion of the paper will be based on this draft with respect both to the requirements and the tools of product modelling. Anyway, the opinions contained in this paper should by no means be interpreted as views of the respective ISO working groups.

STEP is different from the previous neutral exchange standards in the following major points [4]: STEP is to cover far more than geometric data: among others, it must be able to deal with topology, features, tolerances, materials. It will be further extended to cover the whole life cycle of products (e.g. technology, maintenance). Applications of STEP include engineering branches as far from each other as mechanical engineering and architecture. At the same time, the applications use common resources (e.g. draughting). STEP has to provide means of long-term archives and aims at the automatic generation and interpretation of the product models, so no human experts may be involved in the exchange.

The aim of this paper is to discuss issues which must not be overlooked if one would like to build product model exchange on a firm ground. The first part of the paper discusses several issues related to the use of types, especially subtyping. The second part exposes problems with the transformations of schemas that are needed for subSTEPs, updated and localized STEPs. A much more detailed discussion of these issues can be found in [9, 10].

2. Issues related to EXPRESS subtyping

EXPRESS is a model specification language, developed for the purposes of STEP [5]. Although the features of EXPRESS address the definition of static object instances (once created, instances are not supposed to change), the language includes facilities for writing procedural code that can be used to specify the conditions of creating acceptable instances (by local rules attached to the definition of object instances) and acceptable models (by global rules that are attached to one or more object types).

Subtyping is a key feature of EXPRESS: subtypes can be composed of one type or of several types. For defining subtypes one may add further attributes and/or constraints, and compose instances out of several components. EXPRESS applies assignment compatibility of instances but there is no strong type checking in the language, e.g. integers typed as lengths and as temperatures can be freely mixed.

The aims of using subtypes can be classified into the following groups: *Refinement*: Given a type, add further attributes in order to get subtypes that contain objects with more detailed descriptions. *Constraining*: Given a type, add constraints in order to get subtypes that contain more tightly constrained objects. *Composition*: Given more than one types, mix them up in order to get subtypes that contain objects including several components. *Combination of constraints*: Given more than one types, mix them up in order to get subtypes that contain objects that fulfil the combination of constraints. (In order to have an effective assignment compatibility rule, combination of constraints is actually restricted to conjunction.)

2.1. Subtyping with attributes, constraints or components?

A crucial question in developing product model schemas is the proper understanding of the difference between subtyping with additional attributes versus subtyping with additional rules:

ENTITY unit_vector; fi : REAL; END_ENTITY;

ENTITY vector SUBTYPE OF unit_vector; r : REAL; END_ENTITY;

The other version is subtyping with rules:

ENTITY vector; x, y : REAL; END_ENTITY;

ENTITY unit_vector SUBTYPE OF vector; WHERE $x * x + y * y = 1.0$; END_ENTITY;

While each approach is correct in itself, the choice should depend on the purpose of the modeller. Fortunately, these two kinds of subtyping suggest a coherent assignment compatibility rule: the left hand side of any kind of assignment has to belong to the same type or to a supertype of the type of the right hand side of the assignment. The soundness of this type checking rule is obvious, since it requires that the right hand object should have the same or more attributes (or more tightly constrained attributes) than the variable on the left hand side.

Another modelling issue is the selection between refinement and composition: *type1 with additional attributes defines type2* , versus *type1 combined with type2 defines type3*. The approach we propose is that composition should be favoured in those cases where the set of the additional attributes is a conceptual unit in itself, i.e., wherever at least a good name can be found and when the additional part may be used several times when defining some further entity types of the schemas. Altogether, it should be clear that composition of subtypes is a modelling technique much different both from using attributes containing pointers to other instances and from embedded instances.

2.2. What subtype structures are acceptable?

The only restriction imposed by the rules of EXPRESS is that no cycles are allowed in the subtype graph. Unfortunately, as it is known from knowledge representation folklore, this rule might not be strict enough when multiple inheritance is present:

ENTITY a_type; END_ENTITY;

ENTITY b_type SUBTYPE_OF a_type; END_ENTITY;

ENTITY c_type SUBTYPE_OF (a_type AND b_type); END_ENTITY;

The problem is whether an instance of c_type is more closely related to b_type or to a_type. In order to answer such questions one needs an ordering of the components of the type [2]. There are two apparent ordering principles: (1) Ancestor types are less specific. (2) A type definition sets the specificity order of its direct ancestors. Subtype structures should be considered legal if and only if such an ordering exists.

2.3. How to resolve collisions of attribute names?

What should be done when a type has an attribute with the same name as an attribute in one of its (immediate of indirect) supertypes? Actually, when an entity type has several components it is hardly avoidable that some attribute names collide. The collision of attribute names should be legal: if it were forbidden, it would decrease the modularity of the development of schemas. The intention behind using colliding names might be that the subtype has more specific information about the object.

Our suggestion is that name collisions among attribute lists should be resolved in favour of the attribute within the more specific type. (This is one more reason of requiring an ordering of the types according to specificity.) An argument against this suggestion may be that it hurts the integrity of the supertype. In our opinion this is not so: whenever this kind of integrity is important, do not use subtyping: use pointers or local instances instead: then each of the colliding attributes will be present.

2.4. How to create families of types?

Schema developers of STEP have early recognized that the use of multiple components of entity types supports modular and coherent program development, so EXPRESS facilitates the definition of intricate subtype structures as this (N.B., the definition should be completed with other entities that are omitted for the sake of brevity here):

ENTITY cooking_appliance
    SUPERTYPE OF ((gas_cooker XOR electric_cooker) AND
                          (cooking_appliance_with_time_indicators OR
                          cooking_appliance_with_heat_indicators))
END_ENTITY;

Composite instances belong to a variety of new subtypes: the pseudo-logical XOR, AND and OR supertype expressions say that, within each instance of cooking_appliance, (1) according to the XOR relation, exactly one of the gas_cooker vs. electric_cooker pair of components has to be included, (2) according to the OR relation, one or both of the components cooking_appliance_with_time_indicators, cooking_appliance_with_heat_indicators have to be included, (3) according to the AND relation, both kinds of inclusion have to be considered.

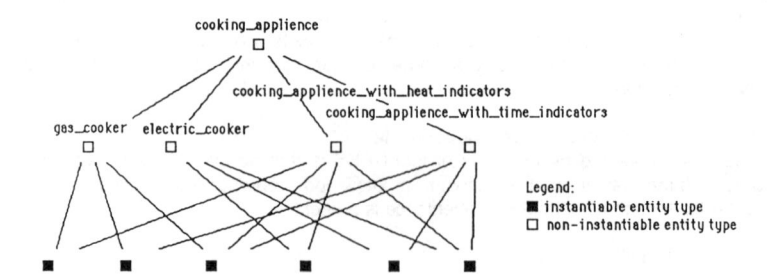

Fig.1. Subtype structure with multiple levels

Our main concern here is that this structure can be made more tractable if the modeler uses components that, in contrast with the previous structure, *aren't subtypes of cooking_appliance* (N.B., some definitions have been omitted here, too):

ENTITY cooking_appliance SUPERTYPE OF
    ((gas_cooker XOR electric_cooker) AND (heat_indicators OR time_indicators));
END_ENTITY;

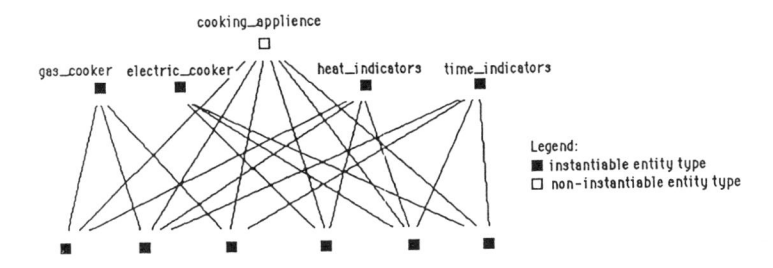

Fig.2. Subtype structure with mixed in components

The second version offers several advantages: (1) it is built only of two kinds of types, a basic type and a variety of add-on components, (2) it has fewer non-instantiable entity types, and (3) the components may be used alone or in other composed types. Further discussion is needed whether the components of an instance may depend on the actual values of the attributes of the instance.

2.5. Further Issues

How to represent structures of instances that constrain each other? Constraints among related instances can be expressed by global rules or distributed among the local rules of the instances. Both approaches have their own problems: When using global rules an additional structure may be required to specify the scope of the tests. When using local rules, no matter whether the checking of the constraints works as a greedy or lazy process, there may arise intricate timing issues.

How to define generators of a canonical representation? Using different generators of an entity type is a common requirement in product modelling. Although subtyping seems to be an appropriate tool to fulfill this need, several problems emerge here, too.

3. Remarks on the formal definition of application protocols

There is a widespread opinion among experts working on STEP that the ability to define Application Protocols (APs) is a necessary condition of the practical use of a product model exchange standards [3,7,8]. APs are supposed to fulfil a very broad range of requirements by providing the means of (1) isolating several smaller, streamlined subSTEPs; (2) the coordinated improvement of STEP (compatibility of versions, support during the whole life cycle of products); (3) using STEP as a common base of regional, national, industry or company-specific standards.

In the typical situation of product model exchange a program system (an application) sends a product model to another one. These applications have been developed independently, by using different programming tools and with no respect to the product model exchange facility. They need not share data definitions and algorithms, some of them may apply full type checking, others may deal with typed object instances but variables with no concept of type. Moreover, no one-to-one correspondence exists between the instances in the pools of the applications, and those of the neutral product model file.

Assume that the sending system, supported by the information represented within the schema and by its in-depth knowledge of the application itself, is able to isolate a proper representation of the product from the sender's actual state. On the receiving side, in order to integrate the content of the product model file into the current state of the application, the postprocessor of the receiving system has to consider a schema and a mapping. Once again, assume that the postprocessor is indeed able to make this integration.

Whenever there are differences between the schemas and/or mappings used on the two sides, the product model transfer enters the realm of APs: *Transformations of mappings* might be needed if different media are involved, or if the pre/postprocessors work under different processing priorities [1]. *Transformations of schemas* should be able to cope with limited subSTEPs, with different, updated and local versions.

Schema transformations should be transposable onto PM files, i.e. APs should deal with two layers of problems: one layer concerns the transformations of the schemas, the other concerns the schema transformations having applied to PM files. Although transformations may be invertible, existence of inverse transformations is not guaranteed. Last but not least, product model transfer needs a concept of the validity of schemas and product model files, i.e. there should be given a formal and operative criterion to be applied on them.

The variants of using these transformations can be studied on Fig. 3. (Sch0 is the ancestor of two other schemas, Sch1 and Sch2; Mf0, Mf1 Mf2 are product model files; the arcs of the diagrams are pre- and postprocessors and transformator programs.)

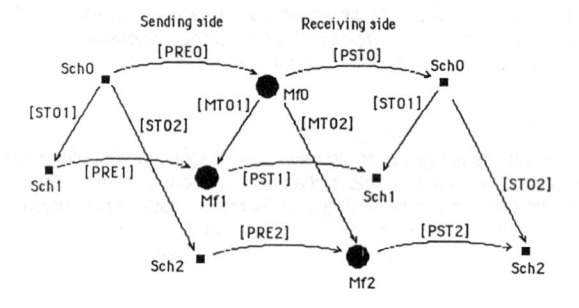

Fig.3. Schema and product model transformations.

Several variants of sending / receiving configurations may arise, e.g.: *Send by Sch0 --- Receive by Sch1* (product model sent according to the basic schema, received by a system that knows only a simpler schema); *Send by Sch1 --- Receive by Sch0* (product model sent according to a descendant schema, augmented with company-specific details, received by a system that knows only the basic schema); *Send by Sch1 --- Receive by Sch2* (product model exchange with using two different updated versions or company-specific extensions of the core standard).

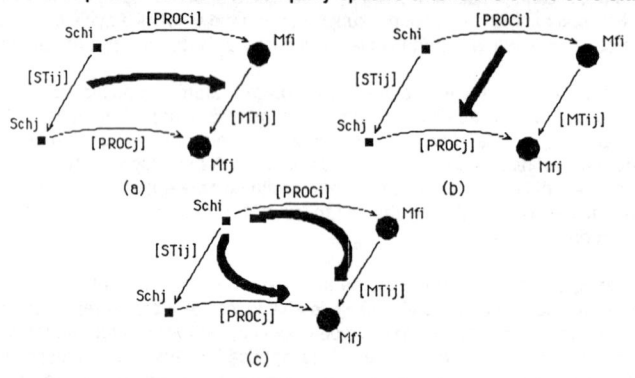

Fig.4. Transpositions and commutativity of transformations.

Looking up the previous settings, several questions recur: *Preserving the validity of schemas:* What kind of transformations keep the validity of the input schema? (Example: a transformation is surely invalid if it deletes an entity type without deleting all the attributes of that type within other entities). *Transposition of a transformation to the model file* (Fig.4a): What kind of transformation can be transposed so that the transposed version will properly work on the model files? *Transposition of a processor to work with a transformed schema* (Fig.4b): What kind of schema transformation can be used with an automatic update of a pre- or postprocessor? *Commutativity of processing and transformationn* (Fig.4c.): Under which conditions is it true that processing of the product model according to a transformed schema can be substituted by the transformation of a processed product model file?

Further questions arise with respect to inverses: What kind of transformation of schemas is invertible? What kind of schema transformation, having transposed to the model files, is invertible? When can be found a subrange where a transformation is invertible?

It seems to be most probable that the checking of the validity of the transformations could not be an autonomous process: STEP has to formalize the criteria of the validity of its components, then specify, a maybe interactive, process of transforming valid components into other valid ones.

4. Conclusion

The first part of the paper has exposed several, loosely coupled issues related to the requirements of advanced product modelling tools. The conclusion of these investigations is that schema development is an intricate process and the misuse of the tools should be avoided by careful design and selection of the features of the tools. The understanding of application protocols needs more formal means and this field deserves more attention: the paper has tried to shed light on some of these problems.

Acknowledgment Discussions with K. Klement, D. Schenk, B. Wenzel and G. Krammer have helped much in the clarification of the ideas presented here.

References

[1] Fowler J.: Latest Version of STEP Data Access Interface Specification National PDES Testbed Memo, October 1990
[2] Keene S.E.: Object-oriented programming in COMMON LISP: A programmers' guide to CLOS. Addison-Wesley 1988.
[3] N.N.: State of the art of interfaces for the exchange of product model data in industrial aplications. CEN/CLC/AMT/WG STEP N16, March 1990
[4] Owen J.: STEP Part 1: overview and fundamental principles. ISO TC 184 / SC 4 / WG1 Document No 467, March 1990.
[5] Schenk D.: Exchange of Product Model Data - Part 11: The EXPRESS Language ISO TC 184 / SC 4 / WG1 Document No 496, July 1990.
[6] Smith B.M.: External representation of product definition data. ISO TC 184 / SC 4 / WG1 Document No 433, August 1989.
[7] Wenzel B.: A Software Development Architecture for STEP. ISO TC 184 / SC 4 / WG1 Document, September 1989.
[8] Woodall A., Pearson M., Owen J.: Critical Issues for Application Protocols. ISO TC 184 / SC 4 / WG1 Document No 431, October 1989.
[9] Markus A.: Proposals for extensions of EXPRESS and the format of the product model file. ISO TC 184 / SC 4 / WG7 Document No 2, October 1990.
[10] Markus A.: Toward the formal definition of application protocols. ISO TC 184 / SC 4 / WG7 Document No 3, October 1990.

Strategic Material Flow Analysis Using Simulation

V B Patange
School of Engineering, Sheffield City Polytechnic,
Sheffield, England

U S Bititci
Strathclyde Institute, Glasgow, Scotland

Summary :

Computer Simulation can be an effective tool for studying the material flow in a production line. A case study at BUKO Ltd, the manufacturer of supermarket trolley is discussed. The problem is focussed on increasing the production volume at minimum cost and consistent quality to meet customer demand. Various simulation models were developed on PCModel software using different manufacturing strategies. The bottlenecks in each case were identified and the effect of increasing the workload capacity on the overall performance of the line is analysed.

Introduction :

Recent trends in Production Systems have been automation of shop floor activities (FMS, Robotics etc) and computerisation of supporting engineering and management functions (CAD, CAM, MRP). Considerable effort has also been spent on integration of various engineering and management functions with each other as well as with the shop floor activities in order to ensure the availability of the right information at the right place at the right time.

Shop floor disciplines such as JIT and OPT have been developed to rationalise and synchronise shop floor activities. However, many organisations fail to relate manufacturing strategy to appropriate place of automation and try to achieve improvements in their production performance by introducing automation [1].

This paper outlines such an experience with a company based on the east coast of Scotland, which manufactures supermarket equipment. A brief introduction to the company is given and the project carried out is discussed in detail.

Introduction to Organisation :

Buko Ltd is a medium sized concern located at Glenrothes and employs nearly 300 employees. At present, Buko produces supermarket equipment like trolleys, shopping hand baskets, roll containers and wire meshed shelvings.
Virtually, all products are made from either stainless steel or mild steel wires. These wires are available in rolls which are first straightened and cut into required size. This process takes place in "Cutting and Bending" section. The finished material from this section is transferred to the stores where it is held until all the components of that order are complete and ready for assembly. This takes nearly a week.

The order is loaded in the welding shop in the following week, where the mesh is produced by spot welding the cross wires. All the welding operations are completed in this section and goods are assembled to semi-finished stage. These goods are then transferred to the electroplating shop where they are zinc plated to give a smooth and bright finish.

Initial Findings :

Buko found low production rates in their trolley welding shop. Buko's real needs were to increase the production output per day of these trolleys, to meet ever increasing market demands. Therefore, Buko requested a study of their cart line with a view of developing a company wide strategy towards automation of these operations.

There are four cart lines in the shop in which 13 different types of carts are produced. Two of these lines have robots performing certain welding operations and the other two lines are fully manual. Only one manual line was considered for this exercise.

The manual operations involved activities like:

- Loading the jig with pieces of wire/wire components
- Feeding the jig between the electrodes for spot welding
- Removing the welded mesh from the jig

Table 1 shows the sequence of operations in the manual welding line, where five operators attend nine work stations. The decision for the individual operator assignment is taken by the shopfloor supervisor based on the demand and the priority of the operation.

While defining the problem of improving the production rate of carts, it was necessary to reduce the operation time. So, in the above manual operations, the problem was focussed to improve the performance of the workcentre with high content of repetitive labour and high throughput. Therefore, study of complete cart line was planned so that the bottlenecks could be identified and areas in which considerable improvement is required can be located [2].

The objectives were set for the study. The higher production rate virtually means higher output or higher material flow in the shop.

In order to create real and dynamic models, use of computer simulation was made [3]. "PCModel", a commercial simulation software was used to construct these models and study the material flow.

Simulation Models and Analysis of Results :

To construct realistic models of the cart line, information regarding layout of the line, sequence of operations, operation/activity time, handling time, current shop loading methods and production order size was collected [4].

All models were built by using the workscheduling rule for maximum utilisation of resources except Model 5. The first model (Model 1) was constructed to study the behaviour of the present production line. The complete manual cart line was developed showing all the nine workcentres (see fig 1). This model was examined under present shop loading, handling and batching methods. The simulation study showed that station 3 (welding sides of the basket) was a bottleneck.

Following suggestions were made to increase the production capacity of station 3,

(a) creating additional spot welding machine parallel to station 3, capable of performing similar operation at the same speed.

(b) replacing the manual welding station 3 with a robot which could weld the sides as well as the fronts of the basket and thus completing operations 3 and 4 in one set up.

Considering the above alternatives, models 2 and 3 were constructed and subsequent bottlenecks were identified. This exercise was carried out not only to increase the material flow in the shop, but also to smoothen this flow by balancing the production line.

Model 2 attempts to flush the material through station 3 by increasing the capacity (fig 1) of this workcentre. The effect of increasing the capacity at station 3 was studied, keeping all other parameters unchanged. It was observed that the bottleneck at station 3 was shifted to station 5 which is a classical behaviour of any unbalanced line.

The robot in Model 3 eliminates operation 4. It was assumed that robot takes approximately same time to do operation 3 & 4 as in other robotic cart lines. The simulation study shows that this robotic operation is inefficient and was found to be a bottleneck. The reason being that the time taken by the robot to weld sides and fronts in one set up was significantly more than welding the sides alone.

In Model 2, it was found that the bottleneck was shifted from station 3 to station 5, therefore another configuration of the line (Model 4) was developed. The concept of creating parallel manual stations 3A and 3B together with 5A and 5B was used in this model. It shows the effect of changes at stations 3 and 5 when made simultaneously.

Table 3 shows 20% reduction in throughput time for a given combination of cart baskets. This model showed a large pile up of WIP at station 9 (CO_2 welding). Further improvements in this model would have been made possible, if alternative production methods are found to improve the capability of station 9 where the bottleneck is shifted now. It is also evident that station 1 regulates the material flow in the line.

Model 5 is very similar to Model 1 as regards the configuration of the line is concerned (fig 1). No changes were made in equipment or the production facilities in the configuration, except the loading methods. This model runs on Just In Time (JIT) philosophy.

This can be achieved by defining buffer values to each workcenter. During the simulation run when the buffer capacity reaches its maximum value, the workcenter preceding this station stops producing the jobs. This balances production rate of individual workcentres to the final assembly rate and reduces value added Work In Progress(WIP). The WIP level of the production line can be maintained by setting different buffer values to individual workcentres.

The buffer values greatly influences WIP of the system. If more buffers are allocated to each station, WIP increases proportionally. On the other hand, if small buffer values are assigned to the stations, WIP can be reduced, but there is a possibility that some workcentres may remain idle. Therefore, in order to maintain an optimum WIP level, proper buffer values should be given to all workcentres. This is a very important prerequisite to any production system operating on JIT principle.

With this strategy, it was found that it takes the same time to produce given number of cart baskets as it would in Model 1, but reducing the WIP by nearly 33% (ref. table 3).

Conclusion :

The utilisation statistics (table 2) clearly show that workcentres located in the rear end of the line are underutilised particularly from station 3 onwards.

The performance of stations 3A and 3B were found to be satisfactory where the utilisation of the resources are concerned. But, the effect of creating station 3B on the overall performance of the system was insignificant.

Model 4 showed a significant improvement not only in terms of the weekly production rate but also the utilisation of workcentres. When the utilisation statistics of this model is compared with other models in table 2, a remarkable achievement is observed. The cost associated with each model is shown in table 3.

One of the limitations of PCModel is that it does not provide a facility to generate random numbers for various statistical distributions except the uniform distribution. As a result, the effect of large number of probabilistic variables on the performance of the line could not be filtered out. Therefore, this exercise gives only a rough guideline to form a strategy towards automation of the cart line.

It can be concluded from the simulation results that Model 4 shows the reduction in manufacturing lead time by 26 hours, which is 36 % less than Model 1 (see table3). It was found that stations 3, 5 and 9 were the bottlenecks and considerable improvement is required in designing better production methods to increase the throughput of the system.

Table 1

Workstation Number	Description of Operation
1	Welding the panel
2	Forming the panel
3	Welding sides of the panel
4	Welding fronts of the panel
5	Welding the side handle and rear frame
6	Welding the bottoms
7	Welding the top rail
8	Welding the side plates
9	CO_2 welding of loops.

Table 2

UTILIZATION STATISTICS					
Work Station Number	MODEL NO.				
	1	2	3	4	5
1	99 %	99 %	99 %	99 %	55 %
2	28 %	28 %	31 %	30 %	17 %
3A	99 %	68 %	99 %	67 %	99 %
3B	-	68 %	-	70 %	-
4	52 %	81 %	-	81 %	45 %
5A	92 %	99 %	62 %	72 %	82 %
5B	-	-	-	70 %	-
6	32 %	34 %	22 %	49 %	27 %
7	68 %	72 %	47 %	95 %	59 %
8	34 %	35 %	24 %	46 %	28 %
9	85 %	91 %	61 %	99 %	78 %

Table 3

ANALYSIS OF RESULTS-

SYSTEM VARIABLES	MODEL 1	MODEL 2	MODEL 3	MODEL 4	MODEL 5
TOTAL LEAD TIME OF THE BASKET OR THROUGHPUT (IN HOURS:MINS)	17:47	14:48	31:06	14:20	17:47
MAXIMUM WORK IN PROGRESS (WIP) ALLOWABLE IN THE SYSTEM (IN NO. OF BASKETS)	1500	1500	1500	1500	1500
MAXIMUM WIP ATTAINED BY THE SYSTEM DURING THE STEADY STATE (IN NO. OF BASKETS)	1500	1400	1500	1300	1000
QUANTITY PRODUCED IN UNIT TIME OF 85 HOURS FOR TYPICAL COMBINATION OF (A=1500 + B=1000 + C=1500) = 4000	2700	2950	1950	3500	2700
TOTAL TIME CONSUMED BY THE SYSTEM TO PRODUCE THE ABOVE QUANTITY	136:56	125:17	181:40	110:35	137:13
BOTTLENECK STATIONS	3	5	3	9	NO BOTLNK
ADDITIONAL COST OF THE PROJECT	-	£12,900	£130,000	£18,900	-
ADDITIONAL LABOUR REQUIREMENT	-	1 SEMISKILL	-	2 SEMISKILL	-

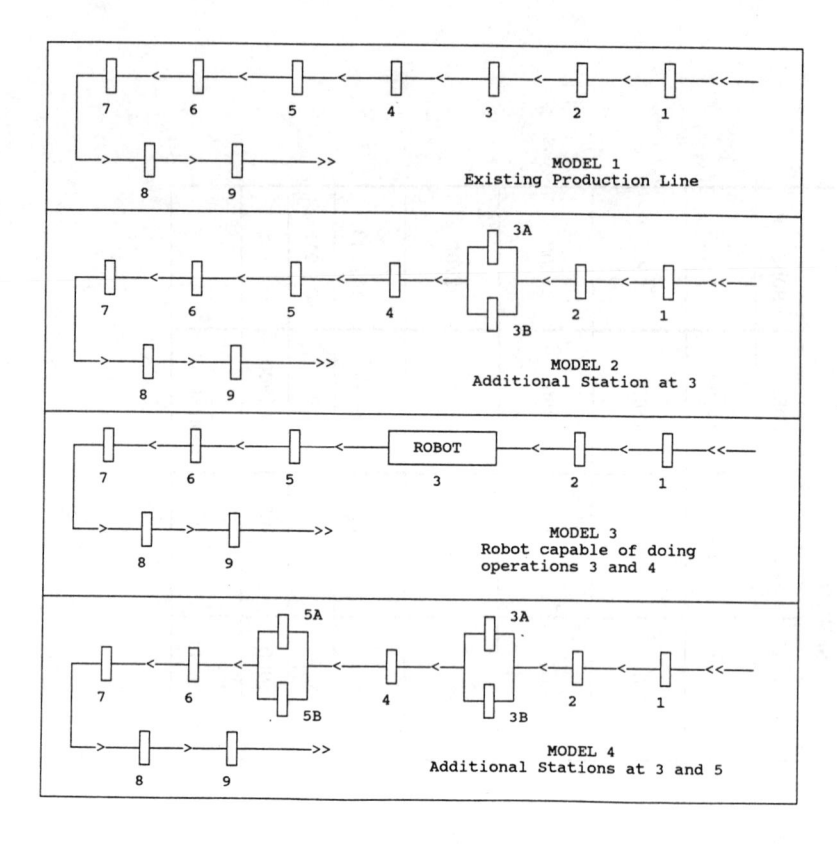

Figure 1

References :

[1] Gardner, L.B.; 'Automated Manufacturing', pp 2-12.

[2] Patange, V.B.; 'Automation of Spot Welding Operations', MSc Thesis, University of Strathclyde, Glasgow (1989).

[3] Carrie, A.S.; 'Simulation of Manufacturing System', 1989.

[4] Chaharbhagi K.; Rahnejat H.; 'Application of Simulation in Planning the Design of Flexible Assembly System', Proceedings of Institute of Mechanical Engineers 1986. pp 95-109.

[5] 'PCModel', User's Guide by David White. Simulation Software Systems, 2470 Lone Oak Drive, San Jose, CA 95121, USA.

A Lambda Calculus Model for the Officer-Cell Phenomenon in Manufacturing Systems

G. R. LIANG

Department of Industrial Engineering & Management,
National Chiao Tung University,
Hsinchu, Taiwan, R. O. C.

Abstract
Hierarchical policies are frequently used for implementing automated manufacturing systems. From a physical viewpoint, a hierarchical control model is usually interpreted through a shop-cell-workstation-equipment hierarchy. From a personifying viewpoint, a hierarchical control is often typified like a general-officer-sergeant-troop hierarchy in military. Both approaches interpret the same reality. In contrast with a cell consisting of several workstations, a coherent examination enforces an argument of an officer consisting of several sergeants. A new model based on lambda calculus is proposed to interpret this puzzling phenomenon.

Introduction
Hierarchical policies [1,2,3,4] are generally treated as promises for constructing factories of the future. Essentially, these policies are to pass commands from higher levels down to lower levels. Many models were built to describe their behaviors. For example, the Automated Manufacturing Research Facility (AMRF) [4,5] in National Bureau of Standards (now NIST), U. S. A., used a policy called hierarchical control model. Gershwin [2] used a frequency spectrum to describe the hierarchical behavior in manufacturing system. One application is the Miniline 1300 [6] in IBM. The mathematical representation for their approaches is based on the concept of state space in control theory. Weber and Moodie [7,8] used the frames in artificial intelligence to construct the hierarchical structure. A hybrid model using the frames and the AMRF model is in [9]. Biemans [10] proposed to model manufacturing systems using reference model method. Naylor and Volz [11] suggested an approach using the concept of assemblages of software components. Liang [12,13,14] introduced lambda calculus for modeling the hierarchical control structure. Vallespir [15] gave a GRAI method.

Lambda calculus was initiated in 1940s [16]. Since the advent of electric computer age, it was mainly applied to the design of functional programming languages. For the application of lambda calculus to functional programming, a major consideration is the parallel reductions of lambda terms because the final reduction result is unique according to Church-Rosser theorems [17]. For the application of lambda calculus to manufacturing system, the reduction of lambda terms are corresponding to the activities in manufacturing system. Naturally parallel reductions can not be done arbitrarily. The concept of time has to be added into reduction processes because timing sequence is significant in manufacturing.

Officer-Cell Phenomenon

The management of a traditional factory usually is under a hierarchy like general manager, managers, supervisors, workers, etc. For an automated factory, a popular way to save manpower is simply to replace individuals by computers. Consequently, a hierarchy of men is replaced by a similar hierarchy of computers. One major reason to support this approach is the reduction of complexity. For example, a single factory which has 85 orders, 10 operations, and only one substitutable machine holds over 10^{800} schedules [18] that one could create. It was indicated in [1] that a two-level hierarchy will reduce the complexity from e^{NH} down to $10e^{NH/10}$ by grouping an average 10 entities of the global model. Here N is the number of entities, and H is the horizon considered. So hierarchical policies are generally suggested for constructing computer-integrated manufacturing systems.

From a physical viewpoint, the AMRF has a five-level structure: facility, shop, cell, workstation, and equipment. A cell consists of several workstations, and a workstation consists of several equipment. Decisions from cell level are passed down to workstation level, then to equipment level. The Miniline 1300 has a four-level structure: top level, middle level, lower level, and machine level. From a personifying viewpoint, to trace a hierarchy like cell, workstation, or equipment may reveal an original hierarchy of manager, supervisor, or worker. Synder [19] typified this hierarchical control like an order-passing procedure in a military hierarchy. For example, the generals pass orders to the officers; the officers pass orders to the sergeants; the sergeants pass orders to the troops.

Either a physical viewpoint or a personifying viewpoint, both approaches interpret the same reality. It is possible to find a connection between them. For example, the troop, sergeant, and officer levels can be assigned to be corresponding to the equipment, workstation, and cell levels, respectively. However, a coherent examination enforces an argument of an officer consisting of several sergeants because a cell consists of several workstations. If there exists a real connection between them, then this puzzling officer-cell phenomenon must be answered first. In the following, a lambda calculus model is proposed for solving this puzzling officer-cell phenomenon.

Lambda Calculus Model

Lambda calculus was proposed in [12,13,14] for modeling the hierarchical structure in manufacturing system. The theme is to use the reductions of lambda terms to represent the activities in manufacturing system. Then the lambda terms form a hierarchy through substitution processes [17]. A lambda term [17] is generated from two basic operations repeatedly: abstraction and application. For example, a lambda term $(\lambda x.\lambda z.z)ab$ is generated through two abstractions and two applications. Through two β reductions [17], this is represented as $(\lambda x.\lambda y.y)ab \rightarrow_\beta (\lambda y.y)b \rightarrow_\beta b$.

For the reason of easy explanation, some basic concepts for this approach are explained in the following through a functional language FP [20]. The relationship between lambda calculus and FP can be found in [21]. First, each functional unit in a hierarchy is treated as a function or a lambda term which can be applied to input data. Then β reductions are done to represent the activities. For example, a functional unit for making a drill of machine M1 down in equipment level is shown in Fig. 1. If input data are sent to this function, then this unit produces a "drill down" output to control the drill. Second, a hierarchy is constructed as a series of substitution processes. For example, a drilling process of machine M1 in workstation level, illustrated in Fig. 2., is represented as a sequence of operations: loading, computing, workpiece moving, drill down, depth control, drill up, and unloading. The lower parts in Fig. 2 are imbedded into higher parts. In Table 1, their corresponding programs in FP are shown on the left, and their operations are shown on the right.

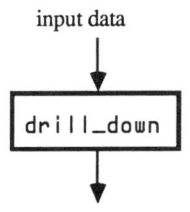

Fig. 1 A functional unit for controlling a drill down in equipment level

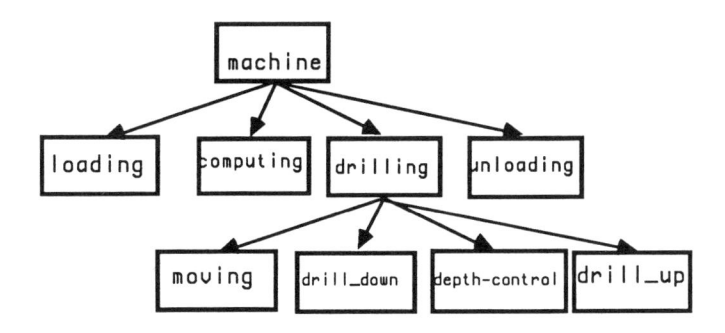

Fig. 2 A drilling process of machine M1 in workstation level

Here the workpiece <1.0,2.0,3.0,4.0> means four ordered drilled holes with depths 1.0, 2.0, 3.0, and 4.0. When this workpiece is sent to machine M1, it is loaded at first. Then some computations are done before drilling. The drilling process involves four steps: move the workpiece forward, make the drill down, control the depth, and make the drill up. The last operation is to unload the workpiece. From a personifying viewpoint, the order-passing procedure is from higher levels down to lower levels. From a physical viewpoint, the higher level consists of the components in the lower levels.

Programs	Operations
	machine:<1.0,2.0,3.0,4.0>

{machine [loading, drilling@computing, unloading]}
 {loading out@%"workpiece loading"}
 {computing distl@[/@[%10.0,!+@&%1],id]}
 {drilling (null -> out@%"finish drilling";
     [[moving@1@1,drill_down,depth_control@2@1,
     drill_up],drilling@tl])}
 {moving &out@[%"move workpiece forward", id]}
 {drill_down out@%"drill down"}
 {depth_control &out@[%"depth of drilling:", id]}
 {drill_up out@%"drill up"}
 {unloading out@%"workpiece unloading"}
	<"workpiece loading" <<<"move workpiece forward" 2.5> "drill down"> <"depth of drilling:" 1.0> "drill up"><<<"move workpiece forward" 2.5> "drill down"> <"depth of drilling:" 2.0> "drill up"> <<<"move workpiece forward" 2.5> "drill down"> <"depth of drilling:" 3.0> "drill up"><<<"move workpiece forward" 2.5> "drill down"> <"depth of drilling:" 4.0> "drill up">"finish drilling">>>> "workpiece unloading">

Table 1 The programs and their operations

Towards a higher level viewpoint, assume machine M1 and two other machines M2, M3 form cell C1; machines N1, N2, and N3 form cell C2; machines L1, L2, L3, and L4 form cell C3. With a similar extension upwards, assume cells C1, C2, and C3 form shop S1. They are illustrated in Fig. 3. More precisely, this hierarchy in lambda calculus model is like a superstring twisted among levels in Fig. 4.

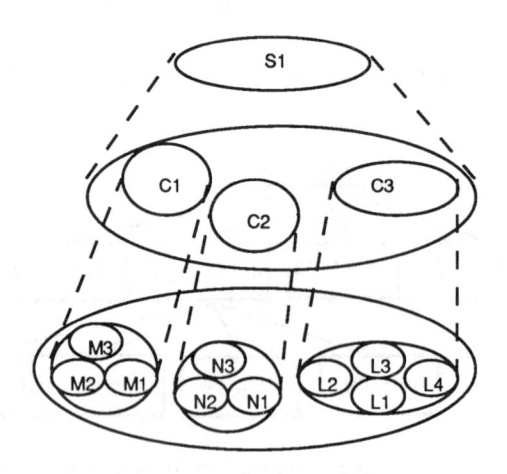

Fig. 3 A three-level hierarchy

From the analysis above, there is a direct interpretation within a lambda calculus model for the officer-cell phenomenon in manufacturing system. If the components in Fig. 3 are extracted from the lowest level to form a hierarchy, then this vertical approach reflects a personifying viewpoint. In other words, it looks like an order-passing procedure that officers give commands to sergeants if the orders are passed from higher levels to lower levels. In terms of lambda calculus terminology, this extraction is a series of substitution processes. If the components are synthesized within a level, then this horizontal approach reflects a physical viewpoint. In other words, a cell consists of several workstations on the shop floor. In terms of lambda calculus terminology, this synthesis is a series of applications and abstractions to form a twisted lambda term. Lambda calculus gives a satisfied interpretation for the puzzling officer-cell phenomenon in manufacturing system.

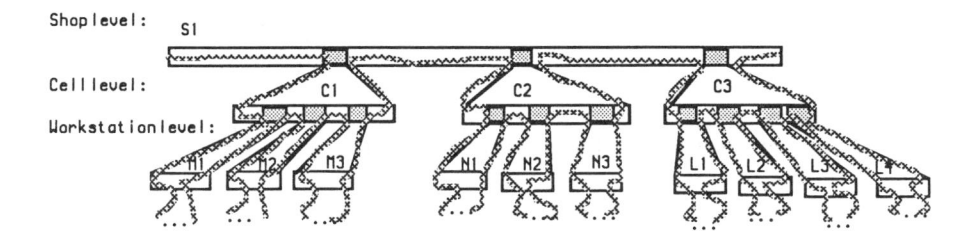

Fig. 4 A twisted lambda term in a hierarchy

Conclusions

A lambda calculus model is proposed to interpret the officer-cell phenomenon in manufacturing system. Under the lambda calculus model, the tangled shop-cell-workstation-equipment and general-officer-sergeant-troop hierarchies become separated. The former is a result from the vertical approach, and the latter is a result from the horizontal approach. Here lambda calculus explicates this phenomenon through functional language FP. However, FP only reflects partial features for the application of lambda calculus to manufacturing system. For example, the timing concept in Table 1 is interpreted from left to right in square brackets. Moreover, methods about distributed and parallel reduction processes for modeling activities in manufacturing system are still absent. The more advanced study requests the implementation of manufacturing-oriented lambda calculus language.

Acknowledgement

This study was supported by the National Science Council, Republic of China, under contract number: NSC79-0415-E009-02.

References
1. Proth, J. M. and Hillion, H. P., Mathematical Tools in Production Management, Plenum Press, 1990.

2. Gershwin, S. W., "Hierarchical Flow Control: a Framework for Scheduling and Planning Discrete Events in Manufacturing Systems," Proceeding of the IEEE, Vol. 77, No. 1, pp. 195-209, Jan. 1989.

3. Libosvar, C. M., "Hierarchies in Production Management and Control: a Survey," MIT Laboratory for Information and Decision Systems, Rep. LIDS-P-1734, Jan. 1988.

4. Jones, A. T. and C. R. McLean, "A Proposed Hierarchical Control Model for Automated Manufacturing Systems," Journal of Manufacturing Systems, Vol. 5, No. 1, pp. 15-25, 1986.

5. Albus, J., Brains, Behavior, and Robotics, BYTE Publications, Inc. 1981.

6. Akella, R., Y. Choong, and S. Gershwin, "Performance of Hierarchical Production Scheduling Policy," IEEE Trans. on Components, Hybrids, and Manufacturing Technology, Vol. 7, No. 3, pp. 225-240, Sept. 1984.

7. Weber, D. M. and Moodie, C. L., "A Knowledge-based System for Information Management in an Automated Integrated Manufacturing System," Robotics & Computer-integrated Manufacturing, Vol. 4, No 3/4, pp. 601-617, 1988.

8. Weber, D. M. and Moodie, C. L., "An Intelligent Information System for an Automated, Integrated Manufacturing System, Journal of Manufacturing Systems, Vol. 8, No. 2, 1989, pp. 99-113.

9. Huber, A., "Knowledge-based Production Control for a Flexible Flow Line in a Car Ratio Manufacturing Plant," Proc. 6th Int. Conference Flexible Manufacturing Systems, pp. 3-20, Nov. 1987.

10. Biemans, F. P. and Vissers, C. A., "Reference Model for Manufacturing Planning and Control Systems," Journal of Manufacturing Systems, Vol. 8, No. 1, pp. 35-46, 1989.

11. Naylor, A. W. and Volz, R. A., "Integration and Flexibility of Software for Integrated Manufacturing Systems, Design and Analysis of Integrated Manufacturing Systems, (Ed. Compton, W. D.), 1987.

12. Liang, G. R., "Functional Manufacturing Systems," Journal of Mechatronic Industry, pp.182-189, May, 1988.

13. Liang, G. R., "Functional Approach to the Implementation of Hierarchical Production Controllers," Proceedings of CIES, pp. 147-157, Taichun, Taiwan, R. O. C., Nov. 1990.

14. Liang, G. R., "The Reconfiguration of Manufacturing Systems Using Transformation Methods," Proceeding of the 3th National Workshop on Automation Technology, pp. 695-704, Kaohiung, Taiwan, R. O. C., July, 1989.

15. Vallespir, B., "Hierarchical Aspect of Production Management System Associated Modeling Tools and Architecture," The First International Conference on Automation Technology, pp. 635-641, Taipei, Taiwan, R. O. C. July, 1990.

16. Church, A., The Calculi of Lambda Conversion, Princeton University Press, Princeton, 1941.

17. Barendregt, H. P., The Lambda Calculus - Its Syntax and Semantics, North-Hilland, 1984.

18. Fox, M. S., "Industrial Applications of Artificial Intelligence," Computer Integrated Manufacturing (Ed. Jurksen, I. B.), Springer-Verlag, 1988.

19. Synder, W., Industrial Robots: Computer Interfacing and Control, Prentice-Hall, 1985.

20. Baden, S., Berkely FP User's Manual, Rev. 4.1, 1986.

21. Revesz, G., Lambda-Calculus, Combinators, and Functional Programming, Cambridge University Press, 1988.

Simulation Design Techniques in Network Analysis

Chandra S. Putcha, Professor and Jesa H. Kreiner, Professor
California State University, Fullerton, CA 92634, USA

Magdi H. Mansour, Associate Professor
Dalhousie University, Halifax, N.S., Canada B3H3J5

Abstract:

Simulation techniques are applied for the network analysis problems in this paper. The network analysis problem as it pertains to the present research work involves determination of project completion time of various networks. This concept has many applications in diverse fields such as the manufacturing, construction, or transportation industries etc. The project completion time depends on the path with the longest total completion time of all the given paths, which again, is a function of various activity durations according to known activity precedence relationships. The total project completion time is a probabilistic quantity, and rigorous mathematical techniques are required to find the 'exact' value. Hence, approximate techniques such as Monte Carlo simulation are used to find the project completion time for the network with input variables having normal and uniform distribution. The results are also compared with Rosenblueth's two point estimate scheme which is being widely used for the last several years for the solution of practical problems. Probabilities of the project completion time greater than a given level are calculated using the above method, modified suitably for the purpose of probability calculation and the results are compared with those obtained using simulation methods.

Introduction and Objectives

In many engineering problems uncertainty in data and in theories is considerable and hence, a probabilistic treatment of the problem becomes almost mandatory. Closed form solution of such problems is almost impossible. Hence, simulation methods are used. As is well known, simulation is the process of duplicating the real world based on a set of assumptions and conceived mathematical models, which is used to predict or study the response of a system. If a problem involves random variables with known or assumed probability distributions, Monte Carlo simulation is required. Monte Carlo simulation methods (1) and Rosenblueth's two point estimate method (2,3) are applied to a limited number of network problems. The necessary data are taken from available literature (4,5).

The aim of this project was to determine the project completion time of a given network with several activities where each activity duration is a probabilistic quantity. This kind of work has many applications in different industries like manufacturing, transportation, construction, etc. In general, in a network there are various paths from the starting point to the finishing point which in turn are functions of activity durations - all probabilistic in the present study. To determine the project completion time, one has to first express the total completion time as a function of all the possible states of nature. The second phase consists of expressing each state of nature as a function of activity durations according to the known activity precedence relationships, and then find the probability that the project completion time is greater than a given specified time. Hence, the project completion time can best be expressed in terms of probabilities.

Work in this direction of mathematical modeling has been done by Putcha and Kreiner (6,7). Also, Putcha and Rao (8) applied some of these concepts to network analysis problems and modified the existing Rosenblueth's method for calculation of probabilities.

In the present study, the effects of precedence relationships (each one can be considered as a state of nature), keeping the starting and finishing point same, is studied. In addition, the effects of duration of activities, statistical distribution of various activities on the project completion time are also studied. The distributions of random variables are taken as uniform and normal to start with. The results are obtained using Rosenblueth's two-point estimate method and then compared with Monte Carlo simulation and the 'exact' method which was being developed as part of the earlier study (8). The data for various activities are taken from the existing literature. A computer program was developed as part of this study which uses Rosenblueth's two-point scheme for a given set of activities in a network and compares the results with those obtained from Monte Carlo simulation and the exact method. The results are compared both in terms of statistical parameter, i.e, mean value, and standard deviation, and also the final values of probability of the event being considered.

Methodology

As stated earlier, two methods, namely, Monte Carlo simulation and Rosenblueth's two-point estimates method are used to evaluate the probability of project completion time greater than a given quantity. Monte Carlo simulation is discussed in detail elsewhere (1,5).

The results obtained using Monte Carlo simulation are compared with Rosenblueth's two-point estimate method. The details are discussed elsewhere (2,3). A brief description of the method is presented here:

Rosenbleueth's method mainly consists of obtaining the moments of y (where y is a function of a real random variable x) using the first two or three moments of x which are supposed to be known. The moments of y are derived as linear combinations of powers of the point estimates $y(x^+)$ and $y(x-)$ where x^+ and $x-$ are some specific values of x. The following expressions are specifically used in the present study:

$$m_y = 1/2 \ (y^+ - + y-) \tag{1}$$
$$\sigma_y = 1/2 \ (y^+ - y-) \tag{2}$$
$$V_y = (y^+ - y-)/(y^+ + y-) \tag{3}$$

The terms m_y, σ_y and V_y represent the expected value, standard deviation and coefficient of variation of y respectively.

Once two-point estimates are obtained for each random variable, the Rosenblueth's procedure then simply consists of combining these and getting the mean value and standard deviation of the final variable (which in turn is a function of several variables) for various precedence relationships as is the case for the given problem. As a final step, the probability of the required event is calculated by modifying the Rosenblueth's method.

Formulation of the Problem

The project completion time T, of a network shown in Fig. 1a and 1b is a function of D_1, D_2, D_3, and D_4 which are the activity durations, i.e.

$$T = f(D_1, D_2, D_3, D_4) \tag{4}$$

The activity durations D_1-D_4 are assumed to have uniform distributions on the intervals $[L_{L1}, L_{u1}]$-$[L_{L4}, L_{u4}]$ respectively where the L_{LL} and L_{uL} represents the lower and upper limits of the duration of various activities. Since the project completion time will depend on the path through the network with the longest total completion time, the above equation can be rewritten as:

$$T = \mathrm{Max} \begin{array}{c} D_1 + D_4 \\ D_1 + D_3 + D_5 \\ D_2 + D_5 \end{array} ; \ \mathrm{Max} \begin{array}{c} D_1 + D_4 \\ D_2 + D_3 + D_4 \\ D_2 + D_5 \end{array} \tag{5}$$

The first expression in the above equation represents the network in Fig. 1a and the second expression represents network in Fig. 1b. Similarly, for the network in Fig. 2, the project completion time T can be expressed as:

$$T = \text{Max} \begin{array}{l} D_1 + D_4 \\ D_3 \\ D_2 + D_5 \end{array} \qquad (6)$$

Results, Discussions and Conclusion

Three examples of network analysis problems shown in Fig. 1a and 1b and Fig. 2 are solved for probability for project completion time being greater than 20 days. In example 1, network is used with the activity durations D_1-D_4 having a uniform distribution [2,4], [4,5], [1,2], [2,3] and [2] respectively. The activity duration D_5 is considered deterministic. In example 2, network 1, is used again as shown in Fig. 1b with parameters of uniform distribution for various activities D_1-D_4 being [0,20], [0,10], [0,5], [0,15], [0,8] respectively. In example 3, network 2 as shown in Fig. 2, is used where the parameters for the uniform distribution are same as in example 1, but the states of nature used for project completion time are different. Example 1, (Fig. 1a) has been solved using simulation with sample size 15 and 30 in Ref. 5, and the mean values have been obtained as 0.26 and 0.27 for 18 repetitions, and the corresponding standard deviations are 0.10 and 0.07 respectively. The ranges are found to be 0.07-0.44 and 0.10-0.37 for sample size 15 and 30 respectively. The same problem has been solved using Rosenblueth's two-point estimate method. The mean value and standard deviation for various states of nature compared quite well with the corresponding simulation results, considering the fact that the simulation used only a sample size of 15. The probability of the event-project completion time being less than or equal to 6 is found to be 0.3974, which is in the same range as established by Simulation Method.

The results for examples 2 and 3 indicate a good correlation between the statistical parameters of all the posssible project completion times obtained using Monte Carlo simulation and Rosenblueth's method for uniform and normal distribution. For example, for the activity precedence relationship ($D_2 + D_3 + D_4$) and for uniform distribution, the mean value and standard deviation using Monte Carlo simulation are 14.98 and 5.39 respectively, whereas the corresponding values using Rosenblueth's method are 15.25 and 5.44.

Table 1 shows the final probabilities of the event--the project completion time greater than 20 by various methods--Monte Carlo and Rosenblueth's method. Again, as before, good correlation has been obtained between the results.

Table 1: Comparison of Probabilities By Various Methods

Event	Monte Carlo (N = 1000)		Rosenblueth (Modified)		"Exact"	
	Uniform	Normal	Uniform	Normal	Uniform	Normal
	Example 2					
$P(T>20)$	0.481	.459	0.500	.500	0.467	0.465
	Example 3					
$P(T>20)$	0.389	.389	0.385	.385	0.398	0.401

The efficacy of Rosenblueth's method has been shown for the solution of a general non-linear network problem and the results compared very well with the simulation results.

Acknowledgements

The authors are thankful to the Computer Center at California State University, Fullerton for the use of computing facilities.

References

1. Warner, R.F. and Kabalia, A.P., "Monte Carlo Study of Structural Safety", <u>J. Struct. Div. Proc. ASCE</u>, Vol. 94, 1968.

2. Rosenblueth, E., "Point Estimates for Probability Moments", <u>Proc. Nat. Acad. Sci.</u>, Vol. 72, No. 10, pp. 3812-3814, Oct. 1975.

3. Rosenblueth, E., "Two-Point Estimates In Probabilities," <u>Applied Mathematical Modeling</u>, Vol. 5, pp. 329-335, Oct. 1981.

4. Au, T., Shane, R.M., and Hoel, L., "<u>Fundamentals of Systems Engineering Probabilistic Models</u>". Addison-Wesley Publishing Company, 1972.

5. Ang. A.H-S., "<u>Probability Concepts In Engineering Planning And Design</u>," Vol. II, John Wiley & Sons, 1975.

6. Kreiner, J.H. and Putcha, C.S., "Modeling of Failure Probabilities of Shafts Subjected to Reverse Bending," <u>Proceedings of 21st Israel Conference on Mechanical Engineering</u>, Technion City, Haifa, 1987.

7. Putcha, C.S. and Kreiner, J.H. "Reliability Factor Analysis of Shafts", <u>Proceedings of Canadian Conference of Applied Mechanics</u>, Ottawa, May 28-June 3, 1989.

8. Putcha C.S. and Rao, S.J.K., "Network Analysis for Project Completion Time--A Probabilistic Study" communicated to <u>Civil Engineering Systems: Decision Making and Problem Solving</u>, Butterworth Publications, 1991.

9. Hart, G.C., "<u>Uncertainty Analysis, Loads, and Safety in Structural Engineering</u>", Prentice-Hall, 1982.

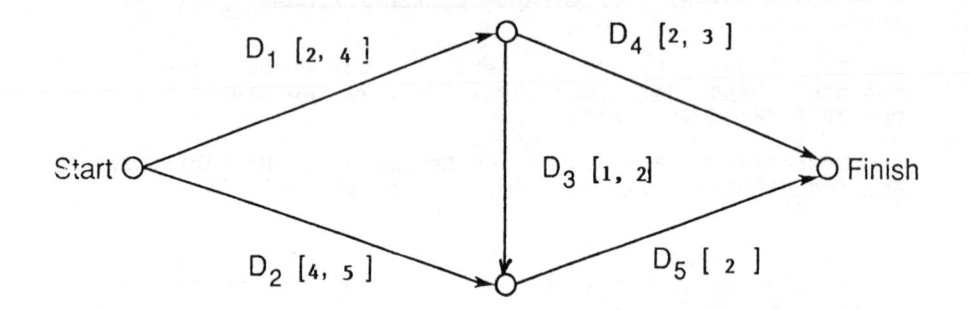

Fig. 1 (a) Network – For Example 1

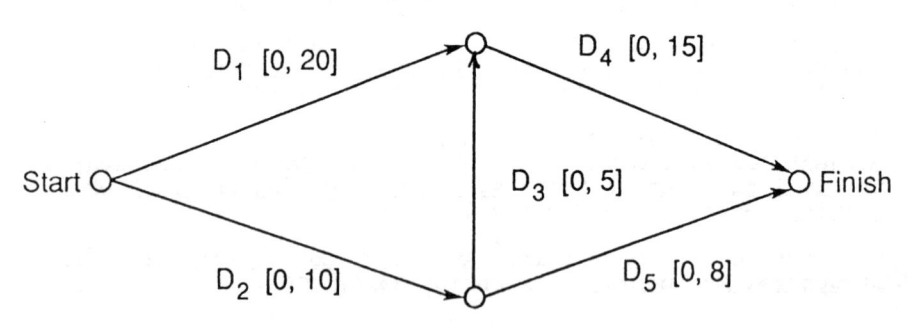

Fig. 1 (b) Network – For Example 2

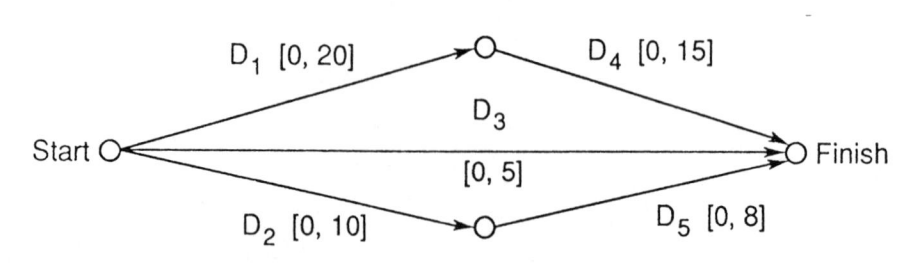

Fig. 2 Network – For Example 3

Manufacturing Cell Formation with Identical Machines: A Simulated Annealing Approach

T. HAMANN, J.-M. PROTH and X. XIE
INRIA, Technopôle Metz 2000, 4 rue Marconi, 57070 Metz, FRANCE

Abstract : This paper addresses the cell formation problem for a manufacturing system containing identical machines. It consists of grouping machines into cells so as to minimize the inter-cell traffic while satisfying the production capacity constraints. An important feature of this problem is that identical machines may be assigned to different cells. Two cases, the uncapacitated case and the capacitated case, are considered. Simulated annealing based algorithms are proposed to solve both cases. Numerical experiments show that the proposed algorithms lead to near-optimal solutions.

1. Introduction

Cellular manufacturing layout consists of grouping machines which manufacture nearly the same parts in the same cell. An optimal manufacturing layout is the one which minimizes the total inter-cell traffic. The cell formation problem has been considered in the literature ([1-10]). Most of these works assume that a unique routing is available for each part type (i.e. each part has to visit a unique sequence of machines in order to be manufactured). However, a real manufacturing system usually contains identical machines. Intuitively, identical machines should be assigned to different cells so as to reduce inter-cell traffic when several part families require the same type of machine.

This paper addresses the cell formation problem with identical machines. It consists of partitioning machines into cells of limited size so as to minimize the total inter-cell traffic while meeting the demand. We distinguish two cases: the uncapacitated case and the capacitated case. Simulated annealing based algorithms are developed to solve both cases.

This paper is organized as follows. Section 2 provides the problem formulation. Section 3 addresses the uncapacitated case while Section 4 addresses the capacitated case. Section 5 presents some numerical results. The last section is the conclusion.

2. Problem formulation

In this paper, we consider a job-shop with identical machines. More precisely, the manufacturing system is composed of a set M = $\{M_1, M_2, ..., M_m\}$ of m machines and a set P = $\{P_1, P_2, ..., P_n\}$ of n part types. Let K be the maximal number of machines in a cell and let f_i

be the inter-cell transportation cost of a P_i-type part. There are I machine types. Let $T_i \in \{1, 2, ..., I\}$ denote the type of machine $M_i \in M$ and let $O(k)$ be the set of k-type machines.

The production process of a part type P_i is determined by a sequence of operations and the related processing times. Each operation can be performed on any machine of a specified type. Let N_i denote the number of operations required to produce a P_i-type part. Let a_{ik} be the type of machines which can perform the k-th operation required to manufacture P_i-type parts and let τ_{ik} be the time required to complete this operation. Let d_i be the demand rate of the P_i-type parts.

Let $Y_{ik}(j)$ be the rate at which P_i-type parts use machine M_j for their k-th operation. Of course, $Y_{ik}(j) = 0$ if $M_j \notin O(\alpha_{ik})$. The production capacity constraint of machine M_j can be expressed as follows:

$$\sum_{i=1}^{n} \sum_{k=1}^{N_i} \tau_{ik} Y_{ik}(j) \leq 1 \qquad \forall M_j \in M \qquad (1)$$

Let $y_{ik}(j,l)$ be the rate at which P_i-type parts use machine M_j for their k-th operation and then use machine M_l for their k+1-th operation. The flow conservation rule implies that:

$$\sum_{j=1}^{m} y_{ik}(j,l) = Y_{i,k+1}(l) \qquad \forall M_l \in O(\alpha_{i,k+1}), P_i \in P, k < N_i \qquad (2)$$

and

$$\sum_{l=1}^{m} y_{ik}(j,l) = Y_{ik}(j) \qquad \forall M_j \in O(\alpha_{ik}), P_i \in P, 1 < k < N_i \qquad (3)$$

Furthermore, the demand should be satisfied, i.e.

$$\sum_{j=1}^{n} Y_{i1}(j) = d_i \qquad \forall P_i \in P \qquad (4)$$

Let $C = (C_1, ..., C_q)$ be a partition of the set of machines M. G_j indicates the cell containing machine M_j. The number of machines in each cell is limited by K, i.e.

$$\text{card}(C_r) \leq K \qquad \forall C_r \in C \qquad (5)$$

Let $F_i(C)$ be the inter-cell traffic cost of P_i-type parts. We have:

$$F_i(C) = \sum_{k=1}^{N_i} \sum_{(Mj, Ml) \in M \times M} f_i y_{ik}(j,l) 1\{G_j \neq G_l\} \qquad \forall P_i \in P \qquad (6)$$

where $1\{x\} = 1$ if the argument x is true and $1\{x\} = 0$ otherwise.

Finally, the cell formation problem consists of finding a partition C of the set of machines and the production ratios $y_{ik}(j,l)$ so as to

$$\text{minimize} \sum_{i=1}^{n} F_i(C) \tag{7}$$

subject to constraints (1) to (6).

The formulation of the uncapacitated case can be obtained by removing the production capacity constraints (1). Problem (7) is a NP-hard problem and below we propose heuristic solutions to this problem.

3. Uncapacitated case

This section presents a simulated annealing algorithm for computing the optimal cells. It is very close to the simulated annealing algorithm presented in [5]. An efficient algorithm is developed for computing the optimal part routings and the total inter-cell traffic costs when the cells are given.

3.1. A Simulated Annealing Algorithm

The simulated annealing algorithm can be summarized as follows:

1. Generate an initial partition C_0 and set $C = C_0$
2. Choose an initial temperature $T_0 > 0$ and set $T = T_0$
3. **While T > T_f do**
   3.1. **Repeat** L times 3.1.1 to 3.1.4
       3.1.1. Generate randomly a partition C' in the neighborhood of C
       3.1.2. Compute $\Delta F := F(C') - F(C)$.
       3.1.3. Generate randomly a real number $R \in (0,1)$.
       3.1.4. If $R < \exp(-\Delta F/T)$, set $C := C'$.
   3.2. Set : $T := g\,T$.

A neighbor of a given partition C is obtained by either: (i) removing a machine from one cell to another cell, or (ii) permutating two machines located in different cells, or (iii) taking out a machine from a cell and creating a new cell with this machine.

3.2. Inter-cell traffic computation with predetermined cells

This subsection provides an efficient algorithm to compute the inter-cell traffic F(C) when the manufacturing cells are given. This algorithm is based on the following properties whose proofs are omitted for lack of space.

Property 1: The partition of manufacturing system into cells being given, the assignment of parts to machines can be determined independently for the different part types and there exists at least one optimal solution in which a unique routing is used for each part type.

Property 2: For each part type, the optimal routing is designed as follows: (i) the first operation is performed in a cell in which a maximal number of consecutive operations can be performed; (ii) the selection of the subsequent cell is made following the same rule as in (i); (iii) the algorithm stops when all operations are assigned to machines. The inter-cell traffic cost $F_i(C)$ is equal to $f_i d_i (V_i - 1)$ where V_i is the number of cells visited by a P_i-type part. Note that the same cell can be visited several times.

4. Capacitated case

In the capacitated case, we start from the cells and routings of the uncapacitated case and derive a three step heuristic solution as described below.

4.1. Machine assignment

First, the cells are modified according to the production capacity requirement in each cell so as to satisfy as well as possible the capacity constraints. Let $W(r, j)$ be the workload of type j machines in cell C_r, i.e.

$$W(r, j) = \sum_{i=1}^{n} \sum_{k=1}^{N_i} d_i 1\{C^*(i, k) = C_r \text{ and } \alpha_{ik} = j\}$$

where $C^*(i,k)$ denotes the cell in which the k-th operation of a P_i-type part is performed.

The machines are assigned iteratively to cells. At each iteration, we choose a cell r^* and a machine type j^* respectively among cells which are still not full (i.e. $\text{card}(C_r) < K$) and among machine types containing unassigned machines with the maximal load. A type j^* machine is assigned to cell C_{r^*} and $W(r^*,j^*):=\max\{W(r^*,j^*)-1,0\}$

4.2. Production ratio determination

Starting from the modified cells, the second step defines the production ratios $Y_{ik}(j)$ in order to minimize the total inter-cell traffic. The inter-cell traffic minimization problem becomes a linear programming problem with real variables when the cells are known. It can be solved by a classical LP algorithm.

4.3. Inter-cell traffic minimization

At the third step, we assume that the production ratios $Y_{ik}(j)$ are given. The simulated annealing algorithm presented in subsection 3.1. is used to partition the manufacturing system into cells so as to minimize total inter-cell traffic. The only remaining problem is the computation of inter-cell traffic costs.

The rate at which P_i-type parts are sent to cell C_r for their k-th operation can be calculated as follows:

$$u(r, i, k) = \sum_{M_j \in C_r} Y_{ik}(j)$$

Thus, among parts sent to cell C_r for their k-th operation, at most u(r, i, k+1) parts stay in the same cell for their (k+1)-th operation. The other parts should leave the cell and thus introduce an inter-cell traffic cost. Following this reasoning, the total inter-cell traffic costs is given by :

$$F(C) = \sum_{i=1}^{n} \sum_{k=1}^{N_i - 1} \sum_{C_r \in C} f_i \text{Max}\{0, u(r, i, k) - u(r, i, k+1)\}$$

5. A Numerical Example

The manufacturing system considered hereafter is composed of 19 machines of 10 different types. There are 2 type-1 machines, 3 type-2 machines, 2 type-3 machines, 2 type-4 machines, 1 type-5 machine, 3 type-6 machines, 1 type-7 machine, 2 type-8 machines, 1 type-9 machine, and 2 type-10 machines.

There are 18 part types. The demand is of 1 part per time unit for each part type and the inter-cell traffic costs are equal to one. The production processes are the following (the values in parenthesis correspond to machine types and processing times respectively):

P 1: (9, 0.302), (8, 0.230), (8, 0.164)
P 2: (6, 0.472), (2, 0.258), (6, 0.012), (8, 0.102)
P 3: (8, 0.359), (8, 0.412)
P 4: (9, 0.422), (2, 0.472)
P 5: (3, 0.344), (10, 0.203), (1, 0.242), (1, 0.437)
P 6: (10, 0.405), (3, 0.110), (4, 0.145), (3, 0.131)
P 7: (1, 0.106), (4, 0.027)
P 8: (3, 0.468), (4, 0.024), (3, 0.309)
P 9: (4, 0.451), (4, 0.160), (2, 0.324)
P 10: (2, 0.125), (5, 0.068)
P 11: (5, 0.162), (6, 0.148)
P 12: (6, 0.438), (5, 0.037), (5, 0.127)
P 13: (6, 0.416), (4, 0.379), (2, 0.425)
P 14: (6, 0.172), (5, 0.173), (6, 0.479)
P 15: (2, 0.014), (2, 0.142), (2, 0.098), (1, 0.415)
P 16: (1, 0.366), (1, 0.112), (10, 0.370)
P 17: (10, 0.125), (2, 0.169), (2, 0.264), (2, 0.201)
P 18: (10, 0.211), (6, 0.485)

This is a randomly generated example whose optimal inter-cell traffic has been chosen to be equal to zero when the maximal number of machines in a cell is 5.

The following table gives the total inter-cell traffic costs of both uncapacitated and capacitated cases starting from 5 randomly generated initial solutions. Note that the maximal traffic (i.e. the traffic between machines) is equal to 35.

Trial	Max. cell size = 5 uncapacitated	capacitated	Max. cell size = 3 uncapacitated	capacitated
#1	0	0.379	3	3.314
#2	0	0.671	3	3.314
#3	0	2.554	3	3.314
#4	0	0.510	3	3.314
#5	0	1.733	3	3.314

6. Conclusion

In this paper, we address the manufacturing cell formation problem with identical machines. A mathematical formulation is given. Two cases, the uncapacitated case and the capacitated case, have been considered. Both cases are a generalization of the well-known inter-cell traffic minimization problem with unique routing. Simulated annealing based algorithms are proposed to solve both cases. Numerical results show that these two algorithms lead to fairly good solutions.

References

1. Askin, R.; Subramanian, S.B.: A Cost-Based Heuristic for Group Technology Configuration. Int. J. Prod. Res., 25 (1), 1987.
2. Burbidge, J.L.: The Introduction of Group Technology. John-Wiley, 1975.
3. Garcia, H.; Proth, J.M.: A New Cross-Decomposition Algorithm: The GPM Comparison with the Bond Energy Method. Control and Cybernetics, 15 (2), 1986.
4. Harhalakis, G.; Nagi, R.; Proth, J.M.: An Efficient Heuristic in Manufacturing Cell Formation for Group Technology Applications. Int. J. Prod. Res., 28 (1), 1990.
5. Harhalakis, G.; Proth J.M.; Xie X.: Manufacturing cell design using simulated annealing: an industrial application. J. of Intelligent Manufacturing, 1 (3), 1990.
6. Hilger, J.; Proth, J.M.: Manufacturing Cells and Part Families: Generalization of the GP method. Information and Decision Technology, 12 (1), 1991.
7. King, J.R.: Machine-component grouping using ROC algorithm. Int. J. Prod. Res., 18 (2), 1980.
8. Kusiak, A.: EXGT-S: A Knowledge-Based System for Group Technology. Int. J. Prod. Res., 26 (1), 1988.
9. McAuley, J.: Machine Grouping for Efficient Production. The Production Engineer, Feb. 1972.
10. McCormick, W.T.; Schweitzer, P.J.; White, T.E.: Problem Decomposition and Data Reorganization by a Cluster Technique. Operation Research, 20, 1972.

Robotics

The Study of Micro-Drive Robotic End Effector

Cai Hegao, Liu Hong and Zheng Quan

Robot Research Institute,
Harbin Institute of Technology, Harbin,150006, P.R. China

Summary

As to enhance robotic effective positioning accuracy, the paper describes a new force controlled micro direct-drive end effector, which can realize delicate adjustment according to the information from force sensor. The active adjustment is actuated by 2-dimensional planar linear stepper motor which can move along x- and y- axes in the range from -2mm to +2mm with the resolution $0.78 \mu m$. The paper presents the modified mechanical structure, the independent force feedback control system, and the macro / micro manipulator system. The theoretical study proves the high dynamic stability of the macro / micro manipulator system. At last, the experiment results of using this system to finish the peg-in-hole assembly are given.

Introduction

The carrying out of delicate assembly tasks with close tolerances is complex because of the interactions which exist between the robot's end effector and its environment.

There are two approaches to solving this problem:
- a quasi-static approach which exploits force information in a logical way in order to performe trajectory corrections;
- a dynamic approach which exploits force information continuously in a servo loop

It is this latter approach which is used in this paper as it seems the best adapted to delicate assembly tasks. The most classic consists in equipping the manipulator with a force sensor and a compliant system. This system, even though used frequently, does not allow a good level of performance to be obtained as the stability problems it creates do not allow very good bandwidths to be obtained.

The other solution, which we propose in this paper, is based on the concept of a macro / micro manipulator.It consists in associating a micro-manipulator to a main robot. This idea has already been presented in certain publications.[1][2][3]

The kernel of the macro / micro manipulator system is the development of micro manipulator,which should have some sensing ability, high frequency response and positioning accuracy[3].This paper presents a micro manipulator which can realize not only active adjustment but also passive compliance.

The following parts of the paper describe the micro manipulator design, the macro / micro manipulator system and its stability analysis,and at last, the peg-in-hole assembly experiment results are given.

Micro manipulator design

The micro manipulator has two parts, one is the passive compliant RCC constructured by double-deck laminated plates, which can realize passive adjustment about x- and y- axes, the other is the force-controlled precision micro-driving wrist (FPMW-II), which is the further development of FPMW-I[3].

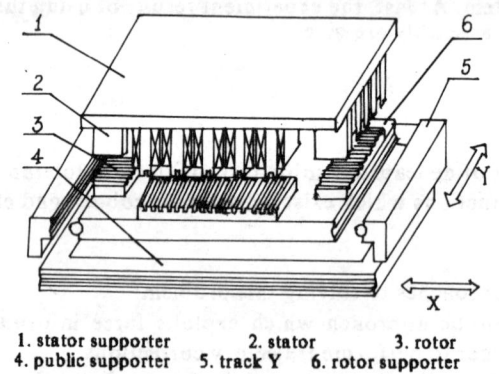

1. stator supporter 2. stator 3. rotor
4. public supporter 5. track Y 6. rotor supporter

Fig.1 Diagram of the FPMW-II structure

The design of FPMW-II has the following characteristics:
- the rotor can move along x- and y-axes independently, the range of motion is from -2mm to +2mm;
- the air gap between the stator and the rotor can be more easily adjusted from 100 μm to 150 μm;
- the highest velocity is 30 cm / sec and high resolution (0.78 μm);
- the largest force output can reach 25.8 N;
- high positioning accuracy,compact structure, light weight and high fre-

quency response.

The main part of FPMW-II is a 5-phase 20 beat hybrid linear stepper motor. It is based on the Sawyer Linear Motor Principle and can realize two dimensional motions along x- and y-axes. The Mechanical structure with a public supporter is shown in Fig.1 . This kind of composition has virtues of light weight, low friction and compact structure. Also, in order to regulate the air gap between the stator and the rotor, we design the track y as an adjusting track. Using the regulating screws and flexible beam, the air gap can be precisely regulated.

The macro / micro manipulator system

As shown in Fig.2, the macro / micro manipulator system is composed of a general purpose commercial robot PUMA-562, the macro manipulator, and its controller, a direct-drive end-effector and a RCC wrist, the micro manipulator, a F / T 30 / 100 force sensor, a pneumatic pump, a pneumatic gripper and an IBM-PC computer. The whole system is constructed to finish the most common peg-in-hole assembly as a typical application of fine manipulation.

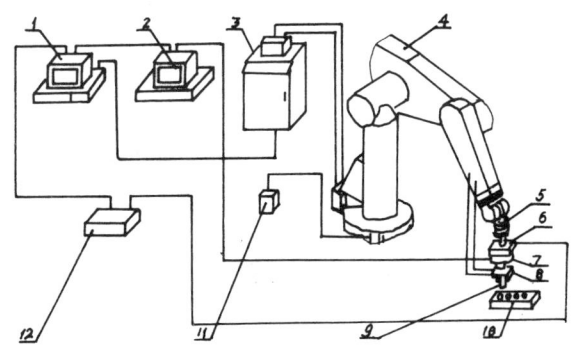

1. IBM-PC   2. F / T sensor controller   3. PUMA-562 controller   4. PUMA562
5. RCC wrist   6. FPMW-II   7. F / T 30 / 100 force sensor   8. pneumatic gripper
9. workpiece   10. hole block   11. pneumatic pump   12. motor driver

Fig. 2 Macro / micro manipulator system

The micro manipulator FPMW-II has been mounted on a Unimation PUMA-562 six axes robot. The robot controller and the FPMW-II microcomputer controller are interfaced through an RS 232 serial link. Communication at each ends is driven by hardware interrupts, allowing buffering of incoming commands without aborting the currently executing process.

The FPMW-II control procedures are invoked by commands received from the robot controller. Several library routines have been implemented in VAL-II. the robot programing language, that can be called by applications program. The FPMW-II microcomputer is essentially "slave" to the robot actually. It will remain in a wait state until a command is received from the robot, at which time the appropriate FPMW-II control routine is executed until another complete command is received. After most commands a status byte is returned to the robot, indicating the correct status of FPMW-II, the status byte can then be used in decision logic at the robot controller level.

The stability analysis of macro / micro manipulator system

The dynamic macro / micro manipulator system model can be simplified as shown in Fig.3. The driving force is actuated on the micro manipulator.

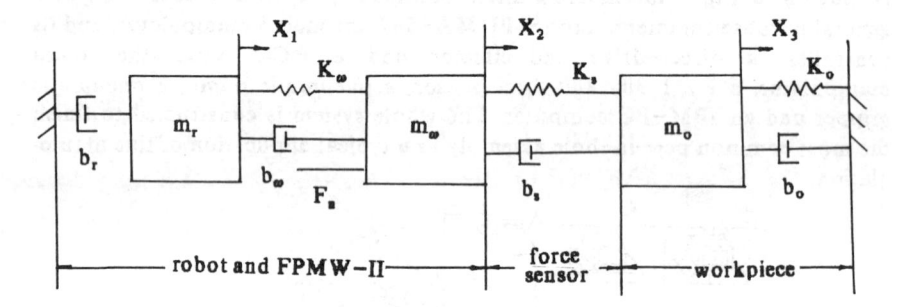

Fig.3 Dynamic model of macro / micro manipulator system

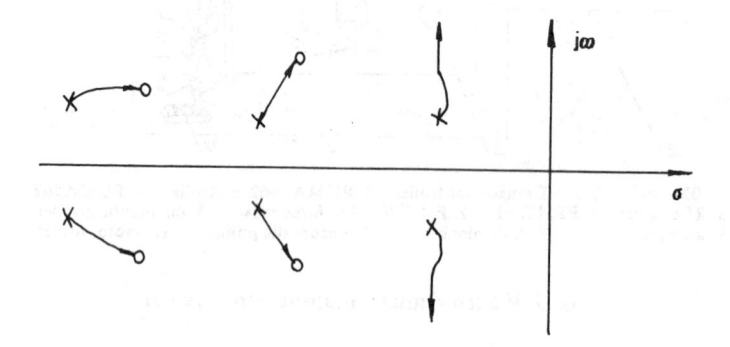

Fig.4 The root locus of the system

The open loop transfer functions relating X_2 to F_a and X_3 to F_a are:

$$\frac{X_2(S)}{F_a(S)} = \frac{D_o D_r}{D_\omega D_r D_o - K_\omega^2 D_o - K_s^2 D_r} \tag{1}$$

$$\frac{X_3(S)}{F_a(S)} = \frac{(b_s S + K_s)D_r}{D_\omega D_r D_o - K_\omega^2 D_o - K_s^2 D_r} \tag{2}$$

where: $\quad D_\omega(S) = m_\omega S^2 + (b_\omega + b_s)S + K_\omega + K_s$

$D_r(S) = m_r S^2 + (b_r + b_\omega)S + K_\omega$

$D_o(S) = m_o S^2 + (b_s + b_o)S + K_\omega + K_o + K_s$

The force Fs is detected by force error:

$$F_s = K_s(X_2 - X_3) \tag{3}$$

$$F_a = K_j(F_d - F_s), \quad K_j \geqslant 0 \tag{4}$$

Synthesizing the formulas (1),(2),(3),(4), There are four open loop zeros and six open loop poles. The angles between the real axis and two asymtotes are ± 90°. Fig.4 shows the root locus of the macro / micro manipulator system. From the diagram of the system root locus, we can find that,with the increase of K_f, the root locus doesn't appear in the right half plane and the system keeps constant stable.

The physical explanation for the stability of the system is that the contact force is adjusted directly by micro manipulator not by robot itself, whose dynamic characteristic is the source leading to system unstable.

The experiment results with peg-in-hole assembly

The crux of the peg-in-hole assembly is the high speed adjustment of the position and the posture. Using macro / micro manipulator system described forehead, the peg-in-hole assembly work with close tolerance and relatively great initial positioning error has been finished.

Based on the force analysis,the system strategy is divided into two periods after the macro manipulator finishes the coarse positioning. The first stage is the approach phase. The FPMW-II can realize quick search along an optimum path. Once the value of the Fz has a sudden decrease, the second stage, i.e., insertion phase begins. At this stage the FPMW-II proceeds high speed adjustment according to the force information from force sensor, and the compliant RCC completes the passive posture adjustment.

The force F_x, F_y, and F_z in the peg-in-hole process are shown in Fig.5. The clearance ratio of the peg and hole is 0.0015 and the whole work cycle is 2.5 second.

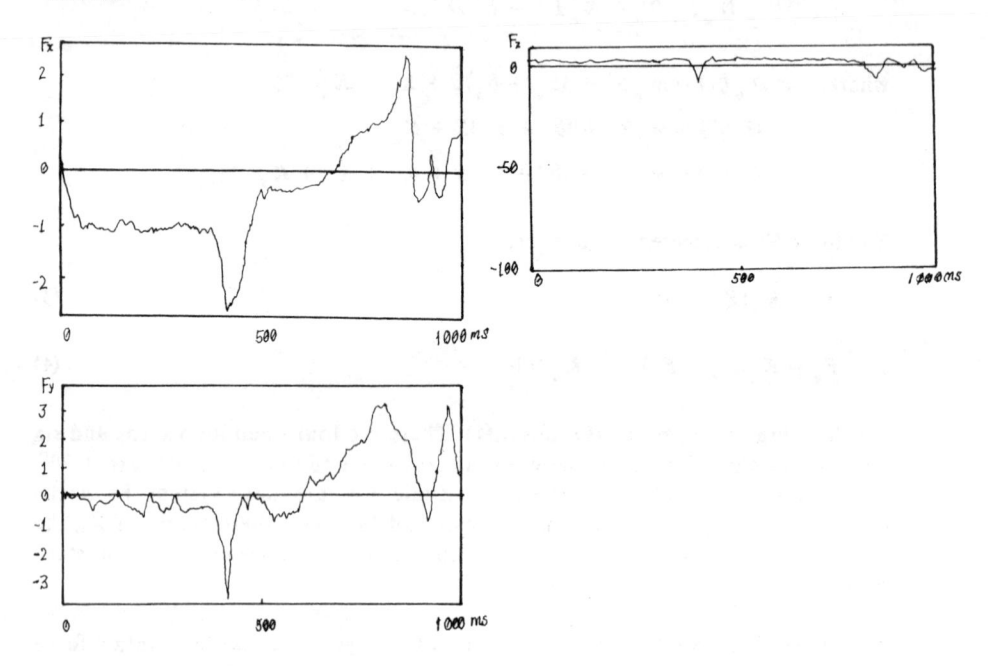

Fig. 5 The force information in the process of peg-in-hole insertion

References

1. Reboulet C. and Robert A."Hybrid control of a manipulator equipped with an active compliant wrist",3rd ISRR,Oct.1985,Gouvieux.

2. Sharon A.,Hogan N. and Hardt D."High bandwith force regulation and ineritia reduction using a macro/micro manipulator system",IEEE Int. Conf. on Robotics and Automation,Philadelphie,USA,April 1988.

3. H.G.Cai,H.Liu,etc.,"A new kind of force-controlled micro driving robotic end-effector(FPMW-I)",1990 Japan-U.S.A. Symposium on Flexible Automation, pp. 249-252.

Development of Excavation Robot for Underground Space Construction

T. Arai, T. Nakamura, K. Homma, R. Stoughton, and H. Adachi
Robotics Department, Mechanical Engineering Laboratory, AIST, MITI
Namiki 1-2, Tsukuba, Ibaraki 305, Japan

Summary
Japan is a mountainous country and her plains are densely populated, especially in large cities such as Tokyo, Osaka, etc. The MITI has started a National Research and Development Program on Underground Space Development Technology to establish a new frontier underground. In this program, an automated excavating robot will be developed to help construct an underground dome. The Mechanical Engineering Laboratory is engaged in the development of key technologies for the robot and has started basic studies on 1) parallel link manipulators, 2) walking vehicles, 3) remote position sensing, and 4) task planning for excavation. This paper describes our concept of an automated excavating robot and a newly developed parallel link manipulator applied to the excavation task.

Introduction
The Japanese Ministry of International Trade and Industry(MITI) has started a National Research and Development Program on Underground Space Development Technology to establish a new frontier at depths of greater than 50 m below ground level [1]. In this program an automated excavating robot will be developed to help construct an underground dome of approximately 100 m diameter and 30 m height. The construction will be planned so as not to effect activities at the ground surface. In the preliminary concept, the area is carefully surveyed and evaluated by using computer tomography techniques and data obtained from several bore holes. Next, the ground around the dome is reinforced with a spiral tunnel and rock-bolts. The inside part is then excavated and the wall is lined without pumping out the underground water. Since the excavation task will be executed in underground water, it must be automated by robots. Unfortunately, automatization technology lags behind in this area. One exception is an underwater bulldozer that will not be applied to this project purpose because of its poor performance record.

The Mechanical Engineering Laboratory is engaged in the development of key technologies for the robot and has started basic studies on a) parallel link manipulators, b) task planning for excavation, and c) walking vehicles [2]-[4]. In our preliminary design the robot system consists of a powerful manipulator on a walking vehicle with a dedicated task planner as shown in Fig.1. Since the manipulator for the excavation task is required to produce large forces and to work in water, a parallel link mechanism is expected to be applied and prototype mechanisms have been developed. The robot is required to move around and execute the excavation task effectively on unstructured terrain. Legged locomotion is expected to support the whole machine stably as well as allow it to walk around on such a terrain. Appropriate task planning including real-time prediction of soil and rock behavior must be developed for automated excavation.

Manipulator description
Two prototype manipulators were constructed to explore the basic characteristics of parallel link mechanisms. A serial link mechanism is employed in most conventional industrial robots and experimental manipulators. On the other hand, a parallel link mechanism, composed of 6 links connecting an end effector with a base in-parallel, has many advantages over conventional serial link manipulators, such as, large load capacity, high positioning accuracy, simple inverse kinematics and durability. Since these characteristics are desirable for its application to the excavation task, a prototype manipulator was built as shown in Fig.2 to check the basic movement and controllability in 6 axis joy stick control and force control tasks. The

manipulator was designed to generate forces and moments at the end effector as isotropically as possible. It is equipped with a passive serial link mechanism in the center to get the position and orientation of the end effector directly for real-time control, since it is difficult to derive the forward kinematics in closed form.

Based on experience with the prototype, a more practical manipulator with large power was developed to be applied to the actual excavation task. Fig.3 shows the configuration of the manipulator. It mainly consists of 6 prismatic links, a round end effector plate, and a base plate. Each link has a DC brushless motor with a reduction gear and a ball-and-screw joint as an actuator. Each link is connected to the base with a universal joint, and to the end effector with a gymbal mechanism and a rotary joint. All the links are allocated in the same configuration as in the prototype to get the homogeniety of force transmission between the actuators and the end effector. Principal mechanical specifications are listed in Table 1.

Currently the following control software is being developed for application to the excavation task.

(i) Trajectory control in task coordinates

Based on discrete data given as a trajectory for excavation, a path with smooth velocity and acceleration is generated, and the end effector is moved along the path. The inverse kinematics, interpolations of position and velocity and acceleration, and joint servos are included.

(ii) Reference control to a given time series of data in task coordinates

The end effector is moved to follow the referenced data in task coordinates in real time. This may be extended to master slave teleoperator control.

(iii) Force control

Force control enables the end effector to produce any force and/or moment in task coordinates or in end effector coordinates. A 6 by 6 set of linear equations using the Jacobian matrix should be solved to derive inverse statics and get the required torques of the actuators.

(iv)Position/force hybrid control in designated coordinates

The excavation task requires the manipulator to move the end effector along the pre-planned trajectory keeping the appropriate excavating force. This is not done by a simple trajectory control or force control. Some forces and positions should be controlled at the same time in the designated coordinates.

Walking mechanism

Locomotion capability is essential for the excavation robot to carry out the excavation task for constructing the huge dome. The locomotion mechanism must be capable of supporting the whole robot stably under any foreseeable conditions. The following requirements must be addressed in designing the locomotion system.

(i) Omni-directional locomotion

The robot should move around work sites quickly to execute the excavation task efficiently. It should be able to move in any direction smoothly from the current position without changing its orientation [5]. This is called "omni-directional" locomotion and is useful for locomotion while doing some manipulation task.

(ii) Adaptability to rough terrain

There is a mixture of fragile rock, earth and sand at the work site, and the land surface is complex and rough. Since there is also muddy underground water, the land surface may not provide good support for the robot. The robot must move around under such bad conditions.

(iii) Supporting capability

The excavation task produces a large reaction force between the robot and the land surface.The locomotion mechanism should actively produce this reaction force and help the manipulator work efficiently.

(iv)Stability

The locomotion mechanism carries the manipulator which has a large motion range while working. The mechanism should stably support the whole robot system even when the manipulator changes its center of mass.

(v)Motion for dexterity

The excavation cutter should be controlled to realize appropriate positioning and orienting for the task. Although this is basically done by the manipulator, a redundant motion would be preferable to perform the task efficiently. The locomotion mechanism should be capable of offering additional motion for this purpose as well as for locomotion.

There are wheel, crawler, and legged mechanisms currently employed for locomotion of robots. These three types of mechanisms should be evaluated in terms of the above mentioned requirements. First, as to omni-directional capability, the wheel cannot attain it, and the crawler can only change its orientation. The leg can realize this function completely. Second, the wheel cannot run around rough terrain. The crawler and the leg mechanisms can do it. Third, while the wheel and the crawler can realize the supporting capability only by increasing their own mass, the leg can produce a force and a moment in any direction actively. Fourth, both the wheel and the crawler have fixed supporting patterns and thus the mechanisms must be made larger to increase their stability. The leg mechanism is capable of adapting its supporting patterns to its environments and thus can achieve the same stability with a smaller size. Fifth, while the wheel and the crawler mechanisms are capable of realizing redundant motion for the robot with an additional suspension mechanism, the leg mechanism can realize this function without modification. From all these points of view, the leg mechanism is expected to be applied to the locomotion part of the excavation robot in spite of several problems to be solved.

Configuration and number of legs should be discussed to design an appropriate leg machine for the required task. In leg configurations there are two main types classified, a mammal type in which each leg extrudes down from a body, and an insect type in which each leg goes up first. Since stability is an important factor in the excavation task, the insect type is desirable to get a wide supporting pattern. As to number of legs, four or six are widely used in many robots. In insect type, having fewer legs is preferable for dexterity and controllability. A leg mechanism as depicted in Fig.1 is proposed for the locomotion of the excavation robot.

<u>Measurement of location and orientation of the robot</u>

Location and orientation of a target object can not be measured using a ruler or optics or other non-contact methods in underwater, underground environments. In this project a location and orientation measurement method based on inertial navigation is proposed for application to the robot working in the excavation task. Since measuring accuracy is generally dependent on the elapsed time, an error is accumulated in a conventional method where a gyro system is on board the robot if the robot works for a long time. In the proposed method a flying object that has a sensor, such as a gyroscope, runs inside a guide tube between the base and the robot, and measures location and orientation at some interval as in Fig.4. A configuration for the sensor and a measuring algorithm will be discussed in the following.

Sensor system

Fig.5 shows a configuration of a flying sensor system in the guide tube. The flying sensor has wheels to reduce a friction between the tube and itself, a driving tether, gyroscopes and accelerometers. Defining sensor coordinates $x_b y_b z_b$ fixed on the flying sensor, three accelerometers and three gyroscopes are set on the coordinate axes to measure the acceleration vector $d^2\mathbf{r}/dt^2$ and the angular velocity vector $\mathbf{w}_b$. In this method gyroscopes and accelerometers are fixed on the flying sensor and it is called a strap-down method.

Measurement of orientation

Let T be a homogeneous transformation matrix between inertia coordinates (or base coordinates) and the sensor coordinates. T is defined as follows.

$$T = \begin{pmatrix} l_x & m_x & n_x \\ l_y & m_y & n_y \\ l_z & m_z & n_z \end{pmatrix} \quad (1)$$

where l, m, n represent unit vectors of each sensor coordinate direction. The orientation of the sensor is defined by the matrix T and this satisfies the following equation.

$$\frac{dT}{dt} = T \begin{pmatrix} 0 & -\omega_{zb} & \omega_{yb} \\ \omega_{zb} & 0 & -\omega_{xb} \\ -\omega_{yb} & \omega_{xb} & 0 \end{pmatrix} \quad (2)$$

Thus, the orientation can be derived by integrating the equation. The orientation is expressed by using roll, pitch, and yaw angles in the actual application.

Measurement of location

Defining vectors **r** and **r**$_b$ which represent a position of a certain point on the sensor in the base and sensor coordinates respectively, then the following equation can be derived.

$$\frac{d^2\mathbf{r}}{dt^2} = \frac{d^2\mathbf{r}_0}{dt^2} + \frac{d^2\mathbf{r}_b}{dt^2} + 2\omega \times \frac{d\mathbf{r}}{dt} + \frac{d\omega}{dt} \times \mathbf{r}_b + \omega \times (\omega \times \mathbf{r}_b) \tag{3}$$

where $\times$ is the vector cross product, and $\omega = (\omega_{xb}, \omega_{yb}, \omega_{zb})^T$. The position **r** can be found by integrating the equation.

Excavation task planning

In underground space development, the working space is located so deep that it is difficult for an operator to observe the excavation process and give detailed command instructions. Therefore, autonomous motion planning from a roughly described goal given by the operator is needed.

As a precondition, we assume that the progress of the work is given. Work is executed with the following process.
(i) Excavation planning : first, the robot senses its position, and the terminated position of the day before and the goal of the day are given by the operator at the surface. With these data, the robot plans the route to the work site and the process of the work.
(ii) Motion to the work site : after that, the robot begins movement to the starting point.
(iii) Recognition of object : the robot moves to the initial working position and makes use of tactile sensors to get information about the surface of the soil. This recognition of the local working area uses tactile sensing only, without measuring tools.
(iv) Motion planning : from the surface information obtained in (iii), a model of the soil is made. Based on this model, precise motion planning such as the end effector trajectory and estimation of generated forces is made, and then motion of the manipulator and legs is executed.
The model represents the shape and physical characteristics of the soil. It is formed by expansion from datum of each point on the surface to the state of the whole surface. To express the fall down of the soil, fuzziness of information is added to the model.
For surface recognition, it is effective to use tactile sensors attached to the manipulator in addition to vision. Tactile sensing gives information about the sensed point; position, force and torque. To expand this to information about the surface, a method of surface modeling such as a Bezier surface can be used. Such an expression is for a definite surface such as the surface of a rigid body. To model the surface with fall down, an N-value probability function is added to the formula. When the soil is hard, the probability function becomes like a δ-function. When the soil is soft, it becomes more gradual.

$$z(x,y) = f(z_0(x_1,y_1), z_0(x_2,y_2), \cdots, z_0(x_n,y_n)) \otimes g(N(x,y)) \tag{4}$$

$z_0(x_i, y_i)$ shows the measured value, f shows the function which expresses free surface, g shows the probability function on N-value, and $\otimes$ shows convolution.
To represent physical characteristics, an N-value function expression is introduced. As in the case of shape, surface information consists of the data for each measured point. However, the function is like a step function, because the N-value changes according to the change in the type of soil.

$$N(x,y) = h(N_0(x_1,y_1), N_0(x_2,y_2), \cdots, N_0(x_n,y_n)) \tag{5}$$

$N_0(x_i, y_i)$ shows the measured value, and h shows the function expression of the N-value.
(v) Correction of motion planning : while executing work based on the motion planning, if the situation turns out to be different from the prediction, the model and the motion planning are

corrected. For example, the slope of the area can be calculated from the reactive force and moment on the edge of the tool, and can be used to correct the model.
(vi) Motion to next working position : when work which must be done at the position is finished, the robot moves to the next working position and repeats the process (iii) to (vi).

Conclusions
This paper discusses a research concept and currently on-going research in key technologies required to excavate and construct an underground space. A parallel link manipulator is proposed for the application to the excavation task. A leg mechanism is discussed to enable the robot to move around underwater over rough terrain and to get dexterity. A measurement method for location and orientation is also proposed. As to excavation planning, a procedure and a method are generally discussed.
 In future works, the manipulator will be applied to an actual excavation task, a prototype leg mechanism will be constructed, and a practical planning algorithm will be developed.
 The authors would like to express their appreciations to the office of National Research and Development Program, MITI for the encouragement of our research project.

References
[1] Agency of Industrial Science and Technology, "National Research and Development Program (Large-Scale Project)", 1990, p.14.
[2] Arai, T., et al., "Design, Analysis and Construction of a Prototype Parallel Link Manipulator", *Proc. of IEEE Inter. Workshop on Intelligent Robots and Systems*, July 1990, pp. 205-212.
[3] Adachi, H., et al., "Development of a Quadruped Walking Machine and Its Adaptive Crawl Gait", *Proc. of Japan-U.S.A. Symposium on Flexible Automation*, July 1990, pp. 91-94.
[4] Homma, K., et al., "Spatial Image Model for Manipulation of Shape Variable Objects and Application to Excavation", *Proc. of IEEE Inter. Workshop on Intelligent Robots and Systems*, July 1990, pp. 645-650.
[5] Lee, W-J, et al., "Omni-Directional Supervisory Control of a Multilegged Vehicle Using Periodic Gaits, *Trans. IEEE R&A*, Vol.4-6, December 1988, pp.635-642.

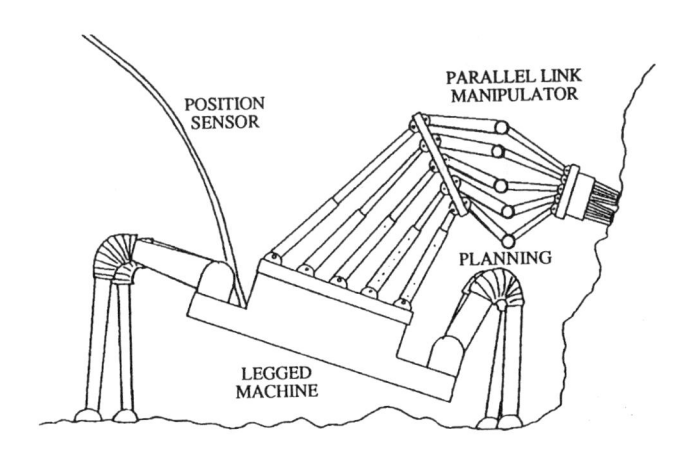

Fig.1. Concept of excavation robot.

Fig.2. Picture of prototype manipulator.

Table I SPECIFICATION OF PARALLEL LINK
MANIPULATOR FOR EXCAVATION

CONFIGURATION	6 PRISMATIC LINKS, 6DOF
MOTOR	100w DC BRUSHLESS
TRANSMISSION GEAR	BALL JOINT (5mm LEAD)
RANGE OF LINK	1,140mm - 1,638mm
NO LOAD MAXIMUM SPEED OF LINK	300mm/sec
PEAK CONTINUOUS FORCE	45.3kgf
PEAK STALL FORCE	125.6kgf
RESOLUTION OF ENCODER	0.00061 (5/8192) mm/pulse
MEASURING LINK	5 ROTARY POTENTIOMETERS
INTERFERENCE DISTANCE	90mm

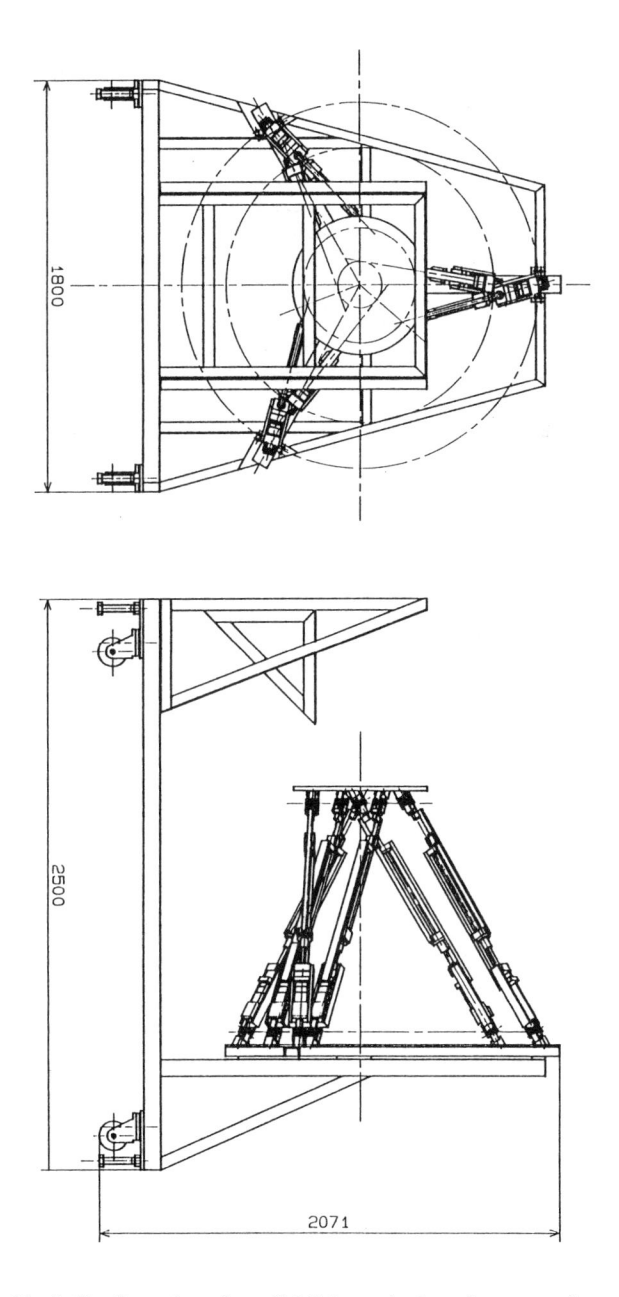

Fig.3. Configuration of parallel link manipulator for excavation.

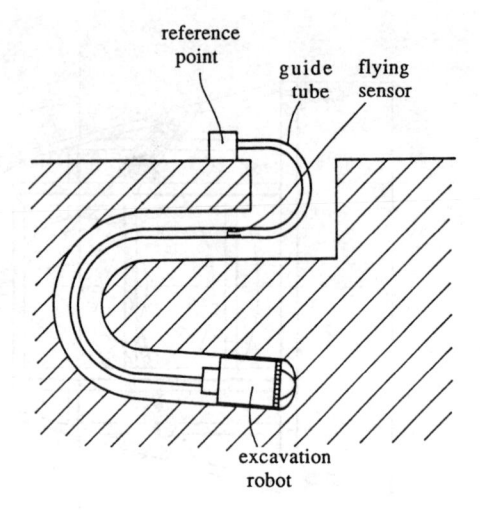

Fig.4. Principle of measurement of location and orientation.

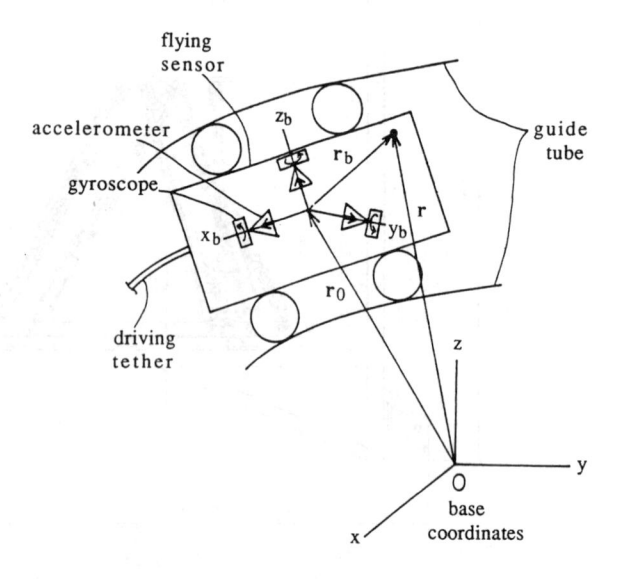

Fig.5. Configuration of flying sensor.

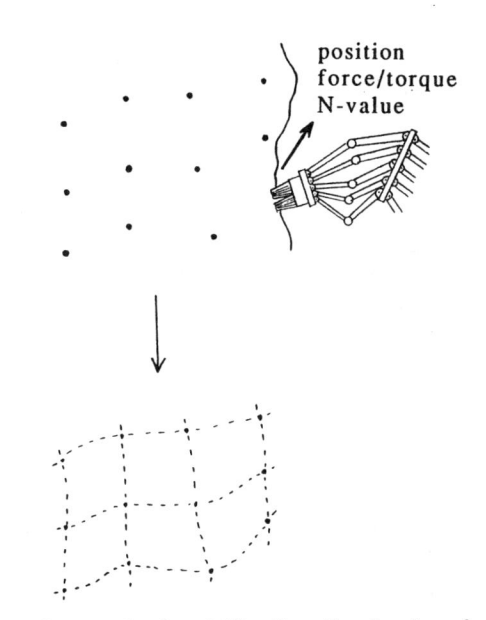

Fig.6. Example of modelling for soil and rock surface.

On a Robot which Plays the Xylophone

On a Robot Which Plays the Xylophone

M. ITOH

Department of Mechanical Engineering
Ashikaga Institute of Technology
Ashikaga, Tochigi, Japan

Summary

This paper deals with the brief description of an apparatus as well as its computer programs which is able to play the xylophone in the range of two octaves. The apparatus is composed of the xylophone with metallic pipes, a controller, an interface and a personal computer. This apparatus, which we call a robot, is operated manually by tapping the key on the keyboard of the personal computer and also can be operated automatically by music programs which are previousely stored in a floppy disk. The mechanism of xylophone unit, the interface and the controller are briefly described.

Introduction

The industrial robots have been developed since the beginning of 1960. At the present, many kinds of robot are manufactured and used in various industrial fields. Concerning the musical instruments, several kinds of robots which can play musics such as playerpiano have been developed. In regard of the robots which can play the xylophone, quite few examples such as the one produced by the Pacific Industrial Co., Ltd., in 1985 [1] have been reported. In this paper, a robot which plays the xylophone in the range of two octaves is described.

Outlines of the apparatus

The block diagram of the apparatus of this robot is shown in Fig.1. The personal computer is composed of a display, a computer proper and a keybord. The xylophone unit is composed

```
┌─────────────┐    ┌─────────────┐
│ Personal    │    │ Xylophone   │
│ computer    │    │ unit        │
│┌───────────┐│    │┌───────────┐│
││ Display   ││    ││ Xylophone ││
│└───────────┘│    │└───────────┘│
│┌───────────┐│    │┌───────────┐│
││ Computer  │┤....├│ Controller││
││ proper    ││    │└───────────┘│
│└───────────┘│    │┌───────────┐│
│┌───────────┐│    ││ Interface ││
││ Keyboard  ││....┤└───────────┘│
│└───────────┘│    │             │
└─────────────┘    └─────────────┘
```

Fig.1. Block diagram of the apparatus

Table 1 Manufacturers of each instrument

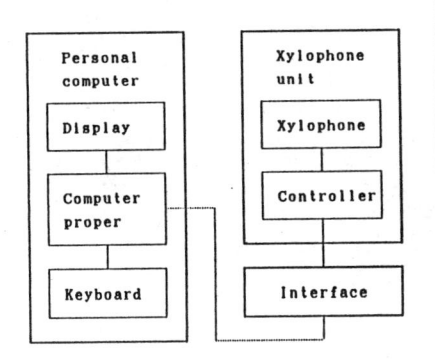

Instrument	Type	Manufacturer
Metallic pipe of xylophone	14K	Kawai Musical Instrument manufacturing Co.,Ltd. Japan
Controller	PZ-AP1	Pacific Industrial Co.,Ltd., Japan
Personal Computer	PC-9801RA	Nippon Electric Co.,Ltd.,Japan

of a controller and the xylophone made of metallic pipes in the range of two octaves. The dotted lines in the figure indicate the electric wirings connecting each instrument. The side and top views of the xylophone unit are shown schematically in Fig.2 and Fig.3. The xylophone unit is composed of four wooden boards F, L, W and S shown in Fig.2. In the figure, the notations F, L, W and S denote the floor, leg, waist and shoulder boards, respectively. The notations Ar and Al represent the right and left arms to play the xylophone. Each board can rotate around the vertical axis of rotation zz'.The leg board L is designed so that it can rotate the upper structure by the length of a half pitch distance of the xylophone pipes in both directions at the positions of the arm tips when the solenoid coil R0 or R1 is excited by alternative currents. Similarly, the waist board W makes the arms rotate by the length of one pitch distance when the solenoid coils R2 or R3 is excited by alternative currents, and the shoulder board S makes the arms rotate by one and a half pitch distance when the solenoid coils R4 or R5 is excited likewise. The arms are rotated in the counterclockwise direction around the axis of rotation yy' when the solenoid coil R6 or R7 is excited, then

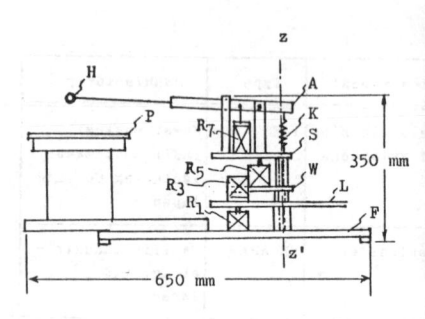

Fig.2. Schematic side view of the xylophone unit

Fig.4. Photo of the front view of the xylophone unit

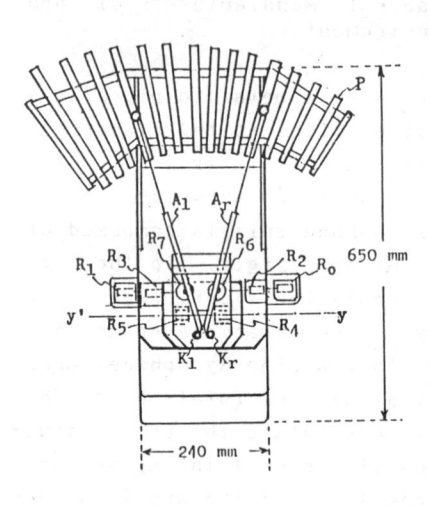

Fig.3. Schematic top view of the xylophone unit

Fig.5. Poto of the top view of the xylophone unit

the hammer Hr or Hl hit the xylophone pipe. The wooden strut ST is fixed on the shoulder board S. The hammers Hr and Hl are pushed up ordinarily by the helical springs Kr and Kl shown in Fig.2 and Fig.3. The photos of the front and top view of the xylophone unit are shown in Fig.4 and Fig.5. The com-

251

ponents of the interface between the personal computer and the controller are shown in Fig.6. In the figure, To $\sim$ T7 are transistors and ro $\sim$ r7 are direct current relays and Do $\sim$ D7 are the diodes to protect the transistors from the impulsive voltage. The manufacturers and types of each instrument are shown in Table 1. The photo of the whole system of this robot is shown in Fig.7.

Fundamentals of movement

As shown in Fig.6, the signals in the address-bus, data-bus and control-bus of the personal computer are connected with the periferal interface chip 8255 in the interface. The terminals of port A, port B and port C of the 8255 are connected to the electromagnetic relays ro $\sim$ r7 through transistors To $\sim$ T7. In this robot, we make the keys on the keyboard of the personal computer PC-9801RA corresponding to the symbols of the musical scale shown in the left side column of table 2. And also, we make the symbols of the musical scale corresponding to the 0 or 1 signal in the column of the solenoid coils Ro $\sim$ R7 shown in Table 2. Here, the notation 0 in the col-

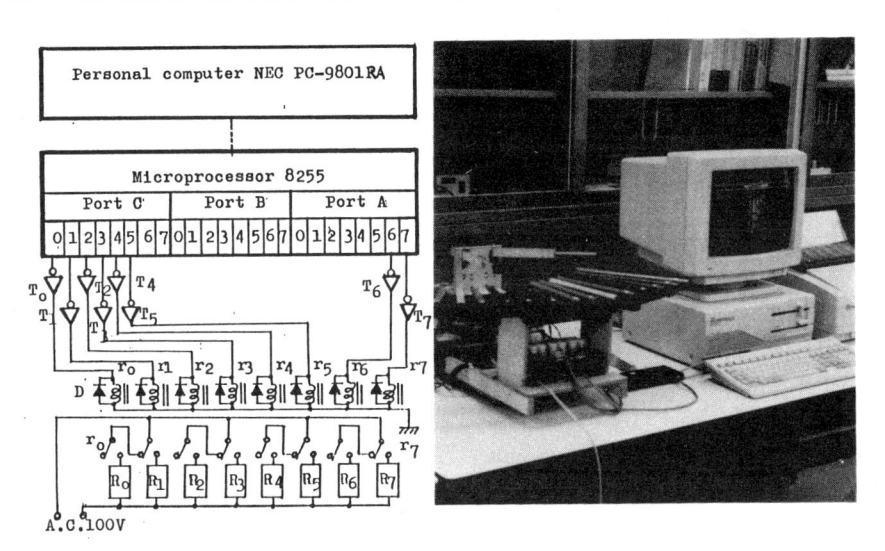

Fig.6. Block diagram of the interface and controller

Fig.7. Photo of the whole system of the robot

umn of Ro~R7 means the solenoid coil is excited and the notation 1 means solenoid coil is not excited. For example, if we hit the key Z on the keyboard of the personal computer, the relays r1, r3, r5 and r7 are excited and as a result, the solenoid coils R1, R3, R5 and R7 are excited. Then the hammer of the left arm hits the pipe G of the xylophone shown in Fig.8 , consequently. In Fig.8, the notation of the small circle means the position of the both hammers.

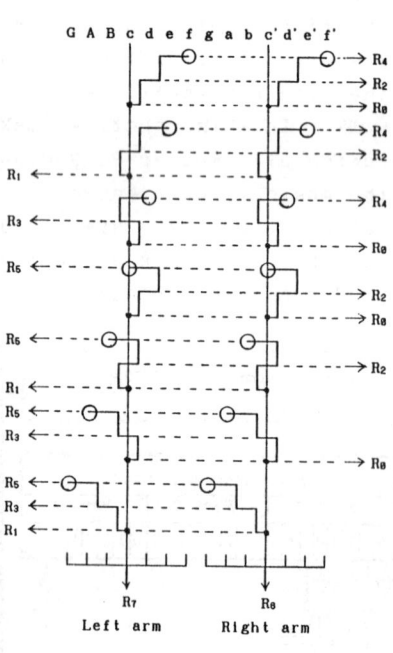

Board	Arm		Shoulder		Waist		Leg		
Port	Port A				Port C				
Terminal	7	6	5	4	3	2	1	0	
Key / Musical scale	2^3	2^2	2^1	2^0	2^3	2^2	2^1	2^0	Hexadecimal
	R_7	R_6	R_5	R_4	R_3	R_2	R_1	R_0	
Z	G	0	1	0	1	0	1	0	1	55
X	A	0	1	0	1	0	1	1	0	56
C	B	0	1	0	1	1	0	0	1	59
V	c	0	1	0	1	1	0	1	0	5A
B	d	0	1	1	0	0	1	1	0	66
N	e	0	1	1	0	1	0	0	1	69
M	f	0	1	1	0	1	0	1	0	6A
,	g	1	0	0	1	0	1	0	1	95
.	a	1	0	0	1	0	1	1	0	96
/	b	1	0	0	1	1	0	0	1	99
□	c'	1	0	1	0	0	1	0	1	A5
A	d'	1	0	1	0	0	1	1	0	A6
S	e'	1	0	1	0	1	0	0	1	A9
D	f'	1	0	1	0	1	0	1	0	AA
	silent	1	1	1	1	1	1	1	1	FF

Table 2 Relationship between musical scale and each sole-

Fig.8. Illustrative diagram of movements of the both hummers

Outlines of computer program

The keys of the keyboard of personal computer and the musical scales have been represented by the hexadesimal numerical values in Table 2 in computer programming. The time intervals of diacritical and rest signs in the musical programs have been represented by the time delay in the computer programming. Several musical programs written in hexadecimal codes have been saved in floppy disks of the computer. The titles and the numbers of the musical compositions saved in the floppy disks are shown on the dislpay of the computer. When an operator selects the number of the musical composition desired and pushes the key of the number on the keyboard of computer, the robot starts its movements.

Conclusions

It has been shown that a robot which plays the xylophone in the range of two octaves was made using a personal computer and a simple controller. The controller is simply composed of electromagnetic relays, transistors, solenoid coils and other subsidiary parts. This robot still needs a minor improvement to play the musical programs which include the sharp or flat notations. In that case, the metalic pipes of the xylopone should be modified to include higher or lower frequency sound.

Acknowledgment

The auther wishes to acknowledge useful advise given by the people in the Pacific Industrial Co. Ltd., and also expresses his gratitude to Mr. F. Ishikawa for his assistance and cooperation in this study.

Reference

1. The Pacific Industrial Co.,Ltd.:Training of the mechanical control by the personal computer PC-9801 (1985) 140-143, in Japanese.

Teach Control Based Off-Line Programming

Emma C Morley, Chanan S Syan, Sue V Grey-Cobb and John R Wilson

Department of Manufacturing Engineering and Operations Management. University Park, Nottingham NG7 2RD, England.

SUMMARY

Off-line programming of robots has helped in increasing the economic application of robots. Despite the many systems available few offer a consistent approach in programming from the shop floor to off-line. This paper explores the development of off-line programming and how consistency may be provided.

INTRODUCTION

The benefits of off-line programming of robots have been recognised for some time. A survey in 1985 identified over 90 languages currently available or under development(1).
In this paper the development of robot programming systems is examined with emphasis on the role of the programmer.

A method of using robot configurations as a means of programming is then explained. As a robot based system it can be applied by users for both on-line and programmers off-line.

TYPES OF ROBOT PROGRAMMING SYSTEM

The principal advantage that a robot has over other automatic
machinery is its flexibility and versatility. However to achieve the most out of this potential a suitable programming system is required.

Programming languages have been classified in a number of different ways. Lozano-Perez(1983) identified three types of robot languages(2);
     1) Guiding Systems
     2) Robot Level
     3) Task Level

While Yong et al (1985) took a different approach and extended this to four levels of languages(3);

    1) Joint Level
    2) Manipulator Level
    3) Object Level
    4) Objective Level

Further distinction can also be made, within some of these levels between textual and non-textual languages (4).

FROM ON-LINE TO CAD SYSTEMS

Early programming systems used teach controls or teach pendants to program a robot. The pendant was used to control the robot's motions and so move the robot to a set position. The location was then recorded via the teach pendant, in this way a program was created. This approach was known as programming by guiding, it was also a joint level type system as the locations would be stored in the form of joint angles of the robot links. This method allowed for good interaction between the programmer and the robot. However a number of disadvantages were present :-

- The robot had to be used during the programming process
    which was an uneconomical use of its time.
- Programs were difficult to debug.
- The programmer was required to work close to the robot
    which posed a potential safety risk.

To help overcome these problems the first off-line programming systems were developed. Initial systems like VAL(4) and AML(5) were textual languages and similar in structure to high level computer languages. Programs could be written off-line and then validated at the robot. Pendants were still required in order to program in reference positions. The main difficulty with these languages was that now the programmer was working away from the robot he had to imagine the robot and its environment whilst writing the programs. In complex tasks errors would become more likely because of the difficulty in visualising the situation.

Though the program could now be written off-line it was still necessary to use the robot for verification and debugging. Hence the amount of robot down time in the programming process was still quite high.

The next generation of robot languages such as VALLII(4) and RAIL(6), however concentrated more on providing additional features to the systems in order to improve their potential. They provided the ability to run subroutines, carry out arithmetic computation and to handle sensor data.

The problem of visualising the robot and its environment has been overcome by the use of CAD systems to simulate the work system. The robot can be modelled in 3-D along with its workplace environment eg. machine tools, conveyors etc. This means that simulations of the robot can highlight potential collisions making possible some debugging of programs an off-line task as well.

The development of these types of systems has gone in two distinct directions. Firstly as a means of programming robots off-line, so helping to reduce robot downtime and secondly as a simulator. CAD off-line systems have not yet become extensively used principally because the costs of such systems are high and so the payback time is too long to be a worthwhile investment. However the simulation systems are being widely used as a means of testing different robot styles and optimising workcell design. Some of these systems are now available to work on PCs. This use of CAD systems in robotics is seen by some as their main task in the near future (7).

As well as the move into graphic systems the nature of robotic languages have also been changing. From the classification of robot languages by Yong et al. most languages to date have been at the manipulator level. A few languages have now started to become available at the object level, such as RAPT(8) and AUTOPASS(9). The emphasis in these languages shifts from the robot to the task that the robot is performing and the objects involved. These languages are aimed particularly at assembly tasks. This is an area of robotic application that has failed to develop as rapidly as had been anticipated in the early eighties. (10).

THE PROGRAMMER

Another important factor is that as the robot languages changed the task of programming has also changed. Programming with the teach pendant involved personnel who were familiar with the robot either from the shop floor or production engineering. The early off-line programming systems(ie VAL) required programming knowledge similar to that required for NC part programming. The later textual systems needed more advanced programming skills. The change to graphic and object level systems has created another shift in programmer from the robot user to the assembly engineer.

The skill level of the potential programmer has an important effect on the nature of a language and should therefore receive human factors attention (11). It may be asked which comes first the programmer or the language ? Also whoever is the principal user of the system others will still require access. This is particularly the case for shopfloor and maintenance personnel. They are still required to make use of the teach pendant and languages must allow for this. (12).

It would therefore be of benefit if the same programming system could be applied at the robot and off-line with the aid of a graphic system. As the task level languages tend to exclude robot operators such a system would preferably be robot based.

FROM TEACH PENDANTS TO OFF-LINE

Whilst off-line programming has a growing influence in the applications of industrial robots, teach pendants will still be a requirement. Their functions are likely to shift from programming to motion control during testing or maintenance. However the ability to debug programs on-line is still likely to remain a requirement for some time. Recent research at Nottingham University(13) has identified the perceptual difficulties faced by an operator in the motion control of a robot. This has helped to identify the importance of the operator recognising the robot status, ie. its configuration and position.

This work involved a PUMA 560 jointed arm. This is a six jointed articulated arm. Using the first three joints (waist, shoulder and elbow) the robot can position its end-effector in most places with a choice of four different configurations. This is determined by the robot being in either a RIGHTY or LEFTY configuration and then with the elbow joint either ABOVE or BELOW the wrist. If the wrist is considered as well then a total of 8 configurations are possible depending on whether the wrist is in a state of FLIP or NOFLIP.

If the robots workplace is divided into four sections and just the first three joints are considered then 16 distinct images of the robot are possible (figure 1). Each could be described as an individual pose. The principle of defining robot positions in terms of poses of configurations has been forwarded before (14). It was then advocated as a means of improving robot accuracy and efficiency in use.

This idea is now being taken a stage further as a potential means of programming robots. A manufacturing process can be described by a number of tasks. The tasks can be simplified into a sequence of poses that the robot assumes. In terms of gross motions this could be described by the 16 poses given in figure 1.

A three digit code is used to describe and identify each configuration. The code is assigned in the following manner.

1st digit: 0-3 Robot location A,B,C or D.
2nd digit: 0,1 Robot configuration, righty or lefty.
3rd digit: 0,1 Arm position relative to the wrist, above or
              below.

Thus a code of 210 represents the robot positioned in quadrants C in a lefty configuration with the arm positioned above the wrist.

It is therefore possible to describe, in the form of gross motions, a sequence of moves by a list of code numbers. On this principle a programming system could be developed using the poses. This could then be applied both on and off-line. An off-line system could provide a menus illustrating the choice of available codes or configurations. On-line with a tech pendant the system could allow the operator to enter the code of a desired location.

A system which allows a similar means of programming both on and off-line has the advantage of a consistent approach. This should allow easier communication between personnel working with the robot at all levels. Similarly training would be easier to provide across the board. Such a method would also be simpler to apply at operator level and thus reduce errors.

CONCLUSIONS

Robot programming systems have developed a long way since the initial teach by guiding

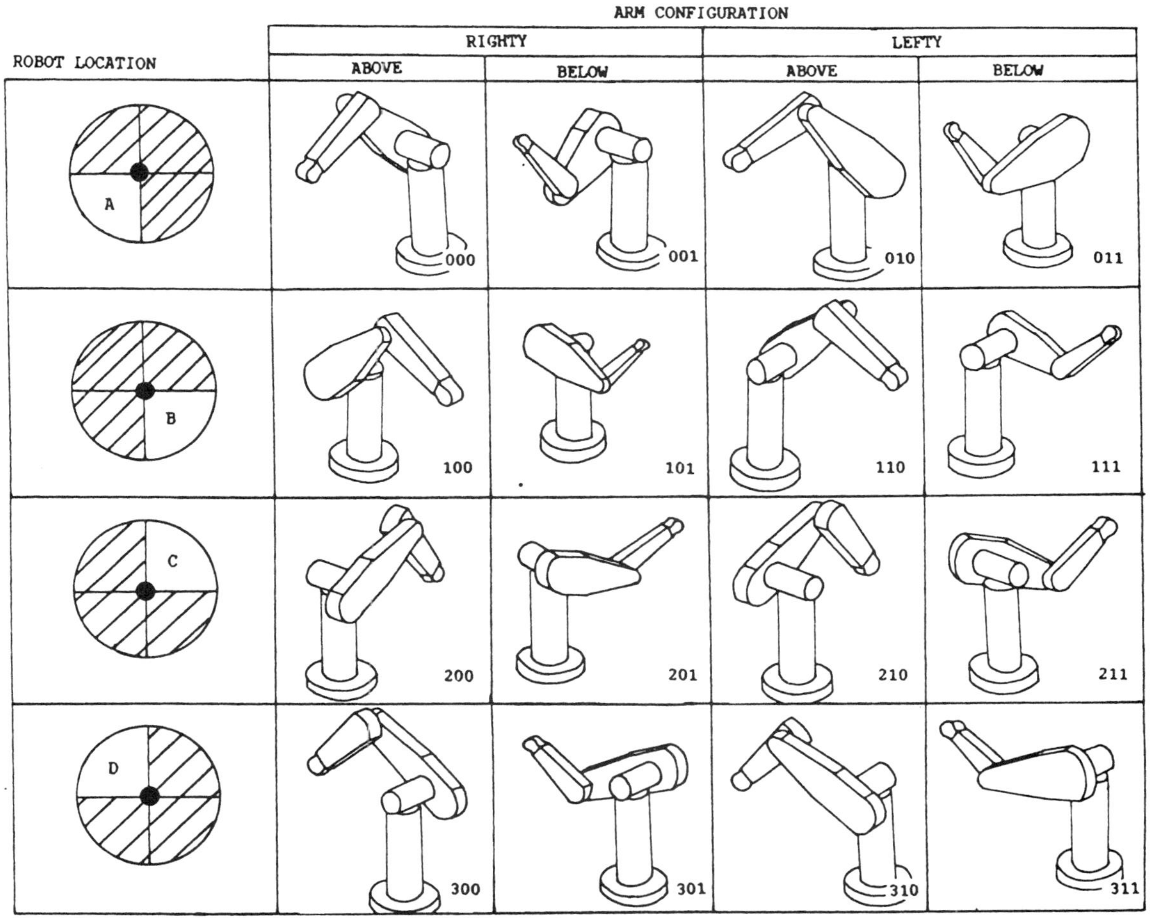

Figure 1. The 16 poses of the PUMA robot with their identifying codes.

approach. Off-line systems can increase the productivity of robots in manufacturing however the low implementation of such systems suggest that improvements can still be made. It is hoped that with the continuing research into the type of system described in this paper a system suitable both on the shop floor and off-line can be developed and that greater acceptance of off-line systems will follow.

REFERENCES

1. Hocken, R.; An overview of off-line robot programming systems. Annals of the CIRP Vol 35/2/1986 pp465-503.

2. Lozano-Perez, T.; Robot programming. Proceedings of the IEEE vol. 71 no.7 July 1983 p821-841.

3. Yong, Y.F.; Gleave, J.A.; Green, J.L.; Bonney, M.C.; Off-line programming of robots. Handbook of Industrial Robotics. Ed. S.Y. Nof. 1985 pp366-380.

4. Grover, M.P.; Weiss, M.; Nagel, R.N.; Odrey, N.G. Industrial Robotics, Technology, Programming and Applications. 1986 McGraw Hill.

5. Taylor, R.H.; AML: A manufacturing language. International Journal of Robotics Research.

6. Franklin, J.W.; Vanderberg G.J.; Programming vision and robotics systems with RAIL. SME Robots VI.

7. Carter, S.; Off-line robot programming the state of the art. The Industrial Robot 14(4) 1987 p213-215.

8. Corner, D.F.; Ambler, A.P. Popplestone R.J.; Reasoning about the spatial relationships derived from a RAPT program for describing assembly by robots. IEE Seminar on UK Robotics Research 1983 p16/1-16/5.

9. Lierberman, L.I.; Wesley M.A.; AUTOPASS: An automatic programming system for computer controlled mechanical assembly. IBM Journal of Research and Development. July 1977 p321-333.

10. Albus, J.S.; Robotics :Where has It been ? Where is it going ?   Robotics & Autonomous Systems 6(1990) p199-219.

11. Parsons, H.M. Robot programming. Handbook of Human-Computer Interaction. 1988   Ed. M. Helander   p737-754.

12. Humrich, A. Problems associated with the off-line programming of robots. Behaviour and Information Technology 1988 v7   no.4   p369-416.

13. Grey, S.V.; Wilson J.R.; Syan, C.S.; Human control of robot motion: Orientation, perception and Compatability. in Human-Robot Interaction. Eds. M. Rahimi and W. Karwowski. Taylor and Francis   1991.

14. Tucker, M.; Perriera, N.D.; Motion planning and control for Improved Robot Performance.   Proceedings of the Controls West Conference. 1985 p1-17.

Introduction of Robots in Docks

Željko Domazet, Ivica Mandić, Tonči Piršić

Faculty of Electrical Engineering, Mechanical
Engineering and Naval Architecture, University
of Split, R. Boškovića bb, 58000 Split, Yugoslavia

SUMMARY

The paper presents a short survey of the previous results in robot introduction on surface cleaning and ship painting jobs in shipyard industry. The possible configurations of the robot for surface cleaning and ship painting in dock are classified in three mean groups and the optimal configuration from several point of view is chosen by using the Analytic Hierarchy Process. According to the chosen manipulators configuration, the preliminary design and calculation of the kinematics and dynamics has been done, and all the characteristic equations and results are represented.

1. INTRODUCTION

The introduction of robots on surface protection jobs in shipbuilding industry represents a big challenge from a few points of view. The great dimensions of work pieces, as well as performances of existing robots seem to be unavoidable obstacle in mechanization of these dangerous and tiring jobs. Obviously, the results depend on the right choice of the new configurations, and because of this, for the first step of robot and/or manipulator introduction in shipbuilding, the surface protection jobs in docks have been chosen. Namely, these operations don't demand so sharp accuracy and the flat ships surfaces have a good approachability, what enables application of nonredundant robots with easily realizable control. The existing solutions can be classified in three mean groups: manipulator on vehicle, self moving manipulator, and manipulator on rails. The priority manipulators choice has to be done gradually, having on mind lot of facts: ships and docks sizes, actual status in robots development, working conditions, prices of human work, safety factors, etc.

2. THE PRIORITY MANIPULATORS CHOICE

Having on mind three quite different types of structures, as well as lot of influences, the problem was decomposed as a hierarchy, in three level, based on the Analytic Hierarchy Process by T. L. Saaty [2]. In the first level, the top of the hierarchy, is the overall goal: "The right choice of manipulator's configuration", Fig. 1. In the second level are five factors of criteria: possibility of installing (PI); manipulators price (MP); control features (CF); working velocity (WV) and energy consumption (EC). In the third level are various configurations of manipulators, as was described in the previous chapter: manipulator on vehicle (MV), self moving manipulator (SM) and manipulator on rails (MR).

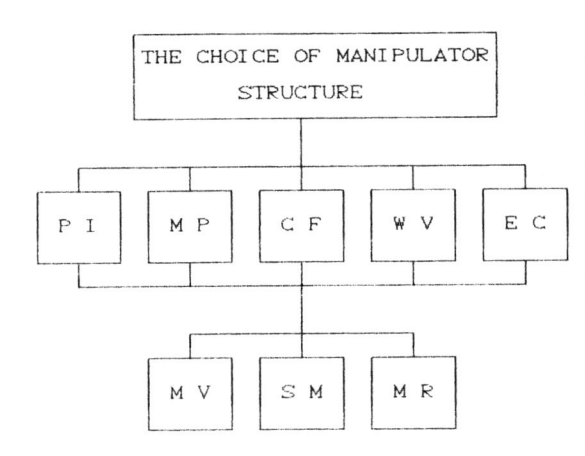

After applying the Analytic Hierarchy Process, as a complex solving framework, the rank - list of priorities of manipulator's structures occured, and the manipulator on rails was on the top of this rank.

Fig. 1. Decomposition of the problem

In order to design a manipulator of our own configuration, we were leaded by results of previous analyze. The suggested structure, represented on fig. 2, has four degrees of freedom (not including the orientation of the end - effector). These four degrees of freedom enable approach to every part of wessel side surfaces and the last two (rotations of top platform of manipulator) enables adequate orientation. Translation alongside the wessel (q_1) is done by transport of whole manipulator onto the fix rails mounted at the top of dock side

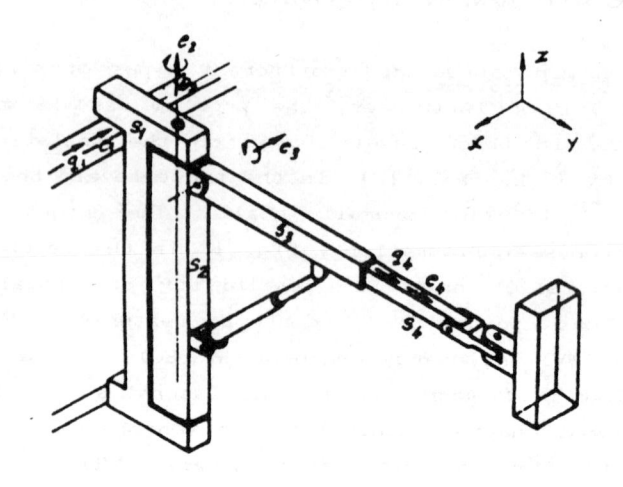

Figure 2. Suggested design of manipulator structure

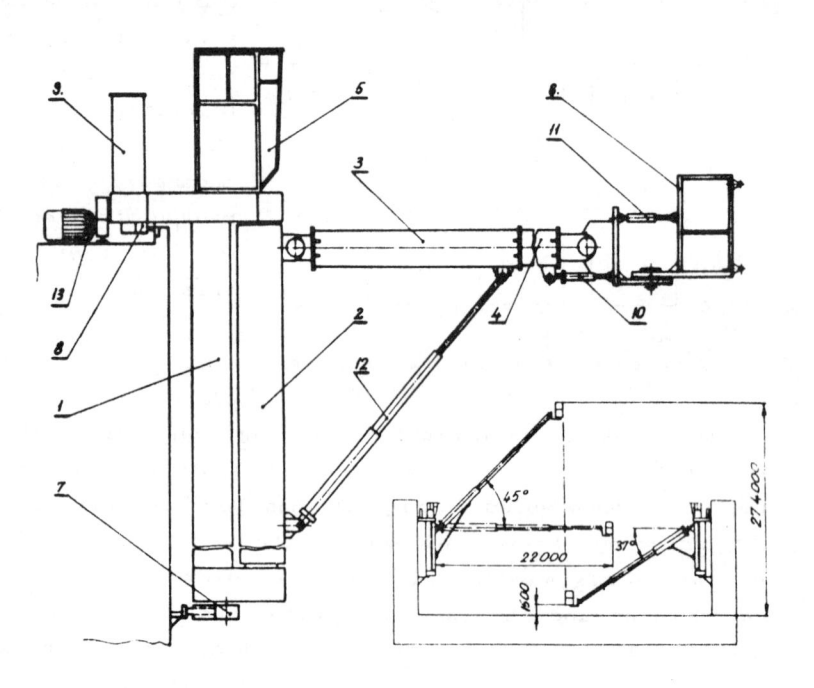

Fig. 3. Preliminary design and working range of manipulator

265

wall. Control is provided by the operator in cabin onto
manipulator's body. His role is working devices control, as
well as cleaning and painting quality supervision. The problem
of the control could be solved with of-line programming, or
even on-line programming, but for the first successful
introduction of robots in our shipyards, it is important to get
a simple structure and control with acceptable price. According
to the previous results, the preliminary design is done, Fig.
3. Manipulator consist of next main part: manipulator segments
S_1, S_2, S_3, S_4 - parts of kinematic chain (1,2,3,4); operator's
cabin (5); gripper equipped with painting and sand blasting
tools (6); lower and upper guides (7,8); cylinders for
orientation (10,11); telescopic cylinder (12); electric motor
for alongside driving (13).

3. KINEMATICS AND DYNAMICS OF MANIPULATOR

Kinematic and dynamic calculation of manipulator is derived by
kinetostatic approach, well known in robotics, which consists
of input data specifications, calculation of external forces,
kinematic chain breaking, setting the equilibrium equations and
calculation of moments and forces in joints. Kinematic and
dynamic parameters are defined by manipulators function and
with aspiration for minimize the dimensions and weight. The
"triangle" law of velocity for relative movements θ_2, θ_3 and q_4
is chosen, as the law of minimum energy consumption, simple for
calculation and control. Because of its function (transporting
the whole manipulator), the "trapezoidal" law of velocity is
chosen for relative movement q_1. Permanent contact between
manipulators end and ship surface is taken into account by
entering the normal reaction force at the manipulator gripper,
so that kinematic chain is opened. In force reduction on the
center of segment mass all the forces of "rejected" part of
kinematic chain, external and inertial forces are taken into
account. The equilibrium equations in manipulator joints are:

$$\vec{F}_r + \vec{F}_\iota + \vec{F}_e + \vec{F}_{rp} = 0 \qquad (1)$$

$$\vec{T}_r + \vec{T}_\iota + \vec{T}_e + \vec{T}_{rp} = 0 \qquad (2)$$

$\vec{F}_r$, $\vec{M}_r$ - reaction forces and moments

$\vec{F}_i$, $\vec{T}_i$ - inertial forces and moments,

$\vec{F}_e$, $\vec{T}_e$ - external forces and moments

$\vec{F}_{rp}$, $\vec{T}_{rp}$ - forces and moments of rejected part

The procedure is derived for simultaneous movements of all segments, during the typical working cycle. As the example, the equilibrium equations and kinematic chain breaking for manipulator segment S_4 are represented: (according to the results of this analyze, it is obvious that inertial and Coriollis forces could be neglected but in the next example the equations are shown in complete form).

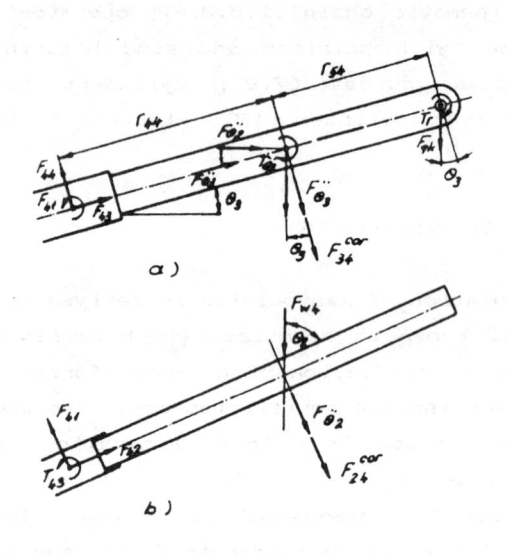

Fig. 4: *Kinematic chain breaking and force reduction for segment S_4*

Equilibrium equations for segment S_4, calculated according to Fig. 3:

$$F_{41} = F_{w4} \sin \theta_2 + F_{\ddot{\theta}_2} + F^{cor}_{24} \tag{3}$$

$$F_{42} = F_{w4} \cos \theta_2 + F_{\ddot{q}_4} + F_{gk} \sin \theta_3 + F_{g4} \sin \theta_3 + \tag{4}$$
$$+ F_{\dot{\theta}_2} \cos \theta_3 - F_{\dot{\theta}_3}$$

$$F_{43} = F_{g4} \cos \theta_3 + F_{gk} \cos \theta_3 - F^{cor}_{34} + F_{\ddot{\theta}_3} + F_{\dot{\theta}_2} + \tag{5}$$
$$+ F_{\dot{\theta}_2} \sin \theta_3$$

$$T_{41} = T_{\ddot{\theta}_3} + T_R + (F_{g4} \cos \theta_3 + F_{\dot{\theta}_3} + F^{cor}_{34} + F_{\dot{\theta}_2} \sin \theta_3) r_{44} + \tag{6}$$

267

$$+(r_{44} + r_{54}) F_{gk} \cos \theta_3 \qquad T_{42} = 0 \qquad (7)$$

$$T_{43} = (F_{\ddot{\theta}2} + F_{24}^{cor} + F_{w4} \sin \theta_2) \cdot r_{44} \qquad (8)$$

where:

F_{ij} - reaction forces in joint, F_v - wind force,
F_q, $F_{\dot{\theta}}$, F_{θ} - inertial forces, T_{θ} - inertial moments
F_{gk} - force caused by gripper weight, F^{cor} - Coriollis force

Similar equilibrium equations are set for all segments, and according to calculated reaction forces in joints, actuators are chosen.

4. CONCLUDING REMARKS

The possibility of mechanization on surface cleaning and painting jobs in docks have been analyzed. The humanization of these dangerous and hard jobs, as well as improvement of quality and efficiency are the main reasons for deriving the preliminary design of such a manipulator. For the optimal configuration of manipulator kinematic and dynamic calculation is derived, actuators are chosen and preliminary design is done. Further steps should be the detailed construction, execution and installation of the manipulator, and test in operational conditions.

R E F E R E N C E S

[1] I. Mandić, D. Stipaničev, Ž. Domazet: Introduction of Industrial Robots on Surface Protection jobs in Shipyards, Elsevier Science Publishing Co., New York, USA, p.p. 447-456, 1986.

[2] T. L. Saaty: Principles of the Analytic Hierarchy Process, University of Pittsburgh, October 1983.

[3] D. Sorić: Analiza kinematike i dinamike manipulatora za poslove površinske zaštite u dokovima, Graduating Thesis, Split, 1990.

[4] W. R. Tanner: A User's Guide to Robot Application, Presented at the First North American IR Conference, October 26-28, 1976.

Autonomous Operation of Multi-Robot-Systems in Factory of the Future Scenarios

E. Freund, J. Roßmann
University of Dortmund
Institute of Robotics Research (IRF)
Postbox 50 05 00, D-4600 Dortmund 50, Germany

Summary

Considering the variety of tasks in Factory-of-the-Future scenarios and the degree of flexibility needed to cope with the problems related to increasing product variances, small batch production and reduced product lifecycles, several means of factory automation are expected to coexist for maximum throughput.

In this paper a multi-robot-system concept, expressed in a hierarchical structure, is proposed that allows the incorporation of different means of automation as there are robots and manipulators as well as hardautomated features on different levels of abstraction.

The feasibility of the proposed approach is emphasized by the description of the CIROS (**C**ontrol of **I**ntelligent **RO**bots in **S**pace) testbed, an experimental multi-robot-system with two redundant robots working together closely in order to autonomously perform experiment servicing tasks in a space laboratory environment. Although the practical implementation in the nationally funded CIROS-project aimed at the service of a space laboratory, the demonstrator has been build up with terrestrial, industrial components and thus the concept and the practical implementation is easily transferable to industrial applications whose needs concerning autonomy and fault tolerance of the involved robotic system are one step in advance of modern CIM-integrated robot-workcells.

1. Introduction

Modern factory automation concepts incorporate automation means with different levels of flexibility and intelligence in order to provide the most economical use of production facilities. Hard automation equipment and fix sequence manipulators, which are tailored to a specific production task, are the least flexible means, but they are economical for great production volumes. Modern single-robot-workcells, expecially if they are integrated in CIM-structures, are far more flexible and of good use to cope with greater product variances and small batch production.

The multi-robot-system concept presented in this paper has been developed and exemplarily *realized* at the IRF in order to improve the performance of robotic systems for state-of-the-art industrial applications but also to allow the robots to conquer working fields that appear not to be feasible with state-of-the-art robot-control technology. Considering the application of robotic systems for more complex factory-of-the-future projects like service and maintenance of production facilities, routine exchange of parts, inspection tasks etc., it is not too hard to imagine a scenario, where several highly sophisticated robots on mobile bases perform these tasks autonomously.

The proposed multi-robot-system concept also appears to be appealing for robotic systems which work in dangerous environments e.g. nuclear power plants or chemical factories. The inherent flexibility and functional redundancy of a multi-robot-system would lead to to a much higher reliability of the automated system because the planning levels, which are always aware of the system's resources, could — in case of failure of one robot — replan the task and instruct another, similarly equipped robot, to complete the task. These replanning capabilities have been implemented and demonstrated in the CIROS testbed.

2. The Multi-Robot Control Structure

In order to provide true flexibility for factory-automation in the future, a unifying structure and design methodology has to be developed, which gives an aid for the top-down development and leads to a coherent automation system. The components of such a system with different levels of activity and intelligence have to cooperate in the factory-of-the-future. This implies the necessity to provide a versatile system-structure to cope with the different communication-, scheduling- and security-problems that will arise.

The proposed structure which is composed of different layers with exactly determined functionality, contains all vital parts of a multi-robot-control-system is shown in figure (1).

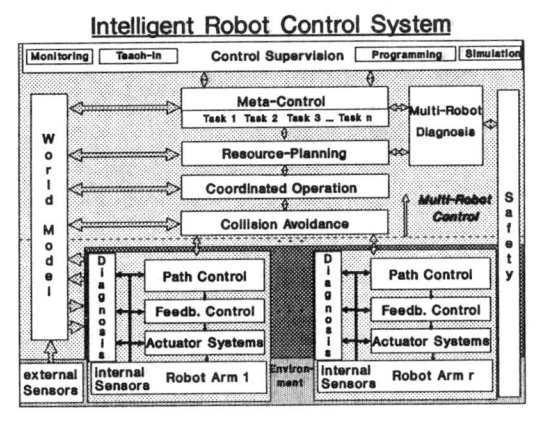

Figure(1): The Multi-Robot-Control-Structure

Basis of a multi-robot-system are the robots, logically connected by superimposed control levels. In figure (1) the robots are represented by blocks containing the different functional components allowing the integration of a robot into a multi-robot-system. Besides their kinematics (robot arm) and their feedback controlled actuator systems, the vital components of the robots are their internal sensors which provide position and velocity informations for the local feedback controller. Essential for sophisticated applications that depend on a very high degree of autonomy as e.g. the autonomous servicing of a space laboratory module, is an individual diagnostic component which can determine irregular behaviour of each single robot. In the exemplary realization, standard MANUTEC-R15 in a gantry configuration have been used, together with their control, which were enhanced by a suitable interface to the computer on which the multi-robot-control specific software-layers are run.

Superimposed on this layer of individual robots are the levels which distinguish the multi-robot-control from state-of-the-art single-robot-controls.

Collision avoidance features, that are considered further below, are placed directly above the single robot layer. They can be seen as a new quality of security feature, which is able to actively avoid collision between robots and their environment by changing the path fed to the single robot's path control. The desired path for each robot is transmitted to the collision avoidance level by the coordinated operation level, which contributes different indespensible features to the multi-robot-control and allows the robots to be run in three modes of operation as shown in figure (2).

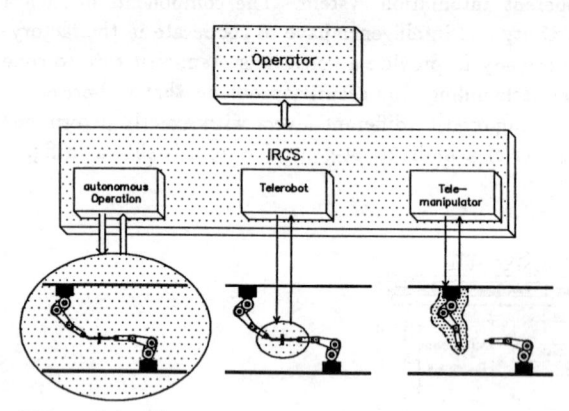

Figure(2): The modes of operation supported by the coordinated operation layer

Its main task is to interpret and run the tasks scheduled by the resource-planner and described in terms of *coordination primitives*, a new concept, introduced to allow the full exploitation of the inherent flexibility of this multi-robot structure. Besides allowing the description of single-robot tasks, the coordination primitive concept permits the *description of multi-robot tasks*.

The lifting of a pipe by two robots for example is described by a single coordination primitive, providing the starting point, the endpoint and the type of path for the movement in terms of an object coordinate system. This information, together with the ability of the coordinated operation level to access the global world model will sufficiently describe the task.

Superimposed on the coordinated operation level are the planning-levels. The resource-planner performs a scheduling of the different tasks fed into the system considering the availability of the necessary resources to complete the tasks. It is based on precomputed plans for single task execution whose starting conditions concerning resources (availability of robots, grippers, probes etc.) are checked at runtime. A major purpose of the resource-planner is the resource handling and the optimization of synergetic effects of concurrent plan execution.

The meta-control facility represents the highest level of the autonomous runtime system. It has to decompose high level plans and user commands into different tasks which can be performed by the underlying layers. The meta-controller is also in charge of the world model management, the communication with the multi-robot diagnosis and the control-supervision layer to provide an easy to use interface to the user.

The world model in this concept can be seen as an information node for the control, where relevant system information is collected on different levels of abstraction and bandwidth to be distributed again on demand.

2.1. The group concept

The complexity of a multi-robot-system increases exponentially with the number of robots involved. Hand in hand with the system's complexity, the software-complexity and the computational burden for the robot controller increases. To overcome these difficulties and to allow the decentralization of computational power, the group concept is introduced. The central issue of the proposed concept is, that the groups are *not configured statically* but are reconfigured by the resource-planner depending on the actual task to perform.

The multi-robot structure itself remains mainly untouched by the group concept, only a substructure for the coordinated operation and collision avoidance level are defined. Each of these levels will distinguish internally between a system and a group level, each with different tasks to accomplish.

Figure (3) elucidates the interdependencies of the collision avoidance and coordinated operation level with respect to the group concept.

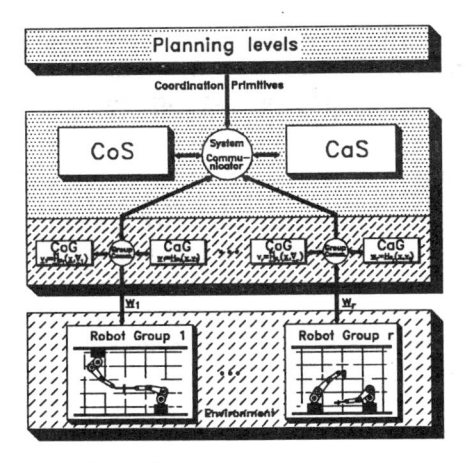

Figure(3): The Group-Concept

Both system levels, that of the coordinated operation (CoS) and that of the collision avoidance (CaS) only exist once in the system and perform configuration and initialization tasks with the help of their global system overview. They generate group processes which only consider the robots in the group they are configured for und thus can satisfy the strong realtime-constraints in order to grant the highest degree of flexibility. Furthermore, the separation of the robots into different groups supports a natural decentralization of the computational burden.

2.2. Collision Avoidance

In state-of-the-art industrial single-robot-workcells, collision avoidance is restricted to the avoidance of an impact between the robot and its a priori known environment by careful manual teaching of the desired paths. The collision free path such becomes an integral part of the robot's program and so the path-planning and teaching has to be redone whenever parts of the environment are subject to changes.

An autonomous multi-robot-system necessarily requires sophisticated collision-avoidance measures which are capable of detecting collision danger and *avoiding collision in realtime* because the pre-taught paths are of no use in environments with independently operating robots with overlapping working areas.

Besides being able to protect such a simple system from serious damage, online collision avoidance also provides several merits for the overall system-performance in far more complicated multi-robot-systems and even allows a great improvement of user-support:

- Tasks can be planned for each robot or robot group individually, without taking care of the position of other robots. This facilitates online-planning very much.
- Independent tasks can be started at any time without being synchronized with each other as far as overlapping working areas and potential obstacles are concerned.
- The manual teach-in support is also enhanced in several ways:
  - The operator can fully concentrate on the robot's TCP, while all other parts of the robot including the elbow, which is — as experience shows — endangered the most, are completely protected.
  - In case of collision danger with for example the elbow, those degrees of freedom of the robot's TCP will be blocked by the collision avoidance that — if the robot would be moved in one of these directions — would lead to an increase of the collision danger.

3. The CIROS project - a practical implementation

The multi-robot control structure as it has been described in this paper served as a basis for the practical implementation in the CIROS-project. Besides realizing the testbed, the development and validation of the proposed versatile multi-robot control architecture was one of the key issues.

The CIROS testbed is equipped with two redundant robots with 6 revolute and one translational axis each. The configuration is shown in figure (4).

The layout of the laboratory as the robots working environment is similar to that of the German Spacelab and is built in a modular manner, so that it is possible to incorporate the latest experiments and to adapt the environment to reality as precisely as necessary. A *tool exchange capability* and torque/force-sensors have been included to allow the robots to operate drawers and flaps for experiment processing, exchange printed circuit boards and feed a heater.

In order to attain the desired functional redundancy, that is the chance to have most of the tasks completed by either the one or the other robot, a configuration had to be chosen that granted widely overlapping working areas of the two robots.

On the one hand, this assures functional redundancy but on the other hand the robots are almost permanently in collision danger. The collision avoidance level prevents the robots from colliding with each other and with the laboratory's walls thus allowing the robots to work in all three modes of operation presented above.

Figure(4): Redundant two robot configuration in a space laboratory environment

4. Conclusion

Summing up the experiences that have been made while implementing CIROS, it has to be stated first that the most natural representation of complex automatic systems for control purposes is to break down the problem hierarchically. The hierarchical breakdown supports implementation efforts in several ways e.g. it allows to implement the multi-robot-system in different teams and only a minimum of communication necessary. Because of the clear interfaces, parts of the control can be revised, so that the system can evolve over time to take advantage of advances in technology. The strict realtime conditions in the lower levels can be satisfied by introducing the group concept and thus decentralize the computational burden.

Essential parts in a realization — besides the single-robot controls which can be state-of-the-art — are the collision avoidance level to provide the necessary security measures for teaching- and autonomous operation, the coordinated operation level to allow the robots to work together in a cooperative manner and the resource-planner to optimize concurrent plan execution.

This work was supported by a grant of the "Bundesminister für Forschung und Technologie (BMFT)" of Federal Republic Germany.

References

1. Ch. Bühler, P. Kaever, U. Kernebeck, J. Roßmann:
   Strukturierung intelligenter Robotersteuerungen für Weltraumanwendungen
   IRF-Reports, C-Z-289, August 1989.

2. E. Freund, J. Roßmann:
   Teleoperated and Automatic Operation of Two Robots in a Space Laboratory Environment
   41st Congress of the IAF, Dresden, FRG, 10.1990.

Sensors, Control and Signal Processing

Interactive Motion Specification Using Splines

M. Caulfield-Browne, Dr B.L. MacCarthy, Dr C.S. Syan.
Dept. Manufacturing Engineering and Operations Management
University of Nottingham, U.K.

Summary

This paper outlines the main principles of a general motion specification method using spline functions and describes an interactive graphics-based CAD system called MODUS, which utilises the method. The spline method has many advantages over traditional techniques and is applicable to a wide range of engineering design problems involving motion specification. The MODUS system is now operational on a SUN 3/80 workstation. Development work is currently being carried out to further enhance its capabilities.

Introduction

In many engineering areas, including mechanism design, robotics, automated guided vehicles, conveyor systems and highway design, precise motion constraints must be satisfied. The fundamental design problem is the specification of a motion which satisfies the constraints and has acceptable dynamic characteristics. Satisfying kinematic constraints often generates motions with unacceptable dynamic characteristics which limit the performance of the system.

Motion curves may be constructed in various ways. They may be derived from the solution of a dynamic model or, more frequently, obtained from using a set of rules and heuristics for specific applications. The problem may be illustrated using a simple radial plate cam.

Consider a cam-follower system, (figure 1). The follower is required to have a periodic dwell-rise-return-dwell motion (figure 2). Standard kinematic curves may be used to specify the follower motion [1]. However, when other than simple motions are required complex blending techniques have to be used [2]. Even when it is possible to use these techniques to satisfy the motion constraints their inflexibility gives limited or no control of the motion. This often results in discontinuities in the derivatives of the motion curve. These discontinuities can lead to systems that exhibit excessive vibration, wear and noise.

Most planar motion specification problems can be reduced to a general interpolation problem. It has been shown [3,4] that solutions based on spline functions have many advantages for a whole range of motion specification problems. Spline functions are piecewise polynomials with continuity between the pieces. Solutions based on spline functions can satisfy exact kinematic constraints and
  - generate an infinite family of solution curves for most problems,
  - allow high order smoothness,

- are computationally stable and efficient,
- allow interactive control and modification of the motion characteristics.

This paper outlines the basic principles of the spline function method and describes an interactive computer-aided design system called MODUS, which is currently under development at the University of Nottingham.

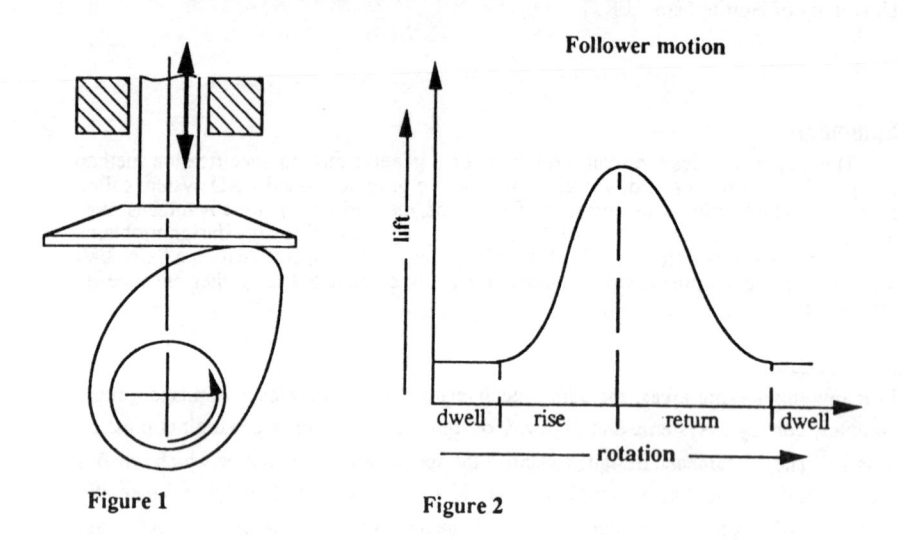

Figure 1

Figure 2

Spline solutions to motion specification problems

The problem is to satisfy a general set of discrete kinematic constraints. This may be expressed in the following manner. Let the set of n interpolation constraints be defined by $\{x_i, y_i\}$, $i = 1,2,...,n$, where x_i is the constraint abscissa (with $x_{i+1} \geq x_i$) and y_i is the corresponding constraint value. The form of the constraint at x_i is given by a corresponding sequence of integers $\{r_i\}$, $i = 1,2,...,n$. For example, if $r_i = 0$ then a value is specified, if $r_i = 1$ a first derivative is specified, if $r_i = 2$ a second derivative, etc. An interpolating function $F(x)$, is required such that

$$F^{(r_i)}(x_i) = y_i, \qquad i = 1,2,...,n. \tag{1}$$

The problem may be solved easily and conveniently with polynomial spline functions.

First consider any closed interval [a,b], partitioned such that

$$a = t_0 < t_1 < t_2 < .... < t_{g-1} < t_g < t_{g+1} = b \tag{2}$$

A polynomial spline $S(x)$ of order k is defined as a function that is a polynomial of order k (degree $\leq k-1$) on each of the subintervals $[t_i, t_{i+1})$, $(i = 0,1,2,...,g-1)$ and $[t_g, t_{g+1}]$ and which has continuity across the break points t_i, $(i = 1,2,3,...,g)$. These breakpoints have an associated knot sequence $\{\lambda_i\}$, $(i = 1,2,...,l \quad l \geq g)$ such that at any breakpoint t_i, μ_i knots may be placed, where $\Sigma \mu_i = l$. Each knot at t_i is said to have multiplicity μ_i. The polynomial spline $S(x)$ has continuous derivatives up to $k - 1 - \mu_i$ at t_i, $(i = 1,2, \ldots ,g$). In the simplest case where a single knot is placed at each breakpoint, $\mu_i = 1$ (for all i) and $g = l$ and a spline function of order k will have $k-2$ continuous derivatives on the interval [a,b].

B-Splines provide the most convenient computational basis for the linear space of polynomial splines on the interval [a,b], (see [5,6]). Any polynomial spline can be represented as a combination of B-Splines.

$$S(x) = \sum_{i=1}^{l+k} c_i B_{i,k}(x) \qquad i = 1,2, \ldots ,n. \qquad (3)$$

where $B_{i,k}(x)$ is the ith B-Spline of order k, and c_i is the ith coefficient. This representation requires knots to be specified exterior to the interval [a,b], but this is purely a computational problem. The solution to the interpolation problem in (1), using functions of the form (3) reduces to solving the linear system of equations of order n $(=l+k)$ expressed in matrix form

$$Ac = y \qquad (4)$$

where the i,jth element of A is given by $a_{ij} = B_{j,k}^{(r_i)}(x_i)$, and c and y are the vectors of spline coefficients and specified values respectfully. It can be shown [4] when no more than $k-1$ interpolation conditions are defined at any point the matrix A is non-singular if and only if the interior knots λ_i satisfy

$$\lambda_i \in (x_i, x_{i+k}). \qquad (5)$$

In practice this means that if n interpolation conditions are specified, then all polynomial splines of order k with $n-k$ interior knots satisfy the conditions provided the knot positions are consistent with (5).

Example
A more specific cam-follower motion may have to satisfy the constraints in table 1

Table 1

ROTATION (degrees)	LIFT (mm)	VELOCITY	ACCELERATION
50.00	0.00	0.00	0.00
70.00	20.00		
140.00	120.00	0.00	0.00
230.00	0.00	0.00	0.00

From the table it can be seen that there are ten constraints ($n = 10$). Thus

$$\begin{aligned}
x_1 &= 50, & y_1 &= 0, & r_1 &= 0. \\
x_2 &= 50, & y_2 &= 0, & r_2 &= 1. \\
x_3 &= 50, & y_3 &= 0, & r_3 &= 2. \\
x_4 &= 70, & y_4 &= 20, & r_4 &= 0. \\
x_5 &= 140, & y_5 &= 120, & r_5 &= 0. \\
x_6 &= 140, & y_6 &= 0, & r_6 &= 1. \\
x_7 &= 140, & y_7 &= 0, & r_7 &= 2. \\
x_8 &= 230, & y_8 &= 0, & r_8 &= 0. \\
x_9 &= 230, & y_9 &= 0, & r_9 &= 1. \\
x_{10} &= 230, & y_{10} &= 0, & r_{10} &= 2.
\end{aligned}$$

If we wish to interpolate this data using quintic splines ($k = 6$), then we must specify $n-k$ ($= 4$) internal knots consistent with (5). Figures 3, 4 and 5 show the resulting displacement, velocity, and acceleration curves using a uniformly spaced knot sequence. However, for this set of imposed constraints, an infinite set of spline interpolation curves can be constructed. Different solutions may be generated by :
- moving the knot positions,
- changing the order of the spline,
- altering the interpolation constraints.

The ability to modify the form of the motion curves, while still satisfying all the imposed constraints, gives the potential for an interactive motion design system.

MODUS - Motion Design Using Splines.

MODUS is a general interactive motion specification package. Spline functions are used to interpolate a set of kinematic constraints. MODUS has been developed on a SUN 3/80 workstation in the Department of Manufacturing Engineering and Operations Management at Nottingham University, U.K. **

For general planar motion problems, spline functions may be used to generate an infinite set of solution curves that exhibit different characteristics. In order to define a motion curve, the following must be specified :
- the interpolation constraints,
- the order of the spline function,
- the knot positions.

The system allows :
- arbitrary combination of interpolation conditions, including non-consecutive specification of derivatives,
- choice of order of spline,
- freedom of knot positions, with the restriction indicated in (5).

The core element of the system is the motion generation module. Applications modules can be 'built' upon this. The structure of the motion generation module is shown in figure 6. Curve modification is achieved by changing the parameters that uniquely define the spline function. Using interactive graphics, the interpolation constraints may be changed, knot positions moved, order increased or decreased, etc. Combining changes in design parameters is a particularly powerful tool. The effect of a change can be viewed through 'windows' which simultaneously show the motion curve and any characteristics of interest. This almost instant change and effect attribute of the system gives the user the ability to 'tune' the motion curves to suit the application. Modifications can be carried out on any of the displayed characteristic curves.

Consider the example given earlier. Given that the follower must satisfy the constraints in table 1, the initial spline motion curves formed with a uniformly spaced knot sequence (knot positions - 86, 122, 158, 194), are shown in figures 3, 4 and 5. Using MODUS the motion is modified by introducing a tuning velocity constraint (geometric velocity of 2.40 at rotation 81.0 degrees) and altering the position of the knot sequence (new knot positions - 70, 100, 125, 125, 200). The effect is shown in figures 8, 9 and 10. The object was to reduce the maximum peak velocity for the motion. This was achieved, and the maximum geometric velocity reduced from 3.04 (figure 4) to 2.40 (figure 9).

The motion generation module has been constructed to allow application modules to be 'built' around it (figure 7). Specific applications modules can then interact with the motion generation module. The first application module developed is for cam design.

For example the cam design module can take as an input the motion curves in figures 8, 9, and 10. Further system inputs are required. The precise inputs depend on the selection of the follower type and dimensions. The system then generates a cam profile and any other required output information, such as velocity, acceleration and jerk curves, pressure angle, or curvature. The cam system can then be studied for excessive pressure angles, undercutting, etc, and if unacceptable the problem segment of the motion curve can be modified. This can be repeated until a satisfactory cam-follower system is produced.

Conclusion and future developments

This paper has outlined the principles of the spline function method for engineering design problems involving motion specification. The method has many advantages over traditional techniques and much potential for a whole range of motion design areas. An interactive graphics-based CAD system called MODUS has been developed, based on the spline function method. The system has been developed in a modular fashion with a core module for motion generation and applications modules based upon it.

Further work is continuing to enhance the capabilities of the system. Additional design handles are being added to the motion generation module to explore, tune and optimize the motion characteristics. Feedback loops are being developed to link the motion generation module and the applications modules. The development of further applications modules is underway.

Figure 3 — Rotation (degrees), Displacement

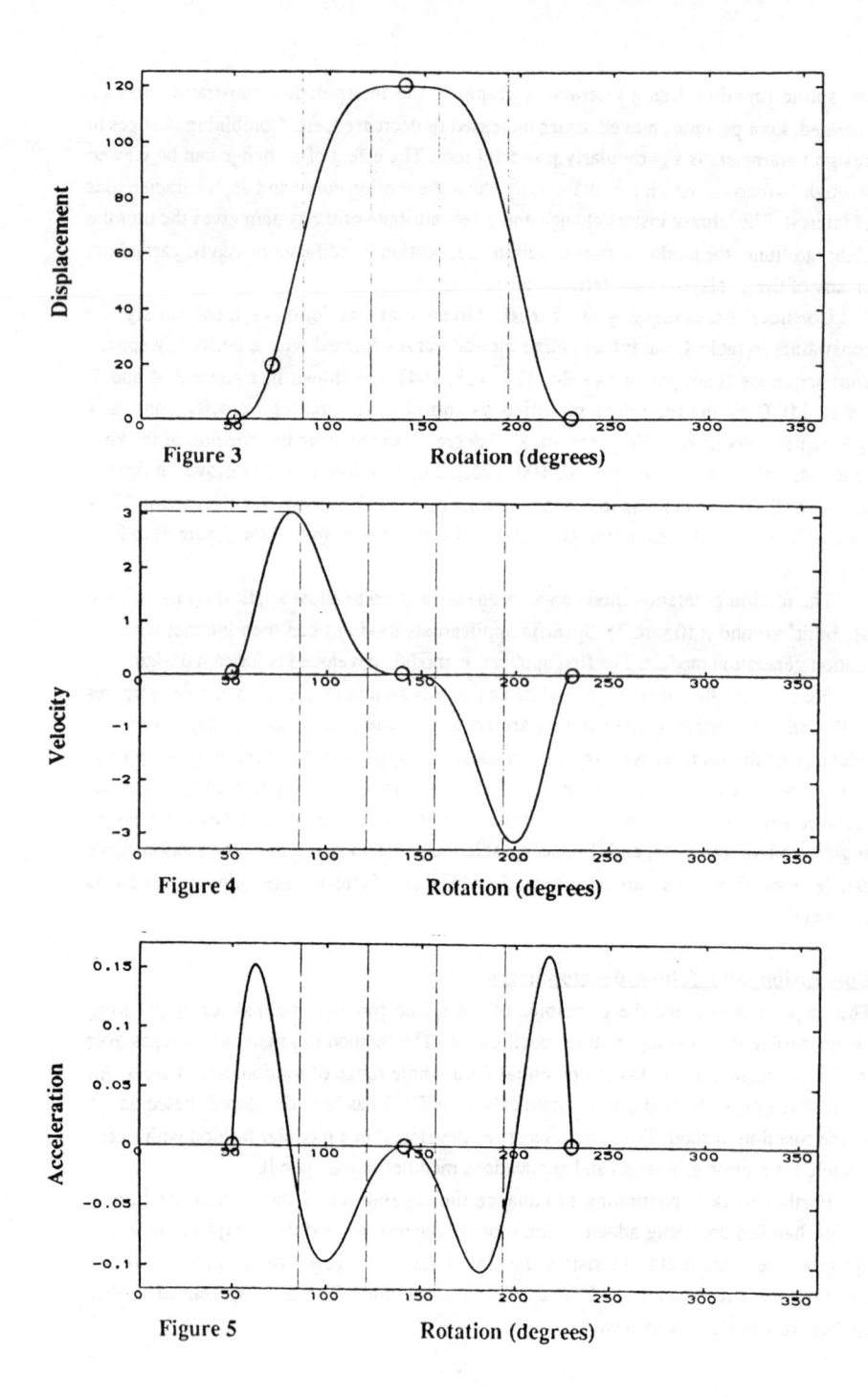

Figure 4 — Rotation (degrees), Velocity

Figure 5 — Rotation (degrees), Acceleration

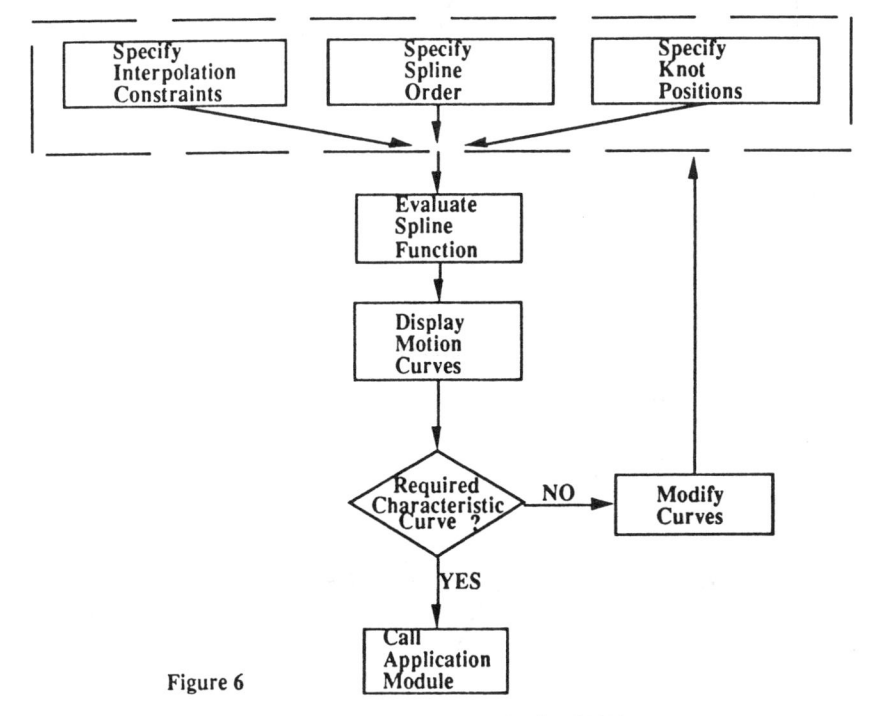

Figure 6

Structure of the motion generation module of MODUS.

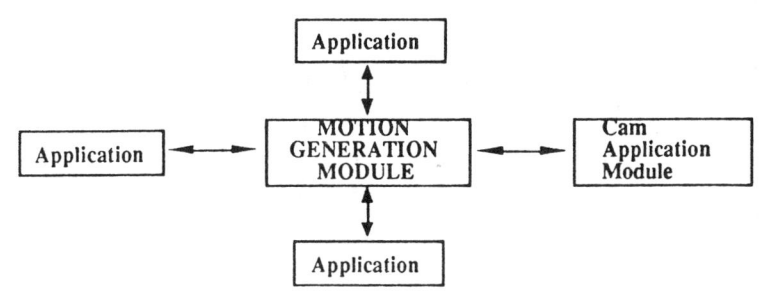

Figure 7

Applications modules built round the motion generation module

Figure 8

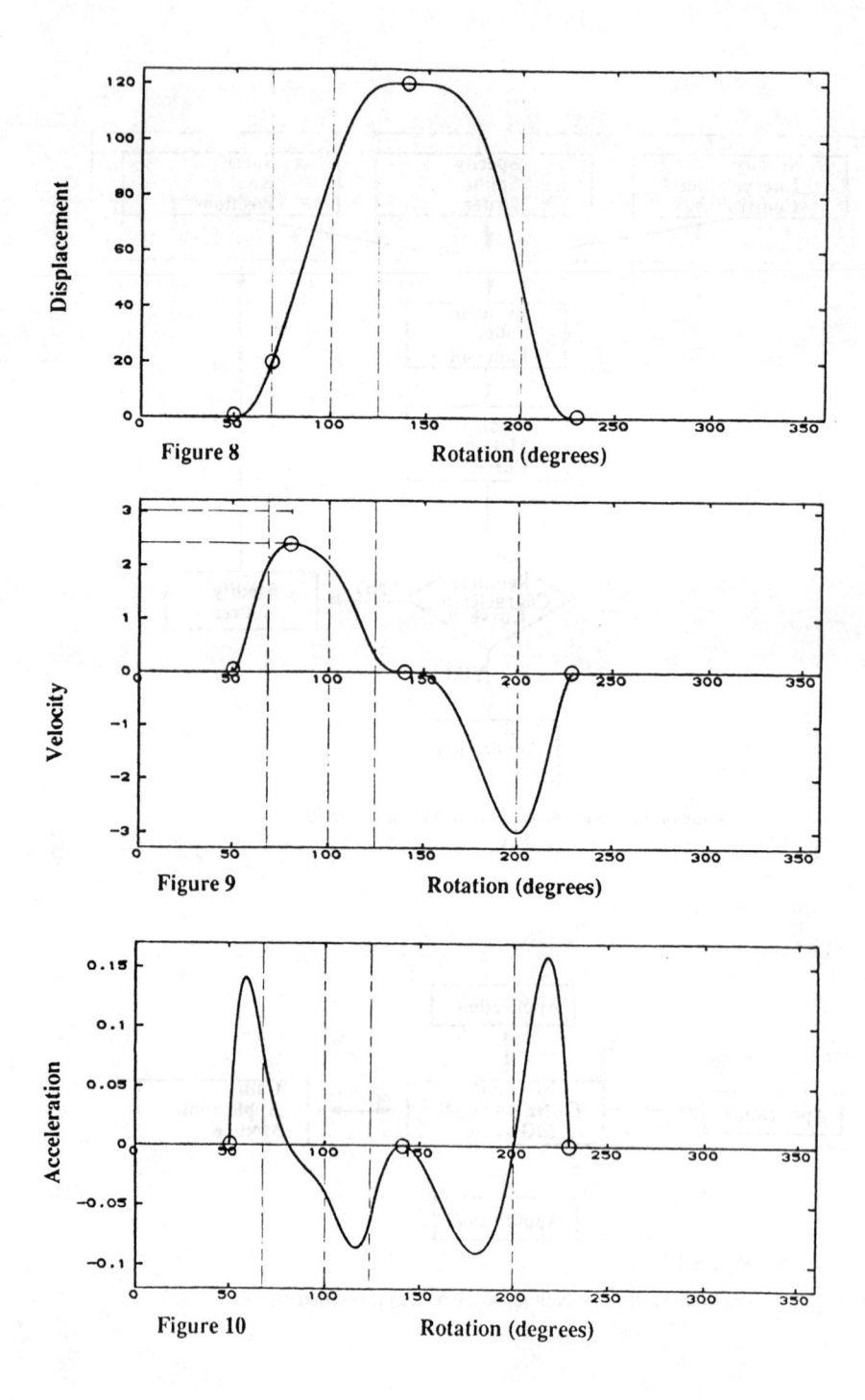

Figure 9

Figure 10

** Acknowledgements

The authors gratefully acknowledge the support of the Electro-Mechanical Engineering Committee of the SERC (GR/F07743) for this work.

References

1. ESDU Item, 82006, "Selection of DRD cam laws".
   Engineering Sciences Data Unit, 1982.

2. Reeve, J. E., "Mechanisms, cam laws 4-blending".
   Engineering, July 1979, pp 759-765.

3. MacCarthy, B. L., Burns, N. D., "An evaluation of spline functions for use in cam design".
   IMECH E. vol. 199, No. C3, 1985, pp 239-248.

4. MacCarthy, B. L., "Quintic splines for kinematic design".
   Computer-aided design, vol. 20, No. 7, Sept. 1988, pp 406-415.

5. deBoor, C., "A practical guide to splines", Springer, 1978.

6. Schumaker, L. L., "Spline functions: Basic theory", John Wiley, 1981.

Fast Algorithm for the Calculation of Generalized Torques in Robotics

B. LEVESQUE and M.J. RICHARD

Department of Mechanical Engineering
Laval University
Quebec, Canada G1K 7P4

SUMMARY

This paper describes an efficient dynamic algorithm for the analysis of manipulators. Appell's principles are exploited in this work where four different methods are analyzed and tabulated in terms of their efficiency. The analytical step by step approach proposed here in reduces the number of necessary operations for the calculations of the generalized forces of a 6 degrees of freedom manipulator by as much as a factor 3 with respect to other conventional methods.

INTRODUCTION

Most of the formulations of the dynamics of a manipulator are derived from the Newton-Euler equations or the Lagrange's equations. The best results, in terms of efficiency of computation, are the ones based on Newton-Euler equations [1,2], which in their general form applied to a six degrees of freedom manipulator gives 738 additions and 852 multiplications [3]. The methods derived from Lagrange's equations give results that are less interesting in the point of view of the number of operations [3], although it has been demonstrated that the two methods could be equivalent with some transformations in the formulation of the inertia tensor [4]. An efficient formulation based on Lagrange's equations has also been proposed by Renaud [5] (as less as 367 additions and 352 multiplications), but unfortunately is function of n^3, where n is the number of degrees of freedom [6].

On the other hand, very little work has been done to improve the efficiency of the formulations based on Appell's equations. The recursive method developed by Vukobratovic and Potkonjak gives 2431 additions and 2929 multiplications, and depends on n^3 [6]. In this paper, a derivation of the complete dynamics of a mechanical manipulator from Appell's equations is presented which leads to a recursive algorithm depending linearly on the number of degrees of freedom, and giving better results than the Newton-Euler's formulation.

APPELL'S EQUATIONS

Appell's equations are very similar to Lagrange's equations. They are based on the computation of the acceleration energy which is the equivalent of the kinetic energy in the Lagrange's equations. For a given rigid body in space, the acceleration energy is defined by the equation :

$$G = \frac{1}{2} \int_m \ddot{z}^2 \cdot dm , \qquad (1)$$

where G represents the acceleration energy of the body, $\ddot{z}$ is the absolute acceleration of the mass element dm, which may be expressed as :

$$\ddot{z} = \dot{v}_p + (\dot{\omega} \times r) + \omega \times (\omega \times r) , \qquad (2)$$

where $\dot{v}_p$ is the linear acceleration of any reference point P of the body, ω is its angular velocity, $\dot{\omega}$ its angular acceleration and r the position vector of the infinitesimal element of mass dm with respect to the reference point P. The evaluation of the integral (1) for a body i gives :

$$G = \tfrac{1}{2} \dot{v}_p^2 \cdot m + \dot{v}_p \cdot (\dot{\omega} \times r_c) \cdot m + \dot{v}_p \cdot [\omega \times (\omega \times r_c)] \cdot m$$

$$+ \tfrac{1}{2} \dot{\omega} \cdot \underline{J}_p \cdot \dot{\omega} + [\omega \times (\underline{J}_p \cdot \omega)] \cdot \dot{\omega} + \omega^2 [\omega \cdot \underline{J}_p \cdot \omega] , \qquad (3)$$

where G_i represents the acceleration energy of the body i, m_i is the mass of the body, r_c is the position vector of the mass center with respect to the point P and $\underline{J}_p$ is the inertia tensor of the body. For a system composed of n rigid bodies, the total acceleration energy of the system is given by the sum of the acceleration energy of each of the bodies. The Appell's equations are defined by the derivative of G, in equation (3), with respect to the generalized acceleration i. One obtains after some algebra and simplifications the n resulting generalized forces τ_i :

$$\tau_i = \sum_{j=1}^{n} \left[m_j [\dot{v}_{p_j} + (\dot{\omega}_j \times r_{c_j}) + \omega_j \times (\omega_j \times r_{c_j})] \cdot \frac{\partial \dot{v}_{p_j}}{\partial \ddot{q}_i} \right.$$

$$\left. + [m_j (r_{c_j} \times \dot{v}_{p_j}) + \dot{\omega}_j \cdot \underline{J}_{p_j} + \omega_j \times (\underline{J}_{p_j} \cdot \omega_j)] \cdot \frac{\partial \dot{\omega}_j}{\partial \ddot{q}_i} \right] , \qquad (4)$$

where the symmetry of the inertia tensor $\underline{J}_p$ has been exploited. In order to simplify equation (4), the following variables represented in figure 1 will be defined; e_i is a unit vector representing the reference axis for the joint i, r_{ji} is a vector going from the origin of the reference system j-1 to the mass center of link i, p_{ji} is a vector going from the origin of the coordinate system j-1 to the origin of the system i, s_i is an indicator used to determine if the joint is revolute ($s_i = 0$) or prismatic ($s_i = 1$) and C_i represents the center of mass of body i. With this notation, the derivatives of the accelerations appearing in equation (4) are easily obtained and one finds for the generalized forces :

$$\tau_i = e_i \cdot \sum_{j=1}^{n} [(1 - s_i)[(p_{ji} \times F_{p_j}) + N_{p_j}] + (S_i)(F_{p_j})] , \qquad (5)$$

where: $F_{p_j} = m_j [\dot{v}_{p_j} + (\dot{\omega}_j \times r_{c_j}) + \omega_j \times (\omega_j \times r_{c_j})]$, (6)

and $N_{p_j} = m_j(r_{c_j} \times \dot{v}_{p_j}) + \underline{J}_{p_j} \cdot \dot{\omega}_j + \omega_j \times (\underline{J}_{p_j} \cdot \omega_j)$. (7)

Equation (5) is the general equation for the computation of the generalized force at joint i, for a revolute or prismatic joint. The terms F_{p_j} and N_{p_j} are the resulting inertial force and torque acting on the mass center j.

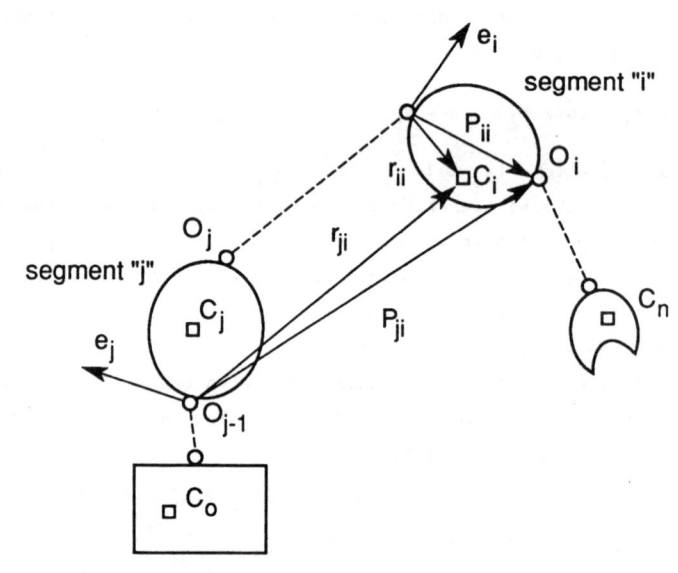

FIGURE 1 - Relations Between Segments

COMPARISON OF THE EFFICIENCY OF THE FORMULATIONS

One must first define a comparison index. This index is called the Theorical Minimum Number of Operations (TMNO) and is the sum of the number of operations needed to compute each of the n generalized forces supposing that all the quantities necessary are expressed in the same coordinate system, and without the loss of any generality. However, for efficiency reasons, the equations presented above are divided into two sets of equations, one set for prismatic joints and the other for revolute joints. With these two variables and the general procedure presented above, the TMNO are 80 additions and 91 multiplications for each revolute joint, and 77 additions and 94 multiplications for each prismatic joint. If one wants to use a single set of general equations that will treat indifferently prismatic or revolute joints, the TMNO would be 92 additions and 115 multiplications, which is about 21 percent more operations for each link. In a search for a better and an efficient dynamic algorithm, the equations of the step by step method, presented in this work, [7] has been examined for a six degrees of freedom robot in two different ways : 1) elimination of zeros and ones from the general algorithm, 2) complete elimination of all zeros and ones. Table 1 compares the performance of several methods including some results from Hollerbach [5] and naturally results from the step by step approach.

TABLE 1 - Comparison Between Methods

METHODS	ADDITIONS	MULTIPLICATIONS
TMNO	100% (498+)	100% (540*)
Uicker,Kahn (Lagrange's)	10,350% (51548+)	12,270% (66271*)
Hollerback 4x4 (Lagrange's)	720% (3546+)	813% (4388*)
Hollerback 3x3 (Lagrange's)	345% (1719+)	406% (2195*)
Luh,Walker,Paul (Newton-Euler's)	148% (738+)	158% (852*)
Renaud: minimum (Lagrange)	73.7% (367+)	65.2% (352*)
Appell's Equations		
Vukobratovic	488% (2431+)	542% (2929*)
Method 1: general	194% (966+)	234% (1266*)
particular	173% (864+)	208% (1125*)
Method 2: general	136% (678+)	154% (834*)
particular	130% (647+)	143% (771*)
Method 3: general	133% (660+)	158% (852*)
particular	124% (617+)	143% (771*)
Method 4: normal	64% (320+)	82% (442*)
minimal	36% (177+)	49% (262*)

In method 1, the vectorial and tensorial quantities are expressed with respect to inertial coordinates with the inertia tensors expressed with respect to mass centers. For example, let us consider the six (revolute) degrees of freedom robot. Applying the above conditions to our calculations, the algorithm must execute 966+ and 1266* which represent, respectively, 194% and 234% of the TMNO. One may introduce some basic simplifications. It should be noted that the articulation axis are always represented by the preceding z axes, in other words, e_i = [0 0 1]. Also, if some initial conditions are introduced such as fixing the base of the manipulator, the number of total operations for our 6 degrees of freedom manipulator would reduce to 864+ and 1125* which represent 173% and 208% of the TMNO.

In the second approach, all quantities are defined with respect to body-fixed or local coordinates which simplifies the calculations by introducing constant inertia tensors, expressed with respect to the mass centers. Reexamining our example, 678+ and 834* are required which means 136% and 154% of the TMNO. As explained earlier, it is possible to introduce particular conditions associated with the mechanism and, in effect, reduce the TMNO to 647+ and 771*. This representation method corresponds to that proposed by Luh, Walker and Paul [1] but, by exploiting Appell's principle in this work, it was possible to reduce their number of additions by 12% and their number of multiplications by 8%. However, if the forces and torques on the end effector must be transformed, the performance of this approach reduces considerably. At best, 719+ and 891* are required which leads toward results of the first method. The third method is very similar to the second method except in the representation of the inertia tensors. In the third method, local coordinates are exploited with inertia tensors defined with respect to some point P, usually taken at the origin of the local coordinate frame. For the case of 6 revolutes, 660+ and 852* must be executed (133% and 158% of the TMNO). This method is very appealing since one can add an object to the end effector, which changes the center of mass of the body, and can calculate the new center of mass with few operations (3+ and 8*).

Finally, the fourth method takes into account the actual structure of the manipulator. This approach is considered an analytical method where the procedure to follow is established step by step searching for structural simplifications along the way. Note that this method is not "general" in the sens assumed in this work. However, the procedure to obtain the final dynamic model is identical for all the possible cases. If one considers again the six revolute manipulator, in a general case (base acceleration, external forces with respect to inertial system, inertia tensors are not principal), this method yields 320+ and 442* which represents 64.3% and 81.8% of the TMNO. Let us suppose that the system has the best possible conditions where the base is fixed, external forces are expressed in local coordinates (body 6) and the local axes allow for the principal moments of inertia, then the step by step method will execute 177+ and 262* which results in a minimum of

36% and 49% of the TMNO. Eventhough, this method is very efficient, its application is limited to particular robot simulations since it relies heavily on the architectural nature of the manipulator.

CONCLUSION

A new derivation method for the dynamical model of manipulators based on Appell's equations has been derived in this work. The greatest attributes of this method lies in its simplicity and directness. A systematic analysis of various formulation methods with respect to vectorial and tensorial representation was conducted. A notable difference was observed in the efficiency of the algorithms whether the vectors and tensors where expressed along inertial, base (0) or local coordinates. The latter system of coordinates provided, by far, the best environment for the development of an efficient algorithm. It was shown that the step by step analytical method reduced the number of operations for a single trajectory point by a factor 3 over other conventional simulation techniques as listed in table 1. The method is appropriate for the control of manipulators due to its rapidity.

REFERENCES

1. Luh, J., Walker, M., Paul, R.: "On-Line Computational Scheme for Mechanical Manipulators", Transactions of ASME, Journal of Dynamic Systems, Measurement, and Control, Vol. 102, 2 (June 1980), pp. 69-76.

2. Paul, R.P.: Robot Manipulators: Mathematics, Programming, and Control, MIT Press, Cambridge, 1981.

3. Hollerbach, J.M.: "A Recursive Lagrangian Formulation of Manipulator Dynamics and Comparative Study of Dynamics Formulation Complexity", IEEE Trans. on Systems, Man, and Cybernetics, SMC-10, 11 Nov. 1980, pp. 730-736.

4. Silver, W.M.: "On the Equivalence of Lagrangian and Newton-Euler Dynamics for Manipulators", in The International Journal of Robotics Research, Vol. 1, No. 2, pp. 60-70, 1982.

5. Renaud, M.: "An Efficient Iterative analytical procedure for obtaining a robot manipulator dynamic model", Robotic Research: the First International Symposium, MIT Press, 1984, pp. 748-764.

6. Vukobratovic, M., Kircanski, N.: Real-Time Dynamics of Manipulation Robots (Scientific Fundamentals of Robotics 4). Springer-Verlag, Berlin, New York, 1985.

7. Lévesque, B.: Etude de l'efficacité des algorithmes de simulation dynamique pour les manipulateurs, Mémoire de maîtrise, Université Laval, Québec, 1988.

Adaptive Control of Machining Using Expert Systems

by

Subramanya, P.S., Osman, M.O.M., Latinovic, V.N
Department of Mechanical Engineering
Concordia University, Montreal.

Summary

There has been a steady attempt to use Artificial Intelligence (AI) and Expert system techniques for the control of machining process. A rule based expert system was developed for real time control of machining process and the system was tested for reliability under simulated machining conditions. Acoustic emmission signals and cutting force signals were used to determine the machining condition.

Adaptive Control of Machining

Previous work in adaptive control of machine tools can be classified into two groups. Adaptive Control of Optimization (ACO) and Adaptive Control of Constraints (ACC). Many ACO and ACC systems have been investigated [1,2]. The ACO method faces the problem that a huge database need to be established before the method can be applied. Furthermore, its reliance on Taylor's tool life equation to obtain the optimum working conditions makes it unsuitable for implementing on-line real time control, in the general sense. In the second method it is difficult to approach the global optimum working conditions with an incomplete knowledge of relationship among the machining variables.

Expert System Method

Expert systems techniques are often effective for controlling complicated problems containing a substantial degree of uncertainty. This is done by treating the control of the machining process as a decision making problem and by establishing a decision making framework to be applied in tandem with multiple sensors technique [3].

The System Data Base

In order to achieve on line prediction and identification of the machining process, it has to be represented by mathematical models. These models provide the hypothesis for selection of proper machining data. This can be achieved by suitable adaptive optimization algorithms. The use of sequential 'Maximum a posteriori'(MAP) estimation procedure based on the Bayesian statistical approach was found very useful [3].

Since the optimization for each new set of data obtained during the operation in order to adopt to the machining conditions, tend to destabilize the process because there is always a different signal coming in. More over the optimization consumes time and the response will be delayed. In order to overcome the above problems, new machining parameters are generated only at the request of the rule base. The rule base requests for new set of parameters only when the in coming signals are classified into a different class.

The classification procedure adopted to group the feature vectors of AE signal is based on the Nonparametric clustering criteria [4]. In order to expedite the classification and also to avoid side effects of irregular noise, the signals are analyzed in batches for classification.

The Rule Base

The rule base contains the production rules which are typically described as :"if <situation> then <action>". In order to obtain reflex-action like decision for unforeseen circumstances, metalevel inference is made possible by this type of grouping of rules.

The 'situation' as a part of the rule represents the actual situation in machining, namely signals from the data which re interpreted as chatter or temperature at the tool point or acting force levels etc. The 'action' can represent the activation of a controller or an estimation algorithm.

Since it is a common knowledge that the machining process follows some hypothesis only if the machining operation is chatter free i.e., stable, the priority will be given to chatter suppression rules over other rules. For example when rough machining steels,

IF <chatter>
THEN < get(hyp(stable(v,s))) >
AND <alter(v,s)>

If <group> is <cluster 3>
THEN < get(hyp(Optim(v,s)) >
AND <Optim(v,s)>
AND <idle, 10>

Inference Engine

The purpose of the inference engine is to decide from the context (current data base of facts, evidence, hypotheses and goals) which production rules to select next. The graph method [5] of inferenceing and dividing the inference process among different modules has been proved to be the best method available for high speed inference.

In Graph method of inferencing, all the rules are in a graph in which each rule is represented by a special count node with premises coming in and conclusions leaving as shown in the figure a of figure 1. The true assertions are placed in a linked list just as in any standard method of inferencing. This is shown in figure b of figure 1. True assertion list is traversed only once. At each assertion, the emanating arcs are traversed, adding one to the COUNT in the count node to which each arc points. When the COUNT in the count node is equal to the TRIGVAL, the conclusion, to which the count node points, is added to the end of the assertion list. It is necessary to add the conclusion to the end so that it will eventually be visited, and used to trigger rules in which it is a premise.

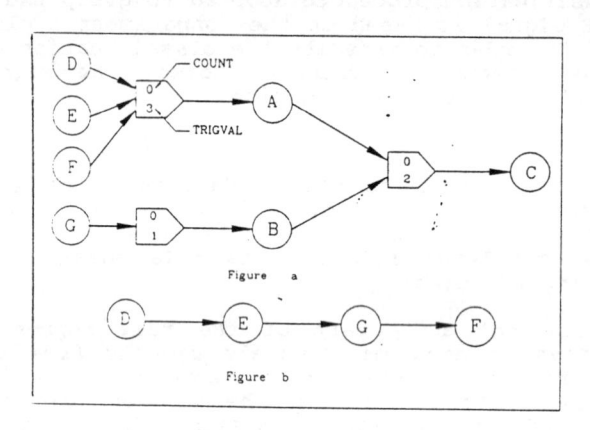

Fig.1 : Graph Method

Application of Expert system for control
Recognition of Machining State

Sensing of the machining state is the key to application of control strategy for machining. Several sensing strategies for tool wear detection have been proposed and evaluated in a number of review articles. Each technique has its own advantages and drawbacks with the result that no single technique has proven to be completely reliable over the entire range of operation.

The most promising tool monitoring techniques is the use of AE and Cutting force information in order to develop an intelligent tool condition monitoring system. The generation of the AE signals directly in the cutting zone makes them very sensitive to changes in the cutting process. Figure 2 to 4 shows the typical AE signal generated during turning of carbon steel with HSS tool. The AE signals do not tell the complete story because they are also sensitive to chip fracture. The energy levels of the AE signal during a chip fracture is very high and probably close to AE level connected with worn tool cutting (in which the chip breaking frequency is lower due to increased ductility of the chip at higher temperatures). Hence it is very likely that if the AE signal is sampled during a chip fracturing event, misclassification will occur. The Cutting force signal on the other hand is not very sensitive

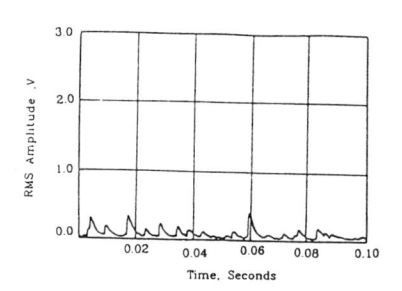

Fig.2: Initial Stage

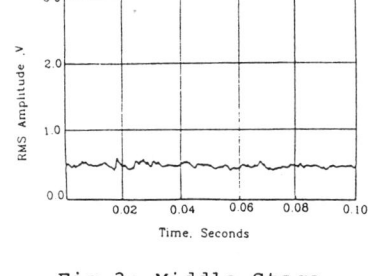

Fig.3: Middle Stage

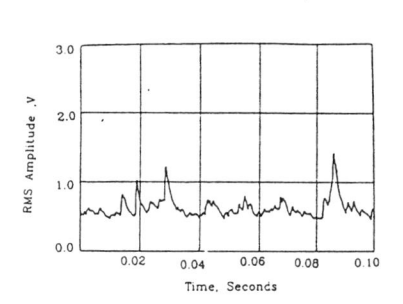

Fig.4: Final Stage

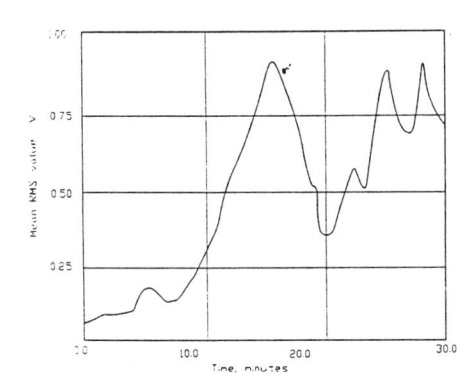

Fig.5: Mean RMS Vs Time

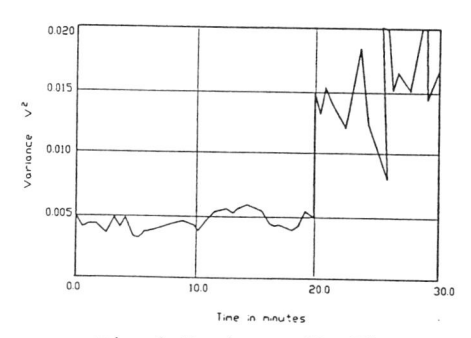

Fig.6 Variance Vs Time

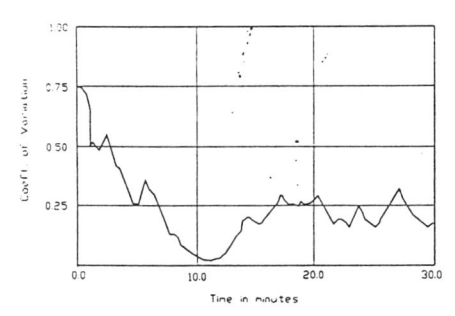

Fig.7: Coeff. of Var. Vs Time

to chip breakage. The inclusion of cutting force as one of the dimensions of the feature vector will tend to reduce the misclassification to a great extent. However, the cutting force information by itself is inadequate for tool wear detection because its magnitude is also dependent on the cutting velocity. Another problem is that although flank wear tends to increase the cutting force, the accompanying crater wear tends to reduce it, so the magnitude of the cutting force may not show any sensitivity to tool wear at all.

The characteristic features of AE signals (Figures 2 to 4) are analyzed and measured by various researchers [6,7]. These characteristic features of AE signals can be summarized as follows (Figures 5 to 7).

1. At the early stage of tool wear, the mean value of the AE signal is small, but the signal contains much of the burst type of AE signal, which results in high value of coefficient of variation.

2. At the middle stage of tool wear, the mean value of the AE signal becomes larger, but the coefficient of variation becomes small because the variance is kept small.

3. At the final stage of tool wear, both the mean value and the variation increase, and consequently the coefficient of variation also increase slightly as compared with that at the middle stage.

The changes in the characteristic features of RMS signal of the AE provides the basis for classification into three different classes in the system database as shown in Figure 8 using the pattern recognition principle (clustering).

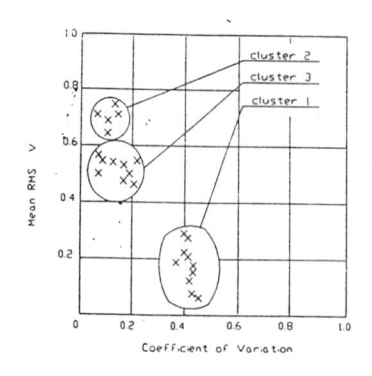

Fig.8: Clustering

Conclusions

The expert system method for controlling the machining operation is found to be very effective when subjected to simulated machining conditions. The Cutting speed and feed generated by the expert system at various times of machining(Simulation) are shown in Figures 9 and 10.

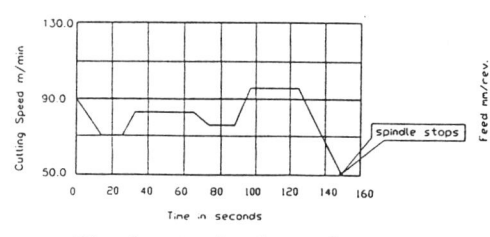

Fig.9: Variation of Cutting Speed.

Fig.10: Variation of Cutting Feed.

It is observed that the system can provide a satisfactory performance even with incomplete and unreliable data and knowledge.

References

1. Kaminskaya, V.V. et al.: Trends in Development of Adaptive Control. Machines and Tooling. 45. 1974. pp 66-70.

2. Wick, C.: Automatic Adaptive Control of Machine Tools. Manufacturing Engineering. 1977. pp 38-45.

3. Subramanya, P.S.; Latinovic, V.N.; and Osman, M.O.M.: Expert Control of Turning Process. 5th International Conference on CARS & FOF. Virginia, USA.

4. Devijver, P.A.; and Kittler, J.: Pattern Recognition - A Statistical Approach. Prentice hall. New Jersey. 1959.

5. Neapolitan, R.E.: Forward-chaining Versus a Graph Approach as the Inference Engine in Expert Systems. Proc. Applications of Artificial Intelligence III. SPIE vol. 635. April 1986. pp. 62-69.

6. Blum, T.; Insaki, I.: A Study of Acoustic Emmision from Orthogonal Cutting process. Jr. of Engg.for Industry. Vol. 112. August 1990.

7. Moriwaki, T.; Tobito, M.: A New Approach to Automatic Detection of Life of Coated Tool Based on Acoustic Emission Measurement. Trans ASME. Vol. 112. August 1990.

A Reinforcement Connectionist Path-Finder

José del R. MILLÁN, Marc BECQUET
Institute for System Engineering and Informatics
Commission of the European Communities. Joint Research Centre
Building A36. 21020 ISPRA (VA). ITALY
e-mail: j_millan@jrc.it, m_becquet@jrc.it

ABSTRACT— *In this paper, the problem of path-finding is tackled with the use of a reinforcement connectionist system able to find and learn feasible paths. The learning phase is a stochastic trial-and-error procedure from performance feedback. These concepts are described and applied with the objective of real-time decision. The paper presents the problem formulation, the system architecture and the learning algorithm. Experimental results are given, including robustness aspects and generalization capabilities.*

1. Introduction

Traditional approaches to *robot path finding* rely on *planning*. However, planning is unsuitable for real-time decision making, since it is conceived as an off-line activity that acts on a perfect model of the environment. Therefore, it cannot be used in cases where the environment is (partially) unknown or dynamic and when it is necessary to take into account a continuous set of robot configurations and robot actions. Moreover, a path-finder navigating in a real environment must be able to cope with unexpected events. An alternative to planning is *to transform each perceived situation into the proper action, and iterate this mapping until the goal is reached.*

In this paper we present a *reinforcement connectionist system* able to find and learn suitable *situation-action rules* as to generate feasible paths for a mobile robot in a cluttered 2D environment. The criterion used for evaluating the quality of a path is a compromise between minimizing the path length and maximizing the distance to the obstacles.

2. Problem Formulation

An intelligent agent interacting with its environment is said to improve when it learns to react with a more and more useful action —where *usefulness* is defined in terms of the agent's goal— to each situation. The *associative reinforcement learning (ARL)* problem [1, 2, 3] offers a simple and general framework for developing this kind of systems. Simply stated, the ARL problem is that of learning to associate with each stimulus **X** the action **Y** that maximizes reinforcement z —either present, future or cumulative.

The approach to the ARL problem adopted here is *to estimate the gradient of the reinforcement with regard to the system's actions by measuring the correlation between variations in actions and variations in reinforcement.* Once this gradient is available, it is used for modifying the agent. This learning paradigm is known as *reinforcement learning.*

The *input* to the path-finder consists of an *attraction force* and several *repulsion forces* exerted respectively by the goal configuration and the obstacles on the current configuration. The robot does perceive only obstacles in its immediate surroundings.

Let the *shortest path line (SPL)* and the *shortest path vector (SPV)* be, respectively, the line and the vector that connect the robot current and goal configurations. The direction of the attraction force is that of the SPV, its intensity being an inverse exponential function of the distance between the current and goal configurations.

Each repulsion force represents the resistance of an obstacle toward the robot SPV following process. Each such force has the direction of a bisector of the SPL and its perpendicular, starting at the current configuration and heading the opposite quadrant to that where the obstacle lies. Because the directions of the repulsion forces are specified with respect to the direction of the attraction force, they are implicit in the codification and therefore are not included in the input to the system. The intensity of each repulsion force depends on three factors. The first factor is aimed at avoiding obstacles in the proximity of the SPV. The second allows to avoid obstacles near to the robot current configuration. The third acts in case the SPV intersects an obstacle: the deeper the penetration into the obstacle the more distant from the SPV will be the next robot motion.

The SPL and its perpendicular divide the workspace into four quadrants. Repulsion forces in a quadrant try to deflect the robot move toward the center of the opposite quadrant. Thus all the information related to the repulsion forces can be reduced to four signals: *intensity of the environmental repulsion from each quadrant.*

In short, *the number of input signals* to the path-finder *is independent of the environment*, and these signals are five: the intensity of the attraction force and the four environmental repulsion intensities. Since the output of the path-finder is computed with respect to the direction of the attraction force (see below), this direction does not need to be included in the input of the system. Each input signal is codified as a real number in $[0, 1]$.

The *output* of the path-finder represents the step taken by the robot and is codified as a move in *relative cartesian coordinates* with regard to the SPV. That is, the axis X is the SPL. Both coordinates take real values in $[-1, 1]$. The maximum distance the robot can cover in a single step is limited by the *perception range*: the robot could collide with obstacles "not completely" perceived since the robot concentrates on its immediate neighbourhood.

The *reinforcement signal* is a measure of how good is the answer of the system to a particular stimulus. It is calculated on the basis of the quality of the configuration reached by the robot —a combination of the attraction and repulsion factors— and the way in which this configuration has been reached. It is a real value in $[-1, 1]$.

The determination of both the input and goodness degree is reminiscent of the *potential field approach* to path planning [4].

Two are the benefits of this codification scheme. First, since the goal is codified in the input information and the input is independent of the environment used during the learning phase, it allows to transfer the knowledge acquired for a situation to a different one. Second, it reduces the complexity of the task to be solved. In the robot path finding problem, the consequences of an action can emerge later in time. Thus, actions must be selected based on both their short- and long-term consequences. Since the environmental reinforcement signal is computed using *global information* —i.e. it is based not on the current robot configuration but on the SPV—, the path-finder gets a measure of the short- and long-term consequences of an action only one time-step after it has been executed. The task is reduced to learn, for each stimulus, the action which maximizes the environmental reinforcement signal.

3. System Architecture

The path-finder is made of two elements, namely the *step generator* and the *critic*, and interacts with its environment as depicted in Figure 1. At each time t, the environment provides the path-finder with the input pattern $\mathbf{X}(t) = \big(x_1(t), x_2(t), x_3(t), x_4(t), x_5(t)\big)$, together with the *environmental reinforcement signal* $z(t)$. The input pattern is fed to both the step generator and the critic. Nevertheless, the step generator does not receive directly the environmental reinforcement signal but the *heuristic reinforcement signal* $h(t)$ elaborated by the critic. The latter is an enrichment of the former based on past experience of the path-finder when interacting with the environment. The step generator produces instantaneously an output pattern $\mathbf{Y}(t) = \big(y_1(t), y_2(t)\big)$ that it is the output of the path-finder. The environment receives this action $\mathbf{Y}(t)$ and, at time $t+1$, sends to the path-finder both an evaluation $z(t+1)$ of the appropriateness of the action $\mathbf{Y}(t)$ for the stimulus $\mathbf{X}(t)$ and a new stimulus $\mathbf{X}(t+1)$.

To illustrate the usefulness of the critic, let us consider the two following situations. First, the output of the step generator at time t $\mathbf{Y}(t)$ is the best action possible for a given stimulus $\mathbf{X}(t)$, but the associated environmental reinforcement $z(t+1)$ is -0.5. For example, if the current and goal configurations are located at opposite sides of an obstacle, then the best move could be to send the robot away from the goal. Alternatively, the step generator receives an environmental reinforcement signal $z(t+1)$ of $+0.5$ as a consequence of producing the action $\mathbf{Y}(t)$ for the stimulus $\mathbf{X}(t)$, when it could have generated a better action $\mathbf{Y}'(t)$ which allows to reach the goal configuration from the current one. These two situations show that the environmental reinforcement signal is not sufficient. To solve the path finding problem, what the step generator needs is high reinforcement in the first case, and low, perhaps negative, reinforcement in the second case. The goal of the critic is to transform the environmental reinforcement signal into a more informative signal. The resultant signal is the heuristic reinforcement signal.

4. Learning Algorithm

Among the different kinds of reinforcement learning algorithms proposed in the literature, we adopt the *associative search, AS,* algorithm [1, 2, 5] because its appropriateness toward the specifications of the input, output and reinforcement signals we have.

A central issue for any reinforcement system is to explore alternative actions for the same stimulus. *Stochastic units* provide this source of variation. But the stochastic behavior of the path-finder should tend to be *deterministic* with learning. Otherwise, the path-finder could not generate *stable solution paths* after it eventually discovers them. By using a suitable algorithm for transforming the stochastic units into deterministic ones [6] the path-finder not only discovers the suitable situation-action rules, but it also does it in a very short time.

4.1 The Basic AS Algorithm

Continuous stochastic units compute their output in three steps [7], as expressed in Equations (1) through (5). The only difference with respect to Gullapalli's formulation is the way in which the variance is computed.

Since the signals we are interested in are continuous, a separate control of the location being sought (*mean*) and the breadth of the search around that location (*variance*) is needed. The first step is to determine the value of these parameters. The mean μ should be an estimation of the optimal output. A simple way is to let $\mu_i(t)$ equal a weighted sum of the inputs $s_j(t)$ to the unit i plus a threshold $\theta_i(t)$:

$$\mu_i(t) = \sum_{j=1}^{n} \left(w_{ij}(t) s_j(t)\right) + \theta_i(t). \tag{1}$$

The standard deviation σ should be small if the expected output of the step generator is close to the optimal, and it should be high in the opposite case. Since the heuristic reinforcement signal provides this comparative measure of the output goodness, $\sigma(t)$ should depend on the *expected heuristic reinforcement*, $\widehat{h}(t)$:

$$\sigma(t) = k_\sigma * \widehat{h}(t), \tag{2}$$

where k_σ is a constant and $\widehat{h}(t)$ is a trace of the absolute value of past heuristic reinforcement received:

$$\widehat{h}(t) = \xi * \text{abs}\big(h(t)\big) + [1 - \xi]\widehat{h}(t-1), \tag{3}$$

with ξ being a constant in $[0, 1]$.

As second step, the unit calculates its *activation level* $a_i(t)$ which is a *normally distributed random variable*:

$$a_i(t) = N\big(\mu_i(t), \sigma(t)\big). \tag{4}$$

Finally, the unit computes its output $s_i(t)$:

$$s_i(t) = f\big(a_i(t)\big) = \frac{2}{1 + e^{-\beta a_i(t)}} - 1, \tag{5}$$

where β is a constant in $[0, 1]$.

A deterministic unit computes its output as a weighted sum of its input:

$$s_i(t) = f\left(\sum_{j=1}^{n}\big(w_{ij}(t) s_j(t)\big) + \theta_i(t)\right), \tag{6}$$

where $f(\cdot)$ is the same function as in Equation (5).

In the AS family of algorithms, the weights are modified according to the following general expression:

$$\Delta w_{ij}(t) = \alpha h(t) e_{ij}(t-1), \tag{7}$$

where α is the *learning rate* and e_{ij} is the *eligibility factor* of w_{ij}. The eligibility factor of a given weight is a measure of how influential that weight was in choosing the action. Each particular version of this general algorithm differs from the others in the way h and e_{ij} are calculated.

In the experiments reported here, we use the version of the AS algorithm that best suits the robot path finding problem. This version has been identified after a comparative study on twenty five versions of the basic AS algorithm:

$$h(t) = z(t) - \widehat{z}(t-1), \tag{8}$$

$$e_{ij}(t) = \frac{\partial \ln N}{\partial w_{ij}}(t) = s_j(t) \frac{a_i(t) - \mu_i(t)}{\sigma^2(t)}, \tag{9a}$$

$$e_{ij}(t) = s_i(t) * s_j(t), \tag{9b}$$

where $\hat{z}$ is the *expected primary reinforcement* calculated by the critic. In order to undertake this prediction task, the critic is built as a second network out of deterministic units, and since it is provided with input/output pairs, a *supervised algorithm* is used. e_{ij} is calculated in two different manners according to the kind of units of the step generator. If the units are stochastic, it is computed in such a manner that the learning rule corresponds to a *gradient ascent* mechanism on the expected environmental reinforcement [5]. If the units are deterministic, it is the *Hebbian rule*.

4.2 A Strategy for Discovering Stable (Quasi) Optimal Paths

It has been stated above that in order to obtain stable solution paths, stochastic units should become deterministic as learning proceeds. The way in which σ is computed guarantees that the breadth of search will diminish asymptotically to zero as the path-finder learns. But this is not acceptable to solve the problem efficiently.

A mechanism for accelerating the generation of stable solution paths is the following. When the path-finder discovers, after a certain experience with the environment, *acceptable* paths, it might be more suitable to search the solution paths in a more "controlled" manner by *transforming the stochastic units of the step generator into deterministic units*. In addition, the weights are not updated after each step, but after the path has been generated and only if the path is not *(quasi) optimal, qo-path*. The acceptability and optimality criteria are defined by the boolean expressions:

$$qo\_path = \left(length < k_{qo_{length}} * \text{dist}[conf_{start}, conf_{goal}]\right) \wedge \left(minimum\_clearance < k_{qo_{clear}}\right), \quad (10)$$

$$acc\_path = \left(length < k_{acc_{length}} * \text{dist}[conf_{start}, conf_{goal}]\right) \wedge \left(minimum\_clearance < k_{acc_{clear}}\right), \quad (11)$$

where $k_{qo_{length}}$, $k_{qo_{clear}}$, $k_{acc_{length}}$, and $k_{acc_{clear}}$ are workspace-dependent constants.

The weight updates are delayed and not applied when a qo-path is generated because, otherwise, the changes to the weights could probably prevent the path-finder from reproducing it. Since the output signals are continuous and the optimality criterion is very severe, even little weight modifications could alter the step generator as to produce actions sufficiently away from the optimal ones.

The process for discovering a stable qo-path requires a further refinement. The fact of obtaining an acceptable path does not mean that the qo-path is nearby. The discovered acceptable path could be close to a local optimum. Even more, if collisions happen while deterministic units are used, the acceptable paths could be lost because of the penalty received. For these reasons, if after consuming a fixed quantity of computational resources the path-finder does not discover a qo-path, then the deterministic units should turn back to being stochastic again. Another acceptable solution should then be obtained. Figure 2 shows the final algorithm for discovering a stable qo-path.

5. Experimental Results

The system has been implemented in Common Lisp on a VAX station 2000. The *step-generator* we have used to carry out the simulations has three layers of units, all the units in a layer being connected to all the units in all the layers above. The hidden layer is made of four units. As usual, the functionality of an input layer consists simply in forwarding the signal it receives. The hidden and output units of the step-generator are stochastic units that become deterministic as learning proceeds. The *critic* has a hidden layer of four deterministic units and, obviously, one output deterministic unit.

In the simulations, the robot moves inside the workspace until either it reaches the goal or it collides. The learning phase is carried out in an incremental way. Initially, the path-finder has to learn to generate a feasible path from the first starting configuration in the training set. Then, it tries to generate a feasible path from the second starting configuration. As the necessary knowledge required for solving this new situation is discovered and codified, part of the previous knowledge could be lost. So, the path-finder must learn again to generate a feasible path from the first starting configuration. The iteration "learn to generate a feasible path from the new starting configuration—recover the ability to generate feasible paths from the previous starting configurations" is repeated until the path-finder is able to generate feasible paths from all the starting configurations.

Figure 3 shows the behavior of the path-finder once the learning phase is finished. Panel A depicts the paths generated by the path-finder from every starting configuration in the training set. Obstacles are shown as circles and every initial configuration and the goal are respectively represented by a triangle and a square. In panel B, instances of the situation-action rules discovered by the path-finder are illustrated; for every starting configuration considered —little circles— the move to be taken by the robot is shown. The path-finder is able to produce stable collision-free paths from almost all the initial configurations in the training set. The number of steps required to reach this state of the path-finder has been 77315. The only situations not properly handled by the path-finder are a subset of the most difficult ones, that is those in which an obstacle lies in between the goal and the current robot configuration and is very close to this configuration. These situations may be handled by getting appropriate guidance from a symbolic path-planner [8].

Until now, we have assumed that the robot can perfectly perceive the workspace. Nevertheless, a robot navigating in a real environment is subject to noisy and inaccurate measurements. To test the ability of our path-finder to work under real conditions, a 20% of white noise is added to each input signal. Figure 4 depicts the resulting behavior. Comparing this figure with Figure 3, panel B, it is evident that the path-finder demonstrates a *strong robustness*, since both "maps" are very similar.

Figure 3, panel B, illustrates some of the generalization abilities of the path-finder. It tackles many more situations than those perceived during the learning phase. In addition, the path-finder is also able to navigate in workspaces different from that used during learning. Figure 5 shows the behavior of the path-finder when both the goal and the number and location of the obstacles have changed. The result of this experiment proves that the *situation-action rules learned are workspace-independent*.

6. Future Work

The simulations carried out in this paper demonstrate the adequacy of a reinforcement connectionist learning approach to implement local obstacle-avoidance capabilities. The formulation of the problem used to test this approach is a difficult one, since the input and output are continuous, the environment is partially unknown, and the optimality criterion is severe. The problem, however, has been simplified by assuming a point robot and circular obstacles. To overcome these limitations, we are currently extending the prototype for dealing with a 2D mobile robot and polygonal obstacles.

Acknowledgements

José del R. Millán acknowledges his indebtedness to Carme Torras. This work is the result of a long and close collaboration with her. He thanks Thomas Barbas for his support in the beginning of this project at the JRC. The authors also thank Jacques Locquet for creating and maintaining the suitable computing environment.

References

[1]. Barto, A.G., Sutton, R.S., & Brouwer, P.S. Associative search network: A reinforcement learning associative memory. *Biological Cybernetics*, 40, 201–211. 1981.

[2]. Sutton, R.S. Temporal credit assignment in reinforcement learning. Ph.D. Thesis, Dept. of Computer and Information Science, University of Massachusetts, Amherst. 1984.

[3]. Barto, A.G., & Anandan, P. Pattern-recognizing stochastic learning automata. *IEEE Transactions on Systems, Man, and Cybernetics*, 15, 360–374. 1985.

[4]. Khatib, O. Real time obstacle avoidance for manipulators and mobile robots. *The International Journal of Robotics Research*, 5, 90–98. 1986.

[5]. Williams, R.J. Reinforcement-learning connectionist systems. Technical Report NU-CCS-87-3, College of Computer Science, Northeastern University, Boston. 1987.

[6]. Millán, J. del R., & Torras, C. Reinforcement learning: Discovering stable solutions in the robot path finding domain. *Proc. 9th European Conference on Artificial Intelligence*, 219–221. 1990.

[7]. Gullapalli, V. A stochastic algorithm for learning real-valued functions via reinforcement feedback. Technical Report COINS-88-91, Dept. of Computer and Information Science, University of Massachusetts, Amherst. 1988.

[8]. Millán, J. del R., & Torras, C. Learning to avoid obstacles through reinforcement. In L. Birnbaum & G. Collins (eds.) *Machine Learning: Proceedings of the 8th International Workshop*. San Mateo, CA: Morgan Kaufmann. 1991.

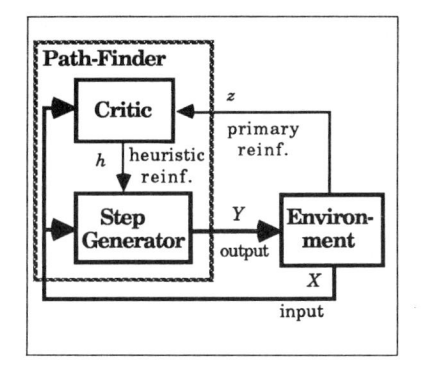

Figure 1. Connectionist path-finder according to the ARL architecture.

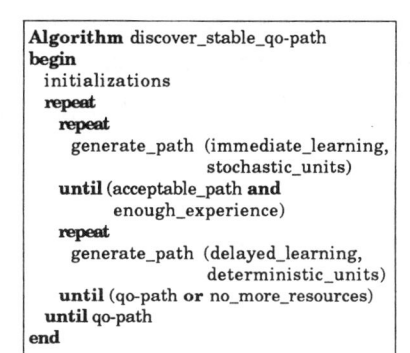

```
Algorithm discover_stable_qo-path
begin
  initializations
  repeat
    repeat
      generate_path (immediate_learning,
                     stochastic_units)
    until (acceptable_path and
           enough_experience)
    repeat
      generate_path (delayed_learning,
                     deterministic_units)
    until (qo-path or no_more_resources)
  until qo-path
end
```

Figure 2. Algorithm for discovering a stable QO-path.

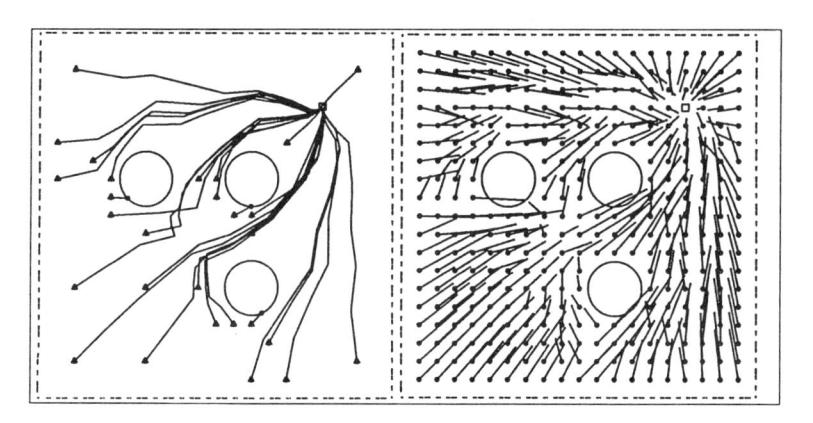

Figure 3. Behavior of the path-finder.

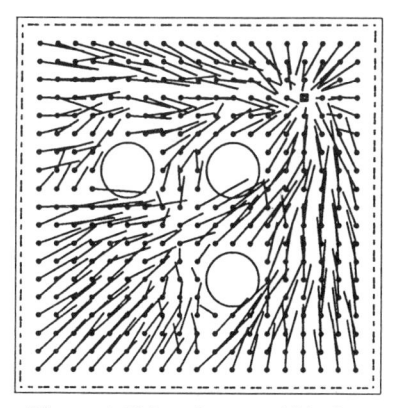

Figure 4. Noise tolerance exhibited by the path-finder.

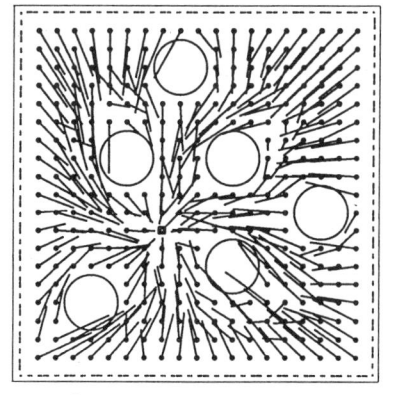

Figure 5. Generalization abilities: Experiments with a new goal and more obstacles.

'True' Surface Determination Using a Novel Algorithm of Directional Search through a Point Table Sampled by a Contact Probe

Mahnaz. Akbary_Safa, Ibrahim.I.Esat*, Colin.B.Besant.
Department of Mechanical Engineering,Imperial College
Exhibition Road, London ,UK
*Department of Mechanical Engineering, Queen Mary College
Mile End Road, London, UK

ABSTRACT

A coordinate measuring machine with a contact probe may introduce errors in surface measurement due to the contact point not coinciding with the measured point. The magnitude of the error depends on both the surface and probe geometry. A typical measuring operation involves a touch trigger probe moving in a vertical plane in equal steps, generating a uniform distribution of data in the xy plane. The sampled data, can then be processed by fitting parametric surface patches to the regular grids. The surface geometry together with the probe shape provide all the necessary information to obtain the true surface. There are two problems with this approach; the computational inefficiency and, the loss of geometrical information due to surface patch interpolation. The method proposed in this paper avoids using surface patches. It extracts the 'true' surface directly from the sampled data using a very efficient directional search technique.

1. Introduction

The geometry of manufactured workpieces or components made from various materials is described by geometry, dimension and tolerances in technical drawings or by CAD data on computer systems. Geometry and dimension are fundamental for the proper function of a single workpiece (e.g. a turbine blade) [1] or for the function of two or more matching workpieces such as nuts and bolts, etc. Thus, geometric quality control is important in the manufacturing process. Advanced manufacturing technology based on automation [2], decreasing tolerances and the need for very fast feedback of data [3] for production control all require effective and very precise computer-controlled flexible workpiece testing equipment. This has led to the development of computer-aided coordinate measurement [4-5] which requires new generation of measuring devices as well as new and more general geometric models and theories [6].

In some measurement processes the data points from an object profile are taken and compared with the nominal set of those points which are prior knowledge. However in many instances the nominal model or the approximate orientation of the object being measured might not be known. In which case the surface geometry is developed using the measured data points collected from a spherical tip probe. Regardless of any prior information measurements on sculptured surfaces with a spherical tip probe could result in contact radius error. That is: The data measured is the probe tip point while the data required is the surface contact point. Therefore the measured surface may not produce the 'true' surface. Before any analysis can be made on the measurements, a method must be developed to compensate for this error.

In general once the surface is generated, then using suitable parametric surface patches together with a mesh generation programme, one can create a functional description of the surface from which the normal to the surface can be calculated. If the normal vector to the measured surface is known, then the 'true' surface geometry can be obtained. This approach has serious weaknesses due to a number of reasons. The surface patch approach will eliminate surface discontinuities which may be a functional and an important part of the surface. For example the intersection of two planes can be detected from the measured point table and yet once the

surface patches are fitted the detailed information on the discontinuity will be lost.
The algorithm presented in this paper avoids using parametric surface patches. It generates the 'true' surface from measured data using a very efficient directional search technique. A 21/2 axis measuring probe is considered, a coordinate point table generated by the measuring system represents the position of the probe tip. The measuring increments is controlled in the xy plane and the z coordinate is captured.

2. Definitions

The following definitions relate to figure 1. r_c is a PC (probe contact) point. The unit normal vector at r_c is n and the pair (r_c, n) are PC (probe contact) data. r_o is the centre of probe sphere and is given by:

$$r_o = r_c + Rn$$

with R as the radius of the ball, r_m is the probe tip and is expressed by:

$$r_m = r_o - Rk$$

where k is its unit normal vector, the pair (r_m, k) are PL (probe location) data. r_s is the projection of the measured point on to the 'true' surface, ie where the tip of the probe should be recording, and is expressed by: (e is the error in the z coordinate between r_m and r_s).

$$r_s = r_o - (R+e)k$$

A sequence of line segments obtained by connecting PL points is called PL path, and a sequence of line segments obtained by connecting PC points is called PC path. PL path is always a straight line in the (xy) plane and is independent of the probe contact point (see figure 4). PC path may not be a straight line in the (xy) plane and is dependent on the surface.

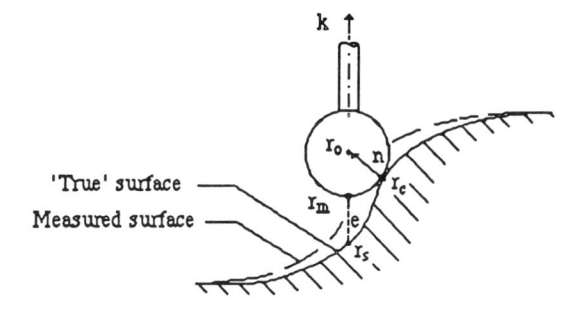

Figure 1. Probe in contact with a three dimensional surface.

Digitising three dimensional surfaces using a mechanical probe is performed by collecting the surface detail as a series of probe points across the surface. These points are captured by the probe tip (PL data). Due to the finite size of the probe and the slope of the measured surface, the tip of the probe may not be in contact with the surface. This introduces an error in the measurement (Fig.1).

In this work, PL data is generated by controlling the probe in the x,y direction and reading the z value. This will form a uniform distribution of data in the x/y plane. In order to obtain the same uniformity in the 'true' surface data, we consider r_s and r_m, on the 'true' and measured surfaces respectively, to be the corresponding points. Since the x,y coordinates of r_s and r_m are equal, the error is in the z coordinate. Hence z_s represents the 'true' z value and is expressed by $z_s = z_m - e$. The algorithm developed to carry this analysis is presented here.

3. Data capture by controlled probe movement

The movement of the probe is controlled in the xy plane resulting in a uniform distribution of the captured PL data. This movement involves the probe covering a distance 'a' in the positive z direction, followed by a distance 'b' in the x or y direction see figure 2. Once 'b' is covered, it is lowered to make contact with the surface. However if the probe makes contact with the surface before 'b' is achieved then the movement described above is repeated until it has. For this several attempts may be needed, and points are recorded only when a uniform distribution is achieved. Therefore an xy plan view of this motion gives a uniform grid as shown in Figure 4.

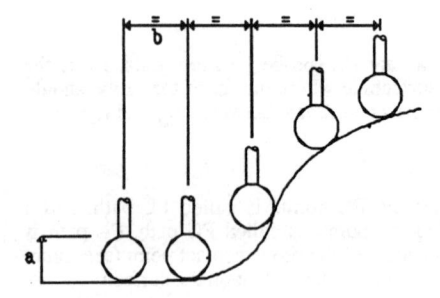

Figure 2. Probe movement (side view)

4. Generating the contact data

The normal vector to the 'true' surface at PC point is the same as the normal to the measured surface at PL point. Since the probe involves the translation of a rigid body with no angular rotation, slopes at r_{cs} and r_m are the same, and hence so are the normal vectors. In figure 3 this is shown in 2D. r_{cp} is a point on and fixed to the probe coinciding with the contact point, r_{cs} is a point on the surface coinciding with the contact point, ρ is a vector between r_{cp} and r_m.

$$r_m = r_{cp} + \rho$$

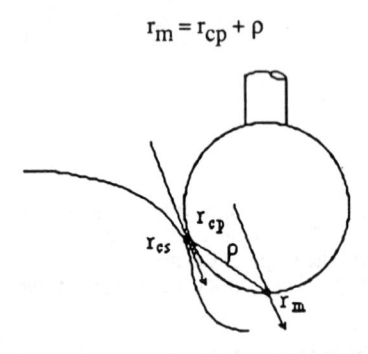

Figure 3. Slopes of the part surface and the measured surface

The probe is controlled in the xy plane and the z coordinate is measured. No explicit information is available about the surface. The only information obtained are the PL points. From three neighbouring PL points r_{m1}, r_{m2}, r_{m3} the unit normal vector (n) shown in figure 1 can be calculated. r_{m1} represents a point with suffices i,j and r_{m2}, r_{m3} are any two consecutive neighbouring points to r_{m1} in anticlockwise direction, which divide the neighbourhood area in to eight triangular area. There are eight pairs of r_{m2} and r_{m3}, see figure 4(B).

$$V = (r_{m2} - r_{m1}) \times (r_{m3} - r_{m1})$$

$$n = V / |V|$$

$$r_c = r_o - n$$

Where

$$r_o = x_m i + y_m j + (z_m + R) k$$

Having then calculated all the PC points, a corresponding table of PC points to PL points is generated. Then a search routine is carried out to find which set of x_m, y_m is equal to x_c, y_c to generate a table of the projected points as shown in table 1.

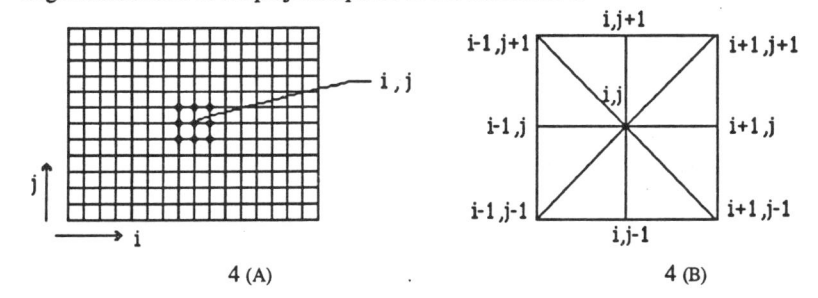

Figure 4.(A) PL data in the xy plane (plan view) ; (B) The eight neighbouring points (the data structure)

5. Directional search algorithm

One of the main steps in the algorithm for generating the 'true' surface is based on the recognition that the value of the projected z coordinate is available in the contact ('true') surface data. This can be seen in Fig. 1, where r_s as the projection of the probe tip is on the contact surface. Therefore the search routine involves comparing a measured point against the contact points for an equal set of x,y coordinates. In previous work the search was reduced from comparison against all the contact points to a search in the vicinity of the require point. This was done by reducing the area of search, since the maximum distance between the corresponding points can not be greater than the probe radius. Although this proved to be much more efficient, it was found that there was need for a further improvement on the search technique of this algorithm.

	measured points	contact points	Projection points
i,j	$x_m y_m z_m$	$x_c y_c z_c$	$x_m y_m z_c$
	----------	----------	----------
	----------	----------	----------
I,J		$x_c y_c z_c$	

Table 1.

Consider a point with indices i,j shown in table 1, the number of steps between a measured and corresponding contact point in the x,y direction is determined by k1 and k2. where k = INT(k) and dx, dy the increments in the x,y direction.

$$k1 = (x_m - x_c) / dx \quad \text{and} \quad k2 = (y_m - y_c) / dy$$

$$I = i + k1 \quad \text{and} \quad J = j + k2$$

Adding the number of steps k1 and k2 to i,j will then take the search to a starting point in the contact table of indices I,J. This point is very close to the required projection point, and the area of search is then a rectangular patch with diagonal indices (I,J) and (I+1,J+1) in which the required point lies, see figure 5. This method avoids a blind integer search in the vicinity of the contact point. Instead it performs a directional search, taking the steepest slope towards the required tangent and aiming to the rectangular area containing the point.

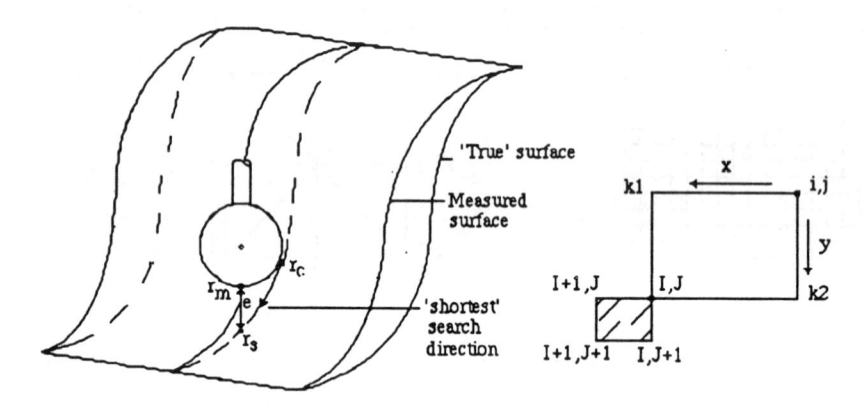

Figure 5. Shortest search direction towards the required (shaded) area.

6. Complexity

In a previous publication [7] it was recognised that if the search routine was carried out by comparing a measured point against all the contact points, the complexity of this search would be proportional to n^2. Such a search routine was not necessary since the maximum distance between a contact point and its corresponding measured point would not be greater than the radius of the probe. Therefore the maximum number of steps of search would be (R / dx) in +- x and (R / dy) in +- y directions, that is (2R / dx)(2R / dy) number of search is needed for each sampling point. The number of steps of search would then equal to $(4R^2 / dxdy)n$. Therefore computational complexity is based on a search in the vicinity of a contact point corresponding to the measured point. The search routine proposed in this paper reduces the number of steps of search much further, this is due to a considerable reduction in the area of search. In an example where R/dx is approximately equal to 5 then the search routine can be almost 100 times faster.

7. Discussion

The analysis carried out showed that although the algorithm is very efficient, a number of calculations are required to find the projection point in the search process. This is due to the fact that a measured x,y coordinate may never be equated exactly with an x,y coordinate of the contact points, instead a suitable interpolation would be used to determine the best match. This may have major effects on the computational efficiency and as a result defeat the aim of the algorithm. Therefore the improvement on the search routine presented in this work was found to be of considerable importance.

8. References

1. Cardew-Hall, M "Inspection of Turbine blades" PhD thesis, 1987

2. Hales, A M "Automated Inspection for Flexible Manufacturing" Manufacturing systems, pp 15-19, December 1985

3. Lee, M K "Data feedback in an Integrated Design to Manufacture System" PhD thesis, November 1990

4. Loney, G C and Ozsoy, T M "NC machining of free form surfaces" Comput.-Aided Des. Vol 19, No 2, pp 85-90, 1987

5. Choi, B K, Jun, C S "Ball_end cutter interference avoidance in NC machining of sculptured surfaces" Comput.-Aided Des. Vol 21, No 6, pp 371-378, 1989

6. Lenz, K J and Merzenich, W "Achievements of accuracy by error compensation of large CMMs" Precision Engineering, pp 228-230, October 1988

7. Akbary_safa, M, Esat, I I, Besant, C B "An efficient algorithm for error elimination from surface measurement generated by a mechanical probe" SPIE/SPSE Symposium on Electronic Imaging Science & Technology, February 1990

Robotic Deburring of Parts with Arbitrary Camber Using Hybrid Position/Force Control

H. G. CAI, Y. ZHENG, Q. TAO and H. LIU

Robot Research Institute
Harbin Institute of Technology, Harbin 150006 P.R.China

Summary

The design and implementation of a multi-microprocessor based system to control the interaction forces between a six-axis robot and a workpiece is described. The force controller worked in parallel with a robot controller by calculating position corrections that allowed force to be controlled in a desired manner. Force control was implemented in deburring experiments, in which the commanded normal force ranges from 1N to 10N and rms force errors is less than 1N as the tangential speed is increased from 5mm/s to 100mm/s.

Introduction

Recently, much interest has been paid to a third class of jobs -- deburring to which robots might profitably be applied [1]. Traditionally, manual deburring is the only deburring method available, and represents a time-consuming and expensive solution. Deburring costs for some cast parts can be as high as 35 percent of the total part cost. This is a major reason for the development of automated deburring operation [2].

In this paper, a hybrid position / force control system based on three-grade processor control system is introduced. The schematic of the hybrid position / force control system is shown in Fig.1. Force control system was integrated a supplemental force control architecture with existing robot motion controls. On the basis of analysizing the software and hardware of PUMA 562 and making use of the function of real-time path control, the hybrid position / force control system was created without any modification of robot original position controller. The one merit of the system is that there is no modification in the original position controller, which reduce the amount of labour greatly. Another one is that we can make full use of the rich software and hardware resource of PUMA robot which greatly increase the flexibility of programming.

The rule of the adjustment of the orientation

As shown in Fig.2, considering the geometry of arbitrary camber, cylinder rotary file was used as grinding tool. As shown in Fig.3, there may be three situations of contact between tool and parts. The orientation of the tool must be revised according to the geometry of the camber. The rule of the modification of the tool's orientation is relationship between the torque and force imposed on end effector. In accordance with three contact situations, three force arm can be got respectively. Obviously, L_{t1} is bigger than L_{t3} and smaller than L_{t2}.

As to the different state in Fig.3 (b), (c), the orientation of end effector should be adjusted differently. The direction of adjustment is signed respectively. We can get β by

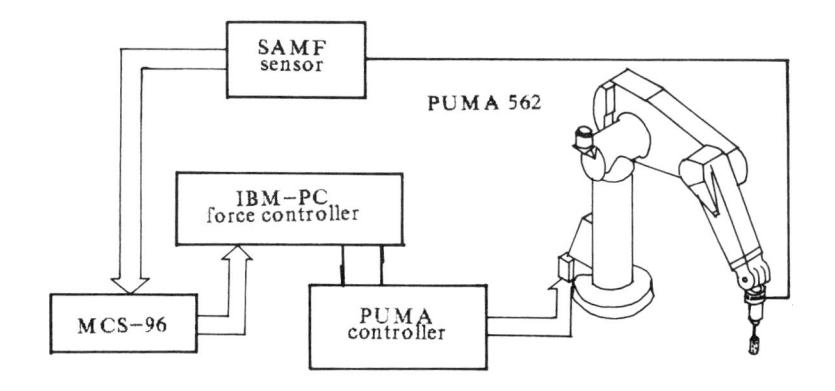

$$\beta = arctg\frac{F_{zt}}{F_{yt}} \tag{1}$$

Fig.1. The schematic of the system

Strategy for robotic deburring

Considering the deburring of a surface by a robot manipulator, the objective is to use an end effector to smooth the surface down to the command trajectory, depicted by the bashed line in Fig.2. In accordance with two findings in [2],[3]:(1) when the cutting tool is moving with constant speed along surface, the tangential force imposed on end point varies significantly due to the variation of burr size. If the burr is large enough, the contact force increase until a seperation of the tool from the the part occurs. Therefore, when the phenomenon happens, the contact force increases and robot must decrease the tangential component of its speed to maintain a constant tangential force. But, to simplifize the design of the force controller, the tangential speed is limited in a small range in our system. In the light of finding (2), the normal force should be maintain constant during the deburring process.

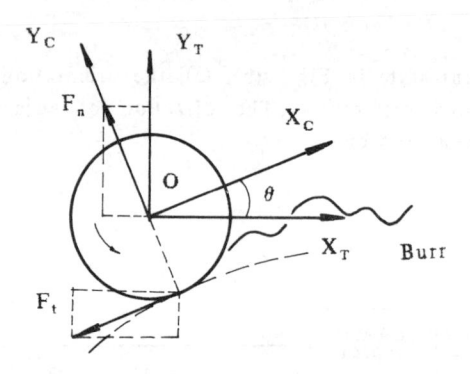

Fig.2. Camber deburring

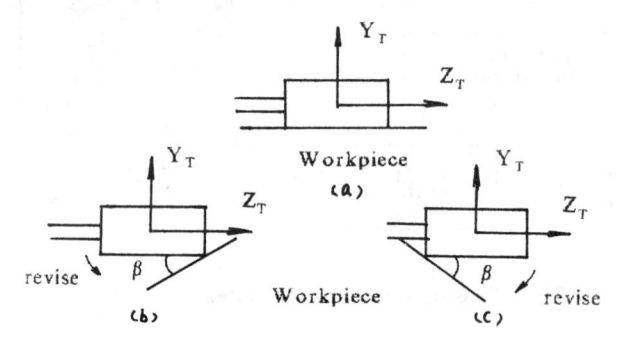

Fig.3. The situation of contact between tool and workpiece

Force controller design

Fig.2 shows a grinding wheel contacting with a workpiece where $OX_t Y_t Z_t$ stand for TOOL frame and $OX_c Y_c Z_c$ for compliance frame $\{C\}$. Both Z_t and Z_c axes were in coincidence. The target of the control system is to maintain constant normal force and tangential speed when deburring parts.

There exist a angle between the TOOL and {C} frame. Defining a 6 × 6 matrix H which stands for the rotation transformation from TOOL to {C} frame,

$$H_\theta = \begin{bmatrix} Rot(Z,\theta) & I \\ I & Rot(Z,\theta) \end{bmatrix} \quad (2)$$

The force sensor frame was coincidence with TOOL frame by suitable mounted. If the modifaction of the orientation of end effector is considered, the $H_{\theta,\beta}$ is expressed as

$$H_{\theta,\beta} = H_\theta \times H_\beta \quad (3)$$

where the angle β is mentioned in the section of the adjustment of the orientation. According to the analysis of statics, we have

$$F_{xt} \cos\theta + F_{yt} \sin\theta = \frac{M_{zt}}{R} \quad (4)$$

the angle θ can be calculated as:

$$\theta = 2 arctg \left(\frac{R \cdot F_{yt} \pm \sqrt{R^2(F_{xt}^2 + F_{yt}^2) - M_{zt}^2}}{R \cdot F_{xt} + M_{zt}} \right) \quad (5)$$

Where : F_{xt}, F_{yt}, M_{zt} can be measured by force sensor,

R is the radium of grinding wheel,

M_{zt} is the torque around z axis in TOOL coordinate.

We chose the one which is close to zero. Incements in TOOL frame can be achieved from

$$\delta^T X = H^T \cdot K_j \cdot {}^e E_j \quad (6)$$

The increments in TOOL frame can be sent to PUMA controller to correct the norminal path to maintain desired contact force.

Experiment result

A series of robotic deburring experiments have been made. The force error, the difference between desired and actual normal force, was the target variable in these tests. The experiments were performed over a range of tangential speed, 10 to 100 mm / s, and normal force, 1 to 10 N, that were realistic for deburring. The camber to be deburred was aircraft engine parts, located in WORLD frame. Fig.4 is the normal force history during deburring.

Fig.5 is a plot of the rms force error vs. travel speed at normal force of 5 N.. Increasing travel speed introduce disturbance with high frequency content. The normal force was limited to 10 N, because at higher levels the force was sufficient to stall the grinder wheel. Note that the deburring process. The flat centered near 60 mm / s indicates that certain speeds were more desirable than others from a force error point of view.

Conclusions

A robotic deburring system was developed in this paper using PUMA 562 manipulator with unmodified controller and external force controller. Using VAL – II real – time path control ability, the force controller was interfaced to the original position control loop. A hybrid controller for deburring operation was designed and implemented on the external computer. Rms force error of deburring experiments increased with increasing tangential speed. In future work, the control of the robotic deburring should be improved. As mentioned in the section of strategy of robotic deburring in order to prevent the seperation of the grinder wheel and workpiece, the tangential speed was limited in a small range. This limitation was bound to increase the operation time and decrease the productivity in factual manufaction. We will seek to develop a self – tuning strategy in case the phemenon happens.

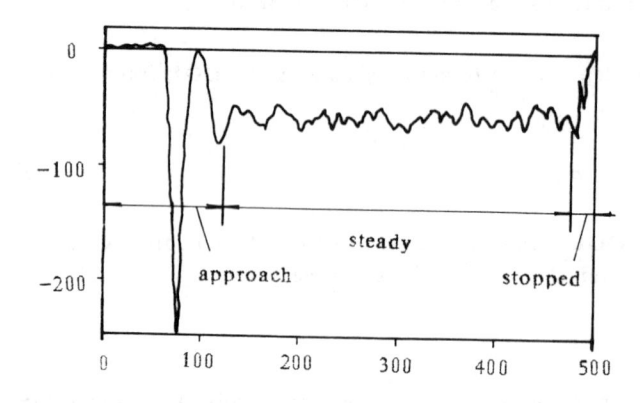

Force   uf (1uf = 0.11 N)

Time   T (T = 14 mm)

Fig. 4 The normal force history during deburring

$F_d = 50 uf$

Force   uf (1uf = 0.11N)

tangential speed (mm / s)

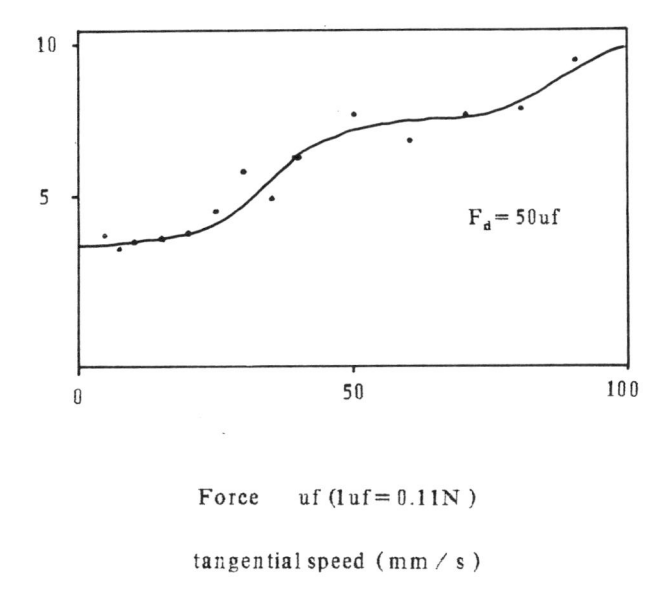

Fig 5 Rms force error vs.tangential speed

References

1. Steven, A.; Hewit, J.R.: Hybrid position and force controlapplied to robotic polishing of turbine. Towards Third Generation Advanced Robotics ICAR'87.

2. Kazerooni, H.: Automated Deburring Using Impedance Control. Conference on Robotics and Automation ,April 3, 1987

3. Kazerooni, H.: An Approach to Automated Deburring by Robot Manipulator. Tournal of Dynamic System Measurement, and Control 108 pp, 354-9 (1986)

Design of Block Networks

A.A.RUBCHINSKY

Moscow institute of Steel and Alloys
Moscow, USSR

Summary
A new notion of a block network is introduced. A block formally represents an object transforming input resources into output products. The notion of a block network describes the collaboration of such objects. The essence of the design block network problem is in the following: how to obtain the required products from given resources with the use of some blocks, transforming certain input resources into some products which act as resources for other blocks. This article suggests one of the possible approaches to this problem.

Basic definitions and notations.

Let us introduce some necessary formal definitions (in the case of splittable resources). Block Z_i is characterized by a pair of set $\langle A_i, B_i \rangle$, $A_i = \{a_1^i, \ldots, a_{ni}^i\}$, $B_i = \{b_1^i, \ldots, b_{mi}^i\}$ where a_l^i ($l=1, \ldots, ni$) are the initial resources and b_k^i ($k=1, \ldots, mi$) are output products. Elements of A_i are called also block Z_i inputs and elements of B_i – block Z_i outputs. Block Z_i converts the set A_i into the set B_i.

Define the operation of two blocks Z_1 and Z_2 (in above order) as connection. Let

$$b_1^1 = a_1^2, \ldots, b_k^1 = a_k^2 \quad (k \leqslant \min\{m_1, n_2\}). \tag{1}$$

Resulting network obtained by blocks Z_1, Z_2 connection implements the conversion of set $A = \{a_1^1, \ldots, a_{n1}^1, a_{k+1}^2, \ldots, a_{n2}^2\}$ into set $B = \{b_1^1, \ldots, b_{m1}^1, b^2, \ldots, b_{m2}^2\}$. This connection means that the types of product $b_1^1, \ldots, b_k^1$, manufactured by the first plant are the resources for the second plant.

Block network is some connection of block $Z_1, \ldots, Z_L$ that complies with the above rules (in any order). Such network implements the conversion of set $A=\{a_1,\ldots,a_n\}$ into set $B=\{b_1,\ldots,b_m\}$ which is a result of joint functioning of connected blocks.

The resource "splitness" means the possibility of the feed of one block output to several other blocks inputs and to the network outputs. Block inputs not connected to outputs of other blocks are called <u>network inputs</u>, and all block outputs are called <u>network outputs</u>.

Consider the totality M of all block network with the blocks of set $Z=\{Z_1,\ldots,Z_L\}$. Blocks $Z_1,\ldots,Z_L$ correspond to the available plants. Let

$$H = \bigcup_{i=1}^{L} A_i, \quad P = \bigcup_{i=1}^{L} B_i. \tag{2}$$

Given also is the set of <u>original resources</u> A_O and the set of <u>final products</u> B_F.

Assume S is a block network of M, A is the set of S inputs and B is the set of S outputs. Network S including blocks of the given set Z is <u>admissible</u> if

$$A \subseteq A_O, \quad B_F \subseteq B. \tag{3}$$

Thus the admissible network consisting of given blocks $Z_1,\ldots,Z_L$, permits one to have the required final products B_F (or something more) produced from available initial resources A_O (or some part of these resources).

Formal statements and algorithms.

The formal problem statement is the following. Given $\langle Z, A_O, B_F \rangle$, $Z=\{Z_1,\ldots,Z_L\}$. It is required to design a block network S using only blocks of Z which satisfies conditions (3).

To solve the design problem, introduce the notations:

$$R = A_O \cup H \cup P \cup B_F. \tag{4}$$

Assume that for some $i \in \{1,\ldots,L\}$

$$A_i \not\subseteq P \cup A_O. \tag{5}$$

Then the block Z_i is not included to any admissible network (because it is impossible to "use" its inputs). Assume all these blocks are removed. For all other blocks the inclusion

$$H \subseteq P \cup A_0 \qquad (6)$$

holds; which follows from the conclusions $A_i \subseteq P \cup A_0$ ($i=1,\ldots,L$).

Statement 1. The condition

$$B_F \subseteq P \qquad (7)$$

is necessary and sufficient for the existence of design problem solution.

To prove the necessity, it is obvious: that if $b \in B_F$ and $b \notin P$, then the second condition in (3) does not hold. The proof of the sufficiency is constructive; that is, it implies the design of an admissible block network.

Algorithm 1.
1. For each $b \in B_F$, find block Z_i such that $b \in B_i$, which exists by virtue of (7).
2. Each input of blocks chosen at step 1 is either contained in set A_0 or coincides with the output of some other block by virtue of (6). The corresponding resources and block are included in the network.
3. The process of inclusion of blocks and resources goes on until the entire network is constructed all inputs of which are connected with the original resources of A_0. The termination of the process is guaranteed by inclusion (6).

Algorithm 1 permits one to find a single solution of the design problem. However, it may practically be important to find all possible solutions, i.e. all feasible block networks.

Enumerate all elements of the set R (see (4)) as

$$r_1,\ldots, r_n ,\ldots, r_{N-m+1},\ldots, r_N \qquad (8)$$

so that $r_1,\ldots,r_n$ be the original resources, and $r_{N-m+1},\ldots, r_N$ be the final products. Conditions of design problem $\langle Z, A_0, B_F \rangle$ may be written with the help of the following notations:

$$a_{ij} = \begin{cases} 1, & \text{if } r_i \in A_j \\ 0, & \text{if } r_i \notin A_j \end{cases} \quad (i=1,\ldots,N;\ j=1,\ldots,L) \quad (9)$$

$$b_{ij} = \begin{cases} 1, & \text{if } r_i \in B_j \\ 0, & \text{if } r_i \notin B_j \end{cases} \quad (i=1,\ldots,N;\ j=1,\ldots,L) \quad (10)$$

Let S be an arbitrary network with blocks of $\{Z_1,\ldots,Z_L\}$. Introduce parameters

$$z_j = \begin{cases} 1, & \text{if block } Z_j \text{ is included in network } S, \\ 0, & \text{otherwise}; \end{cases} \quad (j=1,\ldots,L) \quad (11)$$

$$u_i = \begin{cases} 1, & \text{if original resource } r_i \text{ is used in network } S,\ (i=1,\ldots,n) \\ 0, & \text{otherwise.} \end{cases} \quad (12)$$

Statement 2. If network S is admissible for problem $\langle Z, A_0, B_F \rangle$, then the following conditions hold:

$$\sum_{l=1}^{L} b_{il} z_l + u_i - a_{ij} z_j \geq 0 \quad (i=1,\ldots,n;\ j=1,\ldots,L), \quad (13)$$

$$\sum_{l=1}^{L} b_{il} z_l - a_{ij} z_j \geq 0 \quad (i=n+1,\ldots,N;\ j=1,\ldots,L), \quad (14)$$

$$\sum_{j=1}^{L} b_{ij} z_j \geq 1 \quad (i=N-m+1,\ldots,N). \quad (15)$$

The search of all admissible networks is based on the statement which in some sense is inverse to Statement 2.

Statement 3. Assume $z_1,\ldots,z_L$, $u_1,\ldots,u_n$ is an arbitrary solution of the inequalities system (13)-(15). Then the admissible block network exists such that block Z_j is included into it only when $z_j=1$ ($j=1,\ldots,L$) and resource r_i is included into it when $u_i=1$ ($i=1,\ldots,n$).

The proof of statement 3 is constructive.

Algorithm 2.

Without any loss of generality we may assume that $z_1 = \ldots = z_p = 1$, $z_{p+1} = \ldots = z_L = 0$; $u_1 = \ldots = u_q = 1$, $u_{q+1} = \ldots = u_n = 0$.

1. Consider input r_1^1 of block Z_1. Two cases are possible: r_1^1 coincides with some output r_i^j of block Z_j ($j \leq p$); r_1^1 does not coincides with any output of block Z_j ($j \leq p$). Then by virtue of inequality (13) r_1^1 coincides with one of the resources r_i ($i \leq q$). In the first case, go to step 2; in the second case go to step 3.

2. Connect input r_1^1 with output r_i^j of block Z_j and go to step 4.

3. Connect input r_1^1 with resource r_i.

4. Repeat steps 1 thru 3 for all the inputs of block Z_1.

5. Repeat steps 1 thru 4 for all blocks $Z_1, \ldots, Z_p$.

The outputs of the constructed network S include all the required products, which follows from inequalities (15). Network inputs are original resources r_i with $u_i = 1$.

Thus, to find all the different solutions of the design network problem, it is necessary to find all solutions of inequalities system (13)-(15).

If the costs of the original resources of A_0 are $\alpha_1, \ldots, \alpha_n$ and the costs of block $Z_1, \ldots, Z_L$ are $\beta_1, \ldots, \beta_L$, then it is possible to state the optimal network design problem:

$$\sum_{i=1}^{n} u_i \alpha_i + \sum_{j=1}^{L} z_j \beta_j \longrightarrow \min \qquad (16)$$

with restrictions (13)-(15).

Unsplittable resources.

To be able to recognize some quantitative indices of the converted products, block Z is defined almost similar to the above, with one important difference: inputs and outputs do not form sets; rather, they form multisets. A <u>multiset</u> is a totality of ordinary set elements (<u>carrier</u>) with nonnegative integer numbers, element <u>weights</u>. Block Z converts input multiset $A = \{v_1 a_1, \ldots v_n a_n\}$ into output multiset $B = \{\mu_1 b_1, \ldots, \mu_m b_m\}$ (v_i, μ_j are elements a_i, b_j **weights**.

In connections of blocks each unit of a resource may be moved only as a whole from a single block output to a single block input. Unsplitness of a resource implies exactly this.

The conditions of the design problem are described by set $\langle Z, A_O, B_F \rangle$, where $Z = \{\alpha_1 Z_1, \ldots, \alpha_L Z_L\}$ is a multiset of block, A_O is a multiset of original resources, B_F is a multiset of final products. The consideration of block multiset Z means that given are numbers α_i of the same type block Z_i ($i=1,\ldots,L$).

The problem of the block network design is stated similar to the above; A_O and B_F are the given multisets. Multisets H, P, R are defined by the analogy with the above (see (2), (4) and (8)).

Denote by $\nu(x, X)$ the weight of element x in multiset X. Denote

$$a_{ij} = \nu(r_i, A_j) \quad (i=1,\ldots,N;\ j=1,\ldots,L),$$
$$b_{ij} = \nu(r_i, B_j) \quad (i=1,\ldots,N;\ j=1,\ldots,L),$$
$$c_i = \nu(r_i, A_O) \quad (i=1,\ldots,n),$$
$$d_i = \nu(r_i, B_F) \quad (i=N-m+1,\ldots,N).$$

Assume S is any network with block of multiset Z. Let z_i be equal to the number of Z_j type block, included into network S. Define variable u_i by formula (12).

Statement 4. If network S is admissible for problem $\langle Z, A_O, B_F \rangle$ then the following conditions hold:

$$\sum_{l=1}^{L} b_{il} z_l + c_i u_i - \sum_{l=1}^{L} a_{il} z_l \geq 0 \quad (i=1,\ldots,n), \qquad (17)$$

$$\sum_{l=1}^{L} b_{ij} z_{ij} - \sum_{l=1}^{L} a_{il} z_l \geq 0 \quad (i=n+1,\ldots,N), \qquad (18)$$

$$\sum_{l=1}^{L} b_{il} z_l - \sum_{l=1}^{L} a_{il} z_l \geq \alpha_i \quad (i=N-m+1,\ldots,N), \qquad (19)$$

$$z_j \leq \alpha_j \quad (j=1,\ldots,L). \qquad (20)$$

Statement 5. Assume $z_1,\ldots, z_l, u_1,\ldots, u_n$ is some so of inequalities system (17)–(20). Then the admissible block network exists, such that block Z_j is included into it with weight z_j ($j=1,\ldots,L$) and resource r_i is included if $u_i=1$ ($i=1,\ldots,n$).

An On-Line Control System of Tool Status in Continuing Cutting

E. Ceretti, G. Maccarini, L. Zavanella, A. Bugini

Dipartimento di Ingegneria Meccanica-Universita' Degli Studi di Brescia

ABSTRACT

More importance is being given to the problems of quality and productivity in manufacturing industry today. In order to achieve these aims the correct use of the cutting tool becomes fundamental.

By opportunely checking tool wear a product of higher quality can be obtained together with a decrease in the scrapping of pieces.

The present paper deals with the realization of a checking system able to monitor the tool status by means of a signal derived from the cutting forces; in this way abnormal situations during cutting can be identified, and the appropriate human intervention made. The system is based on a dynamometer made up of a load cell. The axial component of the drilling cutting force is recorded while studying its behaviour as a function of the cutting parameters and of the cutting edge wear.

INTRODUCTION

The economic convenience in machining processes depends also on the useful cutting tool life and its operational conditions. Thus far monitoring of tool working conditions becomes fundamental mostly with regard to tool replacement, in order to prevent unexpected failure which can cause damage to both machine tool and work piece. In recent years more importance has been given to research activity in order to create a sensor that would enable the monitoring of the cutting process, so using the full potential of unmanned machining and cutting tool [1-5].

To this end, tool replacement must take place either when tool wear has damaged the cutting edge or due to unexpected tool failure (chipping). Common industrial practice implies tool replacement after a maintenance interval, which is determined by the economic optimization of the machining process: this fact determines an under-utilization of the cutting tool, whose nature is stochastic.

This study focuses on some experimental results obtained through an 'on line' monitoring system able to control tool conditions and to directly interact with the machining process.

The research study is organized as follows:
- in the first section attention is focused on the description and realization of the equipment (sensors and related measurement line, data acquisition and manipulation) which enables the working job to be controlled;

- in the second section it is focused on finding several variables (to be correlated with tool wear), on the definition of the signal conditioning and finally on the realization of models correlating these variables with tool status [6];
- the third and last section is the experimental one and results are compared with theoretical models.

THE MONITORING SYSTEM

Of the various cutting operations the drilling ones are widely diffused so the monitoring of drilling conditions is an important economic task. Our study focuses on this field and particular attention is given to evaluation criteria for tool life; i.e. tool wear, fracture and hole surface quality.

A C40 steel block is machined without lubrication through a 4 axis numerically controlled machining center with horizontal spindle. Centering, drilling, boring and fine boring operations are made for every hole. The cutting force during the machining process is monitored, in fact this value is correlated with tool wear, thus achieving tool status control [7-9].

The initial intent was to measure the feed force in drilling jobs, because of the relevant differences of the forces between tool and work piece during the processes under examination: these forces increase while tool wear grows.

Commercial products are not able to detect directly the cutting forces (such as absorbed current in working): often they require a considerable alteration of machine components if the machine itself is not suitably equipped. As an example, consistent with our experimental results, the need to equip the machining center so as to avoid the measurement of axial thrusts implies an alteration in the machine structure. As a consequence, our attention focuses on the realization of a load cell to clamp between work piece and work table: this solution implies some constraints in work piece dimensions. Market research into load cells to position between work piece and pallet showed how most of them were limited by the maximum tangential static load (the piece weight in our case).

The correct load cell, to be clamped onto the work table of the machining center, was realized and used for the research.

LOAD CELL DESCRIPTION

The load cell constitutive elements are:
a) four rings, each equipped with four strain gauges (full bridge connected), supplied with a strain gauge exchange also able to measure unbalancing;
b) two anchor plates holding the rings and enabling the load cell to be

fastened to the table and the work piece to the cell.

Such a load cell meets the following fundamental requirements:
- detection of feed force in drilling which compresses the rings, because of the relative position between the machine tool spindle and the ring itself;
- minimization of torque and shear force effects, which may affect measurement during the metal cutting.

In order to achieve these aims the strain gauges are positioned in the axial plane, equally distanced from the constraint point and from the load application point (i.e. the generating line where the sum of deformations due to shear load becomes zero). In this way it's possible to detect the deformations on the rings caused by radial loads, thus evaluating the effective loads, once a convenient system calibration is performed. The sum of output signals from the four rings leads to a signal proportional to the total feed force in drilling.

Other cutting force types (such as tangential or axial cutting force in boring and feed force in fine boring) may be recorded with the load cell by processing and manipulating the output signals.

The load cell obtained has been characterized for different load situations and positions between the plates. These tests are fundamental for characterizing the load cell attitude, thus verifying the validity of the initial hypothesis and determining the calibration coefficient (9 mV/kg) which enables the correlation between the electric signal acquired (mV) and the force (kg). The calibration chart is represented in figure 1: α) is the response of the load cell to compression, β) is the response to shear load.

The results obtained show a linear signal and the small influence on the shear load to the value acquired in static compression.

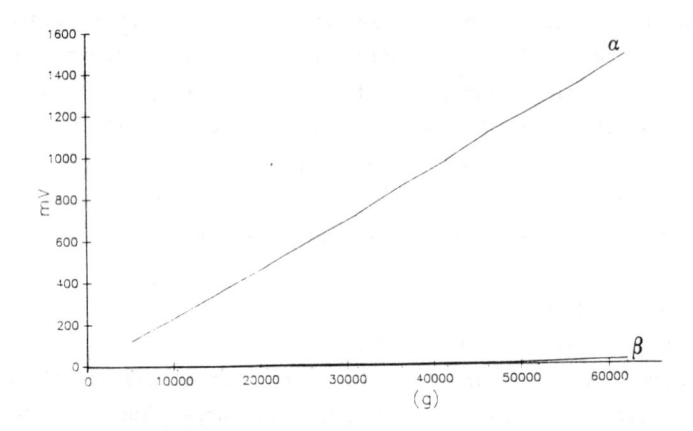

Fig. 1. Load cell calibration

EXPERIMENTAL RESULTS:

A) The load cell was planned so as to measure the feed forces in drilling jobs and this machining process was analyzed first. Before the drilling operation, centering was made without recording or analyzing the signal.

The cutting parameters are shown in the following table:

STEP	S	F
1	500 rpm	100 mm/min
2	500 rpm	50 mm/min
3	500 rpm	25 mm/min

Drilling tool
$\phi = 19$ mm
$\beta = 120°$

A time domain analysis of the data acquired has shown a common signal consisting of an offset component, proportional to the feed force, and an oscillating part (see figure 2). Figure 3 shows the average of the sum of the values recorded by the four channels.

When considering different positions of the hole (some examples are given in Table 1), the single ring output obviously changes but the sum of the four signals remains almost constant (maximum deviations ± 5%).

The next phase compares the acquired data with other experimental results. In this section, attention is paid to Kronenberg's studies, based on an experimental analysis of the chip formation phenomena.

From [14] we can derive the equation determining the feed force:

$$P = C \, d^{\alpha} \, a^{\beta} \qquad (1)$$

where d is the tool diameter, a the feed per revolution and α and β convenient parameters. From [10] $C = 57.5$, $\alpha = 1$, $\beta = 0.78$.

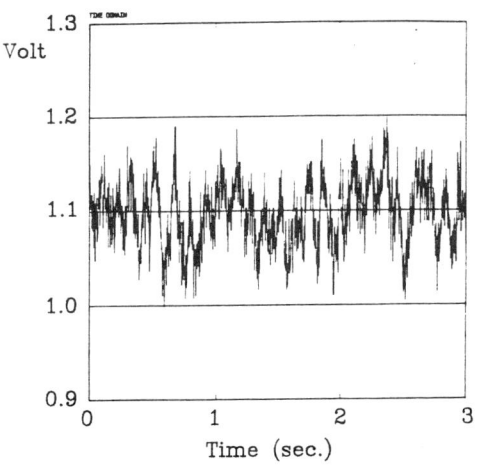

Fig. 2. Single ring signal

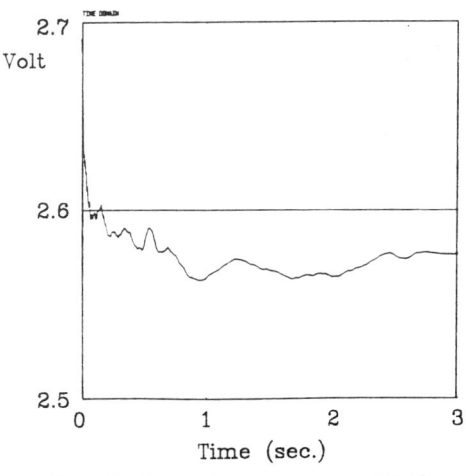

Fig. 3. Four rings average signal

The following table shows the comparison with the experimental values obtained from our research (Ps).

a	P	Ps
0.2 mm/rev	311 daN	294 daN
0.1 mm/rev	181 daN	184 daN
.05 mm/rev	106 daN	115 daN

The agreement of the values obtained is quite good (deviation of ± 2 %).

Figure 4 shows the trend of the average feed force while increasing cutting time. Only after 100 minutes working was there a noticeble increase in the feed force which infact lead to breakage.

B) The second process studied was a boring job, modifying the hole dimensions from 19 mm to 24 mm in two steps. The tool utilized was a widia insert boring bar and the machining process was carried out with 710 rpm spindle speed, 39 mm/min feed and 2,5 mm depth of cut.

A time domain analysis of the signal acquired performed on a single ring, is shown in figure 5. A sinusoidal no zero average signal whose frequency is about 12 Hz expresses the spindle rotation frequency its average is the axial force, while the sinusoidal component is due to ring deformation generated by the shear torque. The axial force was retrieved by an average on every ring signal and by adding up the obtained values.

The shear force values were obtained after an evaluation of the cutting process forces and of their application point.

Refearring to [10] we obtain for the shear force this expression:

$$F_t = C \ A^n \ HB^m \ \beta/50^p$$

where: $C = 19.8$, $n = 0.862$, $m = 1/2.2$, $p = 1/1.36$, $A=0.11 \ mm^2$ (chip section), $HB=148 \ daN/mm^2$ and $\beta=70°$.

The calculated value and the experimental one are:

$$F_t = 37 \ daN, \qquad F_{ts} = 38 \ daN$$

If the data acquired were insufficient to determine a time-force diagram we were able to detect the breakage of the tool which shows an increase in the shear force of 100 % while in the axial one of 300 %.

CONCLUSION

A correlation between force and wear is pointed out and a monitoring system was realized; the next step consists of determining the cutting force over which the cutting process had to be stopped because of a great decrease in the machined surface quality or the tool breakage.

The sensor realized allowed the evaluation of cutting forces and the correlation with tool wear.

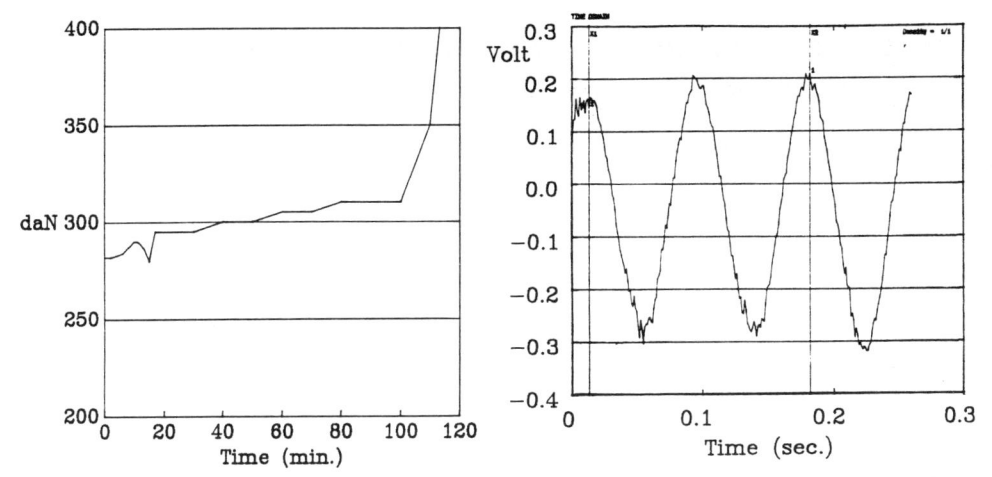

Fig. 4. Force-time diagram

Fig. 5. Single ring signal (boring)

Pos.	ring 1	ring 2	ring 3	ring 4	Tot.(mV)	Tot.(Kg)
C2	828	266	381	1106	2581	286.8
A3	1097	-62	1391	150	2576	286.2
B3	838	240	1082	412	2572	285.8

Table 1: values from the single rings and their sum for different position of drilling (step 1)

BIBLIOGRAPHY

1. J. Tlusty, G. C. Andrews: "A critical review of sensor for unmanned machining" Annals of the CIRP Vol. 32/2/1983.
2. L. Dan, J. Mathew: "Tool wear and failure monitoring techniques for turning: a review" Int. J. Mach. Tools Manufact. Vol. 30 n. 4 1990.
3. R. Teti, G. F. Micheletti: "Tool wear monitoring through acoustic emission" Annals of the CIRP Vol. 38/1/1989.
4. S. R. Hayashi, C. E. Thomas, D. G. Wildes: "Tool break detection by monitoring ultrasonic vibration" Annals of the CIRP Vol. 37/1/1988.
5. F. Giusti, M. Santochi, G. Tantussi: "On-line sensing of flank and crater wear of cutting tools" Annals of the CIRP Vol. 36/41/1987.
6. H. K. Tönshoff, J. P. Wulfsberg, H. J. J. Kals, W. König, C. A. Van Luttervelt: "Developments and trends in monitoring and control of machining processes" Annals of the CIRP Vol. 37/2/1988.
7. E. Brinksmeier: "Prediction of tool fracture in drilling" Annals of the CIRP Vol. 39/1/1990.
8. A. Thangaraj, P. K. Wright: "Drill wear sensing and failure prediction for untended machining" Robotics & Computer-Integrated Manufacturing Vol.4 N.3 1988.
9. K. Subramanian, N. H. Cook: "Sensing of drill wear and prediction of drill life"; Trans. of ASME, J.of Eng. for Ind., May 1977.
10. M. Kronenberg: "GRUNDZÜGE DER ZERSPANUNGSLEHRE"; Springer-Verlag, 1969.

Temperature Compensation of Sealed Pump-Gauge Zirconia Oxygen Sensors Operated in the AC Mode

M. Benammar and W.C. Maskell

Energy Technology Centre, Middlesex Polytechnic, Bounds Green Rd., London N11 2NQ.

ABSTRACT: Temperature compensation of an instrument, incorporating a fully-sealed zirconia pump-gauge oxygen sensor operated in the AC mode and used for oxygen partial pressure measurement, was suggested and implemented with relatively simple hardware. The compensation was achieved by measuring the temperature of the sensor using a thermocouple and including the temperature parameter in the signal processing. The system operated well in the temperature range 650-800°C.

1 INTRODUCTION

Oxygen sensors are used in a variety of applications including monitoring and control of air-to-fuel ratio in combustion systems, life support breathing systems and general environmental management. A miniature zirconia pump-gauge oxygen sensor operating in the AC mode [1] in the temperature range 650-800°C has been developed. The operating temperature was maintained using integral thick film platinum heaters. Theory predicted and experiments confirmed large temperature dependence of the sensor output. A system for controlling the temperature of the sensor was built based upon the control of the resistance of the platinum heaters using a Wheatstone bridge arrangement [2]. It was found, however, that the resistance at a given temperature of these heaters showed some long term drift [3]. As a result the temperature of the sensor was not constant and consequently temperature compensation of the sensor output was investigated.

In the present communication, analogue-implemented temperature compensation (with thermocouple) of the sensor is described and simplification of the hardware used is discussed.

2 THEORY

With a sinusoidal current applied to the pump of the device

$$I = A \sin x \tag{1}$$

where $x = \omega t$, theory indicates that the oxygen partial pressure inside the device should oscillate at the same frequency and result in an EMF, E, appearing on the gauge [1]

$$E = \frac{RT}{4F}\left[\ln\left(\frac{P_1}{P_0}\right) - \ln\left(1 + \frac{RTA}{4Fv\omega P_0}\cos x\right)\right] \tag{2}$$

where R, T and F have their usual significance, v represents the internal volume while P_0 and P_1 are the mean internal and external oxygen partial pressures respectively. Various methods for signal processing have been described [5] for simultaneous computation of P_0 and P_1. The method of concern in this communication was based on the operation of the device in a controlled-loop system involving application of a bias current to the pump in order to maintain the mean value of the gauge EMF constant. This method has been successfully implemented and will be presented in a separate communication [4]. The following equation may be written

$$\frac{1}{\pi}\int_0^\pi E\,dx = V_r \tag{3}$$

where V_r is a reference DC voltage. It can be shown [5] that the bias signal should automatically adjust P_0 in order to satisfy

$$\frac{P_1}{P_0} = \exp\left(\frac{4FV_r}{RT}\right)\frac{1 + \sqrt{1 - \left(\frac{RTA}{4Fv\omega P_0}\right)^2}}{2} \tag{4}$$

The ratio (P_1/P_0) is best kept close to unity in order to minimise leakage effects [6]. Substituting for P_0 from eqn.(4) in eqn.(2), the expression of the AC component of the gauge EMF (i.e. difference between the EMF, E, and its mean value V_r) is given by

$$E - V_r = -\frac{RT}{4F}\ln(1 + u^2 + 2u\cos x) \tag{5}$$

where

$$u = \frac{RTA}{8Fv\omega P_1} \exp\left(\frac{4F}{RT}V_r\right) \qquad (5a)$$

This AC component may be converted into a DC output from which the external oxygen partial pressure P_1 may be calculated. By using a simple RMS converter consisting of a rectifier followed by a low pass filter, its output may be written

$$V_0 = \frac{1}{\pi}\int_0^\pi |E-V_r|\, dx \qquad (6)$$

No analytical solution has been found for the above integral; however numerical integration [5] indicated that to a close approximation

$$V_0 = \frac{R^2T^2A}{8F^2v\omega\pi}\exp\left(\frac{4F}{RT}V_r\right)\frac{1}{P_1} \qquad (7)$$

The output of the RMS converter was dependent upon T and T^2. Temperature compensation (of V_0) was investigated under two operating conditions: $V_r=0$ and variable V_r. These methods were based on the use of a *type R* thermocouple.

Reference voltage equal to zero

For a zero reference voltage, the output may be written (from eqn.7)

$$V_{01} = \frac{R^2A}{8\pi F^2v\omega}\frac{T^2}{P_1} \qquad (8)$$

Temperature compensation can be achieved by dividing the output voltage by T^2 to produce an output independent of temperature within the operating range. The thermocouple EMF (e), measured with a 20°C cold junction, is to be converted into T^2. The straightforward solution is to use a thermocouple lineariser to convert the thermocouple EMF into temperature (T) and a squarer to obtain T^2. However, within the relatively small range (650-800°C) of operating temperature of the sensor, the linearity between (T) and (e) was excellent and the lineariser was not required. Furthermore, equally good linearity between T^2 and (e) was anticipated, making it possible to convert the thermocouple EMF into temperature square without the use of lineariser and squarer. This is an advantage in analogue implementation where non-linear circuits are often avoided for simplicity and cost reasons. The following equation may be written:

$$T^2 = a_1 e + b_1 \tag{9}$$

where a_1, b_1 are constants calculated using linear regression: a_1=164.5 K² µV⁻¹ and b_1=-141.5x10³ K². Theoretical calculation showed that, within the temperature range 650-800°C, the maximum error introduced in V_{01} by the approximation of eqn.(9) is 0.4%. Theory suggested also that the ratio (P_1/P_0) should depend on the amplitude of the AC component of the EMF and should vary between 1 and ½.

Variable reference voltage

The output of the electronics given in eqn.(7) may be written:

$$V_0 = \frac{R^2 A}{8\pi F^2 v \omega} \frac{Y(T)}{P_1} \tag{10}$$

where

$$Y(T) = T^2 \exp\left(\frac{4FV_r}{RT}\right) \tag{10a}$$

The principle of temperature compensation was based upon automatically adjusting the reference voltage V_r in order to keep Y(T)=constant [Y(T)=C] within the operating temperature range. The output of the electronics (eqn.10) would then be independent of temperature, and may be written

$$V_{02} = \frac{R^2 A}{8\pi F^2 v \omega} \frac{C}{P_1} \tag{11}$$

In the case where the reference voltage is equal to zero the constant C is equal to T² [compare eqn.(11) and eqn.(8)]. A choice of C=10⁶ is equivalent to a temperature T=1000K. From eqn.(10a)

$$V_r = \frac{RT}{4F} \ln\left(\frac{C}{T^2}\right) \tag{12}$$

The expression of V_r as a function of temperature is complicated. However a plot of V_r versus the thermocouple EMF (e) for the range 650-800°C and C=10⁶ indicated that the required V_r for full temperature compensation may be expressed as a linear function of the thermocouple EMF.

$$V_r = a_2 e + b_2 \qquad (13)$$

where a_2 and b_2 are constants. The values for the constants were calculated using linear regression: $a_2=-3.54$ and $b_2=24.56$ mV. It can be shown that the maximum error introduced by the approximation of eqn.(13) is 0.52% (deviation of Y(T) from a constant). This is an important finding because temperature compensation is shown to be possible without using non-linear elements (analogue divider, squarer) making the electronics simple to implement. Theory suggested also that the ratio (P_1/P_0), given by eqn.(4), should vary between 0.425 and 1.3. Thus the introduction of a DC component in the gauge EMF modified the ratio between the mean internal and external oxygen partial pressures; this was expected because a part of the DC component is dependent upon the log of the ratio (P_1/P_0) [see eqn.(2)].

3 EXPERIMENTAL

The temperature range considered was 650-800°C. The oxygen partial pressure was set at 4kPa by mixing air and nitrogen using mass flow valves from Brooks Instrument. The frequency and amplitude of the pumping current were 4Hz and 0.8mA respectively and the internal volume of the device was 0.98mm^3.

Figure (1) shows the schematic diagram of the temperature-compensated instrument for oxygen partial pressure measurement. The pumping sinewave was obtained from an oscillator adopted from Jung [7] and the voltage-to-current converter from Froelicher *et al* [8]; the gauge EMF was amplified (x500) using an INA110 instrumentation amplifier; the DC component of the EMF was obtained from a simple low pass filter, and the bias signal used to control P_0 was obtained from the difference between the reference voltage V_r and the DC component. The thermocouple (Pt-Pt13Rh) was attached to the sensor close to the gauge as recommended by Fouletier [9]. The thermocouple EMF was amplified (x1000) using an instrumentation amplifier. The errors introduced by the variation of the cold junction temperature were small but automatic cold junction compensation could readily be incorporated in appropriate cases. The amplified thermocouple EMF was converted into T^2 and V_r according to eqn.(9) and eqn.(13) respectively. The divider was built around an analog multiplier [10]. The switch allowed the two modes of temperature compensation to be tested.

4 RESULTS AND DISCUSSION

Figure (2) shows the results obtained in the range 650-800°C. The temperature compensation with a reference voltage equal to zero was excellent: while the non-compensated output varied by 31.5% (theoretical error was calculated to be 32.5%), the compensated output (V_{01}/T^2) varied by only 1.2%. The second compensated output (V_{02}) varied by 4.2%: this deviation from an ideal compensation could be caused by a difference between the DC component and the reference voltage, i.e. the bias signal was low but was not equal to zero. The compensation could be improved by some fine tuning of the constants a_2 and b_2. The minor difference in outputs V_{01}/T^2 and V_{02} resulted from small differences in the amplification of the circuit blocks.

5 CONCLUSION

The output of pump-gauge devices operated in the AC mode shows a strong temperature dependence. Techniques for temperature compensation, requiring simple hardware, were suggested and successfully tested; full temperature compensation was achieved over the whole range of operating temperature (i.e. 650-800°C) of the device. The simplicity and good performance of this system suggests that temperature compensation could be used as an alternative for the often cumbersome temperature control.

6 REFERENCES

[1] Maskell, W.C.; Kaneko, H.; Steele, B.C.H.: Miniature oxygen pump-gauge. III. Application of a periodic current waveform. J. Appl. Electrochem. 17 (1987) 489-494.
[2] Benammar, M.; Maskell, W.C.: Temperature control of thick film printed heaters. J. Phys. E: Sci. Instrum. 22 (1989) 933-936.
[3] Ioannou, A.S.; Maskell, W.C.; Proc Symp Deposition and Characterisation of Electronic Materials; Manchester 17-19 July 1991; to be published by the Royal Society of Chemistry
[4] Benammar, M.; Maskell, W.C.; in preparation.
[5] Benammar, M.; (1991) PhD Thesis, CNAA, London.
[6] Kaneko H.; Maskell, W.C.; Steele, B.C.H.: Miniature oxygen pump-gauge. I. Leakage considerations. Solid State Ionics 22 (1987) 161-172.
[7] Jung, W.G.: IC Op-Amp Cookbook. Sams-Macmillan, Third edition. Indianapolis 1974. p.455.
[8] Froelicher, M.; Gabrielli, C.; Toque, J.P.: A high output power (70V, 1.5A) potentiostat-galvanostat. J. Appl. Electrochem. 10 (1980) 71-74.
[9] Fouletier, J.; (1976) PhD Thesis, University of Grenoble.
[10] Clayton, G.B.: Operational Amplifiers. Butterworth 1972. p.115.

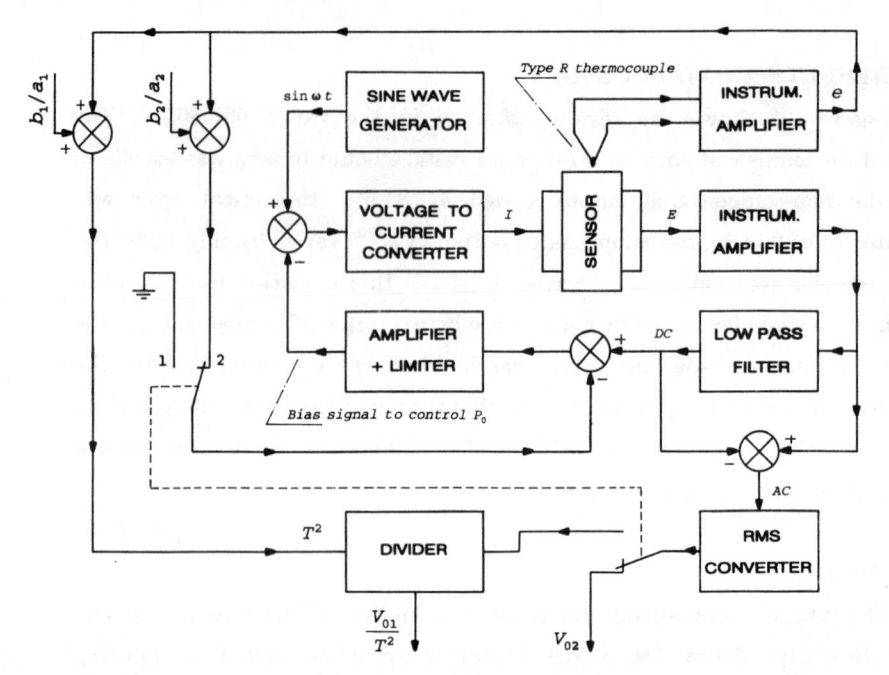

Fig.1. Schematic diagram of the temperature-compensated instrument for oxygen partial pressure measurement. Switch in position 1 and 2: compensation according to eqn.(9) and (13) respectively.

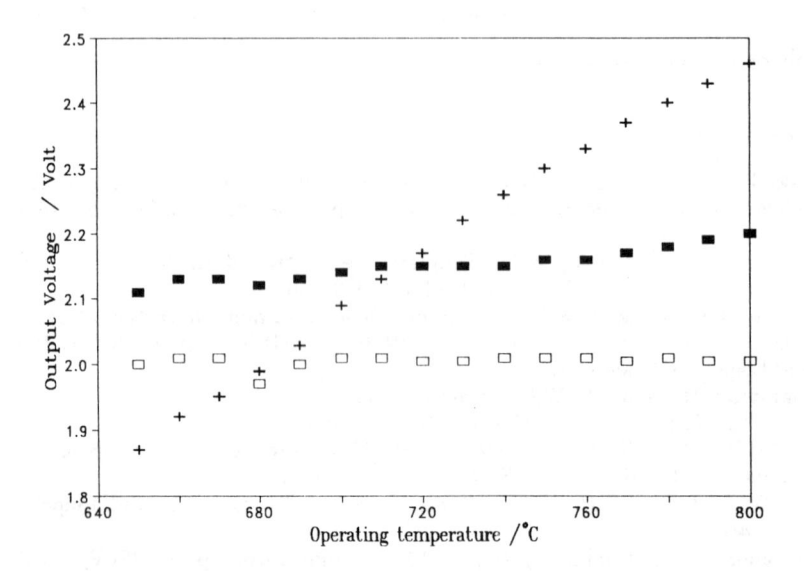

Fig.2. Temperature compensation using the circuit of figure (1). P_1=4kPa, ω=8π rad s^{-1}, A=0.8mA. +, V_{01}; □, V_{01}/T^2; ■, V_{02}.

A Technique for Coupling Non-Compatible CNC Controllers with Touch Trigger Probe Systems

Mrs. Erping Zhou, Dr. David K. Harrison & Mr. David Link

Department of Mechanical and Computer-Aided Engineering, Staffordshire Polytechnic

Summary
The use of in-process measurement in modern CNC machine tools has increasingly become a major feature of machine tool systems. It is perhaps only possible by in-process measurement to take measures in time before component variations, tool wear and other error sources result in rejects. This paper describes a technique for coupling non-compatible CNC machine tool controllers with touch trigger probe systems. The latter is applied in a CNC milling machine controller which is non-compatible with the probe system and the coupling is carried out by a Personal Computer (PC) and relevant hardware and software.

Introduction
In the traditional measuring and gauging of components or setting tools on the machine, manual equipment such as micrometers, vernier callipers, go and no-go gauges etc. are typically utilized. These manual gauging and measuring methods may lead to "down time", i.e. time when the machine is not fulfilling its prime function of cutting material. With the introduction of automated machining and especially CNC systems together with limited manning concepts, the manual approach is decreasing in viability. At the same time, the development of CNC has created completely new possibilities for in-process measurement [1].

The terminology "in-process measurement" refers to the measuring or gauging of a component, while it is clamped in the machining position. This form of measurement may also include component setting, tool setting and tool condition monitoring [2].

A touch trigger probe system is the basic sensor requirement to effect in-process measurement and these have also become widely used for measuring purposes via co-ordinate measuring machines. Many modern CNC controllers are compatible with these probe systems. However, there are still a large number of CNC machine tools which have "non-compatible" controllers with respect to probe systems.

This paper describes a technique for coupling non-compatible CNC machine tool controllers with touch trigger probe systems. The latter is applied to a Bridgeport Series I CNC milling machine controller which is non-compatible with the probe system and the coupling is carried out by a Personal Computer (PC) and relevant hardware and software.

Background

The touch trigger probe system relies on a switching principle which is based on a kinematic location to give the high repeatability required. The trigger signal is transmitted to the interface, which converts the trigger signal into an acceptable form for the machine's controller. CNC control systems are available from many leading manufacturers which incorporate probe-compatible software. This type of controller has the software which includes NC preparatory function codes and process parameter blocks which allow for the integration of a probe system [3]. The terminology "non-compatible controller" with respect to a probe system means that there is no probe compatible software. The probe system can not be used directly with the CNC machine tool to effect tool setting and workpiece dimension measurement. There are many such controllers, especially those found on early CNC machine tools. The Bridgeport Series I CNC milling machine controller is this type.

The Bridgeport Series I CNC Milling Machine Controller
 and Strategies for Coupling Probe Systems

A CNC controller has an internal computer which performs most of the logic functions, i.e., program storage, canned cycles, linear and circular interpolation, loops, macros, etc. In order to perform all these functions the computer must be instructed by a program called an operating system. The operating system is usually built into the machine and is contained in an internal read-only memory (ROM). The operating system in the Bridgeport Series I CNC controller is called BOSS (Bridgeport Operating System Software) [4]. The computer used in the Series I CNC is the LSI-11. Other cards in the system include the XDI (External Data Interface), ERS (Rom and Series Interface), and RCK (Rate Clock). The system is bus-structured. All the functions in the system are connected to a common bi-directional bus and receive the same interface signals. The LSI-11 micro-computer controls the bus. It is obvious that this kind of controller has its own machining and setting system controlled by the software and hardware. The Bridgeport Series I CNC milling machine control system is shown in Figure 1. This consists of the machine table, three directional actuators (stepping motors) and the CNC controller.

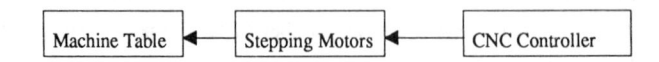

| Machine Table | ← | Stepping Motors | ← | CNC Controller |

Fig.1. The Bridgeport Series I CNC Milling Machine Control System

The first consideration of connecting the touch trigger probe system to the controller may be called the software solution. If a very detailed BOSS source program could be obtained, then it would be possible to rewrite the BOSS to include probe-compatible software. Thus, as soon as the probe touches the workpiece, the probe interface would send the output signal to the LSI-11. Measured values are generated upon the switching pulse by reading the position counters of the machine tool dynamically. Unfortunately, CNC machine tool manufacturers will seldom provide such source programs and rewriting the source program would take an inordinate amount of time and resources.

A hardware solution, which is the second strategy for coupling, is the use of a logic control circuit to control the machine movement (see Figure 2). Here a group of relays are used to regulate the controller control panel switches. These switches are typically the X & Y direction movement switches and the stop switch. When the probe touches the component, the output signal of the probe interface is sent to the relay which controls the stop switch. Thus the machine movement is stopped and the measured value can be taken from the display of the machine.

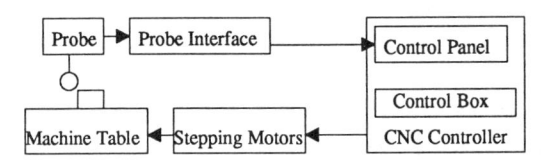

Fig.2. The Hardware Solution for Coupling the Controller with a Probe System

The advantage of using the hardware method to link the non-compatible controller to effect in-process measuring is that it is quite easy and cheap, but it has no support from the software and thus the measuring functions are limited.

The third coupling method could be called remote microcomputer control. This is the way that combines both software and hardware solutions. A diagram of the system is shown in Figure 3. Here the logic control system is used to connect the probe system with a Personal Computer (PC). The PC is employed to control the movement of the machine, instead of the CNC machine tool controller, when in-process measuring. Therefore multi-function measurement can be made through the measuring routines which have been written and stored in the PC. The control will be returned to the CNC machine controller through the logic control system when machining is to be carried out.

Using a PC to couple the touch trigger probe system with a non-compatible CNC controller has more advantages than the others. The PC has flexibility and speed of operation and it is much cheaper and easier than the software solution and leads to a very flexible measuring system with a much wider application area than the hardware solution. Therefore a remote PC has been used with a Bridgeport Series I CNC milling machine to effect in-process measurement in the following case study.

```
    ┌─────────┐      ┌─────────────────┐      ┌──────────────────┐
    │  Probe  │─────▶│ Probe Interface │─────▶│ Personal Computer│
    └─────────┘      └─────────────────┘      └──────────────────┘
         │                                             │
         ▼                                             ▼
    ┌──────────────┐    ┌────────────────┐    ┌──────────────────────────┐    ┌────────────────┐
    │ Machine Table│◀───│ Stepping Motors│◀───│Programmable Logic Control│◀───│ CNC Controller │
    └──────────────┘    └────────────────┘    └──────────────────────────┘    └────────────────┘
```

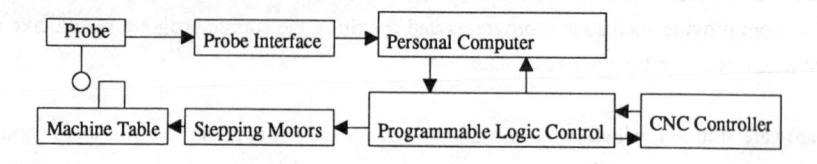

Fig.3. Personal Computer Solution for Coupling the Controller with the Probe System

<u>Hardware Needed for Interfacing the PC to the Machine Controller</u>
Since the BOSS can not recognise the signals from the probe system, the PC is used to control the milling machine's actuators (stepping motors in this case). The PC needs a way to communicate with the peripheral device. Here I/O hardware and its corresponding software provide the interface to transfer data between the PC and the stepping motors.

The stepping motors are the digital actuating devices which drive the machine tool slides. The input to a stepping motor is a train of digital pulses, which can be provided by a digital output port on the PC. Devices such as stepping motors generally employ parallel I/O. Many types of the parallel digital I/O interface models can be obtained commercially.

The Bridgeport Series I CNC milling machine has three stepping motors which separately control X, Y, Z direction movement. Here a 24 bit parallel digital I/O interface model PIO12 (MetraByte Corporation) was selected. This parallel digital I/O card has 24 Transistor-Transistor Logic (TTL) or Diode Transistor Logic (DTL) compatible digital I/O lines which are provided through an 8255-5 Programmable Peripheral Interface (P.P.I.) and consist of three ports, an 8 bit Pa port, an 8 bit Pb port, and an 8 bit Pc port. Each of the ports can be configured as an input or an output by software control according to the contents of a write only control register in the P.P.I. The Pa, Pb and Pc ports can be read as well as written to. In this case the Pa port is used for output ports which supply six trains of digital pulses, three of them are used separately for X, Y, Z stepping motors and the other three are utilized separately for the opposite directions. The Pb port is used as an input port for the signal from the probe system.

Since the PC is only used for controlling in-process measuring, the other functions that the Bridgeport Series I CNC milling machine has, must be kept in operation. Thus if in-process measuring is needed, then the machine would be controlled by the PC and if machining is needed, the machine should be returned to its original status, i.e. controlled by the LSI-11. A programmable logic control system is employed to address this change of control. The system is also controlled by the PC through the parallel digital I/O interface.

Software Design for Coupling
Once the hardware interface had been identified, the next stage was to design the software. There are two sections in the design. The first is to control the pulse output and the probe's input signals. The other is to control the accuracy of the movement.

Although assembly language is usually used in the control of stepping motors, as it uses a symbolic representation of actual binary code that the computer executes directly thus it gives programmers the potential for accomplishing tasks with maximum flexibility and efficiency, it is difficult to use when developing and debugging a program. Lacking of structure makes assembly language programs difficult to read, enhance, and maintain. In this case study the C language was selected as the software program to control the stepping motors as well as to manage the in-process measuring routines. C's inherent efficiency compared with its ability to operate directly on the bits and bytes of the computer's memory allows people to use C in place of assembly language in the control situation. Also the C language has the speed of assembly language and the extensibility and few of the restrictions of the other high level languages.

The In-Process Measurement Sequence
In practice there are three in-process measuring sequences which can be applied. The first measuring sequence is effected after the milling machine has finished its part program. Thus, at the beginning of the procedure, the milling machine is controlled by the BOSS to fulfil the machining functions, then the probe is mounted in the spindle instead of the cutter. The Programmable Logic Control (PLC) system is used to change the control status from the BOSS to the PC and thus measurement functions can be effected. After measuring, the control status is returned to the BOSS through the PLC system to enable the next procedure to begin.

The second measuring sequence is effected during the machining part program. This sequence is typically triggered if some crucial dimensions require measuring before more machining is carried out. This also occurs at a tool change. Firstly the control status rests with the BOSS and the machining part program is inputted to the BOSS. After the crucial dimension has been cut, the tool is returned to the tool change position through the tool change command (M6) which is written in the part program and the probe is inserted at the tool change. Then the control status is set for PC control through the PLC system and the measurement can be effected. Once the measuring is finished and the dimensional tolerance of the component is within the limits, the PLC system receives this information, so the probe is changed for another tool and the control status returned to the BOSS. The next machining operation can then begin. If the dimensional size measured is not satisfactory, then a signal is shown on the PC screen and the PLC system will not change the control status, thus no more machining can begin until the part program is modified or the system is manually over-ridden.

The third sequence happens before machining begins and is used for component setting. In this situation, the control status is at first with the PC. By placing a probe in the spindle, some points may be measured and the coordinate values are stored. These coordinate values are then used as offsets for the part program.

Conclusion

The technique of using a PC and a compatible hardware and software interface for coupling non-compatible CNC controllers with touch trigger probe systems has been demonstrated. In-process measurement can be effected by cheap flexible software written in the C language. PC's are relatively common and it is not difficult or uneconomic to fulfil the measuring process via a non-compatible CNC controller with a probe system. The original CNC machine's manufacturing functions are maintained through the programmable logic control system and it causes no disturbance between the machining and the in-process measuring.

References

1. Link, D.; Farshbaf, M.R. and Ghosh, S.K.; "Developments in Flexible Manufacturing Employing Integrated Technologies",2nd Int. Conf. on Advanced manufacturing Systems and Technology (AMST '90), 19-21 June 1990, Trento, Italy.

2. Sotirov, G.R.; Vitanov, V.I; Ghosh, S.K.; Minkov, N.H. and Harrison, D.K.; "Multicriteria Optimization of Cutting Processes", Int. Conf. Developments in Forming Technology (ICDFT), Vol 2, pp.11.11-11.20, Cascais, Portugal. September 1990.

3. Tlusty, J.;Andrews, G.C.; "A Critical Review of sensors for Unmanned Machining" Annals of the CIRP Vol. 32/2/1983.

4. Anon. M-131 Installation & Maintenance Manual for the Series I CNC Milling, Drilling & Boring Machine (BOSS 4.0/4.1,5.0&6.0 Controls), Bridgeport Textron, December 1980.

Trajectory Planning for Robotic Technology System Control

V. G. Gradetsky, A. M. Ermolaev

Institute for Problems in Mechanics of
Academy of Sciences of USSR
117 526, Moskow, pr. Vernadskogo, 101, USSR

SUMMARY

Methods for solving three main technological tasks for the trajectory generation problem on contour treatment of outside surfaces of solid bodies are considered.

Results can be used in designing new program simulation systems of robot-beam (laser, water jet, gas) technologies and the collision detecting problem on contour treatment with the help of the internal parameters of the model.

The method can be applied for simulating the motion of two coordinated manipulators and the technological manipulator installed on the basis of the wall climbing robot intended for cutting, welding, painting, cleaning on the sloping and vertical plane and on the ceilings.

MAIN EQUATIONS OF THE MOTION

In the paper the problem of the contour controlling in the task of robot-beam technologies is considered. In this way laser, water jet, and gas technologies is considered. This technologies determine the special conditions of the connection between the processing body and the technological beam, where there is not the dynamic influence in the sense of the phisical contact between robots during the motion [1].

The conditions of the connection during the treatment is the constant value of a sloping angle of a character line at a normal vector of a processing surface. It is proposed that the technological beam has a character line and its rotation around this special line maintains the conditions of the connection.

In this paper the solution of the task is built with

useing the approach based on Jacobian matrices (velocity level).

Utilizing the approach based on the homogeneous transformation matrices [2] we can derive the main matrix equation determining the coordinated motion of two robots. This equation is given

$$C^1 A^1 = C^2 A^2 Q \qquad (1).$$

Here, the matrix $C^1 = C^1(q_1^1, \ldots, q_n^1)$ determines the homogeneous transformation from the gripper coordinate system of the first robot with respect to the some refering coordinate system. The matrix $C^2 = C^2(q_1^2, \ldots, q_m^2)$ determines the homogeneous transformation from the gripper coordinate system of the second robot with respect to the same refering coordinate system.

The matrix $A^1 = \text{const}$ defines the homogeneous transformation from the coordinate system of the technological beam with respect to the gripper coordinate system of the first robot.

The matrix $A^2 = A^2(u,v)$ determines the homogeneous transformations from the coordinate system connected with each position of the character line on the processing body with respect to the gripper coordinate system of the second robots. The parameters (u,v) are the parametrical coordinates of the processing surface. The matrix A^2 is determined in the paper [2] and built with the treatment velocity taking into consideration. The matrix $Q = Q(\theta)$ defines the rotation of the matrix A^2 around the character line.

Utilizing the approach of skew symmetric matrices and differentiating the equation (1) with respect to time (t) the following equations are obtained:

$$J^1 \dot{q}^1 = J^2 \dot{q}^2 + J_1^2 (\dot{u}, \dot{v})^T + J_2^2 \dot{\theta}$$
$$J^1 \dot{q}^1 - J^2 \dot{q}^2 - J_2^2 \dot{\theta} = J_1^2 (\dot{u}, \dot{v})^T = \dot{x}$$
$$J\dot{q} = \dot{x} \qquad (2).$$

Here, $\dot{q} = (\dot{q}_1^1,\ldots,\dot{q}_n^1,\dot{q}_1^2,\ldots,\dot{q}_m^2,\dot{\theta})^T$. The matrix $J_{(6\times n)}^1$ is the Jacobian matrix of the first robot. The submatrices $\{J_{(6\times m)}^2, J_{1(6\times 2)}^2, J_{2(6\times 1)}^2\}$ are derived from the Jacobian matrix of the second robot. The functions $u=u(t)$, $u=u(t)$ are the known functions determining the treatment curve on the outside surface of the processing body.

It is necesessary to use the next structure (the regularization method [3]) for solving the equation (2) if $n+m+1 \leq 6$ for three dimensional tasks or $n+m+1 \leq 3$ for two dimensional tasks:

$$\left[\begin{array}{l} \dot{q}_\alpha = \arg[\min_{\dot{q}}\{\|J\dot{q} - \dot{x}\|^2 + \alpha\|\dot{q}\|^2\}] = (J^TJ + \alpha I)^{-1}J^T\dot{x} \\ \alpha = \arg\{\min_{\alpha>0} \|\alpha \frac{d\dot{q}_\alpha}{d\alpha}\|^2\} \end{array}\right. \quad (3).$$

In the other case we have the next task [4]:

$$\min_{\dot{q}} \{0.5\dot{q}^T H\dot{q} + \dot{q}Ab\},$$
subject to $J\dot{q} = \dot{x}$.

Here, $Ab=[\partial s(q)/\partial q]$. The function $s=s(q)$ is minimazed with respect to time [4]. This function allows the existence of the obstacles in the treatment area [5] taking into account. The matrix H is the positive definite symmetric matrix. Useing the Lagrangian function and the regularization method the solution of the problem is given

$$\left[\begin{array}{l} \dot{q}_\alpha = H^{-1}(J^T\mu_\alpha - Ab) \\ \mu_\alpha = \arg[\min_{\mu}\{\|JH^{-1}J^T\mu - (JH^{-1}Ab + \dot{x})\|^2 + \alpha\|\mu\|^2\}] \\ \alpha = \arg(\min_{\alpha>0}\|\alpha \frac{d\mu_\alpha}{d\alpha}\|^2) \end{array}\right. \quad (4).$$

Here, $\mu_\alpha = [(JH^{-1}J^T)^T JH^{-1}J^T + \alpha I]^{-1}(JH^{-1}J^T)^T(\dot{x} + JH^{-1}Ab)$, The infinimal value of the parameter α is 0 (equations (3),(4)) if $\det(J^TJ) \neq 0$ or $\det(JH^{-1}J^T) \neq 0$ [3]. The parameter α

allows to avoid the singular positions during the motion. The solution of equations (3) and (4) is searched by a fourth-order Runge-Kutta method.

SIMULATING RESULTS

Let us considere the task of gas cutting by means of the wall climbing robot with a manipulator installed on its basis. It is proposed that the motion of the wall climbing robot has been generated in the interactive mode with thecnological restrictions taking into consideration before buliding motion trajectories of the manipulator. Two projections of technological equipment is shown at Fig. 1a and 1b. The motion of the manipulator for an each intermediate position of wall climbing robot was built by means of the equation (3) and shown for single position at Fig. {2,3,4,5}. The cutting path is a convex quadrangle.

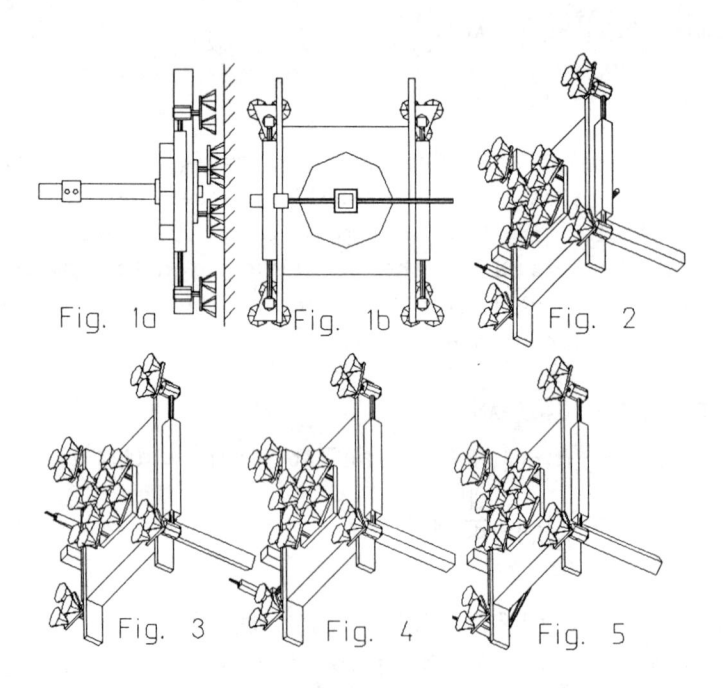

Fig. 1a Fig. 1b Fig. 2

Fig. 3 Fig. 4 Fig. 5

CONCLUSION

It is proposed the method for solving the task of programming the coordinated motion of two robots in robot-beam technologies. The method allows to avoid the singular positions by utilizing the regularization theory. The proposed method was applied to the problem of gas cutting with useing the wall climbing robot and a technological manipulator installed on its basis. The considered method may be used for tasks of the global control.

REFERENCES

[1] J.Y.S. Luh, Y.F. Zheng, "Constrained Relations between Two Coordinated Industrial Robots for Motion Control", The International Journal of Robotics Research, vol. 6, n. 3, p.p. 60-70, 1987.
[2] V.G. Gradetsky, N.N. Rizchov, A.M. Ermolaev, V.V. Semin, "Motion Planning of Technology Robots in Specific Environments", presented at the 5-th ICAR Conference, Piza, 19-21 June, 1991.
[3] A.S. Leonov, "The Substantiation of Choosing the Regularization Parameter by Criterions of Optimality and Relations", The Journal of Calculating Mathematics and Mathematical Phisics, vol. 18, n. 6, 1987 (in Russian).
[4] R.V. Mayorga, A.K.C. Wong, K.S. Ma, "An Efficient Local Approach for the Path Generation of Robot Manipulators", Journal of Robotic Sysytems, vol. 7, n. 1, p.p. 23-55, 1990.
[5] O.Khatib, "Real-Time Obstacle Avoidance for Manipulators and Mobile Robots", The International Journal of Robotics Research, vol. 5, n. 1, p.p. 90-98, 1986.

Vision in Pyramids Object Recognition in Real Time

Ze-Nian Li

School of Computing Science
Simon Fraser University
Burnaby, B.C. V5A 1S6, Canada

ABSTRACT
A hybrid pyramid multiprocessor vision machine has been built for automated object recognition. The machine has 512 one-dimensional single-instruction multiple-data (SIMD) array processors at the bottom and a 63-node transputer-based multiprocessor system on the top. Parallel hardware links are developed to enable massively parallel data communications between the array processors and the transputers. Most of the low level image preprocessing work can be accomplished within tens of milliseconds. Moreover, special pipeline techniques are developed to enable various flexible and effective data flow modes in the 1D pyramid. The 1D hybrid pyramid is shown to be suitable for three types of object recognition tasks: (a) Hough line detection, (b) detection of arbitrary shapes using the LInear Generalized Hough Transform (LIGHT), and (c) object recognition using the Pattern Trees.

1. INTRODUCTION
Pyramid structure is now quite popular and favorable especially in parallel computer vision, because it supports the multiresolution approaches and capitalizes on advantages of both the mesh and the pipeline architectures. There have been many attempts in building pyramid machines. The Image Understanding Architecture at the University of Massachusetts-Amherst is by far the most ambitious pyramid-like machine under development. The large pyramid will eventually consist of three levels with 512×512 processors at the bottom, 32×32 in the middle, and 8×8 on the top, and would achieve roughly a terra-op in performance on 32-bit integer arithmetic [1]. Most researchers, however, are still striving with their "prototype" pyramid machines with a base of 16×16 or 64×64.

We have built a hybrid pyramid multiprocessor vision machine for real time processing of object recognition tasks. The pyramid has the AIS-4000 at the bottom [2] and 63 MIMD transputers on the top in the form of a binary tree of 6 levels. It is thus called a *1D Hybrid Pyramid* (as opposed to common 2D pyramids built with multiresolution 2D arrays). For economic reason, the machine does not employ hundreds of transputers and is thus affordable. This paper will present two aspects of our pyramid vision project: (a) the development of a high speed parallel communication link (PARLink) between the AIS-4000 and the transputers; (b) the development of flexible pyramidal pipeline algorithms that maximize parallelisms in the "half-scale" 1D pyramid. Several examples will be used to illustrate parallel and hierarchical object recognition in the 1D hybrid pyramid. Section 2 describes the pyramid machine. Section 3 introduces special pyramidal pipeline techniques. Brief description of several application examples are presented in Section 4, and a brief conclusion is given in Section 5.

2. THE 1D HYBRID PYRAMID
Figure 1 depicts the 1D hybrid pyramid. The AIS-4000 has fine-grained parallelism with 512 processors arranged in a single-instruction multiple-data (SIMD) 1-D array architecture. It can readily simulate a 2-D array and handle over 3 billion operations per second. Each transputer is a T-800 with 2 MBytes memory. Since the AIS-4000 is basically a 1D array of SIMD processors, the transputer nodes, in most times, are connected into a binary tree of six levels, although the transputer technology does allow a somewhat more flexible and reconfigurable structure. The single root node at level 0 interfaces with a host Sun workstation.

The 32 leaf transputer nodes (at level 5) are connected to the AIS-4000 via 64 custom-made parallel links. Three of the four links of each T-800 node are used to form the pyramid. The remaining single link of each T-800 is connected to a programmable cross-bar switch which provides flexibility for additional desirable connectivity.

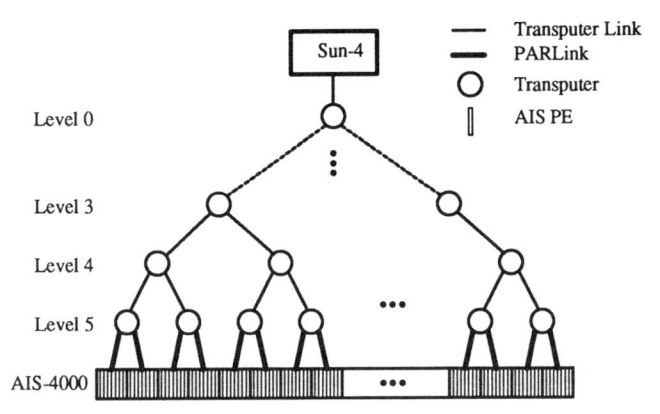

Fig. 1 The 1D Hybrid Pyramid Machine

The key to combining the AIS-4000 and the transputer nodes into a pyramidal architecture is a high speed communication link between the AIS-4000 and the transputers. A parallel link, called PARLink was recently built in our lab. The AIS-4000 has eight SIMD nodes packaged into one custom VLSI chip called a Pixie. Each Pixie chip interfaces to one 32K by eight bit static RAM chip for local image storage. Since the AIS-4000 has 512 SIMD nodes, this means there are 64 clusters of Pixie and RAM chips. This architecture lends itself to the construction of 64 parallel links connecting each cluster up to the transputer pyramid. Two of these links are connected to one transputer node for a binary tree configuration. The tranputers are ordinarily connected with 20 Mbit/sec serial links according to an INMOS protocol. Therefore the hardware requirements for each PARLink are to read eight bit parallel data from the static RAM chips of the AIS-4000 and translate this into the INMOS serial format. This requirement is reversed for communication from the transputers to the AIS-4000. The core of each PARLink is the INMOS C012 chip, and only a few additional buffers and control signals required. All PARLinks are controlled by some custom microcode subroutines supplied by the AISI. Connections are made directly to the data lines of the static RAM chips. Control signals are obtained from the AIS-4000 digital I/O channels as well as direct connections to the microcontroller. A throughput of 1 MByte/sec can be achieved for each PARLink, which allows an image to be passed in 4 msec, with a total bandwith of 64 MBytes/sec for all of the PARLinks.

3. SPECIAL PIPELINE TECHNIQUES FOR THE 1D HYBRID PYRAMID

The 1D hybrid pyramid facilitates a quick and convenient way to input data from the bottom of the pyramid. Live images (512×512) are taken with on-line cameras and input to the AIS-4000 frame buffers. The processed images can then be sent to the transputer nodes via the 64 PARLinks. The massive parallelism characterized by the SIMD array processors is naturally embedded in the pyramid machine, because each level of the pyramid is itself an array. Moreover, much more parallelism is plausible from the pyramidal structure, since the various levels of the pyramid operate in MIMD mode, and the nodes at the upper levels of the pyramid are more sophisticated and operate in MIMD mode within each level as well. This section will focus on several special pipeline techniques that are useful and sometimes unique to our 1D hybrid pyramid.

3.1. Pipelining Among Multiple Levels in a 2D Pyramid

Pipeline technique is suitable for the hierarchical structure of the pyramid. A simple way of pipelining is to assign different functions to the multiple levels which are of progressively reduced resolutions, and the pipelining occurs between these levels. For a real-time system in which image data come in from the bottom continuously, this multilevel bottom-up information flow and abstraction can provide substantial speed-up.

For example, the O(log N) pyramid Hough algorithm for line detection by Jolion and Rosenfeld [3] can readily be implemented in this fashion. Briefly, each processor at the bottom level of the pyramid examines a $2^h \times 2^h$ block of the edge image and merges edge pixels into short lines, longer lines, and so on. A divide-and-conquer process will usually take place in blocks of size 1, 4, 16, ..., $2^h \times 2^h$. The lines are then passed up to the parent nodes and merged recursively up in the pyramid. A practical threshold K for the number of lines reported by each node can be set to avoid excessive number of short lines and noise being detected.

Two major factors will affect the performance of the above pyramidal pipeline technique. (1) For it to function properly, a task must be decomposed hierarchically and assigned to each level. The topologies of any existing pyramid structures usually impose a severe constraint of locality. While locality is the essence/basis for parallelism in low level parallel image processing and some image analysis algorithms, it is not always the case that a whole hierarchical feature/parameter space (conceptual hierarchy) can be mapped level by level, node by node to the pyramidal structure (physical hierarchy). In other words, more flexibility is often needed for the mapping. (2) For it to perform efficiently, each level should have relatively balanced load. For the above example, assume the image has a resolution $2^N \times 2^N$ and the base of the pyramid has $2^M \times 2^M$ nodes, then $h = N - M$. If h is not small, e.g. the pyramid is short ("half-scale") as in our case, then the leaf nodes are overloaded and become the bottleneck. Fig. 2(a) illustrates this situation. For the ease of drawing, we have drawn $N = 5$ and $M = 2$.

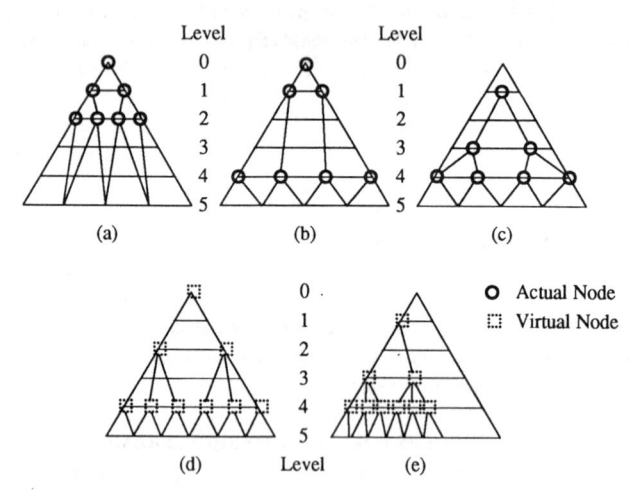

Fig. 2 Basic R-Pipe and C-Pipe Modes

3.2. The Flexible R-Pipe Modes

To alleviate the above two problems, some further pipelining techniques will be introduced. First, we will examine the pipelining along the row dimension, and hence the *R-Pipe*. The reason the leaf nodes are overwhelmed is that each of them is topologically connected to a block of $2^h \times 2^h$ and is responsible for the processing at levels $N, N-1, ..., N-h$. If each $2^h \times 2^h$ block is divided into 2^p subblocks along the row dimension, and the leaf nodes will only process the $2^{h-p} \times 2^h$ subblocks before they send the intermediate results to their parent nodes in 2^p steps in a pipeline fashion, then the processing for the level $N-(h-p)-1$, $N-(h-p)-2, ... N-h$ along the row dimension will be taken over by the parent nodes. Fig. 2(b) shows this

R-Pipe when $p = 2$. Conceptually, the physical pyramid leaf nodes are now being placed at a lower level. Similarly, the parent and grandparent nodes have their own flexibilities in placing themself at proper levels in the conceptual hierarchy shown in Fig. 2(c). In this way, the strict level by level hierarchy is broken which makes the mapping between a hierarchical problem domain and the physical pyramid structure less constrained. Moreover, the system load among the processors at all levels can be better balanced.

3.3. The More Flexible C-Pipe and RC-Pipe Modes

The R-Pipe provides the flexibility for the physical pyramid nodes along the row dimension. Since there is only a 1D pyramid, for many tasks, the nodes in the 1D pyramid (binary tree) can simply sweep along the column dimension by repeating a similar process N times. Hence, identical pipeline modes shown in Fig. 2(a)-(c) can also be realized along the column dimension (and therefore C-Pipe), as long as the nodes are viewed as virtual nodes which correspond to the single actual node at each column. In this way, a 1D Pyramid can trivially emulate a 2D Pyramid. In this section, however, we will point out that 1D pyramid is not merely a degenerated version of 2D pyramid. It actually provides more flexibility in terms of functionality and speed.

3.3.1. The Anisotropic Nature of the 1D Pyramid

As discussed in [4], the AIS 1D array architecture provides an anisotropic feature in that the row dimension is massively parallel and the column dimension is similar to a standard single CPU architecture with the capability of fast random access to its local memory. By arranging the 2D (ρ–θ) parameter space accordingly, our Hough line detection algorithm [4] can actually take advantage of this asymmetric nature of the AIS 1D array architecture, and enable a faster summing process in the column dimension. Now, in the 1D pyramid, each node is emulating a column of virtual processors and has a column of the image/feature space in its own local memory. The virtual nodes do not have to be connected in a binary tree fashion along the column dimension, they do not even have to cover the whole column. In short, the constraint of locality does not exist at all along the column dimension.

3.3.2. The C-Pipe and RC-Pipe Modes

Fig. 2(d) and (e) are two examples of many additionally possible C-Pipe modes. Beside the modes in Fig. 2(a)-(c), Fig. 2(d) shows that the virtual nodes can be connected into binary, ternary, ... trees at various levels. Fig. 2(e) shows that arbitrary size of "region of interest" is supported in the C-Pipe mode. Moreover, the R-Pipe and C-Pipe modes are entirely independent, i.e. any combination of them (*RC-Pipe* mode) is allowed. Fig. 3 depicts a possible RC-Pipe modes in an emulated 2D pyramid.

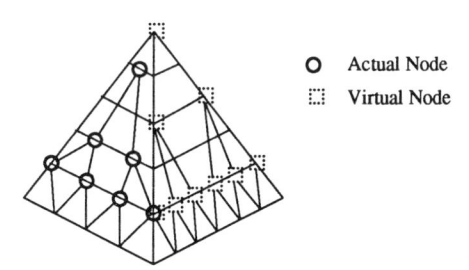

Fig. 3 Combination of R-Pipe and C-Pipe in Simulating a 2D Pyramid

For many parallel vision applications the feature/parameter space has more than two dimensions (e.g. curve detection using the Hough transform). The C-Pipe can readily be extended to handle higher dimensionalities and still offer good parallelism.

4. APPLICATION EXAMPLES

The following is a brief description of some plausible pyramid vision examples. The AIS-4000 and the PARLinks support the 1D hybrid pyramid with a base of resolution 512. However, the timing results were obtained using only 31 (instead of 63) transputer nodes that are in place, 16 of them being leaf nodes at level 4.

The AIS-4000 and the PARLinks provide great efficiency in data transfer and low level image processing, which will not be reemphasized in this paper.

4.1. The Jolion-Rosenfeld Algorithm

This algorithm is implemented using various RC-Pipe modes to illustrate a working example and the potential for speed-up. At the lowest level, edge pixels with identical ρ and θ are first checked and combined before the merging process which has a time complexity of $O(L^2)$, where L is the number of lines to be merged. A simple divide-and-conquer process is employed which starts to merge lines in 16×16 blocks in all cases, and then in larger and larger blocks. The overall time complexity of this implementation is between $O(L)$ and $O(L^2)$. No effort was made to optimize the algorithm in other ways. Since all the lines are accepted (in other words, the threshold $K = \infty$), the speed of this algorithm depends largely on the image content, the more lines in the image, the longer the merging time. To show this, three synthetic test images (512×512) are used. The first image has $32 \times 32 = 1024$ small squares and hence the most lines. The second has 32 wide and short rectangles, and the third has a single big rectangle in it. Table 1 shows the summarized timing result when various granuarities of pipelining are applied. For example, when 8 pipeline steps are taken at level 4, each leaf transputer at this level is responsible for a 32×64 subimage. Conceptually, the physical level 4 transputer is functioning as a level 4 node row-wise and a level 3 node column-wise. We called this R4-C3 mode.

Table 1 Pipeline Timing Result for Three Test Images

Total Pipeline Steps at Level 4	Area Covered by Level 4 Transputers	Pipe Mode for Level 4 Transputers	Total Execution Time		
			Image1	Image2	Image3
1	32×512	R4-C0	4500.3	2627.3	295.3
4	32×128	R4-C2	3012.4	1557.2	280.9
8	32×64	R4-C3	2733.0	1465.5	272.4
64	16×16	R5-C5	2349.5	1393.6	271.5

As Table 1 indicated, the 64-step pipelining yields close to 200% speed-up for Image 1 and Image 2, which is approximately the optimal expectation (think about 15 additional transputers helping 16 transputers). Image 3, however, shows an extreme case where virtually no improvement is achieved. It is because after the preprocess the number of the lines to merge is two at most times regardless of the size of the block.

Table 2.1 and 2.2 show the timing for the last 8 steps of the 64-step pipeline for Image 1. The sum of the partial timing should be obtained by summing up the numbers in the bold font. The subtle difference shown is that the level 3 transputer nodes are operating in R3-C3 mode in Table 2.1, whereas they are in R3-C4 mode in Table 2.2. Basically, more merging work is being passed up to the level 2 transputer nodes in Table 2.2, hence a more load-balanced pipe. This small change realizes a small speed-up of ¯12 msec for this portion, and reveals the fundamental reason for the overall speed-up in the pipeline.

Table 2.1 Partial Timing of a 64-step pipeline (I)

Pipe Step	Level4 (R5-C5)	Level3 (R3-C3)	Level2 (R2-C2)	Level1 (R1-C1)
57	**26.75**	1.15		
58	**26.76**	1.15		
59	**26.76**	1.15		
60	**26.75**	1.15		
61	**26.76**	1.16		
62	**26.76**	1.15		
63	**26.76**	1.15		
64	**26.75**	**1.15 + 49.6**	**18.7**	**57.7**

Table 2.2 Partial Timing of a 64-step pipeline (II)

Pipe Step	Level4 (R5-C5)	Level3 (R3-C4)	Level2 (R2-C2)	Level1 (R1-C1)
57	**26.74**	1.15		
58	**26.76**	1.15		
59	**26.76**	1.16		
60	**26.75**	1.15 + 20.0	18.6	
61	**26.76**	1.15		
62	**26.75**	1.16		
63	**26.76**	1.15		
64	**26.74**	**1.14 + 20.0**	**19.0 + 17.9**	**57.5**

4.2. The LIGHT algorithm

The LInear Generalized Hough Transform (LIGHT) was shown to be capable of detecting arbitrary occluded shapes and suitable for parallel processing [5]. Random patterns in 512×512 images were detected in less than one second on the AIS-4000. Because of the rudimentary operations supported by the SIMD array processor, more complex and heuristic algorithms for inexact pattern matching were not applicable in our AIS implementation of the LIGHT algorithm [5]. In the 1D hybrid pyramid, the AIS-4000 executes the image preprocessing and vote accumulation as before. Afterwards, each layer of the 3D accumulate array is pipelined into the transputer nodes. The transputer receives consecutive pattern entries and applies heuristic tolerance measures for the inexact pattern matching. This application requires the transputers operating in the RC-mode in a 3D parameter space. It also demonstrates the necessity of having more powerful processing elements at the higher levels of the pyramid.

4.3. Object Recognition Using the Pattern Trees

The *pattern tree* described by Burt [6] can be a powerful tool in active vision. A pattern tree model for a particular object consists of a set of nodes and links. Each node represents a particular object feature, while links represent the relative positions of these features. Typically, object representation and search in the pattern tree follow a coarse-to-fine strategy. However, a pattern tree is usually not a balanced and complete binary tree (or quad-tree). The topologically flexible pyramidal structure proposed in this paper alleviates the difficult mapping problem. Some of our preliminary results on this topic will be presented in a coming paper [7].

5. CONCLUSION

This paper reported the construction of a 1D hybrid pyramid. Parallel hardware links were developed to enable massively parallel data communications between the array processors and the transputers. Special pipeline techniques were developed to enable various flexible and effective data flow modes for real-time object recognition in the 1D hybrid pyramid.

The reported activities are a portion of our long term pyramid vision project. It is now well-known that pyramid provides a unique parallel and hierarchical platform. It supports the multiresolution approaches and capitalizes on advantages of both the mesh and the pipeline architectures, and has the great potential for parallel vision. However, the topologies of any existing pyramid structures usually impose a severe constraint of locality. Furthermore, a full-scale 2D pyramid with thousands of nodes is yet to be created, and its availability is going to be limited for the years to come. We have shown that, with the exploration of special pipeline techniques, a 1D "half-scale" pyramid is not merely a degenerated version of a 2D pyramid. It is not only economic, but could also be more flexible, efficient, and adequate for many parallel vision tasks.

6. ACKNOWLEDGMENTS

This work is supported in part by the Canadian National Science and Engineering Research Council under the grants OGP-36726, EQP-42445, EQP-90642 and a Strategic Research Initiative Grant from the Centre for System Sciences at the Simon Fraser University. The author would like to thank Dr. John Ens, who contributed greatly to the design and implemention of the PARLinks, my colleagues Drs. Wo-Shun Luk and Stella Atkins for their kind support and encouragements, Frank Tong for many stimulating discussions and Danpo Zhang for his assistance in developing and implementing the pyramidal algorithms.

REFERENCES

1. E. M. Riseman and A. R. Hanson, Progress in computer vision at the University of Massachusetts, *Proc. DARPA Image Understanding Workshop*, 1990, 86-96.
2. L. A. Schmitt and S. S. Wilson, The AIS-5000 parallel processor, *IEEE Trans. on Pattern Recognition and Machine Intelligence 10*, 3 (1988), 320-330.
3. J. Jolion and A. Rosenfeld, An O(logn) pyramid Hough transform, *Pattern Recognition Letters 9*, (1989), 343-349.
4. Z. N. Li, F. Tong and R. G. Laughlin, Parallel algorithms for line detection on a 1xN array processor, *Proc. IEEE Int. Conf. on Robotics and Automation*, 1991, 2312-2318.
5. Z. N. Li, B. G. Yao and F. Tong, A linear generalized Hough transform and its parallel implementation, *SFU Technical Report CSS-IS 91-03*, abridged version in IEEE Conf. on CVPR '91, 1991.
6. P. J. Burt, Smart sensing within a pyramid vision machine, *Proceedings of the IEEE 76*, 8 (1988), 1006-1015.
7. Z. N. Li, J. Ens and F. Tong, Object recognition in a 1D Hybrid Pyramid, *(in preparation)*, 1991.

Mobile Robot Hydroblast System

Majid K. Babai
Robert M. Rice
Steven A. Cosby

Process Engineering Division
Materials & Processes Laboratory
Science and Engineering Directorate
NASA/Marshall Space Flight Center, AL 35812
 USA

Abstract
High pressure cutting and stripping techniques have a wide range of application in America's space effort. Hydroblasting is commonly used during the refurbishment of reusable Solid Rocket Boosters. This technique allows the stripping of thermal protective ablator materials without incurring any damage to the painted surface underneath, through control of several parameters. The process can also be regulated to strip paint and primer. Automated hydroblasting has obviated the need for personnel to work within a hostile high pressure water environment, since computer-controlled robotic systems can perform this task in a fraction of the time required for manual operations. Mobile Robotic Hydroblast System (MRHS) is a new addition in this field, developed for its flexibility and maneuverability. In addition CAD/CAM technology has played a vital role in the design, optimization, and integration of these processes.

Introduction
Surface cleaning and material removal using pressurized water is not a new technology. Low pressure water cleaning is an every day occurrence from garden hose to kitchen faucet. Water dissipates a large quantity of energy, as dramatically illustrated in the production of electricity at a hydro-electric dam. Pressurized water cleaning utilizes both the impact of water droplets and frictional erosion, as water is forced over the surface material to cut through and drag particulates away. High pressure water cleaning, referred to as hydroblasting, amplifies this power. A pressure of 12,000 - 17,000 psi is typically employed to clean up NASA's Solid Rocket Boosters (SRB's) after their retrieval from the ocean. Water jets under 50,000 - 60,000 psi pressure carrying abrasives, can even cut through the toughest metallic alloys.

Background
Space Shuttle SRB's are retrieved and refurbished after each Shuttle flight. With an average of six shuttle flights per year and two SRB's per Shuttle, SRB refurbishment plays a vital role in keeping NASA's fleet in operation. Hydroblasting is a crucial step in the refurbishment

process. Each SRB is covered at critical locations with protective coatings to shield the base metal against the intense drag and heat generated by high launch velocities, as well as the corrosive ocean environment experienced at splash down. Obviously, these materials (referred to as the Thermal Protection System or TPS), designed to resist hostile conditions and intense heat, are not easily removed. TPS removal at the Kennedy Space Center (KSC) takes place in two stages: (1) stripping around the fasteners for exposure prior to SRB disassembly, and (2) removal of the TPS, paint, and base coat from the SRB segments.

The variation of several parameters control stripping effectiveness of the water spray. The angle of the water to the subject, the water pressure, its volumetric flow rate, as well as nozzle type, stand off, water jet traversing speed, and direction of water jet traverse, all have an effect on the process and must be optimized to produce the desired stripping results. Hydroblast parameters cannot be effectively controlled manually. Manual operation is also inefficient, dangerous, and difficult. Controlling water at a pressure that can slice through the TPS places the operator in constant peril of being himself cut by the water spray. The operator must wear full body protection, a face shield equipped with an air hose, and must fight against 72 pounds of backthrust force generated by the water discharged through the nozzles. To remove the operator from direct contact with high pressure water and flying debris, hydroblasting has been automated in two stages: the first prior to disassembly and the second afterward.

A hydroblast refurbishment facility for processing disassembled SRB segments has been installed at KSC. Developed in the Productivity Enhancement Complex at Marshal Space Flight Center (MSFC), the system utilizes an adjustable frequency controlled Hammelmann pump system and high pressure water discharge nozzles attached to the end of a six degree of freedom, a NIKO gantry robot with 15 foot telescoping z-axis. A turn table located within the robot's work envelope is also integrated into the system, allowing full coverage of the flight hardware by the robot.

Mobile Robot evolution

As above, MRHS, which is presently in the certification and implementation stage at Kennedy Space Center, was developed by NASA and USBI at the Productivity Enhancement Center at Marshall Space Flight Center. The MRHS was designed to solve problems associated with manual refurbishment. The system was developed for automated removal of the Thermal Protection System (TPS) from the Space Shuttle Solid Rocket Boosters prior to disassembly. TPS has to be removed from field joints, the Aft skirt interior, the trailing edge, the system tunnels, External Tank Attach Ring, and other miscellaneous areas in order to expose all fasteners. This proved to be a more challenging development process than experienced earlier for the SRB's segment cleaning. The system had to be designed for minimum impact on the existing facility, yet allow access to geometrically hard to reach areas imposed by the sheer size of the flight hardware. The robot's extensive path programming had to also be developed at MSFC (700 miles away from the flight hardware) without any physical mock-up. This would not have been possible without the availability of state-of-the-art graphic simulation/offline programming technologies.

The MRHS was designed and built as an integrated system consisting of a six degree of freedom articulated robot mounted on a transportable platform (Figure 1). The vehicle can be raised up to 33 inches off the ground by four hydraulic cylinders for extra reachability. The carrydeck cabin was also modified to accommodate two operators and a Karel Robot Controller. Air conditioning and soundproofing were added to the cab's watertight plexiglass, creating an enclosure which protects equipment, operators and the robot controller from heat, humidity and debris.

Figure.1 Mobile Robotic Hydroblast System

The robot itself is a GMF S-420, selected for its 264-lb payload capability and large reach envelope. It was waterproofed and given compatible subsystems such as a robust and adjustable end effector for high pressure water delivery through either a dual-orifice rotary nozzle or a single-orifice nozzle. The rotary nozzle consists of twin jets fixed on a rotating hub which turns by the back pressure of the water jet. Remote controlled diversion valves enable easy selection of either nozzle or of the system's "dump line". Also added were an inclinometer for deck leveling, a pressure transducer to monitor water pressure, and a displacement transducer to input the deck height to the robot controller for verification and added safety. Any parameter beyond acceptable range causes an interrupt so the controller can take appropriate action to safeguard personnel and hardware. Such action could involve diversion of water to the dump line, stopping the robot's motion, or prompting the operator for some verification.

Graphic simulation has been essential for design and development of the MRHS. It has been effectively employed from the robot selection phase all the way to the establishment and orientation of tag point coordinates, as well as verification of arm reachability and evaluation of collision detection (Figure 2).

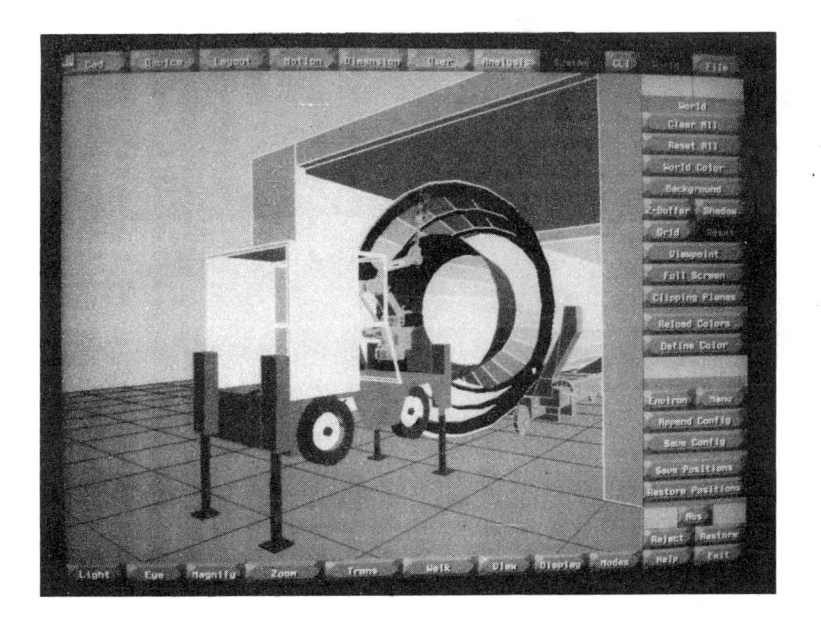

Figure 2. Graphic Simulation Utilization in Robotic Programming

System operation

Once the SRB's are recovered from the ocean, they are placed on transportable dollies guided on rails (Figure 3). After a deionized water shower to rinse off corrosive salt water, hazardous materials safeguarding, and a post flight assessment, the Boosters are ready for TPS stripping. The MRHS is positioned at a set distance from the center line of the track, then the carry deck rises to a set height, elevating to within 0.5 degree tolerance. The booster is rotated to a fixed orientation while the transport moves the hardware such that a low pressure water jet, discharging from the single-orifice nozzle, hits a check point on the booster. The stripping process begins after verification of end effector set up by the operator, and confirmation of deck height, leveling, and high water pressure by the robot controller.

Figure 3. Solid Rocket Booster on Transportable Dollies

Parameter optimization must be performed for each material which the hydroblast is called upon to remove. TPS materials which have effectively been removed by hydroblast include specially formulated Marshall Sprayable Ablatives (MSA), K5NA, Marshall Trowellable Ablator (MTA), PRC - 1422 polysulfide sealant, and cork. The dual-orifice rotating nozzle assembly removes a much wider strip across thinner TPS materials as compared to the single-orifice nozzle. Test results show this nozzle to be effective at removing TPS materials in most areas. Testing must be performed and caution exercised before any critical hydroblasting. The high pressure water spray employed is a potent force. At 15,000 psi, a water spray that is allowed to dwell at the same point on an aluminum plate will cut into the plate, and can eventually cut through it. It is a good habit to start the robot motion before water pressure is increased, or to increase pressure while the spray is being dissipated into the void surrounding the part to be hydroblasted, in order to preclude this possibility.

The MRHS becomes a portable robot which can endure long term operation under harsh conditions, its human occupants safely activating any one of forty-eight robot path programs for a given location, to remove TPS while monitoring the robot's operation. This unit, first of its kind, can be conveniently maneuvered around the boosters while its robotic end effector holding the nozzles can withstand the delivery of 22 gallons of water per minute under 15,000 psi pressure. The nozzle assembly moves through a smooth preprogrammed robotic path while removing TPS with increased accuracy. At the same time, personnel and flight hardware safety is increased while refurbishment time is decreased. It has been conservatively estimated that the system can save 20 hours of serial time per flight set. This equates to a 40 percent reduction in TPS removal time prior to SRB disassembly at estimated savings of $67,000 per launch.

Conclusion

Hydroblast stripping procedures have proven to be an important adjunct to the Solid Rocket Booster's refurbishment process. The fully integrated computer-controlled hydroblast system provides a safe and efficient means of stripping and processing reusable flight hardware. Development of the automated water blast and Mobile Robotic Hydroblast System has already resulted in the savings of millions of dollars and thousands of hours of precious processing time at Marshall Space Flight Center and Kennedy Space Center. As to its untapped potential, the Mobile Robot offers a prototype for a traveling, ergonomic control center with attached robot that can expel media other than water. Already under evaluation at MSFC and USBI is its feasibility for sealant application, paint spraying, and wet ablator application. USBI is considering a major automated system for commercial airliner cleaning, painting and stripping based on the MRHS concept. Applications for other NASA programs and for military and commercial aerospace have yet to be imagined.

References

1. Robert M. Rice, "Process Report on the Automated Hydro Removal of TPS" Prepared for NASA, Contract NAS8-36300, January 1986.

2. Hoppe, D. T. ; Babai, M. K. ; "High Pressure Water Jet Cutting and Stripping", November 1990.

Generalized Approach to the Control of Assembly Process Using Tactile Sensors and Grippers with Soft Fingers

Borovac B., Šešlija D., Stankovski S.

Institute of Industrial Systems Engineering, Faculty of Technical Sciences, University of Novi Sad, 21000-Novi Sad, V. Vlahovića 3, Yugoslavia

Summary

Conventional grippers perform a firm grasp and do not allow any movement of the object within gripper fingers all the time the object is grasped. In this paper it is proposed to use grippers with soft sensored contact surfaces which allow the object motion if an external force on the free object end is applied. The proposed approach is described on the example of the "peg-in-hole" assembly task. The change of pressure between the object and the finger is the basic information for the succesful task realization. A new approach to the control synthesis is suggested.

1. Introduction

The assembly process based on exact measurement of the force acting on the manipulator tip [1-3] requires both very expensive torque-force sensor (usually placed at the robot wrist) and powerful computational device. The control strategy is based on comparison of the actual direction and intensity of reaction force with those defined in advance and which are often called the nominals. The compensation action has to match them. The object is firmly grasped and together with the gripper can be considered as one rigid body. In the case of soft grippers (grippers where contact surface between the object and gripper is not rigid) the reaction force acting on the free end of the object cause small object displacements within the gripper. As a consequence, the pressure portrait between the object and gripper is changed, which is the basic information for the compensation action.

2. Soft Finger Design

The requirements imposed to soft fingers are to increase grasp stability and, at the same time, to sense the contact pressure between the object and finger. The presence of the sensor should not affect the finger's local shape adaptability. These requirements are fulfilled by the design shown in the Fig. 1. The sensored surface of the finger consists of an array of the

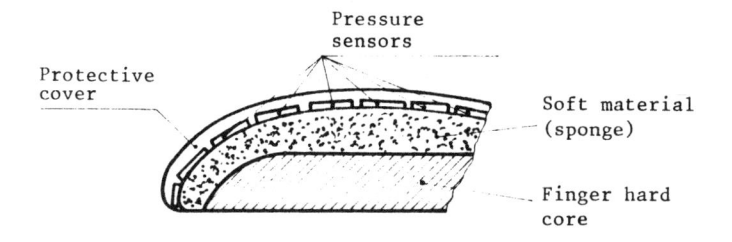

Fig. 1. Soft finger design

separate pressure sensors of FSR type [4]. Each of them is fixed on the hard surface to prevent bending and breaking of the thin film of the FSR material. The sensors are covered with thin layer of silicon rubber which has to ensure protection from external influences. Local shape adaptability is ensured by a thick soft sponge layer placed below the pressure sensors.

When the object is grasped, the finger surface imidiately adapts to the local object shape. Each change in the object position while it is being grasped will cause a change of the pressure portrait of the finger. The information about pressure distribution and its change is the basic information for the object manipulation when it is in the contact with the surroundings.

3) Grasp Interpretation Based on Pressure Information

Grasping can be interpreted as a process of surrounding the object by the gripper where contact occur only at certain points, depending on the chosen grasp. To extend this idea, let us suppose, that the grasped object end is, instead by gripper surface, surrounded by a continuous surface which has the same characteristics as the contact gripper surface. Let further be supposed, that this surface is pressure sensitive and, for the purpose of simplicity, that the pressure distribution in the absence of the external disturbance is uniform over the whole surface. Let us call such a surface the sensorial surface. Any motion of the object caused by an external force will cause a change of the pressure profile on the sensorial surface. A certain external action will always cause the same response of pressure sensors and same type of pressure profile on the sensorial surface will be obtained.

If universal gripper, like Belgrade-USC-IIS Hand [5, 6] is used, a given object can be grasped in various ways in order to realize one and the same task. As a consequence, one pressure sensor of the sensored gripper surface can be in contact with various parts of the object, which means that the information from one particular sensor cannot be used in the same way each time. The sensorial surface represents the entire surface of the object where the gripper fingers can be placed. If the gripper finger with the built-in pressure sensor is placed on the areas where pressure sensors of sensorial surface respond to the object motion caused by an external force, the same pressure will be sensed by the finger's pressure sensor. For an appropriate use of such information, the exact position of each pressure sensor mounted on the gripper has to be known on the sensorial surface.

The sensorial surface can be represented by generalized cylindrical surface swept around the object, as shown in Fig. 2. Let us suppose that the sensorial surface consists of a number of pressure sensors distributed in a matrix form and then swept around the object so that the arrays of the sensors parallel with the axis of the sensorial surface constitute the matrix columns while the horizontal arrays form the matrix rows. Let us, consider further the case shown in Fig. 2.a. The application of force on the object in the direction x' tip will cause a change of the uniform pressure distribution. Let us consider the pressure portrait at one column of the sensors as shown in Fig. 2.a and Fig. 2.b. The role of sensors on the sensorial surface will play the sensors on the gripper which are at these points. In the case of the functional adaptive gripper (e.g. Belgrade-USC-IIS Hand) which can grasp the object in various ways, exact correspodance

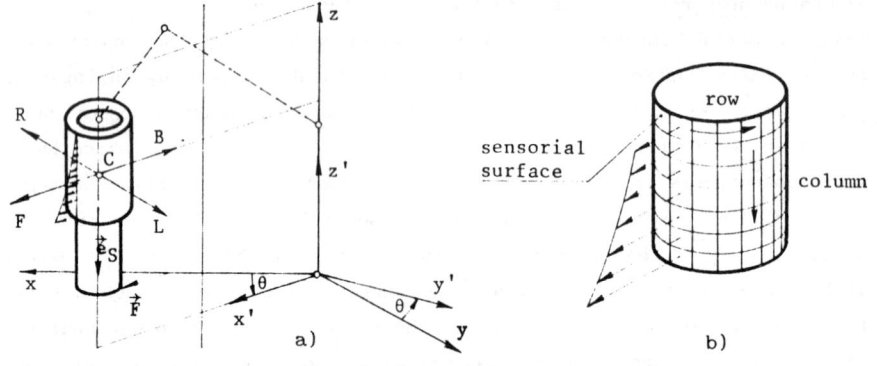

Fig. 2. Sensorial surface

between the points on sensorial surface and gripper depends on the grasping mode employed. If the fingers are sensored as shown in Fig.1. and the object is grasped by a pulp-pinch grasp, the column of sensors from Fig. 2.b. is formed by the array of sensors on a single finger. If the object is grasped by power grasp, so that fingers are swept around the object, each row of sensors on the sensorial surface from Fig. 2.b. will be formed by the same finger's array of sensors. Then, the columns are formed by sensors on the same vertical, each on the "next" finger. The number of rows and columns depends on the number of fingers or sensors on a single finger.

4) Proposed Control Strategy and Experimental Results

The control strategy based on information from pressure sensors mounted on the contact surface between the gripper and the object is different from the conventional ones. The difference is primarily based on the nature of the information which can be obtained from the pressure sensors. The information about the intensity and direction of the external force and torque acting on the gripper is not practically possible to obtain from pressure sensors but it is very easy to obtain the information about the profile of pressure intensity on the gripper and its dynamic behaviour. If the sensored area is covered by a number of separate sensors, then, the dynamic information about the pressure profile and its position on the surface is the basic information needed for control purposes.

The information from pressure sensors alone is not sufficient for a succesful realization of the complete task imposed to the robot. It has to be accompanied by additional sensorial information. For example, vision may participate in the feedback information about spatial behaviour of the manipulator including all phases of navigation before grasping is done, transfer of the gripper to the terminal position and realization of the final task. The role of the pressure sensors vary in dependance of the task imposed. In the grasping and transfering phases, its role is reduced to ensuring a stable and reliable grasp, while in realization of the final task its role can be significantly increased.

Let us consider the simple peg-in-hole assembly task. Let be supposed that the object is close enough to the hole so that assembly does not require additional global navigation. The free end of the object has to be inserted into the other object in the direction of the unit vector $\vec{e}_S$ which is parallel to the sensorial surface axis (Fig. 2). Reaction forces generated during the insertion will cause response of the pressure sensors. The robot has to move the gripper in such a way to eliminate disturbance and to continue insertion. For the spatial case shown in Fig.2.a. there are four main directions for the compensational motion relating robot itself: forward, backward, left and right, as indicated in Fig. 2.a., by F, B, L and R, respectively. A given external disturbance will always cause the same change of pressure portrait which will always require the same compensation action. The situation shown in Fig. 2. will require the forward motion of the gripper each time when it occur. If the coordinate frame $O_{x'y'z'}$ is associated to the robot first link, and rotates together with the first link of the robot, then the motion toward, for example, z axis is always backward motion regardless of the angle θ between the coordinate frames O_{xyz} and $O_{x'y'z'}$. Once defined, the compensational motion with respect to the frame $O_{x'y'z'}$ works in the whole working space of the robot without any additional correction. A dynamic Cartesian control scheme for the control of manipulation robots described earlier [7] is especially convenient for the assembly control strategy described above.

The approach proposed was verified experimentally. The assembly task, reduced to the planar case, consisted of placing a prismatic block into a specially designed hole. A Belgrade-USC-IIS Hand with one figertip sensored in the way shown in Fig.3. was used as a gripper. The hand was mounted on

Fig.3. Sensored fingertip

industrial robot ASEA IRb-L6/2. The block was grasped with pulp-pinch grasp and the sensor array constituted a column on the sensorial surface. Let us consider only the part of the task realization when the assembly begins. The nominal motion of the robot was defined as a vertical downvard. If the relative position of the objects to be assembled is not propper, the compensation has to be performed in such a way to enable a succesful fulfillment of the task imposed. In the initial moment the block axis did not coincide with the hole axis which caused the block inclination from its nominal position. This situation is shown in Fig. 4.a. In this position which differs from nominal, a redistribution of the pressure on the fingertip takes place and according to the control strategy discussed above the robot moves backward to cancel out the disturbance (Fig. 4.b).

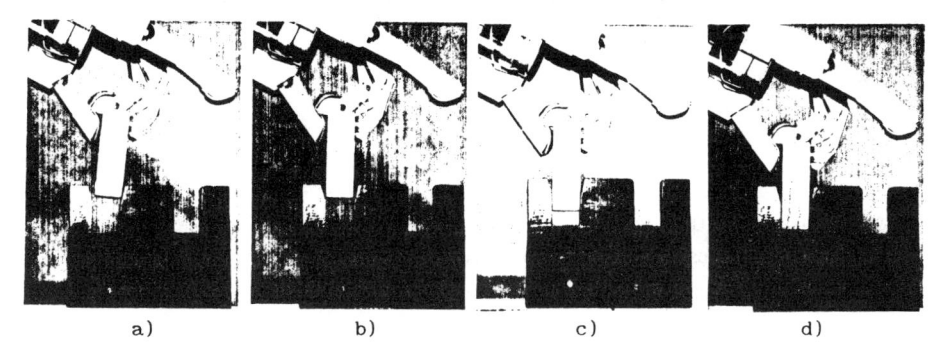

Fig.4. Realization of the assembly task

Afterwards the insertion is continued till its succesful completion. The photos in Fig. 4. illustrate the sequence in the assembly task realization.

5) Conclusion

The paper presents a new approach to the control strategy in those tasks when the contact between the gripper and surroundings occurs. Basic information is the pressure redistribution between the gripper and the object. A new control strategy is developed and tested experimentally.

6) References

1. D. E. Whitney, "Historical Perspective and State of the Art in Robot Force Control",The Inter. Jour. of Robotic Research, Vol. 6, No. 1.,1987.

2. Nevins L.J., D.E. Whitney. "Assembly Research", Automatica, Vol. 16., pp.595-613, 1980.

3. D. Stokić, M. Vukobratović, "Simulation and Control Synthesis of Manipulator in Assembly Technical Parts", Jour. of Dynamic Systems, Measurement and Control, 1979.

4. Borovac B., Šešlija D., "Pressure sensors - Investigation of basic characteristics and design suggestions", Technical Report, FTN-IIS-081--001-MARCH 1990.

5. D. Zelenović, D Šešlija, I. Ćosić, R. Maksimović "On the Flexibility of Robotic Assembly", X-th ICPR, Nottingham, 1989.

6. Bekey B., Borovac B., Tomović, R. "The Belgrade-USC-IIS Robot Hand", Video Proc. of 1991. IEEE Conf. on Robotics and Automation, Sacramento

7. Borovac B., Stokić D., Vukobratović M.,"Contribution to Cartesian Control of Robots",Proc. of Second International Conference on CIM, Troy, 1990.

Monocular Blur Depth Algorithms for an Image-Processing System

Robert P.M. Craven, B.S.Ch.E.
William K. Preece, M.S.M.E.
James E. Smith, Ph. D.

West Virginia University
Center for Industrial Research Applications (CIRA)
Mechanical and Aerospace Engineering Department

Abstract

Distance or depth of field information can be extracted from single frame images. The sharpness of the edge of an object within an image is quantified and calibration curves are produced. Various algorithms are discussed and compared as they apply to microscopic images. This study concludes by discussing potential applications of these range-finding techniques for robotic vehicles and for part recognition in an automated conveyor system.

Introduction

Depth perception is a necessary attribute which allows higher-level organisms to visually interact with a three-dimensional environment. Humans utilize many different strategies, in parallel, to estimate the depth of their surroundings including texture gradient, size perspective, motion parallax, aerial perspective, relative upward locations in the visual field, occlusion effects, outline continuity, and surface shading variations [1], all of which can be emulated by one computer algorithm or another. Computers can further utilize specialized lighting schemes and time-of-flight techniques such as ultrasonics and lasers. Perhaps the most efficient techniques, from a computational time perspective and being free from the correspondence problems associated with multiple image techniques, are monocular techniques. Prominent among the monocular techniques is the practice of quantifying blur.

The range to an object may be determined based on the geometric properties of a lens system. For a given image, the lens system is focused at a certain depth and all points at other distances are blurred as a function of the optics. Figure 1 demonstrates the effect distance has on the focus of an object for an optic system with a constant focal length where U_0 is the distance from the point of perfect focus to the lens and V_0 is the distance from the lens to the image plane. If the distance from the lens to an arbitrary point in the scene U is greater than U_0, the distance from the lens to the point where the image is focused V is in front of the image plane.

Conversely, if an object is located closer than U_0, the image is focused behind the image plane resulting in blur. Large lens apertures are used to increase the resolution of the system by narrowing the depth of focus.

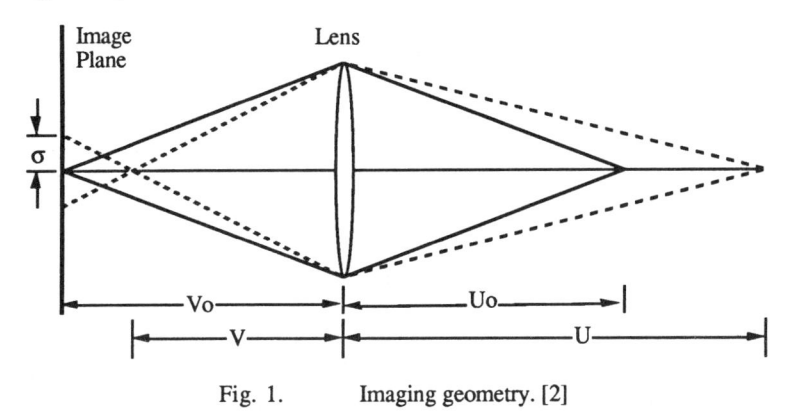

Fig. 1. Imaging geometry. [2]

Many different schemes are used to determine the focus sharpness, or the amount of blur, for an image. Most rely on statistical analysis to determine which len's setting returns the least amount of blur. It is also possible to employ a Fourier analysis to determine image sharpness.

For this study, to minimize computation time, simpler algorithms were tested and then improved upon. Four defocus quantization operators were analyzed to determine their suitability to this application. The first three are Entropy, Variance, and Sum Modulus Difference (SMD) suggested by Jarvis[3]. The fourth is the Edge Intensity Gradient Characterization (EIGC) technique developed by the authors of this study. The following is an example using 20µ microspheres in a microscopic-image processing system. The results are also applicable in the macroscopic world of robots and automation.

Variance

The variance V is a measure of the normalized difference between the average intensity and the intensity at each pixel along the edge intensity gradient. The variance is calculated by

$$V = \frac{1}{N}\sum_{i=1}^{N}[I(x_i) - \bar{I}]^2, \quad \text{where} \quad \bar{I} = \frac{1}{N}\sum_{i=1}^{N}I(x_i). \tag{1}$$

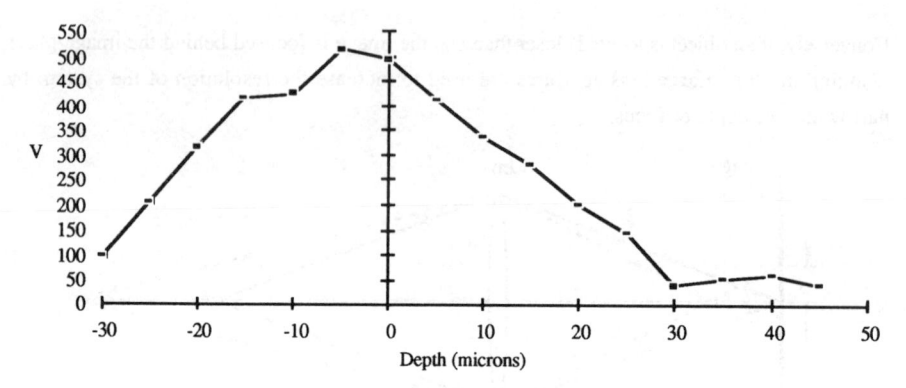

Fig. 2. Variance as a function of depth.

Figure 2 is a graph of variance as a function of depth for the 20 micron microsphere depth data ranging from 30 µm below the plane of focus to 45 µm above the plane of focus. There is a great deal of uncertainty as to the variance value in the -15 to 0 micron range. This would indicate that variance lacks the ability to consistently determine the depth of a particle located just below the plane of focus.

Entropy

Entropy is a quantitative measure of the disorder of a system. In this case, the system under examination is a series of intensity values. As the object is brought into focus, the disorder of the intensity values will decrease. The total entropy of a series of intensity values is calculated by

$$E = - \sum_{x=x_{max}}^{x_{min}} I(x) \ln(I(x)) \quad . \tag{2}$$

The negative sign is used for convention purposes. When the entropy for a group of depth data sets is plotted, the maximum corresponds to the data set most in focus.

The total entropy for the 20 micron microsphere depth data is plotted in Fig. 3. The resulting curve is fairly symmetric up to a depth of 25 µm above the plane of focus. The loss of symmetry is caused by the increased number of data points between x_{max} and x_{min} as the width of the edge slope increases. Perhaps normalizing the entropy with respect to the number of data points will correct this.

$$E_{norm} = - \frac{1}{N} \sum_{x=x_{max}}^{x_{min}} I(x) \ln(I(x)) \quad . \tag{3}$$

The normalized entropy (Fig. 3) failed to produce a well defined trend with respect to depth. For a depth quantification technique to be useful, it must be resistant to the effects of varying data set sizes.

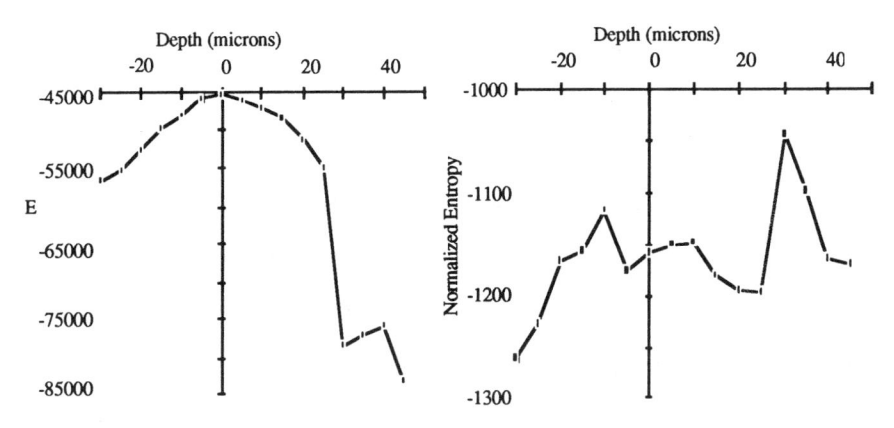

Fig. 3. Entropy as a function of depth.

Sum Modulus Difference (SMD)

The sum modulus difference (SMD) function simply calculates the absolute difference in intensity between two adjacent pixels and sums the differences over the data range or

$$SMD = \sum_{i=2}^{N} |I(x_i) - I(x_{i-1})| \qquad (4)$$

A graph of the SMD for the depth data (Fig. 4) reveals the same inconsistency at the 25 μm depth as the entropy function. Once the SMD is normalized

$$SMD_{norm} = \frac{1}{N} \sum_{i=2}^{N} |I(x_i) - I(x_{i-1})| \qquad (5)$$

and plotted (Fig. 4) the spike disappears. The curve is symmetrical except for a slight depression at -10 μm and a flattening above +30 μm.

The normalized SMD function provides an accurate means of determining the depth of a 20 micron microsphere edge under back lighting conditions. The only disadvantage to SMD is the computational complexity. To perform a SMD analysis of a slope of N pixels wide, approximately 2N+1 mathematical operations are needed, not including the operations needed to obtain the intensity data. The absolute value function also adds to the computation time.

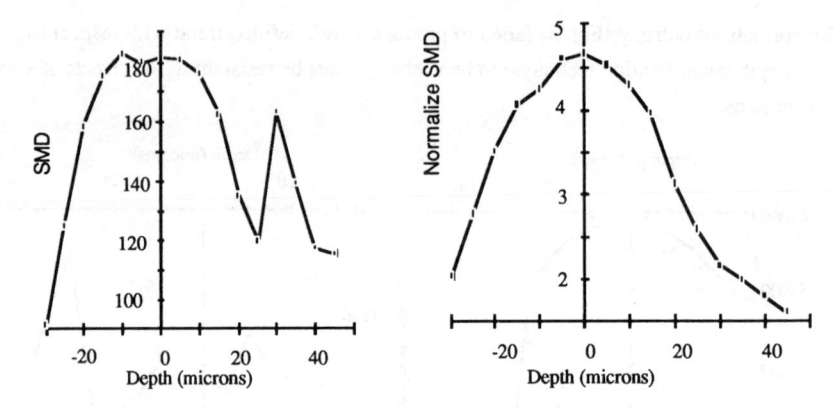

Fig. 4 SMD as a function of depth.

Edge Intensity Gradient Characterization (EIGC)

The basic concept behind EIGC is a quick and efficient method of parameterizing the effects of depth on the particle edge intensity contour slope. The slope $\overline{S_L}$ is a slope estimate of the edge intensity gradient inside a predetermined range (called the slope window W_s) about the position X_{mid} located between the minimum and maximum intensity locations x_{min} and x_{max} respectively. The EIGC slope is calculated by:

$$\overline{S_L} = \frac{1}{W_s}\left[I(X_{mid} - \frac{W_s}{2}) - I(X_{mid} + \frac{W_s}{2} - 1)\right] \quad (6)$$

$$X_{mid} = x_{max} + \frac{(x_{min} - x_{max})}{2} \quad (7)$$

where I(x) is the image eight bit intensity at position x. The EIGC slope calculation is very similar to the SMD calculation except that the sum difference is only calculated near the center of the edge intensity slope to decrease the number of computations needed to define the particle depth. For a typical 20 micron microsphere, the edge width approaches 40 pixels. The normalized SMD function requires 81 mathematical computations including 40 absolute value calculations. The EIGC slope function ($W_s = 5$) requires 8 mathematical computations to quantify the depth of the same microsphere without using the absolute value function.

Figure 5 is a plot of the EIGC slope ($W_s = 5$) as a function of depth for the 20 micron microsphere depth data. The curve is slightly asymmetric as seen in the dip at -10 μm and the curve flattening above 25 μm. These features are also found in the normalized SMD plot. The EIGC slope function provides a comparable level of performance to the SMD with less computational intensity when compared with the normalized SMD function.

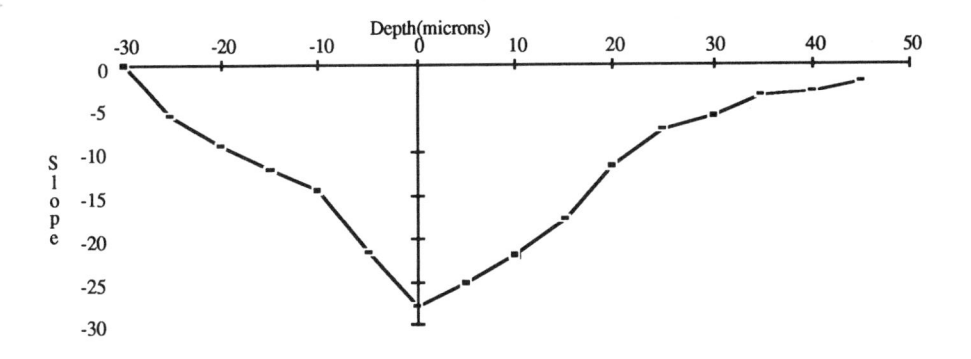

Fig. 5. EIGC slope as a function of depth ($W_s = 5$).

Conclusions

The EIGC technique is not applicable to all vision range finding problems but it is a powerful tool due to it's quick computational time and can be utilized in many situations. If a mobile robot were to use this tool, the optics must be set so that the robot focuses just past the "end of it's nose", thus cutting the calibration curve in half and leaving no doubt as to which side of the focus plane an object lies on. It must also be limited to known objects such as the yellow and black striped poles often bounding such a robot's territory. The technique really excels at part location on a moving conveyor. When one part overlaps another there will be no break to an edge following algorithm but there will be a discontinuity in the depth of the edge corresponding to the overlap. Single frame analysis means faster moving belts due to quicker inspection times, thus improving productivity.

References

[1] R. A. Jarvis, "A perspective on range finding techniques for computer vision," *IEEE Trans. Pattern Anal. Machine Intell.*, PAMI-5, pp. 122-139, Mar. 1983.

[2] A. P. Pentland, "A new sense for depth of field," *IEEE Trans. Pattern Anal. Machine Intell.*, PAMI-9, pp. 523-531, July 1987.

[3] R. A. Jarvis, "Focus optimization criteria for computer image processing," *Microscope*, vol. 24, pp. 163-180, 2nd quarter, 1976.

Real-Time Sensor-Based Approach to the Improvement of Transitions of Robot Manipulator Controls

Y. F. Li
Robotics Research Group, Dept of Engineering Science, Oxford University, UK

R. W. Daniel
Robotics Research Group, Dept of Engineering Science, Oxford University, UK

Abstract
This paper presents a sensor based control approach towards the improvement of transition for robot manipulator when it is required to stop or change its' control laws upon contacting a target. Fibre optic proximity sensor is used for external sensing and the signal filtering algorithms are presented. The *environmentally closed loop* control is achieved. The implementation is carried out on PUMA 560 robot manipulator with a transputer network providing real time computing power.

1 Introduction

Most of today's industrial robots are still *environmentally open loop* positioning systems in the sense that they operate with hardly any or no real time feedback at all from the external environment in which they are working. Such systems cannot provide satisfactory manipulations as both uncertainties about environment and inaccuracies of *internal* controllers may give rise to final positioning errors of the end point. Such positioning errors, even very small, can result in tremendous contacting force which may be damaging to the target or the robot itself [2].

In order to tackle the problem, direct proximity feedback from the end effector is needed and an outer or *external* closed loop is to be formed. Owing to the suitable working range, fibre optic proximity sensor is used here for the range sensing [3]. The nonlinearity of the sensor, However, has presented challenging problems in its' signal processing and therefore has severely limited its' practical application in real time robot control. This paper reports the technique adopted in dealing with the problem and some of the preliminary experimental results in integrating fibre optic proximity sensing with robot control. Special emphasis is placed on the filtering algorithms and their effects on subsequent control actions. The experiments are implemented on PUMA 560 robot, with a transputer network providing real time computing power [5].

Section 2 briefly describes the sensing devices used. In Section 3, filtering algorithms are presented and their similarities and differences shown. Section 4 gives a brief overlook of the control configuration. Section 5 contains a preliminary conclusion.

2 Fibre Optic Proximity Sensor

Fibre optic sensors, being insensitive to electromagnetic interference, small in size and light in weight, have been used in teleoperation [6], intelligent grasping [1] and precision assembly [4]. The reflectance type sensor is used here due to its high accuracy and suitable range.

The sensor output voltage is given by:

$$v = \frac{a}{d^2} + b \tag{1}$$

where a and b are constants and d is the sensor-target distance.

The distance information is contained in the voltage signal, which is nonlinear in terms of the distance and is contaminated by noise. Appropriate filtering algorithm is therefore essential before any attempt can be made to integrate the sensor with control.

3 Filtering Algorithms

Equation 1 represents the nonlinearity in the sensor output. To obtain distance information from such output, which is corrupted by noise, is a nontrivial task and special methods have been adopted towards this end [3]. Proposed and implemented here is an extended Kalman filter approach to the problem.

The system state equation is given by

$$d(k+1) = d(k) + t(k+1) + e(k+1) \tag{2}$$

where $d(k)$ is the sensor-target distance, $t(k+1)$ is the motion step along the direction perpendicular to the target between time (k) and $(k+1)$, and $e(k+1)$ is the motion noise introduced at step $k+1$.

The output equation is

$$v(k+1) = h(d(k+1)) + w(k+1) \tag{3}$$

where $v(k+1)$ is the output voltage of the sensor, $h(.)$ is the response curve of the sensor, characterized by equation 1 but obtained through off-line identification, and $w(k+1)$ is the sensor noise.

Here, we assume $e(k)$ and $w(k)$ for simplicity to be stationary Gaussian and independent and $E[e(i)e(j)] = Q\delta_{ij}$ and $E[w(i)w(j)] = R\delta_{ij}$.

Using extended Kalman filter, we arrive at the following equations:

Prediction:

$$\hat{d}(k+1|k) = \hat{d}(k|k) + t(k+1)$$

$$P(k+1|k) = P(k|k) + Q$$

Estimation:

$$\hat{d}(k+1|k+1) = \hat{d}(k+1|k) + K(k+1)(v(k+1) - h[\hat{d}(k+1|k)])$$

$$P(k+1|k+1) = P(k+1|k) - K^2(k+1)S(k+1)$$

with

$$S(k+1) = P(k+1|k) \left[\frac{\partial h(d)}{\partial d} \bigg|_{d=\hat{d}(k|k)} \right]^2 + R$$

$$K(k+1) = \frac{P(k+1|k)}{S(k+1)} \frac{\partial h(d)}{\partial d} \bigg|_{d=\hat{d}(k|k)}$$

This is the *distance based* extended Kalman filter. It works with acceptable accuracy. However it has proved to be very difficult to increase the speed of convergence experimentally. This is highly undesirable as the 'filtered' information is actually fed into the control law calculations and slow estimator convergence can lead to delayed or wrong decisions by the controllers, which in turn may result in disastrous actions being taken by the robot.

If instead of deriving the distance in the filtering algorithm, we filter the voltage signals first, followed by a voltage to distance conversion as a separate inverse non-linearity outside the filter feedback loop, we then arrive at a *voltage based* filter.

The state equation and output equation now become

$$v(k+1) = f[v(k), k] + e_1(k+1)$$

$$z(k+1) = v(k+1) + w_1(k+1)$$

where $v(k)$ is the sensor output voltage, $f[v(k), k]$ is a nonlinear state transition. $z(k+1)$ is the sensed voltage subject to sensor noise $w_1(k)$.

Taking the same steps as above, we can obtain the following filter equations:

Prediction:

$$\hat{v}(k+1|k) = \hat{v}(k|k) + F[\hat{v}(k|k)]t(k+1)$$

$$P(k+1|k) = \frac{\partial f}{\partial \hat{v}(k|k)} P(k|k) + Q_1$$

where $F[v(k|k)]$ is a non-linear state function associating robot state with the sensor state.

Estimation:

$$\hat{v}(k+1|k+1) = \hat{v}(k+1|k) + K_1(k+1)[z(k+1) - \hat{v}(k+1|k)]$$

$$P(k+1|k+1) = P(k+1|k) - K_1(K+1)S(k+1)K_1(k+1)$$

with $S(k+1) = P(k+1|k) + R_1$ and $K_1(k+1) = \frac{P(k+1|k)}{S(k+1)}$

Compared with the previous filter equations, we note that the *distance based* filter gain contains a nonlinear function of the state in $\partial h/\partial d$ whereas the *voltage based* filter gain does not. Though nonlinearity is introduced to the state equation in the *voltage based* case, this has barely affected the filtering, as its effect represented by the term $t(k)\partial F/\partial v$ is very small in our sensing range. This makes a big difference between the two filters, which contributes to the difference in their convergence speed.

3.1 Experimental results

The robot used is PUMA 560 and a transputer network is used as computing power for real time signal processing and control. The fibre optic sensor is mounted on the end effector of the robot. In this experiment, the robot moves in an *externally open loop* fashion. The robot command, generated off-line, is given in joint space and sent to the joint controller. The proximity sensor readings are taken at the same sampling rate as the inner loop or *internal controller*, 1000Hz.

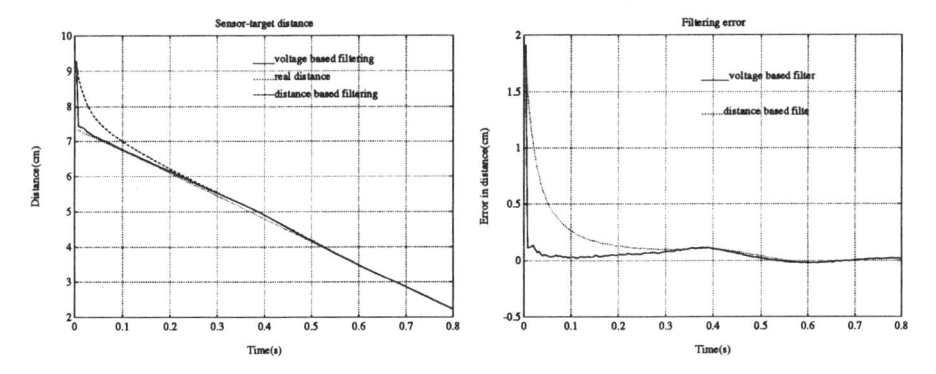

Figure 1: Distance

Figure 2: Filtering error

The performances of the two filters are shown in the experimental results of Fig 1 and 2. The *voltage based* filter outperforms the *distance based* in convergence speed, though their final accuracies are nearly the same.

4 External Control Configuration

The *externally closed loop* control is carried out using the same experiment set up previously except that the proximity sensor signals are now fed back to the *internal* controller in real time. . This experiment is aimed at integrating the proximity sensor into the outer or *external loop* so that end effector-target positioning errors can be compensated for directly. It is also a preliminary investigation into the effects that these filtering algorithms have on the performance of a real robot under *external loop* control.

The control is organized in a hierarchical structure, shown in Fig 3, which takes advantage of the joint sensing and maintains stable joint control all the time.

The dynamics of the robot manipulator can be represented in vector and matrix form as follows:

$$A(q)\ddot{q} + B(q,\dot{q}) + C(q) = T \qquad (4)$$

where : $A(q)$ is the mass matrix, $B(q,\dot{q})$ is the vector of Centrifugal and Coriolis terms, $C(q)$ is the vector of gravity terms, T is the vector of torques, q is the vector of joint angles, $\dot{q}$ is the angular velocity vector and $\ddot{q}$ is the angular acceleration vector.

The independent PD joint control scheme is used in the joint controller design. When an *external* loop containing the proximity sensors and the related processing modules is closed around the *internal* or low level control loops, a controller is needed in order to compensate for errors in the external loop and in order for the closed loop to achieve the desired performance. Proportional control is used in the external loop. The performance of the closed loop in one dimensional case is shown in Fig. 4. Also shown in Fig. 4 is the comparison of the responses with different filters in the *external* sensor signal processing. The small differences in filter performances can mean quite an appreciable difference to the internal control actions, as seen in Fig. 5.

The entire closed loop was realized with a sampling period of 1ms in a single rate fashion. Once the sensor-target distance is extracted in real time, a higher level command can be given at proper moment to instruct the robot to stop or change its' control laws.

5 Conclusion

The integration of proximity sensor with real time robot *external* control is important to its' fast and smooth transition. Fibre optic proximity sensor is suitable for such tasks and its'signal processing problems can be solved by extended Kalman filtering. The convergence speed as well as the accuracy of the filter is very important to robot control actions. Though the *environmentally closed loop* control has been achieved, the influences

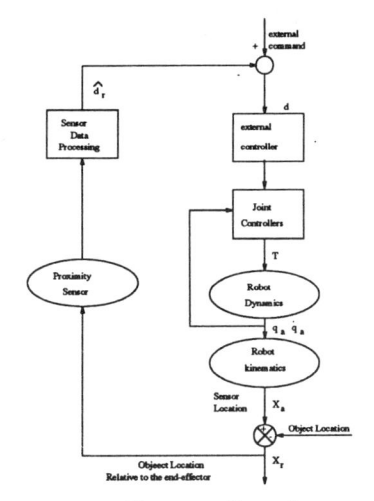

Figure 3: Control structure

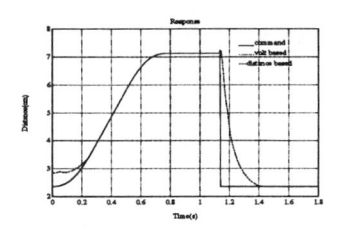

Figure 4: Robot external position

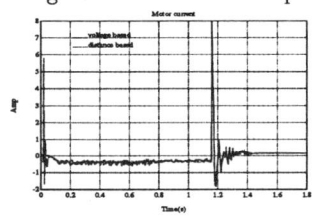

Figure 5: Input motor current

of accuracy of joint sensing on that of the *external* sensing remain to be tackled in the future.

References

[1] D. J. Balek and R. B. kelley. Using gripper mounted infrared proximity sensors for robot feedback control. In *Proc. IEEE Conf. Robotics and Automat.*, St louis, Mo, Mar 1985.

[2] P. Elosegui, R. W Daniel, and P. M Sharkey. Joint servoing for robot manipulator force control. In *IEEE Conf. Robotics and Automat.*, Cincinnati, Ohio, May 1990.

[3] B. Espiau. An overview of local environment sensing in robotics applications. In P. Dario, editor, *Sensors and Sensory Systems for Advanced Robots*. Springer-Verlag, 1988.

[4] P. Karkkainen and S. Pieska. Robot dynamic sensors based on fiber optic for precision assembly. In *7th Int. Conf. Robot Vision and Sensory Control.*, Zurich, Switzerland, Feb 1988.

[5] P. M Sharkey, R. W Daniel, and P. Elosegui. Transputer based real time robot control. In *IEEE Conf. Decision and Control.*, Honolulu, Hawaii, Dec 1990.

[6] C. Wampler. Multiprocessor control of a telemanipulator with optical proximity sensors. *Int. J. Robotics Research*, Spring 1984.

A New Active/Passive Robotic Precision Assembly System with F/T Sensor

Cai Hegao and Yang Ting

Robot Research Institute,
Harbin Institute of Technology, Harbin, 150006, P. R. China

Summary

In this paper, we present a new active/passive robotic assembly system based upon force feedback. It combines active and passive functions in one and the active or passive state of each joint can be set by program as necessary. The test setup we built is a macro/micro manipulation system. It is controlled by an IBM-PC computer through C language and macro-instructions. It is proved that the new system developed is really effective and available.

Introduction

The assembly of discrete parts is one of important aspects in today's industry, which consumes a great deal of time and labour. For years, assembly robot has attracted more and more research workers. So for, researchers have developed many kinds of assembly strategies and assembly-oriented equipments [1]. Generally speaking, they can be devided into three types. The first one is passive method, which usually uses a device, such as a RCC wrist[2], to absorb the position errors of mating parts passively. The second one is active method, which uses a controllable device, such as a HI-T Hand[3], to adjust actively during the parts mating process. And the third one is active/passive method, which is the combination of active and passive methods[4].

To sum up, all the three assembly systems mentioned above have their advantages and shortcomings. The passive system can complete the assembly with a simple system configuration and scheme in very short time. But it can only handle small misalignment between parts with chamfer and often produce large insertion force in the mating process. In addition, it has low adaptability to the changing work environments and can only work in vertical state. The active system can accommodate large initial position error even in case of an unchamfered part. It has higher adaptability in mating conditions and the inserting force can be drastically reduced. But the system is usually complex and expensive, and a comparably long insertion time is required due to long search motion. This hampers its applications in actual assembly processes. The active / passive system has most of the advantages of both the active and passive systems. It is believed to be an ideal method from the view point of practical use and reliability. But the adaptability and reliability of Presently available active / passive systems are still poor and can only work near the vertical state. This is because that the active and passive functions are independent each other in the system and the passive device used in system is usually a RCC wrist which is uncontrollable.

To find a really effective assembly method, we have developed an active / passive assembly system, which employs a specially designed active / passive micro manipulator(APMM) to combine both active and passive functions in one. Detail discription of the system will be given in the following parts.

Test Setup

In order to make the experiments of different assembly strategies. A test setup is built, which has the characteristics of good universality. It can implement different assembly strategies without any hardware modifications.

The active / passive micro manipulator(APMM)

The kernel of the system is the specially designed APMM (see Fig.1), which has four d.o.f.(two for translations and the other two for rotations) and each of them can be set either active or passive state by control as necessary. It is driven by four stepper motors and can move from −8mm to +8mm along X and Y coordinates and rotate from −8 to +8 degrees around the X and Y axses respectively. A special universal coupling is designed for the rotating transmission of X and Y axses, which can absorb the position change between two axses in three dimensions. A rolling track atructure is adopted to reduce the friction. In general, the APMM has mainly the following characteristics:
* Freqency response is 200 Hz
* Force output: $F = 16N$, $T = 1NM$
* The compliancy does not relate to the posture of APMM
* Compact structure and low friction

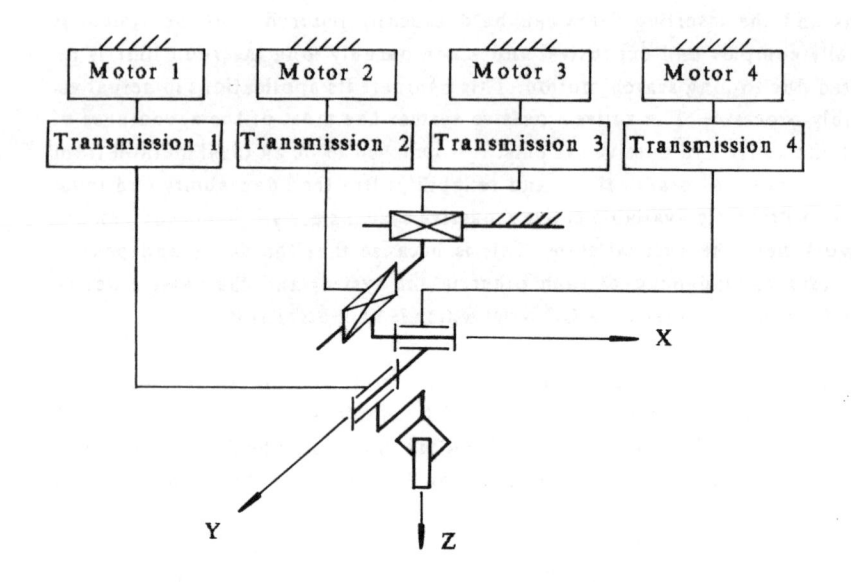

Fig.1 Conventional diagram of APMM

System configuration

The global configuration of the system is presented in Fig.2. The APMM is mounted on the end of a PUMA 562 robot and a F / T 30 / 100 sensor is attached to the bottom flange of APMM. An IBM-PC is connected with PUMA 562 controller, F / T sensor controller as well as the motor driver of APMM through a serial interface and a parallel interface respectively. The PUMA 562 and APMM are controlled by IBM-PC through C language and macro-instructions.

The system is a macro / micro manipulation system. When assembling ,the PUMA robot, the macro manipulator, takes care of the coarse positioning and the insert motion, while the APMM, the micro manipulator, executes the required fine position adjustment, based upon the the force information abtained from the force sensor. This combination makes the system have the quality of high precision and large workspace. In general, the new system has the virtues of:
  * Combining the active and passive functions in one and the active or

passive state is alterable
* Simple system configuration and scheme
* Can carry out assembly operation along any direction
* High frequency response (100Hz)
* High dexterity
* High reliability and good adaptability to the changing work environments
* Larg positioning error is allowable even in the case of unchamfered parts

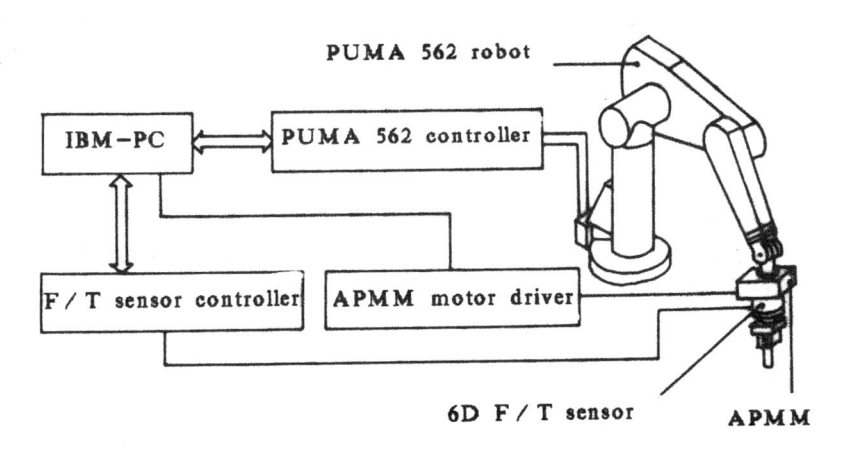

Fig.2   General configuration of system

Interface with the computer

In order to realize active / passive control, special attention has been paid to the interface of IBM-PC, the main controller of system, with F / T sensor controller and motor driver of APMM. An interface board, which has the quality of optical-isolation, has been made as shown in Fig.3. Where U_1 is 74LS245, U_2 is decoder circuit, U_3 is 8253, U_4 is 8255, U_5 is optical isolation and U_6 is motor driver. During assembly, 8253 provides motor's driving pulse, 8255 controls the rotating direction and working state of motors according to the force information from F / T sensor.

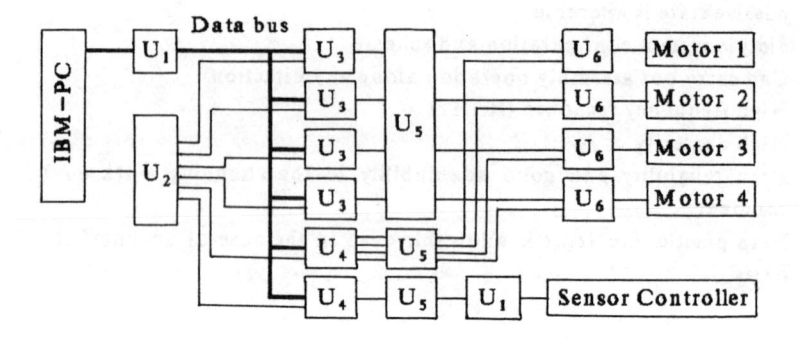

Fig.3 Block diagram of interface between computer and peripheral equipments

Assembly Strategy

Assembly strategy is one of the key problems for achieving successful assembly. Up to now, several assembly strategies have been developed. But, as we know, neither of them can work well enough to be widly used in practice with satisfaction. One of the main causes is that the active and passive operation have not been combined effectively and naturally. As a matter of fact, a really ideal assembly manipulation must be a process of tacit cooperation of active and passive accommodations, which is similar to the assembly operation conducted by human beings. From this point of view, we have developed an assembly strategy imitating human's operation based upon the close study of mating process. In general, the strategy is divided into two phases, hole search and peg insert.

In the first phase, we adopt the strategy similar to human's operation. We subject the peg to an angular rotation in a certain direction and let the lowest point of peg's front edge go into the hole at first. Then rotating the peg in the direction of reducing the angular offset. Meanwhile, PUMA 562 and the other joints of APMM cooperate with the angular modification mentioned above according to the force information and control algorithm in a proper way. By this, we can make the front edge of the peg into the hole.

We begin the second phase with a passive assembly operation, i.e. all joints of APMM are set passive before executing the peg's insert command. During the inserting, we adjust the mating states by proper configuration regulation along the corresponding coordinates based upon the force information and keep the assembly from wedging condition till the signal of "finish" has been received.

In addition to the strategy mentioned above, we have also made some experiments of other assembly strategies such as vibration method. That is to make the APMM vibrate in a certain way during mating process by program. This method has also well performed.

Conclusions

Assembly is one of the key areas for achieving productivity improvements in today's industry. Nowadays, robotic assembly has become a highly important aspect of robot industrial applications. We present here a newly developed active / passive assembly system, which has the virtues of both active and passive assembly methods. Differing from the active / passive systems of the past, it combines the active and passive founctions in one. Hence the system is fully controllable and can conduct assembly operation along any direction. The system we built is a macro / micro manipulation system. It employs a PUMA 562 robot as macro manipulator and a specially designed active / passive micro manipulator(APMM) as the micro one which is mounted on the end of PUMA 562 robot. The whole system is controlled by an IBM-PC computer through C language and macro-instructions. The experiments of peg-in-hole have been done successfully with clearance of 0.01mm within several seconds. It is proved that the new system is really effective and available.

References

1. H. S. Chao; H. J. Warnecke; D. G. Gweon: Robotic assembly: a synthesizing overview. Robotica 5 (1987) 153-165.

2. Daniel E. Whitney: Robot motion: planning and control. The MIT Press 1982.

3. T. Goto, T. Onoyama and K. Takeyasu: Precise insert operation by tactile controlled robot HI-T-HAND Expert 2. Proc. 4th international symposium on Industrial Robots (1974) 154-160.

4. J. Simons, H. Van Brussel: Robotic Assembly. Springer-Verlag 1985

> # Extending the Productivity of Industrial Robots by Efficient Integration of a Novel Force Torque Sensor

Juergen Wahrburg

ZESS (Center of Sensory Systems) and Institute of Control Engineering
University of Siegen, Siegen, Germany

Abstract. Advanced manufacturing tasks such as assembly operations or machinig of complex workpieces have only to a very small extent been solved by use of industrial robots so far. In order to contribute to automated solutions of these problem areas the design of a novel force-torque sensor system and its integration with standard industrial robots are presented. Offering a high flexibility and powerful signal processing capabilities the sensor system proves to be a valuable means to enlarge the features of robotic systems.

INTRODUCTION

Most of all robots presently used in industrial automation are not yet equipped with external sensors, but work as purely position controlled machines. However, the ongoing automation process leads to an increasing number of more complex problems which are not solvable in this way. This holds especially for an enormous variety of different manipulation and machining tasks where the robots make direct contact with other objects. In these cases the robots must be capable to adapt their motions automatically to unpredictable tolerances or positional deviations that may arise in their environment. Furthermore some applications demand the fulfillment of certain production parameters, e.g. of contact forces in a grinding or deburring process, in order to assure a constant quality of the output products.

Although the use of appropriate sensors, particularly of force-torque sensors, suggests itself as a solution to these problem areas, the realization of this idea leads to some fundamental difficulties. Several valuable papers concerning this field have already been published, see for example [1]...[6], but still very few sensor-based robots are used in industrial applications so far. A major problem arises due to unsolved questions concerning the interface between sensors and robot control systems, because often neither sensors nor robots have the necessary features to be connected efficiently.

Focussing on the use of force-torque sensors which are of special importance in assembly and machining tasks, this paper presents some basic ideas to improve robot-sensor integration by a more general approach. They are outlined by describing the design of a smart force-torque sensor as well as its integration into robotic systems. A main objective of our work is to be compatible with standard robots in the sense that using the sensor requires no or only minor modifications of industrial robot controls.

USE OF FORCE-TORQUE SENSORS IN ROBOTIC SYSTEMS

Areas of application and present state of the art

Most of the force-torque sensor designs currently known are 6-component measuring devices which are mounted at the wrist of the robot as a link between arm and gripper. In a threedimensional coordinate system they allow to monitor the respective three interaction forces and torques between the tool of the robot and a fixed workpiece or between a workpiece handled by the robot and a stationary environment. In the following text we concentrate on stiff force-torque sensors which have no compliance and usually use strain gages as transducers. They can enhance the flexibility and productivity of industrial robots in several fields of application, as for example

- assembly operations,
- grinding and deburring,
- contour following,
- adjustment of the absolute robot position,
- monitor and savety functions.

The potential sensor related improvements are opposite to great efforts which have to be made to achieve satisfactory solutions from a technical as well as from an economical point of view. The technical challenges of force-torque sensor applications result from two main reasons.

- It is a demanding task to process the six decoupled force-torque values provided by the sensor in such a way that they change the motion of the robot into the proper direction. To that an appropriate control algorithm has to be developed to transform the sensor information given in the force-torque space into new reference positions and orientations in the robot coordinate system.

- Sensor signal processing takes place in a closed control loop and must be carried out online during the motion of the robot. Therefore only very fast response times can avoid stability problems and possible damages of the workpiece or the machining tool [7].

At present force-torque sensors are still rarely found in industrial automation. Available sensors as well as robot controllers still must be regarded as separate devices, none of which provides the features to implement the additional operations that are obligatory for efficient signal processing. A force-torque sensor supplying six sensor-referenced output signals does not match the limited sensor functions of industrial robot controls, because the latter usually are not fitted with interfaces for multicomponent sensors, cannot accept the input of six single-component analog sensors simultaneously, and only offer proportional amplification to process the input signals. Furthermore the attainable cycle times which for most industrial robots are already in the range of 30 msec to 100 msec without taking into account any signal delays within the sensor nor the signal transmission times, still prevent the performance of fast sensor guided motions.

Basic concepts for an improved robot-sensor interface

As stated above, satisfying solutions of advanced automation tasks which actually do need sensor support, can hardly be found by a direct connection between force-torque sensors and standard robot controls due to the lack of common physical as well as logical interfaces. Corresponding enlargements of existing systems have not been carried out, because on the one hand industrial components are closed systems allowing no user modifications, and on the other hand the support of many different, application dependent product versions is uneconomic and too costly for robot and sensor manufacturers.

A possible outcome can be obtained by introducing an additional system as a connecting link between sensor and robot control, as shown in Fig. 1. Fitted with its own microcomputer and customized interfaces such a

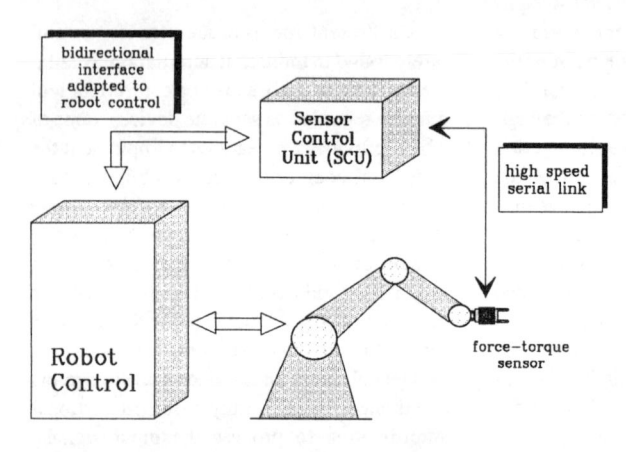

Fig 1. Connecting sensor and robot via a sensor control unit

system which will be called '**sensor control unit** (SCU)' in the following text, is well suited to perform more extensive calculations related to sensor signal processing. Although this approach may look quite simple at a first glance, several points come out in detail which must be considered carefully. As an illustration some essential problem areas are listed in the following.

- Division of sensor related signal processing. It is advantageous to split the arising operations into two layers. We propose to first furnish the force-torque sensor itself with a built-in electronic to carry out the basic functions of signal amplification and adaptation. It is located as close to the transducers as possible and thereby minimizes the influence of disturbances caused by electrically noisy environments. Furthermore it offers a means to optimize the data transmission to the sensor control unit, which forms the second layer to carry out computing-intensive tasks.

- Design of architecture and interfaces of the SCU.

The efficient use of the SCU presumes that it can easily be adapted to different industrial robots as well as to different applications.

- Consideration of the operating instructions of the SCU. Operating of the SCU should be possible on such a level that the same staff who program and operate industrial robots is also capable to use the SCU.

- Achievement of very short cycle times.

As already mentioned, the attainable motion speed of sensor guided robot movements is severely limited due to signal delays within the robot control systems. Therefore the additional computational deadtimes of external sensor signal processing units should be as short as possible.

The next chapters outline the characteristics of a novel force-torque sensor and a SCU which have been designed according to these objectives.

FEATURES OF A NOVEL 6-COMPONENT FORCE-TORQUE SENSOR

The development of the sensor electronic is adapted to a new mechanical force-torque sensor design which is under construction at the swiss firm KISTLER AG. Forces and torques are measured by detecting the deformations caused by the applied load in four elastic cantilever beams. Sixteen strain gages of which four are mounted on each beam serve as transducers. The new mechanical design assures high sensivity and excellent decoupling of the six force-torque components [8]. Cross coupling errors are less than 2 %, linearity errors less than 0.4 % for a measuring range of 200 N (forces) and 7 Nm (torques). A mechanical overload protection for each component prevents damage of the sensor in case of high overloads.

The electronic can be packed on a single printed circuit board by use of SMD (Surface Mounted Devices) - components and fits completely into the sensor, as shown by the photo in Fig. 2. According to Fig. 3 its elements may be divided in an analog part which performs the amplification of the transducer signals, and a digital part, the core of which is an 8051 microprocessor. Its serial port is used to establish a serial link to the SCU according to the RS-485 specifications with an attainable baudrate of 375 Kbaud. This type of connection offers the advantages of a sufficiently low transmission delay, need of only two wires (one twisted pair cable), and bidirectional data exchange enabling the SCU to transmit commands to the sensor electronic. The major functions of the sensor electronic controlled by the 8051 microprocessor inculde the following.

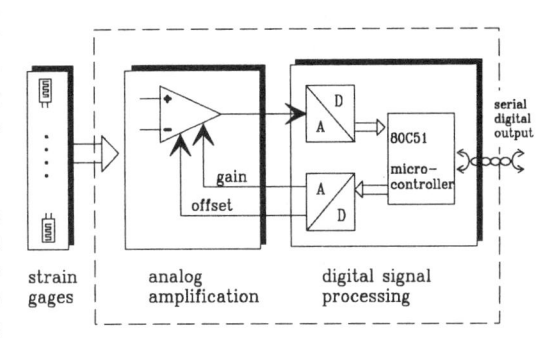

Fig. 3. Hardware of the sensor electronic.

Fig. 2. Photo of the KISTLER force-torque sensor and its integrated electronics

Measuring mode.
- Offset- and gain-setting of the analog amplifier,
- transformation of the transducer signals into digital force-torque values with 12 Bit resolution,
- transmission of force-torque values.

Execution of control commands.
- Compensation of payloads (taring function),
- selftest and plausibilty checks,
- calibration of selected force-torque components.

A high data rate is achieved as the measuring process and the transmission of the results are carried out in parallel. The cycle time to record and transmit all six components is 1 msec for the standard version of the electronic. It can be further reduced if only a selected number of less than six components has to be measured, and if a faster version of the microprocessor is used.

SENSOR CONTROL UNIT (SCU) TO CONNECT SENSOR AND ROBOT

At first the SCU fulfills the basic function of an interface transducer by transforming the sensor information received on the serial line to robot compatible signals. Easy adaptability to specific requirements is achieved by a modular hardware design. As shown in Fig. 4, different I/O-boards can be connected to the CPU through a common system bus. They have been designed to meet well accepted industrial standards, as for example decoupling via optocouplers and 24 V power-supply in case of digital input/output signals.

are incorporated into the motion program of the robot will cause the desired behaviour of the SCU at exactly the proper time. To give an overview only the main functions of the SCU can be listed schematically within the scope of this paper.

- Bidirectional data exchange with the sensor electronic and the robot control,
- dynamical selection of the force-torque components to be actually measured,
- monitoring of given thresholds to which corresponding binary outputs are assigned,
- possibility to record the force-torque signals in internal memory for a detailed analysis of fast transients,
- powerful signal processing capabilities to perform coordinate transformations, filtering, and control algorithms.

The signal processing within the SCU can be accelerated considerably if it is furnished with an optional coprocessor. A TMS 320C25 signal processor reduces the execution times even of extensive calculations down to the microsecond range, for example if a coordinate transformation is performed which transfers the force-torque values measured in the coordinate system of the sensor into a coordinate system used by the robot control. The sum of all processing delays appearing in the sensor electronic and the SCU is at least one magnitude smaller than the deadtime of available robot controls. If in certain applications no way appears to circumvent the time lag of the robot system, the SCU can also be used to control peripheral actuators in the environment of the robot which may have faster response times.

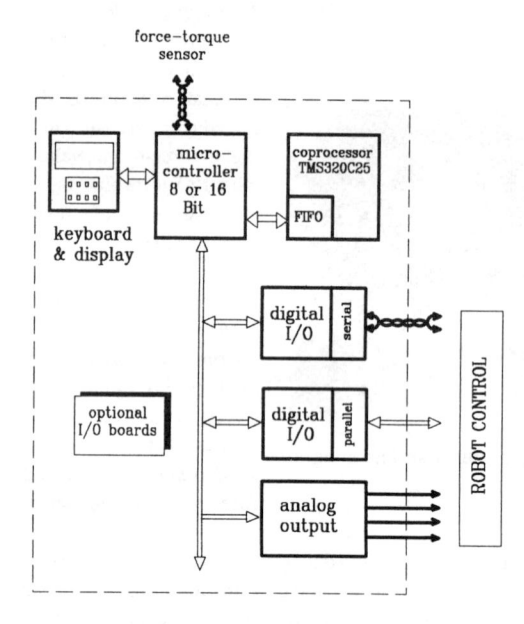

Fig. 4. Hardware of the SCU.

Apart from its flexible hardware configuration the main advantages of the SCU result from the signal processing capabilities implemented in software. It is important to note that all functions of the SCU can be initiated dynamically by the robot control if a bidirectional link to the SCU is established. Corresponding control commands and parameter values which

FIRST EXPERIMENTAL RESULTS

A test installation of prototype versions of the force-torque sensor and the SCU has been carried out with an ABB robot, type IRb 2000.

The bidirectional communication between the S3 - robot controller and the SCU takes place via parallel digital I/O-lines, divided into groups of eight lines which form one-byte ports. As the S3 - system can handle up to three analog sensor inputs, a maximum of three force-torque components can be processed simultaneously. This seems to be sufficient so far because the selected components can be adapted dynamically to the current situation due to the programmability of the sensor system. Furthermore another three components are independently selectable as binary sensors that flag the crossing of programmed thresholds. These outputs are wired to dedicated interrupt inputs of the S3 - system which cause the stop of the robot in the shortest possible time. Signal processing in the SCU has been tailored to the sensor adaptivity functions of the S3 which allow distance and direction searching, speed control and contour tracking.

The functionality of the integrated robot-sensor system has been successfully tested in an industrial environment. At present an installation of the system is in progress to perform a sophisticated deburring task.

CONCLUSION

The experiences which could be gathered up to now turn out that the flexibility of the presented approach can significantly enhance the integration of force-torque sensors and industrial robots. The versatile SCU greatly reduces the interface problems and provides extensive signal processing capabilities which are not offered by standard robot controls. A remaining restriction particularly on applications characterized by fast robot motions results from the still fairly long cycle times of industrial robot systems. However, the next robot generation surely will have improved technical data concerning this point, which will allow to further extend the positive results of previous installations of the force-torque sensor system.

References

1. Whitney, D.E., "Historical perspective and state of the art in robot force control." Proc. IEEE Conf. on Robotics and Automation, St. Louis, 1985, pp. 262-268.
2. Hirzinger, G., "Robot systems completely based on sensory feedback." IEEE Trans. Ind. Electr., vol. IE-33, May 1986, pp. 105-109.
3. Inigo, R.M., and R.M. Kossey, "Closed-loop control of a manipulator using a wrist force sensor." IEEE Trans. Ind. Electr., vol. IE-34, Aug. 1987, pp. 371-378.
4. Tzafestas, S.G., "Integrated sensor-based intelligent robot system." IEEE Control System Magazin, vol. 8, Apr. 1988, pp. 61-72.
5. Freund, E., and Ch. Bühler, "Robot control in manufacturing: Combining reference information with online sensor correction." Preprints 6th Symp. on Inform. Contr. Probl. in Manufact. Technology, Madrid, Sept. 1989, pp. 515-521.
6. Van Brussel, H., "Sensor based robots do work!", Proc. 21st Int. Symp. on Ind. Robots, Copenhagen, IFS Publications, Bedford 1990, pp. 5-25.
7. Wahrburg, J., "Integrating multiple sensors and industrial robots: System architecture and control aspects." Proc. 3rd IEEE Symposium on Intelligent Control, Arlington, Aug. 1988, pp. 165-169.
8. Cavalloni, C., "A novel 6-component force-torque sensor for robotics." Proc. Int. Conf. on Advanced Mechatronics, Tokyo, May 1989.

Expert Systems and Artificial Intelligence

EGRD-Expert Gear Reducer Designer

by
Thomas G. Boronkay, Max L. Brown, Craig Vanderhorst
University of Cincinnati

ABSTRACT:

The design of a Gear Reducers involves the selection of shaft configurations and gear types to be used for the particular application as well as the analytical computations, required for sizing the gears and shafts, and the selection of the bearings. This paper describes the EGRD Expert Gear Reducer System which performs the above functions automatically, with minimum user input.

INTRODUCTION:

A gear reducer is one of the most commonly used mechanical devices for transferring power from a prime mover to the driven machine or mechanism. The design of a Gear Reducer requires several repetitive and tedious calculations in order to achieve functional, size, cost and other objectives. Many of the decisions and choices made by the designer are empirical, based on experience and company standards.

Rule based Expert Systems, originally applied mostly to diagnostic problems (1), are increasingly used in the field of design (2). The knowledge of the designer is captured in the form of "rules" which then direct the logic flow of the design process.

The EGRD Expert System presented in this paper, consists of five modules: Expert, Gear Design, Gear Material, Shaft Design and Bearing Selection and Automatic Drawing modules. All of these modules are interfaced and exchange data and information as required. EGRD generates a design, and produces tabular and graphical outputs of a Gear Reducer with minimum user input.

PROBLEM OVERVIEW:

The design of a gear reducer is a very engineering intensive process. In addition to conforming to the specified geometric configuration, the reducer and its components have to be designed to perform adequately, that is, has to be able to transmit the specified torque, provide the desired speed ratio as well as satisfy cost and size requirements. In addition, gear design must conform to AGMA (American Gear Manufacturers Association) standards. A number of design decisions are based on the experience of the designer and not on analytical computations. Different engineers can generate different designs to perform identical functions, even within the same company.

SOLUTION:

The guiding philosophy was to approach the problem the same way as a human expert. The knowledge used in designing gears, shafts and selecting bearings was captured in knowledge bases. These knowledge bases incorporate rules for design, material selection and automatic drafting.

SYSTEM DESCRIPTION:

As depicted in Fig. 1, EGRD is comprised of the following modules:

Expert -This module is the core of the system. It contains the rules pertaining to the determination of the required number of gear sets, selection of the gear type, calculation of the first estimates of the length of the various steps on the shafts. The other function of the Expert Module is to manage and direct the data flow between the various modules. This module was created by using the *PCPlus Personal Designer* (3), expert system shell. It contains two hundred and thirty six rules.

Gear Design -This module consists of four separate gear design programs. They are used to calculate the various gear tooth parameters for strength, wear and dynamic loads for spur, helical, bevel and worm gears. The gear programs were written in *Basic*.

Gear Material Selection -This module uses the *DBase* data base program to select the gear material compatible with both strength and other user input requirements.

Shaft Design and Bearing Selection-This module calculates the diameters of the various steps on the shafts, selects the bearings from an internal database, calculates the size of the key, if used. The bearing bores are checked against the appropriate shaft diameters and, if needed, recalculated to make them compatible with the bearings. This module is also written in Basic.

Automatic Drawing -This module produces solid models of the assembly as well as the detail drawings of the gears and shafts. It uses script files generated by the *Silverscreen* (4) solid modeling software.

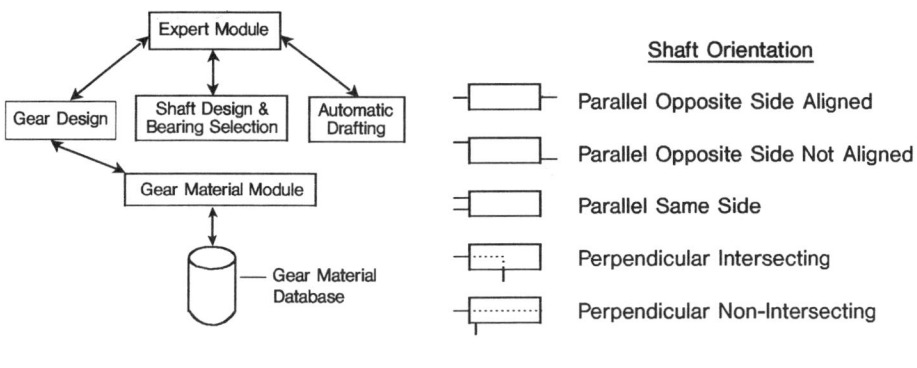

FIG. 1-EDRG System

Fig. 2-Shaft Configuration

SCOPE:

The EGRD system is capable of generating reducers with shaft configurations shown in Fig. 2. Reducers using spur, helical, bevel and worm gears, can be designed by the system. In its current version, the system generates a one or two step reducer. The two step reducer is a reverted gear train. The type of bearings included in the database is limited to ball and tapered roller bearings. The material database contains seventy materials.

GEAR REDUCER DESIGN PROCESS:

The design session starts with the system querying the user for the following inputs:

-Cost Factor
-Size Factor
-Input RPM
-Output RPM
-Acceptable Error in Output RPM
-Shaft Configuration.

Based upon the above input the Expert Module decides on the type of gears (i.e. spur, helical, etc.), the number of steps in the reducer and directs the appropriate gear design module to design the gears. At this time, the user is asked by the Material Selection Module to input his first estimate of the material type and properties to be used. The search of the material database returns a list of materials satisfying the requirements. After the user selects the material, the Gear Design Module proceeds with the tooth design based on both the bending dynamic and wear loading on the gears. In case the wear and/or the dynamic loading conditions are not satisfied, the user is asked to choose either to modify the geometry or the material. In the first case, attempts to reduce the loading are made by modifying the face width, and/or the diametral pitch, iteratively. If the user chooses to change the material and/or the hardness, in order to increase the surface endurance limit, iterations between the Material Database and the Material Selection Module produces a new set of materials for the choice of the user. These steps can be repeated until a satisfactory solution is found or the user decides to try to modify the geometry.

Once the gear design has been completed, the Expert Module passes the gear force data and initial estimated values of the lengths of the various steps of the shafts to the Shaft Design Module. The user is asked to input the type of loading (i.e. cyclic, shock, etc.) as well as the desired life for the bearings. The outputs of this module are the lengths and diameters of the steps of each shaft and the parameters of the bearings. The tabular results for the gears, shafts and bearings are returned to the Expert Module which prints them.

Analysis:

Pinion Shaft Speed	3000	rpm
Gear Shaft Speed	1000	rpm
Diametral Pitch	10	
Number of Pinion Teeth	18	
Pinion Pitch Diameter	1.8 "	
Gear Pitch Diameter	5.4 "	
Face Width	1.0472 "	
Addendum	.1 "	
Dedendum	.125 "	
Whole Depth	.225 "	

Parameters:

Ratio	3
Tooth Form	20_Deg_Per_AGMA_201.02B
Start-Up Horsepower	10
Operating Horsepower	9
Pinion Material	Alloy Steel
Pinion AISI No.	1340
Recommended Heat Treat	Carburizing
Resulting Brinell Hardness	250
Hardenability	Good
Machinability	Good
Pinion Material Basic Stress (PSI)	41667
Gear Material	Alloy Steel
Gear AISI No.	1340
Recommended Heat Treat	Normalized, Quench & Temper
Resultong Brinell Hardness	230

Fig. 3-Gear Train Designer   Fig. 4-Spur Gear Output

Fig. 3 and Fig. 4 depict typical results. Finally, pertinent geometric data is passed to the Automatic Drafting Module which produces the models using parametric design techniques. Fig. 5 and Fig. 6.

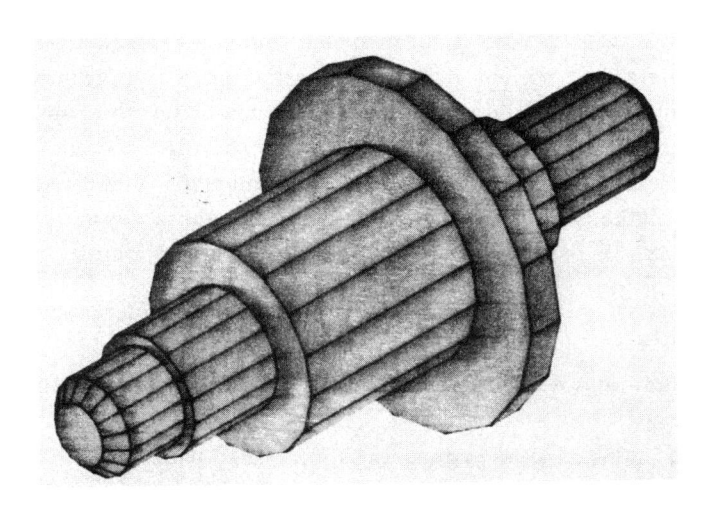

FIG.5-Output Shaft

FIG.6-Assembly

CONCLUSIONS:

Comparing results produced by the EGRD system with ones obtained by traditional methods prove that the concept is valid. The design of a reverted geartrain consisting of four gears, three shafts and six bearings takes approximately four to ten minutes, depending on the number of iterations needed during the design process. Generation of the three dimensional solid models take an additional three to twenty minutes each, depending on the type of rendering desired. These results represent a substantial increase in productivity compared to the traditional methods. Another benefit of such a system is the standardization of the design within an office or an entire company, independent of the engineer using the system.

FUTURE WORK:

Expansion and improvement of this system is being considered in the following areas:
-Increase the number of reduction steps, i.e. gear pairs
-Increase the type of bearings supported by the system and expand the bearing database
-Expand the material database
-Consider manufacturabilty during gear type selection
-Modify the shaft material selection subroutine to allow the shaft materials to be selected from the system's material database.

REFERENCES:
1. Harman & King, Expert Systems; Wiley.
2. Coyne, Roenman at al, Knowledge Based Design Systems; Addison-Wesley.
3. PCplus Personal Designer; Texas Instruments Inc.
4. Silverscreen Manual; Schroff Development Corp.
5. Gear Design Handbook; Buckingham
6. Spotts, Design of Machine Elements; Prentice Hall.
7. Mike Rich, Knowledge Based Engineering Gear Design; Autofact '90 Conference Proceedings.
8. K. Ehrenspiel & P. F. Tropschuh, Anwendung eines wissensbaierten Systems fur die Synthese Beispiel: Das Projektieren von Schiffsgetrien; Springer; Ferlag,1989.

Application of Expert Systems to Process Planning in Manufacturing

Derya Alasya
Engineering Management Department
Old Dominion University

ABSTRACT

The purpose of this project is to illustrate expert system applications and benefits to process planning in a small manufacturing environment. A system is developed for a manufacturing company producing transformers. This system generates a process plan for a transformer given its specifications. The resulting process plan includes an operation sequence, machine types, shuttle sizes, and the processing time for each operation.

INTRODUCTION

To maintain competitiveness in national and international markets, it is mandatory for companies to adopt new technologies in their design and manufacturing functions to be more efficient. Process Planning is one of the most important and complex functions serving as a "bridge" between the design and manufacturing functions, which has been a target area for new technology applications. To date, many computer-aided process planning (CAPP) systems have been developed. However, many of these CAPP systems have fallen short because they are not able cope with the complexity of the problem in a typical manufacturing environment. These traditional CAPP systems require extensive human intervention and are difficult to modify in the case of changes in manufacturing systems. In addition, considering that process planning requires human expertise and heuristics, it creates a great deal of potential for expert systems applications to overcome most of the current problems with the existing systems.

In this project, an expert system is designed and developed for a small manufacturing company called Talema Electronic Incorporation producing toroidal transformers to illustrate the application of expert systems in a job shop environment.

BACKGROUND

Talema produces torodial transformers which differ from conventional transformers in some ways; They are smaller in size and weight (about 50% of conventional transformers), and they have low radiated magnetic field which makes them ideal for compact power suppliers. The operations required for manufacture of a torodial transformer are:

1. Construction of torodial cores by winding a thin electrical steel over a mandril. The cores are annealed at a high temperature for a period of time depending on the size of the core.
2. Windings of primary and secondary levels based on the specifications.
3. Wrapping mylar insulation material between the windings, and after all the winding operations are completed.
4. Attachment of the cores with wire leads.
5. Quality inspection followed by packaging.

Talema has four groups of machines operated manually by the production workers to perform these operations:

1. One machine to construct a torodial core for transformer.
2. Four different types of winding machines in total of 9.
3. Two taper machines for mylar operation.
4. One quality test equipment.

Each winding machine has a shuttle to hold the wires to wind on the cores; type2, type3 and type4 machines have more than one shuttle with different sizes (Table 1). The size of a shuttle is measured by the wire capacity, and the capacity of each shuttle is dependent of the wire size and the number of wire turns.

Machine	Wire Size
type1	- 8
type2	9 - 17
type3	17 - 24
type4	25 - 35

Table 1. Wire Size Ranges for Winding Machines.

Talema is a typical small manufacturing company where most of the manufacturing functions are carried out manually by production supervisors and workers. Determining the process plans for transformers is one of the manual procedures in Talema. The production manager receives a design sheet showing the transformer specifications and a diagram of a transformer indicating primary, secondary levels, and the voltages for input and output. The number of wire turns and wire sizes required for each level are also included on the design sheet. The production manager uses this information and her expertise to:

1. Select the most appropriate machines for all operations and the shuttle sizes to meet the given specifications.
2. Determine a sequence of operations with processing times.

This is a time consuming process, and it becomes complicated in the occurence of changes. As a possible solution, an expert system is designed and developed as a decision making tool to simplify the process planning function in Talema.

DEVELOPMENT OF THE SYSTEM

The system is implemented using an expert system tool called M.1 (Version 2.0, 1986), a product of Teknowledge. M.1 is a rule-based system and uses backwardchaining inference strategy. The knowledge-base of the system consists of facts and rules.

Facts. The facts were developed based on the company documents on the existing machines and processes. The following facts are included in the knowledge-base:

1. Wire size range for each machine.
2. Available shuttle sizes for each machine.
3. Predetermined processing times.

Facts are represented in the following forms (words in capital are used in place of actual numbers):

machine(WIRE SIZE) = MACHINE TYPE   (1)

which states the wire size for a specific machine type.

shuttle(MACHINE#, WIRE-SIZE, TURNS) = SHUTTLE-SIZE   (2)

that assigns the shuttle size for a specific machine type based on the wire size and the number of wire turns.

$$\text{process(MACHINE\#, SHUTTLE-SIZE, TURNS)} = \text{PROCESS-TIME} \qquad (3)$$

where this fact shows the predetermined processing times for a specific machine type based on the shuttle size, and the number of wire turns.

Production Rules. The production rules of the knowledge-base were constructed based on the production manager's expertise. A number of interviews were conducted to model the production manager's problem solving behavior. Many heuristics and rules of thumbs were used in selecting the appropriate machines and the operation sequences. The basic structure of the production rules are as follows:

    if wire-size = S and
      machine = M and
    then machine-number = M

which determines the machine type.

    if machine-number = M and
      wire-size = S and
      wire-turns = WT and
      shuttle(M,S,WT) = SS
    then shuttle-size = SS

which determines the shuttle size.

Consultation. The system provides a highly interactive consultation for the user. The explanation facility provided by M.1 enables the user to ask questions about the system's reasoning process during a consultation. When the system is loaded, the following information is required from the user:
    1. Number of primary and secondary levels.
    2. Number of wire turns for each level.
    3. Wire size for each level.

Once the necessary information is entered, the system automatically starts manipulating its knowledge-base and displays the process plan generated which contains:
    1. The operation Sequence.
    2. Machine number for each operation.

3. Shuttle size.
4. Processing time for each operation.

CONCLUSION

The acceptance of an expert system before its implementation heavily depends on the results of the system evaluation and validation. The evaluation of this system was continously performed throughout the development process when a piece of knowledge added, deleted, or some changes made in the structure. The system's performance was tested for various transformer orders to evaluate the viability, generality and the efficiency of the system. Process plans generated by the system were compared to those generated manually by the production manager. The comparison was made for both conclusions and the reasoning processes to arrive at these conclusions. The evaluation results have shown that the system's performance was at the level of expectations. As a concluding statement, the development of such system for process planning provides two significant advantages:
1. Reduces human intervention considerably.
2. Modifications to the system can be performed easily according to the changes made in the manufacturing environment (e.g. addition of a new machine).

REFERENCES

Alting, L., Zhang, H., "Computer-Aided Process Planning: The State-of-the-art Survey", International Journal of Production Research, 1989, V. 27, pp. 553-585.

Chang, T.C., Wysk, R.A., Wang, H., Computer-Aided Manufacturing, Prentice Hall, New Jersey, 1991.

Chang, T.C., Wysk, R.A., Introduction to Automated Process Planning Systems, Prentice Hall, New Jersey, 1985.

Cohen, P.R., "AI Research Methodology: Three Case Studies in Evaluation", IEEE Transactions on Systems, Man and Cybernetics, May/June 1989, V.19, No. 3, pp. 634-646.

Kusiak, A., Intelligent Manufacturing Systems, Prentice Hall, New Jersey, 1991.

Intelligence Robotic Assembly Cell in CIM Environment

D. Šešlija, S. Stankovski, B. Borovac
Institute of Industrial Systems Engineering, Faculty of Technical Sciences, University of Novi Sad, 21000 Novi Sad, V. Vlahovića 3, Yugoslavia

Summary

This paper presents an approach to the problem of increasing the flexibility of production systems. The approach is based on the integration of classification system and 2D vision module as inputs, expert system for choosing appropriate grasp modes and supervisory control for synchronization of the hand and robot arm movements as outputs. 2D vision module is used in order to react on-line, which is necessary in the industrial environment. The paper shows the transformation of a work order from a higher control level (production control) into the control variables for the actuators.

1. Introduction

Current tendencies in world market indicate a rapid change from the production of a few products in a great number to the production of many product variants in a relatively small number. In order to automate such kind of production systems a lot of flexibility is needed from industrial robots. This is especially emphasized in assembly where many workpieces are contained in a single product. On the other hand, recent developments in robotics and AI like vision, multifingered dexterous hands, sensors, expert systems are only some of the issues that can be incorporated in CIM system. We present here an outline of the system that is currently under development at the University of Novi Sad with the emphasis on the control structure and the efforts to create an environment which is as close as possible to the real industrial environment.

2. Overview of the intelligent robotic assembly cell

Our goal is to develop an intelligent robotic assembly cell (IRAC) that can include the up-to-date achievements in robotics and AI and to serve as a test bed for their applicability in industrial environment. Also, one of our goals is to increase the flexibility of production systems in terms of:

   - ability to change from the assembling of one kind of products to another one in a minimum possible time
   - decreasing the time necessary for assembly without higher investments

that will increase the flexibility in terms of capacity [1].

In order to satisfy these goals, our system uses some elements of the group technology (GT) approach such as classification system, and flexible end effector. One of the starting requirements is also, that the parts to be delivered are not in the predefined position and orientation at the exact pick-up place. The reason is that the devices used for parts feeding and orienting, like vibratory bowl-feeders, are very inflexible. So, our system must have some kind of vision input device that will guide the robot to the part coming via conveyor belt. The proposed system includes a standard industrial robot, vision system, dexterous multifingered hand, position and force (pressure) sensors, as well as the necessary auxiliary equipment: conveyor belt, automatic screwdriver, tools, local buffers etc. as shown in Fig. 1. Our research effort will be limited to the assembly of small and medium mechanical products. Besides, we assume:

- parts are coming to the robotic cell via conveyor belt, one at the time, with previously defined sequence, in random position and orientation.

- there exists a horizontal plane for which the trajectory of the end effector tip holding the largest considered workpiece is above it, and there are no obstacles to manipulation.

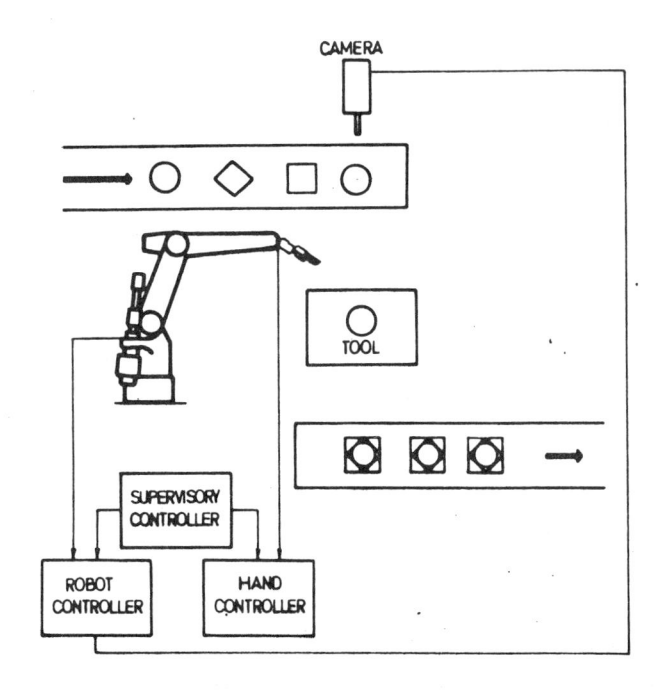

Fig. 1. Overview of robotic assembly cell

- technological procedure for assembly is given in advance.

3. Control structure for the intelligent robotic assembly cell

Our flexible robotic cell should be considered as a part of a broader automated system which in addition to assembly includes machining, transportation, inventory system, production control - all integrated with the aid of computers (CIM). The overall CIM control is hierarchical i.e. the cell receives orders for production of predefined products in some quantity from the higher control level. After finishing assembly, IRAC sends the message to the higher level that is ready to accept a new order. Parts come via conveyor belt randomly oriented either laying down or in the upright position, so that is needed a vision input to detect them. The 3D vision system offer the information about the position and orientation in both planes but they are not, up to now, able to produce the necessary information in, under industrial circumstances, reasonable time scale. We have chosen the 2D vision system in combination with classification system that we have already used as the input for selecting grasp modes [2][3]. The proposed block scheme for control structure of IRAC is shown in Fig. 2.

2D Vision module - The vision module uses simple 2D camera input and information about the object given in form of classification number to obtain the workpiece position -coordinates of the center of gravity (COG), and spatial orientation of the workpiece in vertical plane - upright position or laying down, and in horizontal plane - angle of the main axes. For each geometric primitive [4] in which the workpieces are classified [2] there are some differences in the vision processing algorithm (Fig. 3), i.e. if the workpiece is classified as the primitive rectangle, the vision system should define, beside COG and orientation, the dimensions of the rectangle.

Order processing - This is the place where the work order from a higher control level is entering into the IRAC. The work order in specified form, is coming via computer net. After arrival of the work order, this module is retrieving appropriate classification numbers and sends only significant digits to other modules. For example, the vision module requires only second digit (primitive) except for the rectangle where dimensions are also needed.

Classification - Classification number is determined according to the classification scheme [2][3] in off-line mode based on the shape and other relevant features of workpieces. Classification number for all workpieces that are designed inside the factory is determined when the design phase is finished and for workpieces that are acquired from vendors they are defined after the first delivery. This is currently done manually.

World model - In this module we define the coordinates for all important positions in the robot workspace, like: conveyor belts, camera, tools, obstacles, etc. The grasp depth (Z coordinate) is very important to define properly the grasping position. This is done manually in the interaction with the robot programming unit.

Task description - The information on technological procedure given in the form of technological chart makes the input for this module. The operations specified in the technological chart are decomposed in elementary, predefined, movements and actions of the robot (going down, lifting, transferring, etc..), hand (grasping, squeezing, etc..) and auxiliary equipment. In this way the robot program is obtained as a combination of previously programmed elementary operations.

Feasible grasp modes - This module consists of the expert system for

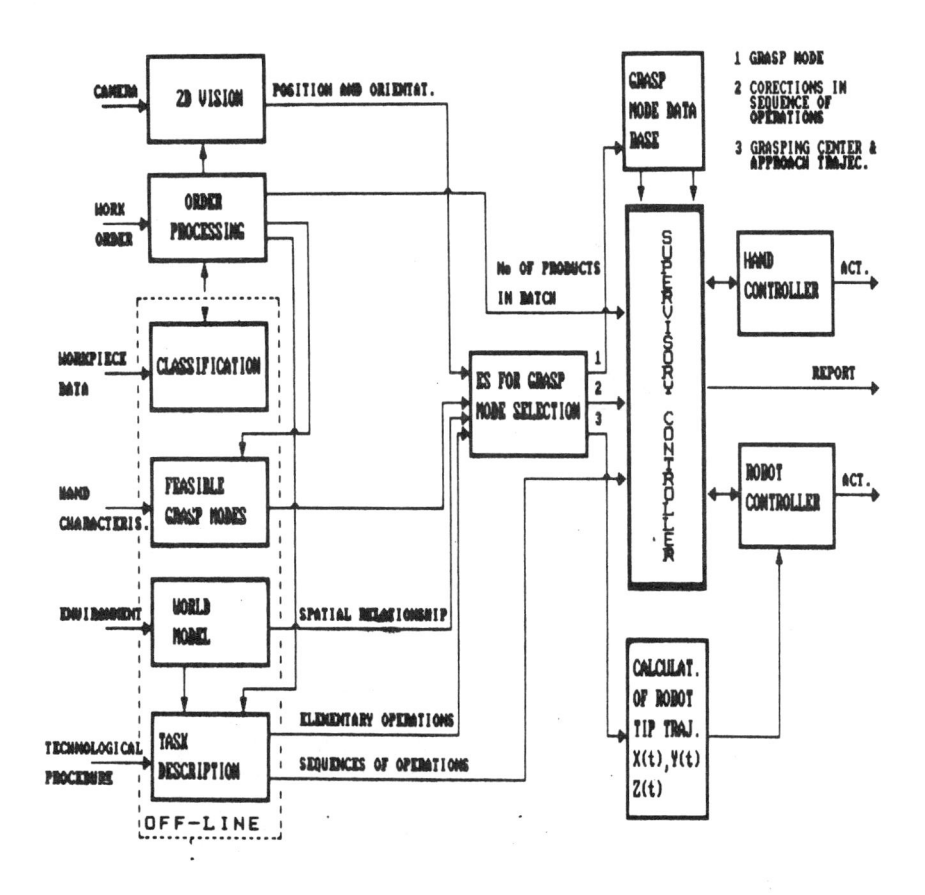

Fig. 2. Block scheme of control structure

deriving a list of feasible grasp modes and data base for feasible grasp modes for workpieces from product range [2]. This module can be realized both in off-line and on-line mode depending on the number of workpieces in IRAC. We have selected off-line mode in order to reduce computational time.

Calculation of robot tip trajectory - Due to the characteristics of workpieces, task and end-effector it may happen that grasping is not taking place around the center of gravity. Also, in some cases the workpieces cannot be grasped with proper orientation required for assembly. In such cases, some corrections such as rotation for $90°$ around Z axis should be introduced in the program previously defined. Data for this corrections should be obtained from the module for selection of grasp mode.

Grasp mode selection - For defining the appropriate grasp mode we suggest the expert system operating in on line mode. In most cases, the grasping is done after the vertical approach trajectory with encompassing of workpie-

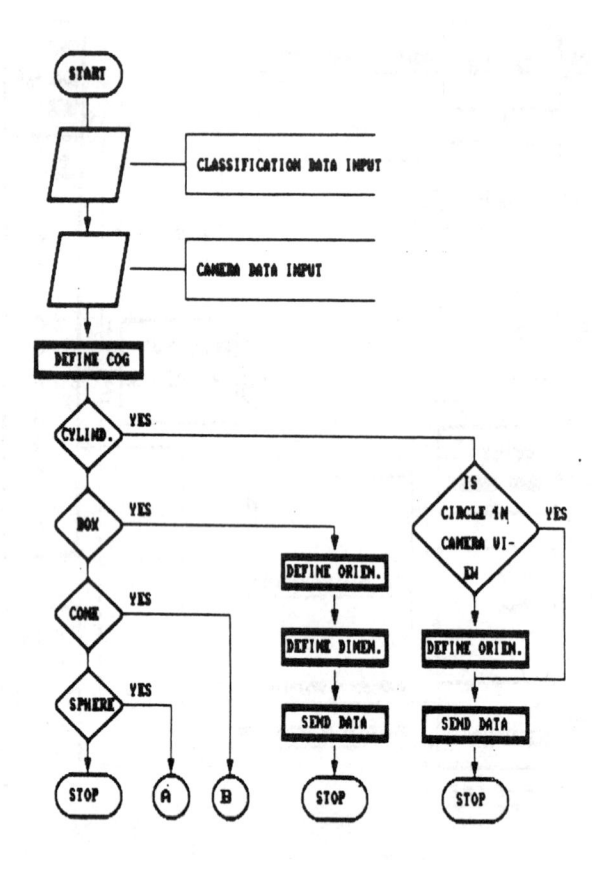

Fig. 3. Vision processing algoritham

ce around the center of gravity. However, there are some other approach trajectories, which combined with various possible grasp modes and requirements from technological procedure, makes the problem much more complicated. This module needs the following information to be obtained: position and orientation of the workpiece, task description - namely, elementary operations and spatial relations between the current robot position and workpiece.

Grasp mode data base - When the grasping mode is defined, the system retrieves appropriate values for control variables from the data base for basic grasps and delivers them to the supervisory controller. The data base is filled with necessary data about basic grasp modes for the applied end-effector. Its size and structure is defined according to the characteristics of the applied end-effector: number of possible grasp modes and number of control variables necessary for defining grasp modes.

Supervisory controller - During a joint operation of two controllers, some kind of synchronization is needed. The two controllers should work in some cases one after another and sometimes at the same time. This module operates like some kind of switchman which overlooks the process and, in right moments, activates the appropriate controllers or auxiliary devices. Based on the information from the order processing module, this module stops the assembly process after the required number of pieces is produced.

4. Conclusions

A conception of an intelligent robotic assembly cell in CIM environment is given. The basic cell structure and control structure are presented. The application of group technology and classification sistems, enabled the use of simple and low cost 2D vision system. Selection of the grasping center and grasp mode, is done by expert system. The application of this system is suggested for flexible, small batch assembly of small and medium mechanical products.

5. References

1. Zelenović, D.M.: Flexibility, a Condition for Effective Production Systems. Intl. Jrnl. of Prod. Research 20 No 3, (1982) 319-337.

2. Šešlija, D.: Investigation of conditions and development of control structure of intelligent robots in flexible assembly. MS Thesis, University of Novi Sad, Faculty of Technical Sciences, 1989. (Serbian)

3. Tomović, R., Zelenović, D., Šešlija, D.: The Role of Intelligent Robots in Flexible Assembly. Computers in Industry 15 (1990) 131-139.

4. Liu, H.: Knowledge Based Grasp Planning for Robot Hands. Ph.D. Thesis, University of Southern California, Computer Science Department, 1989.

The Application of AI and RISC Methodologies to Manufacturing Simulation Tools

I. Astinov*, J.F. O'Kane** & D.K. Harrison**

* Department of Machine Tool Technology, Technical University, Sofia, Bulgaria
** Department MCAE, Staffordshire Polytechnic, Stafford, England.

Summary

Discrete-event simulation is a powerful methodology to aid decision-making and is highly applicable to the manufacturing sector of industry. Many simulation software systems are commercially available but they all suffer from the shortcoming that they require a certain level of knowledge and professional skill for meaningful results to be obtained. This combined with the skill necessary to interpret results has prevented discrete-event simulation from gaining widespread penetration into manufacturing industry.

This paper presents the intermediate results of a collaborative project to harness Artificial Intelligence (A.I.) and Reduced Instruction Set Computing (RISC) methodologies in developing tools for manufacturing simulation. The RISC approach is used in the creation of a kernel for the simulation package. A set of five building blocks are defined which are flexible and powerful enough to build models which encompass a wide variety of manufacturing systems.

Through the use of these five blocks it is possible to simulate a system to a level of detail, the computation being the only constraint. Since there are only five building blocks the simulation kernel is relatively easy to learn, understand and apply. Each of the building blocks is fully explained and the method of model building is illustrated together with case study results. These tools obviate the need to have specialised skills in order to obtain accurate results. Consequently, the A.I. and RISC approach is shown to form a potential step in the development of the next generation of simulation software.

Simulation and A.I.

Over the last decade hybrid systems containing simulation and Artificial Intelligence components have been developed [1,2,3]. Conventional simulation systems are unable to offer advice as to the optimum model to use or the type and extent of experimentation to perform, whereas, Artificial Intelligence can be utilised in the form of knowledge-based systems to automate decision-making via analysing the results of simulation runs and adjusting the model and experimental conditions accordingly until some predefined goal is achieved. An ideal system would be one in which the intelligent component could explain its actions, recommendations and conclusions to the user, since this would inevitably lead to greater confidence in the results it produced. With this in mind the idea of a "Simulators Workbench" has been suggested as one way of achieving this aim [4].

The terms "Intelligent Front End" (IFE) and "Intelligent Rear End" (IRE) have been used to describe support systems offering varying degrees of intelligent decision-making behaviour at either the model building (front) end or the analysis (rear) end. The main advantage of IFE's is the reduction in model development time since the interface can be used by inexperienced persons who only have a superficial knowledge of the particular simulation language and therefore the simulation "learning curve" problem is reduced or in some cases does not exist. IRE's on the other hand could provide assistance to all types of users with facilities to measure the performance of their system, experiment with alternative designs and test "what if" scenarios.

In this paper we describe a system called INMOD [5] which combines the salient features of knowledge-based systems with a simulation kernel for the modelling and analysis of manufacturing systems.

RISC Methodology in Simulation Systems

In general the RISC idea considers the existence of a small set of instructions which are able to carry out the jobs required by the CPU of the computer. This principle is used in defining a set of blocks for modelling and simulating production systems. The aim of this approach is to eliminate some of the disadvantages of the existing simulation software, mentioned earlier and particularly the complexity and the extended professional skills, required for using it.
As a first step towards a definition of a minimum RISC block scheme for modelling, it is appropriate that a simple production system is chosen as an example. Fig. 1. illustrates such a production system. The following observations can be made: -

1. Each production system accepts raw material for processing in specific places, referred herein as inputs. Raw materials are at least of one type, but in general may be several types. The arrival of raw material in the inputs can be strictly scheduled in time, a continuous process or a random event. Since this is one of the most important components of production systems, a separate modelling block is defined, named GENERATOR, with which raw material input is modelled.

2. After processing the raw materials, finished parts are produced and they leave the production system from specific places, named herein as outputs. At this stage of development, it is considered that any quantity of processed parts can leave the production system at the outputs. In this sense, another modelling block, called PIT is defined to represent the production system's outputs.

3. Within the production system, parts are processed on different machines. The word "processing" is used herein to represent the different actions performed on the part - metal cutting, measuring, washing etc. Likewise, the word "machine" is used to indicate the different machine tools and production units utilized in the production system e.g. lathes, milling machines, machining centres, washing machines etc. In other words, each machine performs a particular activity on a particular workpiece. This activity can be described with the time required for the machine to perform it. In general, this time value can be a fixed number or a random variate. An important aspect of machines which should be also considered is breakdowns.

At certain moments machines stop processing parts due to different reasons - breakdowns, maintenance operations, tool change etc. These events are referred herein as "breakdowns". Generally, breakdowns are of two types. Type 1 breakdowns are not related to the process carried out by the machine over the part. A typical example of such a breakdown is a malfunction, electric fuse breakdown etc. Type 2 breakdowns are related to the process performed by the machine, a typical example being tool wear. After a breakdown occurs, the machine should be repaired in order to continue processing parts. Breakdowns can be described with the interval of time they appear. In a similar manner, repairs can be described with the time required to carry out the repair process. Both are expressed as random numbers or variates. All these actions define another modelling block - OPERATION, capable of presenting activities in production systems.

4. Each production system has special places where semi-finished parts are stored due to various reasons e.g. next production facility not available, wait for high priority parts to be machined etc. Within this paper, these places are indicated as "buffers". Examples of buffers are intermediate storages, pallets, or conveyor sections. Since buffers are an essential part of any production system, a separate block called BUFFER is defined, for their presentation and modelling.

5. The elements of a production system interact in a strictly defined order, according to certain predefined logic. This logic is a consequence of the actions that should be made in order to process the parts. Example of such logic is "start the operation if a workpiece is loaded", or, "the robot hand has withdrawn and the shield is closed", or "begin assembly operation, if all sub-assembly components are available". An appropriate block, capable of defining such logic for this purpose is called CONNECTOR.

According to the observations, outlined above, five basic blocks for modelling production systems are defined. Their graphical images and parameters are shown in Fig. 2.

Case Study - A Generalised Manufacturing System

To illustrate some of the features of the INMOD system the concept of a "generalised manufacturing system" (GMS) is used. The GMS is defined as follows :

- the GMS should have n (n>=1) inputs, since each real production system accepts at least one and in general several types of raw materials.

- the GMS should include sequential and parallel processing, as well as intermediate buffering, since any real production system performs a sequence of several different processes. Processing in parallel is applied in the case of bottle-necks or in a set of machines handling the workpieces on rotary tables or similar ;

- the GMS should have m (m>= 1) outputs, since each real production system outputs at least one type of product.

Fig. 3. illustrates the model of a GMS with n = 2 and m = 2.

A model of a production system such as the GMS can be built by describing the system to INMOD in terms of the logic, buffers etc. This description is converted into a formatted file structure which is then used as input to the INMOD processor. The processor executes the model for a predefined period of time. Within this period, each change in the state of the modelling blocks is recorded. This approach is standard to all simulation systems and it enables the subsequent statistical analysis of the model - such as calculating machine utilizations, average and maximum queue lengths, time in system for a workpiece etc to be made. Finally, a summary report of the simulation model run statistics are presented for analysis.

The INMOD 2 System

At present the INMOD system consists of the INMOD PROCESSOR, MODEL EDITOR and INMOD DATABASE. A collaborative project is currently being undertaken to develop the software for an INMOD 2 simulation system. INMOD 2 is an extension of the present structure to include a statistical analysis processor, an object-orientated interface module together with an intelligent front end and an intelligent rear end for model synthesis and analysis.

The intelligent component of the system is an Expert System which contains a knowledge-base of rules/facts pertaining to a range of typical production systems. The proposed INMOD 2 system is shown in Fig. 4.

Conclusions

Based on the experiences of using the INMOD system the following conclusions may be drawn :-

(i) The defined modelling blocks are sufficient enough to create models of over 70% of production systems configurations. INMOD as yet does not include the ability to model material handling features.

(ii) The INMOD system is easy to learn, since it is comprised of only a small amount of modelling blocks which have simple properties to enable inexperienced users to create models quickly and efficiently.

(iii) Different levels of detail in the model can be obtained, depending on the requirements of the simulation study, yet using only five modelling blocks.

(iv) The INMOD 2 software is compatible with any hardware, supporting the ANSI 3.9/78 FORTRAN 77 and ANSI C programming languages, since it is written strictly according to the specifications of the mentioned standards.

It is clear that there is a niche in the market for smaller more user-friendly simulation modelling tools. Systems which provide novice users with an attractive environment to describe, formulate and analyse their production facilities with heightened confidence opens up the usage of discrete-event simulation to a much wider audience.

The brief description of the INMOD system presented in this paper is a small but significant step towards the next generation of simulation software tools and the inclusion of Artificial Intelligence features in such a system provides a powerful and sophisticated decision support system for simulation modelling.

References

1. Lavery, R.G., "Artificial Intelligence and Simulation:An Introduction", Proc of the 1986 Winter Simulation Conference, 448-452.
2. Reddy, R., "Epistomology of Knowledge based Simulation", SIMULATION, Vol. 48, No. 4, April 1987, 162-166.
3. Shannon, R.E., "Models and Artificial Intelligence", Proc of the 1987 Winter Simulation Conference, 16-23.
4. Nolan, P.J., "Intelligent Simulation Interface:5 Year Review", Proc of FAIM Conference, Limerick, Ireland, March 1991, 229-239.
5. Astinov I., "Simulating the operation of machine tools and systems", PhD Thesis, Technical University, Sofia, Bulgaria,1990.

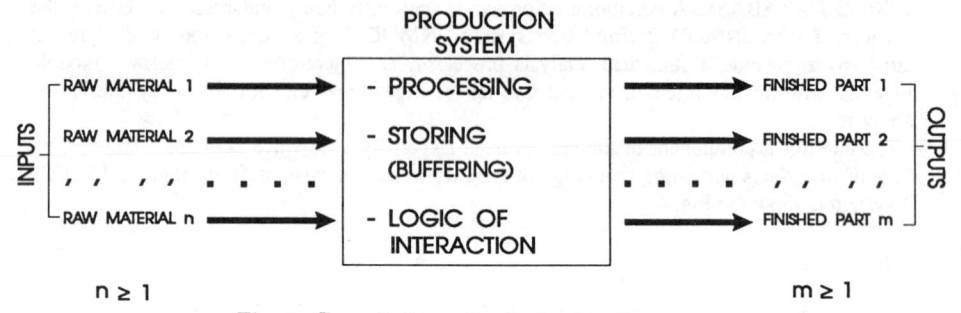

Fig. 1. General view of a Production System

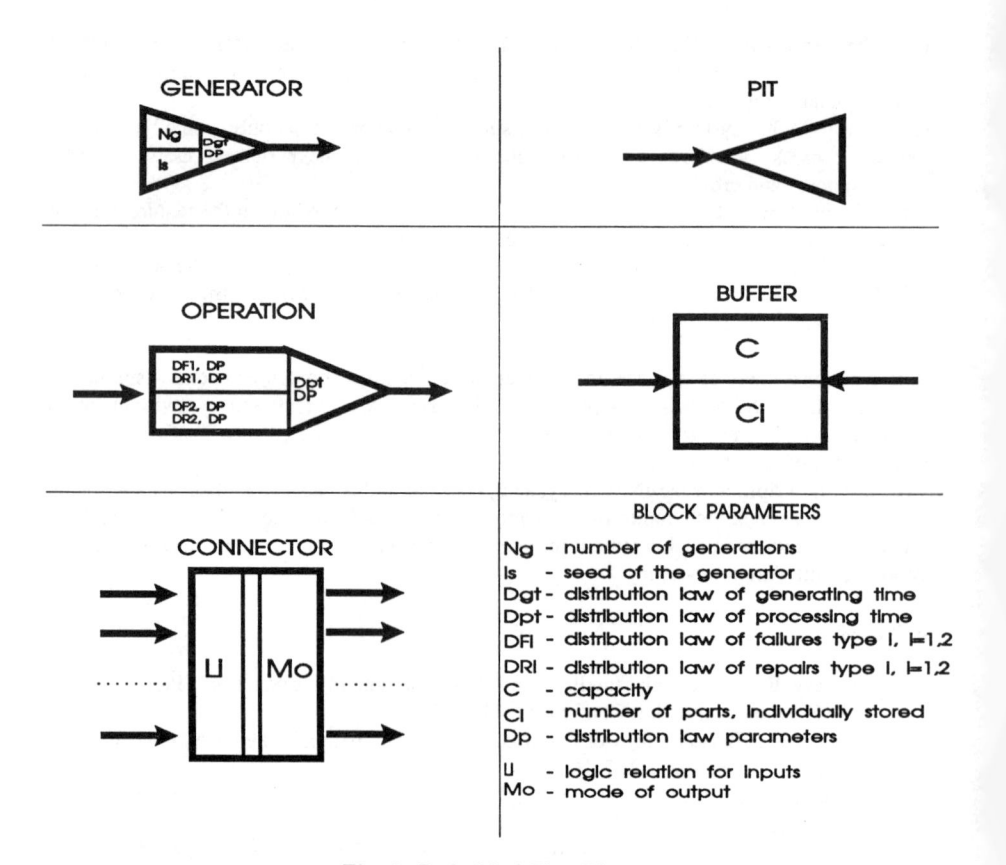

Fig. 2. Basic Modelling Blocks

Control of an Inherently Unstable System Using the Neural Networks

I. I. Esat and T. J. Suh
Department of Mechanical Eng.,
Queen Mary and Westfield College,
Mile End Rd., LONDON E1 4NS.

ABSTRACT

The control theory provides a number of techniques for source allocation in order to optimise a suitable cost function in the manufacturing environment. The control theory widely used in process control and scheduling. This paper assesses the classical feedback control strategy against a neural network based approach by employing an inherently unstable dynamical system. The control of a ball-balancing machine is investigated. It is shown that a neural network based control can provide stability and self organise for an optimum response. This is an important advantage over the conventional control since no explicit formulation of dynamic equations for control is required.

1 INTRODUCTION

The inverted pendulum together and the ball balancing machine are mathematically identical and represent two classic examples of inherently unstable systems. The balancing of these mechanisms has practical applications, one of most important of which is the control of rocket thrusters. There are many examples of classical control design of these systems [1] The control of these systems based on classical feedback theories, requires the precise description of system dynamics. This has to be companied with the desired behaviour of the system which then can be used to form an error or another suitable objective function. Although the linearisation may be acceptable, the systems equations are non-linear in their basic form. In many situations the effective inertia of these mechanisms may not be directly measurable. This is due to fact that these inertias may include many moving parts, such as belts, rotors of driving motors and wheels. The control equations usually exclude non-linear effects of friction. Another problem is associated with the accurate measurement of state space variables. The angular position and velocity measurement of ball on the curved track of the ball balancing machine is specially difficult and prone to error. Therefore, it may reasonably be assumed that, the dynamic equations of the ball balancing machine may not accurately represent the physical system. The neural network based control do not require the dynamics of system equations. Therefore most of the problems associated with accurate definition will not arise in this approach. It is also expected that the neural network approach will tolerate to modest levels of poor signals of the state variables. Some examples of literature on neural networks applied to the ball balancing or inverted pendulum problem are given in the References [2], [3] and [4].

2 BALL BALANCING MACHINE

A typical ball balancing machine is shown in Fig.1. The carrier containing a curved track is constrained to move within the vertical plane with no friction. The system contains two independent variables Y and θ giving four state space variables $(y, \dot{y}, \theta, \dot{\theta})$, the free body diagrams are also shown in Fig.1.

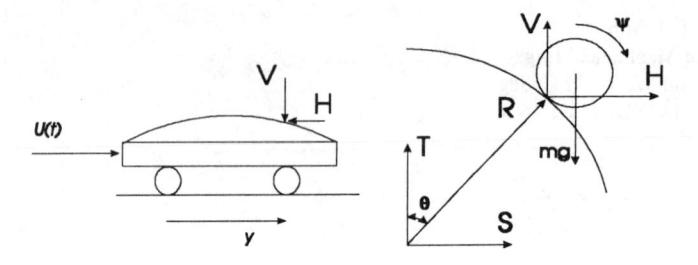

Fig. 1 The ball balancing machine and free body diagrams

In Fig 1, V and H are the reactions at the contact point of the ball and the curved rails. S and T are the horizontal and vertical displacements of the ball relative to carriage respectively. S and T may be written as,

$$T = (R+r)\cos\theta \quad \& \quad S = (R+r)\sin\theta,$$

where R and r are radius of the carriage and ball.
The horizontal force acting on the carriage is given by,

$$U(t) - H = M\ddot{y}$$

The angular position θ of ball on the curved track can be related with the angular rotation ψ of the ball relative to the top of the curved track.

$$\psi = \frac{(R+r)\theta}{r}$$

Vertical forces acting on the ball are V and mg, the equations of motion in vertical direction is given by,

$$V - mg = m\ddot{T}$$
$$V - mg = m(R+r)(-\ddot{\theta}\sin\theta - \dot{\theta}^2 \cos\theta)$$

The equation of motion of the ball in horizontal direction is given by,

$$H = m\ddot{y} + m\ddot{S}$$
$$H = m\ddot{y} + m(R+r)(\ddot{\theta}\cos\theta - \dot{\theta}\sin\theta)$$

And the rotational motion of the ball is given by,

$$Vr\sin\theta - Hr\cos\theta = I\ddot{\psi}$$

$$Vr\sin\theta - Hr\cos\theta = I\frac{(R+r)\ddot{\theta}}{r}$$

If θ is assumed to be small then, equation can be linearised. Eliminating V, H and Ψ, the equations of motion can be written in two independent variables y and θ.

$$\ddot{y} = -\frac{m^2r^2g}{Mmr^2 + (M+m)I}\theta + \frac{(I+mr^2)}{Mmr^2 + (M+m)I}U(t)$$

$$\ddot{\theta} = \frac{m(M+m)gr^2}{(Mmr + I(M+m))(R+r)}\theta - \frac{mr^2}{(Mmr + I(M+m))(R+r)}U(t)$$

rewritting these equations using state space variables x_i, i=1,2,3 4, where

$$x_1 = y, \quad x_2 = \dot{y}, \quad x_3 = \theta, \quad x_4 = \dot{\theta}$$

Now equations may be written in matrix form

$$\dot{x} = Ax + Ub$$

$$A = \begin{bmatrix} 0 & 1 & 0 & 0 \\ 0 & 0 & -a & 0 \\ 0 & 0 & 0 & 1 \\ 0 & 0 & c & 0 \end{bmatrix} \quad and \quad b = \begin{bmatrix} 0 \\ b \\ 0 \\ -d \end{bmatrix}$$

Using classical feedback control strategy, U(t) may be chosen to be,

$$U(t) = k_1x_1 + k_2x_2 + k_3x_3 + k_4x_4$$

Substituting the forcing function into the equation of motion, the final control equations can be obtained. Adjusting the values of the feedback constants conditions of stability may be achieved.

$$\dot{x} = \begin{bmatrix} 0 & 1 & 0 & 0 \\ bk_1 & bk_2 & bk_3 - a & bk_4 \\ 0 & 0 & 0 & 1 \\ -dk_1 & -dk_1 & c - dk_3 & -dk_4 \end{bmatrix} x$$

$$\dot{x} = Ax$$

The characteristic equation may be written as,

$$|\lambda I - A| = 0$$

3 ANALYSIS

Stability conditions based on the Routh stability criterion was obtained. These conditions are given by,

Assuming $k_1 > 0$ and $k_2 > 0$,

$$k_4 > \frac{bk_2}{d}$$

$$k_3 > \frac{(bk_1 - c)}{d}$$

$$d^2 k_2 k_3 k_4 + adk_2^2 + bdk_1 k_2 k_4 > d^2 k_1 k_4^2 + cdk_2 k_4 + bdk_2^2 k_3$$

The stability analysis was carried out, both by using root locus method and by finding values of k_i satisfying the inequilities. The root locus is obtained by plotting the real parts against imaginary parts of eigenvalues of matrix **A**. For k = (5 10 300 20), time response of θ is shown in Fig 3a. The results give an indication of the magnitude of force involved in controlling the system in a stable region.

3 RULE BASED CONTROLLERS

Having established the magnitude of force required to achieve a stable response, a number of rule based control strategies were experimented. It is observed that a set of predetermined simple rules can adequately control and stabalise the system. The state space is divided into 4 regions for the positive and negative sections of the state space variables. The discretised space gives (8x8x8x8) regions each defining a forcing value. Although the initial investigation was carried out by using simple lookup tables, this latter replaced by sixteen simple rules. The lookup tables and latter the rules represent the force required for a stable motion at a given position described by the state variables. The reduction achieved by determining the magnituede of the force based on the angular displacement and its sign by the direction of the angular velocity. Experiments showed that these rules although sufficient to control the ball position the position of trolly, would either move to a positon away from the center or many situation continue to drift. Applying additional, similar rules to the carrier position was not sufficient since the simple rules in the half of the possible positions would contradict each other. A typical rule is given below,

If (x3>.07) and (x3<.2) and (x4>0)   then F = 50.0

A typical response of the angular deflection is given in Fig. 3b. It is interesting to see that a few simple rules could achive a stable solution, ball reaching to the top faster than the example given for the conventional control solution.

4 A NEURAL NETWORK BASED APPROACH

A simple single layer network is employed, the supervised learning method , the most common method, requires a training set of data. The ultimate objective, although known, it is a relatively difficult task to simulate the system from a starting point to the solution (or failure) in order to update the W_i. Most of the papers surveyed, reporting similar research is not clear about this point, although, Anderson does report a method based on discretising the state space [5].   The other important point is the relationship between force and the state space variables. Again the most common approach to training is based exclusively on the input and output state space vectors where the force is not treated as part of the training set. The learning algorithm proposed in this report attempts to train the system to associate the state space variables with the force vector. The diagram shows the single layer network designed for this purpose.

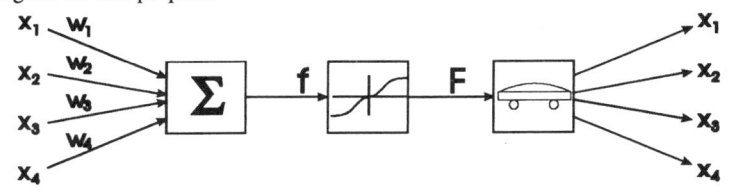

Fig. 2. The single layer neural network for ball balancing problem

The learning algorithm is based on the following formulation,
$$\varepsilon_i = t_i - x_i$$
Where ε_i is the error between the target vector t_i and output x_i
The change in error with respect to weights w,
$$\delta\varepsilon_i = \sum_{i=1}^{4} \frac{\partial \varepsilon_i}{\partial w_j} \delta w_j$$
Here the error vector is described by the Jacobian matrix.
The ij element of the matrix is given by $\frac{\partial \varepsilon_i}{\partial w_j}$
but
$$\frac{\partial \varepsilon_i}{\partial w_j} = -\frac{\partial x_i}{\partial F} \frac{\partial F}{\partial f} \frac{\partial f}{\partial w_j}$$
since
$$f = \sum_{i=1}^{4} w_i x_i$$
$$\frac{\partial \varepsilon_i}{\partial w_j} = -\frac{\partial x_i}{\partial F} \frac{\partial F}{\partial f} x_{i(t-1)}$$
Function F is taken to be logistic (sigmoid) function and represent the force acting on the carrier
$$F = \frac{h-l}{1+e^{-gf}} + l$$
Where h and l represents the highest and lowest allowable force values, g is a suitable gain and f is the force calculated through the learning vector. Therefore the derivatives of F with respect to f can be calculated.
$$\frac{\partial F}{\partial f} = \frac{g(h-l)}{(1+e^{-gf})^2}$$
The remaining partial derivative can be calculated by comparing the values of x and F at two consequtive steps of the numerical integation. That is,
$$\frac{\partial x_i}{\partial F} = \frac{x_{i(t)} - x_{i(t-1)}}{F_t - F_{(t-1)}}$$
With this all the necessary variables are calculated to set up the Jacobian matrix. Therefore the error terms in the weights, δw can be obtained by inverting the Jacobian matrix. This gives the improved weights,
$$w_{i(t)} = w_{i(t-1)} + \delta w_{i(t-1)}$$

a) Classical Feedback   b) Simple Rule Based   c) Single Layer Neural Network.

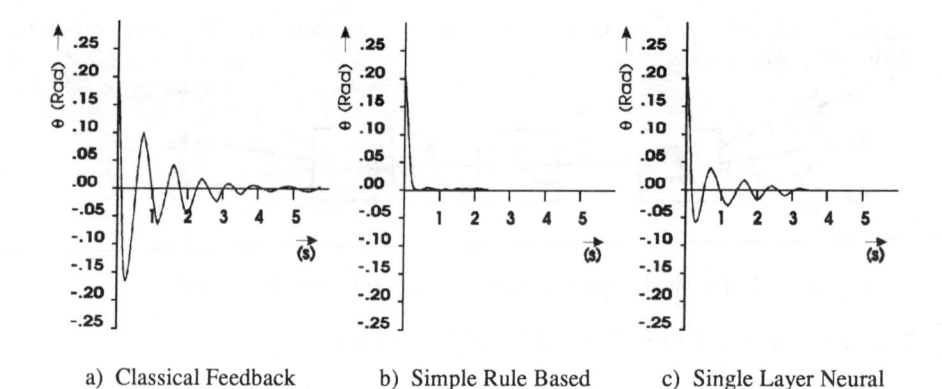

Fig. 3 Time response of the angular position of the ball

5 RESULTS AND DISCUSSION

The subroutine for calculating the driving force replaced with the algorithm described above, that is, the training vector weights were calculated at each step and then used in calculating the force vector. The initial conditions were taken, the same as before. If the weights were taken to be small, then the system required to be simulated a long time. A convergence is not always expected. If the weigths were taken to be 100 with 50% random variation, a convergence is always observed. With these values a successful learning period on a 386 IBM compatible machine, and using a standard Runge-Kutta simulation routine took about 30 minutes. A typical result is shown in Fig 3c. It is interesting to see that the simple rule based system still performed better than both the conventional feedback and the neural network approaches.

REFERENCES

[1] Cheok K. C. and Loh N. K. "A Ball-Balancing Demonstration of Optimal and Disturbance-Accommodating Control". IEEE Contr. Syst. Mag., vol. 7, no. 1, pp. 54-57, Feb. 1987.

[2] Widrow B., "The Original Adaptive Neural Net Broom-Balancer" Int. Symp. Circuits and Syst., pp. 351,357, May 1987.

[3] Tolat V. V. and Widrow B., "An Adaptive Broom Balancer with Visual Inputs", Proc. IEEE Int. Conf. on Neural Networks, San Diego, CA. pp 11-641-11-647, July 1988.

[4] Connell M. E. and Utgoff P. E. "Learning to Control a Dynamic Physical System", Proc. AAA1-87, vol.2, pp.456-460. American Association for Artificial Intelligence, Seattle, WA, 1987.

[5] Anderson C. W., "Learning to Control and Inverted Pendulum Using Neural Networks", IEEE Control Systems Magazine, pp 31-36, April 1989,

Object Location Using an Artificial Neural Network

B. Parsons, M. Stoker, R. Gill

Middlesex Polytechnic
School of Mechanical and Manufacturing Engineering
Advanced Manufacturing Group
Bounds Green Road
London N11 2NQ

Abstract

This paper investigates the use of an Artificial Neural Network in assisting a Robotic arm to locate an object placed in a predefined workspace. The Neural Network was simulated using appropriate computing hardware, and was found to be capable of accepting data from a camera, and providing a suitable output to be supplied to the robot's arm control system.

Introduction

When a human sees and picks up a cup of coffee, there is little conscious appreciation of the bio-mechanics involved. The human is not aware of the exact dimensions of the arm, and has no coordinate system at work which relates the position of the hand to that of the cup. The action is performed with a casualness that results from the years over which the human learnt to perform that and many similar tasks. The learning process was a simple matter of trial and error.

This contrasts the way in which robots have traditionally been programmed. Precise knowledge of the robot's construction and method of control have to be related to the exact positioning of objects in its environment in order to perform useful tasks. There would be obvious advantages if a robot could learn to perform particular tasks; If, once given a goal, a robot could respond correctly to sensory input, as a result of the nature of its design, rather than human knowledge embedded into computer programs.

This task obviously falls into the area of artificial intelligence. However, Expert Systems, which form the basis of the traditional approach to artificial intelligence, still operate by encoding human knowledge in computer programs. Current Artificial Neural Networks, by their ability to learn, seem to offer something closer to that required.

Artificial Neural Networks

Artificial Neural Networks (ANN), as the name suggests, are structures which to some degree are based on our understanding of the behaviour of the biological neuron. There are many forms of Neural Networks, consisting of processing elements organised in a way which may be similar to that of the brain. The more biologically plausible networks offer interest to neurobiologists and the like, but even those networks which are very slightly related to our current understanding of the brain, posses characteristics such as the ability to learn, generalise and abstract. It is these characteristics that have made the networks of much interest, and have spurred research into applications where other more conventional approaches have not sufficed.

ANN's posses the ability to learn. Through a process of adjusting their responses to the environment, they provide consistent outputs to particular inputs. Generalisation is achieved by becoming insensitive to small variations of the input. However, these characteristics are not a result of human knowledge, experience or reason being stored in a computer program, but are a direct result of the nature of the Neural Network.

The Artificial Neuron, Figure 1, is usually designed to implement the process outlined above. The implementation may be by means of software simulation, or interconnection of general purpose devices such as op-amps, or possibly direct silicon implementation. However, due to

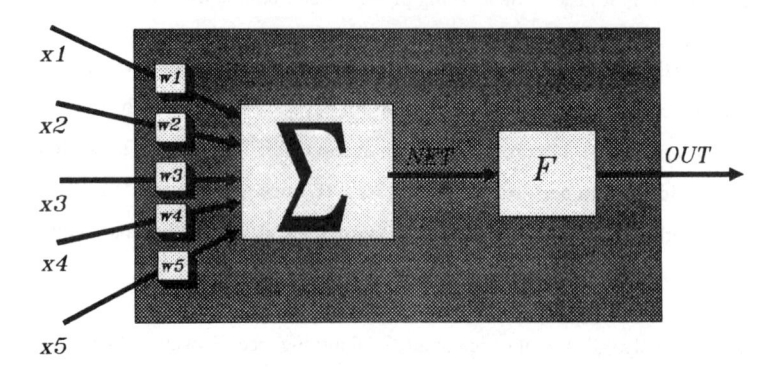

Figure 1 - Artifical Neuron

the large numbers of neurons present in most networks, and the pioneering nature of most of the work in this field, software simulation is usually the only plausible approach.

Each neuron within a network, excluding those in the first or input layer, will receive a set of inputs from other neurons. Each input is multiplied by a weighting value, analogous to the strengths of synaptic connections within the brain, and are then summed. If X represents a set of inputs, and W a set of corresponding weights, then in vector notation NET, the output of a neuron is :

$$NET = \sum XW$$

The NET signal is then usually subjected to an activation function, in order to provide on output signal from the neuron. This often takes the form of a sigmoid function such as:

$$OUT = \frac{1}{1 + e^{-NET}}$$

The use of this function reduces the problem of small signals providing usable outputs, but caters for the possibility of amplified noise or large inputs saturating the outputs of cascaded high gain stages. The result is a neuron that provides suitable gain over a wide range of input signals.

The Power of neural networks results from the interconnection of a number of individual

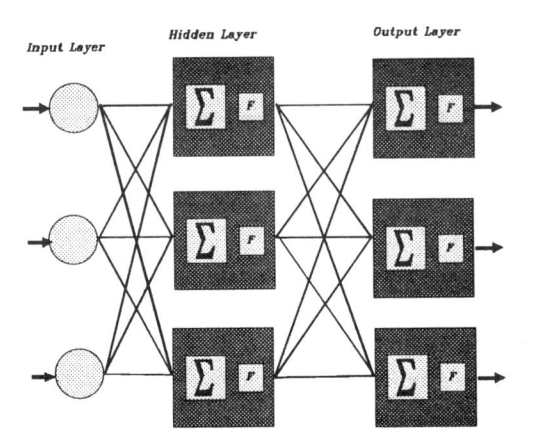

Figure 2 - Multi-layered Network

processing elements. Years of research have led to a myriad of network configurations being implemented. Figure 2 shows a straight forward and common configuration, consisting of layers of neurons, with each neuron directly connected to every neuron in the previous layer.

As varied as the many network configurations, are the various methods used to train the networks. Training usually consists of applying a pattern or a set of patterns to the input of a network. The weight values of the interconnections are gradually adjusted according to the

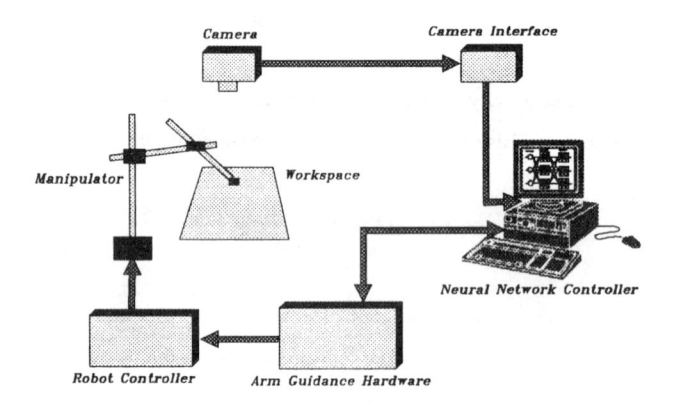

Figure 3 - Schematic of Object Location System

adopted training method. The weights should converge to particular values towards the end of training, such that any future application of one of the input patterns should produce the desired response at the output of the network. Training methods are usually referred to as supervised, or un-supervised. Under supervised training the desired network outputs are already known, with un-supervised training, it is only necessary that different outputs result from different input patterns, as long as these outputs are consistent.

System Outline

During training, an object is moved in a random manner around a pre-defined work area. This may be achieved by attaching the object to the robot arm and either manually moving the arm, or by executing a suitable program. Position sensors are attached to the arm, these take the form of precision potentiometers. The sensors provide a signal, via an interface, which appears at the output of the neural network. This has formed the desired neural network output.

The ANN input is provided by a Dynamic Random Access Memory (DRAM) camera which continuously scans the work area. Thus a training set is formed. The input vector consists of a series of visual representations of an object. The output vector consists of sensory information which relates to the corresponding positions of the robot arm.

The ANN is then trained over a suitable period of time using the training set obtained. Once a network has been successfully trained, an object may be placed in the work area. This will cause the network to provide an output signal which may be used to guide the arm to the object.

The project may be sub-divided into the following categories: Camera Interface, ANN implementation, Sensory input and output, arm guidance hardware. Figure 3 outlines their arrangement.

Implementation

The neural network implemented, was the back propagation network type. Network size was geared towards the application in hand. Twelve neurons were required in the input layer, and to these, binary values representing the scanned image were applied.

Two neurons would be required in the output layer, each one representing the angular position of one of the arm's axis.

The number of layers required in the network was estimated from tests. In the final application, three layers were inserted between input and output, giving the structure shown in Figure 4

Layer Description	Number of Nerons
Input Layer	12
Hidden Layer 1	10
Hidden Layer 2	8
Hidden Layer 3	8
Output layer	2

Table 1 - System Configuration

Conclusions and Further Work

The system developed has shown that a single neural network is capable of providing guidance signals for object location. The system demonstrated that it could operate in real-time. The slowest part of the system was the image acquisition, though here, performance can be improved with more sophisticated hardware.

The use of more powerful computing systems and architectures such as parallel processing could be used to reduce computation time. This area of work is currently being investigated using the INMOS transputer. With an increase in processing power, the system could easily be expanded to perform tasks such as image recognition, or implement more sophisticated neural network paradigms.

References

1. Aleksander I., Morton H., *Neural Computing*, Chapman and Hall, 1990.

2. Caudill M., Butler C., *Naturally Intelligent Systems*, MIT Press, 1990.

3. Fukishima K., Neocognitron: A Self-organizing neural network model for a mechanism of pattern recognition unaffected by shift in position, *Biological Cybernetics* 36(4), 1980, pp193-202.

4. Inmos Ltd, *Transputer Applications Notebook*, 1989.

5. Parsons B,. Control of a Robotic Arm Using an Artificial Neural Network, MSC Dissertation, Middlesex Polytechnic,1991.

6. Wasserman, P. *Neural Computing*, Van Nostrand Reinhold 1988.

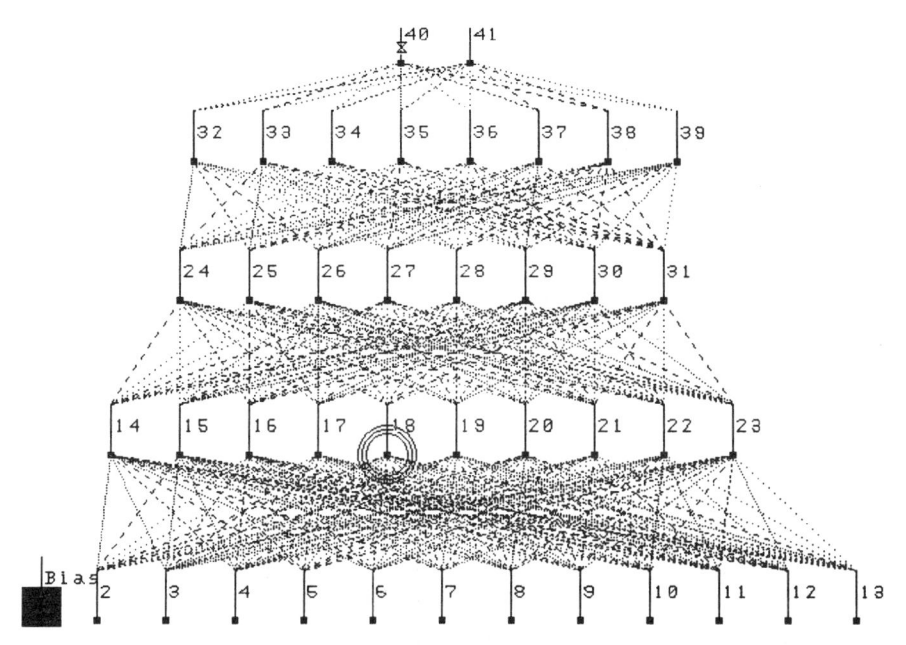

Figure 4 - Neural Network Configuration

An Artificial Intelligence Planner for Assembly Process Planning

Y.P.Cheung, A.L.Dowd
Department of Engineering,
University of Warwick,
Coventry, CV4 7AL,
England

Summary
The needs of manufacturing industry have changed considerably in recent years due to keen competition at home and abroad. To address these changing needs, computers are increasingly used. Where conventional computational methods have been fully exploited and exhausted, Artificial Intelligence (AI) techniques can be applied to meet this demand. AI techniques seek to provide the flexibility that is lacking in conventional methods by attempting to narrow the gap between the human user and the machine.

This paper describes the application of AI techniques to assembly process planning. Various computer-aided process planning systems for machining processes have been reported. Most of these are of the variant-type, i.e. where new plans are made by consulting or editing existing new ones. There are very few truly generative planners (i.e. planners which generate new plans based on existing knowledge or logic) and neither has there been much work on assembly process planning. It is believed that there is an analogy between AI planning and process planning. An assembly process planner based on AI planning is described in this paper.

1. AI Planning

In AI planning, a plan is considered as a possible solution to a specified problem. A problem is often characterised by an initial state description which depicts the start of the problem and a goal state description which indicates what the problem looks like after the execution of the plan. Predicate Calculus, a branch of symbolic logic, was first used to describe these states in a computer. Manipulation of these descriptions using the refutation resolution approach had conveniently produced the required plan as a result of proving that the goal state description is true[1].

The technique of decomposing a problem into sub-problems (if each of these individual sub-problems could be solved in turn, then the whole problem is solved) has been used for planning problems. The problems of using this approach became apparent when researchers failed to solve simple examples such as Sussman's Anomaly in Figure 1.

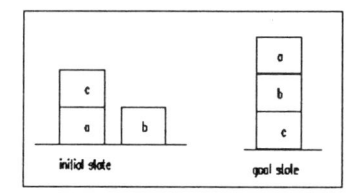

Figure 1: Sussman's Anomaly

The goal state can be defined as: "on(a,b) and on(b,c)" where "a", "b" and "c" represent blocks and "on(a,b)" means that "a" is on "b". If "on(a,b)" is considered first, "c" will have to be placed onto the table first (it is assumed that the table will always have enough space for the blocks). However when considering the next sub-goal, "a" has to be unstacked onto the table first (a requirement of the Blocks World examples is that only one block can be moved at one time) and in doing so the former sub-goal is undone. Similarly, if "on(b,c)" were to be considered first then "c" is on top of "a" which means that "on(b,c)" cannot be achieved first.

This simple example has illustrated the fact that where sub-goals interact with one another, it is not possible to treat each sub-goal as a single independent goal. Little attempt was made to apply AI planning techniques to real world problems which were considered as uninteresting and unchallenging compared with finding a planner that will "do it all". However, it is believed that despite the difficulties of the problem there is an analogy between process planning and AI planning [2].

1.1 Some AI Planners
The first formulation of the planning problem in Predicate Calculus was by Green in 1969 [3]. His system required descriptions (or frame axioms) of what remained unchanged following an action, for the refutation proof to work. Typically the number of frame axioms required is proportional to the product of the number of relations and the number of actions in the problem [4]. Even for the Blocks World examples the number of frame axioms required can be very large.

Frame axioms are implicit in the next significant planner, STRIPS[5]. Its description consists of the goals and actions and a database containing the initial state, pre-conditions (states that must be true before the actions can be performed), add-lists (descriptions that become true as a result of performing the action) and delete lists (descriptions that become false after performing the action). NOAH [6] is another planner which makes use of a set of critics to resolve conflicts. The planning system, TWEAK by Chapman [7] provided a formalism (which has been considered as a "tidying up" of the work on AI planning) for the planning problem. Techniques of these planners were used in the planner described in section 3.1.

2. Process Planning

There are two main approaches to the process plannng problem, variant and generative. In the variant approach a process plan for a new part is created by retrieving previous process plans for a similar part and then making the changes required to the plan to cater for new designs. The generative approach concentrates on synthesizing new plans instead of editing old ones. Very few, if any of such systems are available today.

Some systems based on the knowledge-based approach of implementing manufacturing rules in the form of "if condition then action" statements have been reported [8], [9], [10]. The main problem with such systems is that they were developed for a specific domain (as is the forte of most expert systems) and a completely new system would have to be designed for a slightly different application. In the next section an assembly planner that is based on AI planning techniques is described.

3. Assembly Process Planning

Little work has been done on automating assembly process planning. The main reason is that like process planning in general, very little information on the assembly process has been published as it is usually carried out manually.

The process planning task in general can be sub-divided into two main parts:

i)   Decision on the types of manufacturing processes to be used;
ii) Ordering (or sequencing) of those manufacturing processes.

To achieve i) specific manufacturing databases have to be created as companies differ in the types of manufacturing processes used. As for ii),it is believed that the logic employed in AI planners could be used.

3.1 The Heart of the Planner

The heart of the system performs the sequencing of operations. It is written in Edinburough Prolog on the SUN workstation. The fundamentals of the planner consist of describing the various stages of the problem (i.e. initial, goal and intermediate states) using predicates. For assembly in particular, only the features on the mating pairs concerned need to be considered. Therefore, the basic criteria for an assembly action to occur is to check that it has nothing obstructing its intended action and that its destination is also "clear". The planner was originally based on STRIPS and then revised to incorporate NOAH-like critics. The main problem with this approach is that the logic of the planner, which relies heavily on these critics may not be sufficient to cover all cases. As the TWEAK formalism provided a global view to the planning problem, it was later incorporated into the assembly planner and was able to produce correct orderings of examples taken from some sub-assemblies of the car engine. A more general example of the assembly of the ball point pen was also used to illustrate how it could be tailored for more general cases.

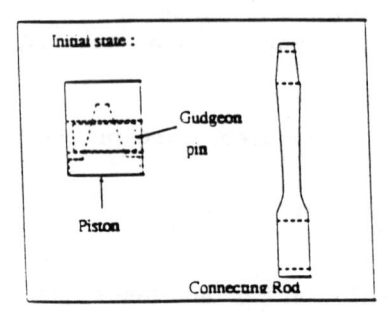

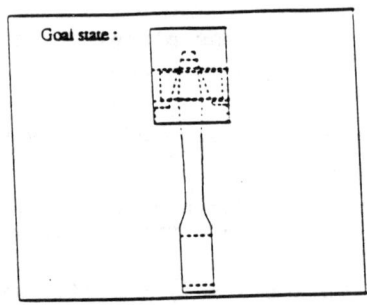

Figure 2. The piston-connecting rod sub-assembly.

In this sub-assembly, the gudgeon pin is inside the piston in the initial state. The initial state of this sub-assembly can be written as:

    assemble(pin,whole,piston,t_hole,nil,nil).
    clear(rod,whole).
    cross(piston,t_hole,axial_hole,ta_area).

The "assemble" predicate represents the fact that the gudgeon pin is in the hole of the piston (called "t_hole"). The next statement shows that the whole of the connecting rod is "clear". The "cross" predicate is used to represent the fact that the two features i.e. t_hole of the piston and axial hole of the connecting rod intersect. "nil" in the above statement indicates that there are no reaction faces which are normally used for pressing actions during assembly. The correct sequence for this assembly is as shown below:

    assemble(rod,whole,piston,axial_hole,nil,nil).
    assemble(pin,whole,piston,t_hole,nil,nil).

At the current stage of development the input to the planner consists of describing the initial and goal states using Prolog which is not desirable for practical use. A pre-processor which can interpret the design data and then translate it into an acceptable form could be added as a front-end to the system. Some comments on product description are given in the next section.

4. Product Description

A popular approach of variant systems is to make use of coding and classification systems such as MICLASS [11], OPTIZ [12], etc. Parts are grouped into families based either on feature or process characteristics. This approach is inadequate in situations where there is much variety amongst products with very few members per part family.

Attempts have been made at interpreting CAD data directly
[13]. The CAD data could be converted into a neutral
format, e.g. IGES (Initial Graphics Exchange
Specification, 1980, NBS in USA). However the functional
and feature information which is not available in the
knowledge extracted directly from IGES is needed for the
process planning stage.

Another approach is to specify explicitly the features of
the parts at the design stage. This requires the direct
design of parts using features which are then stored in
the system's library for process planning at a later
stage. Presently similar CAD systems using the knowledge-
based approach are available but their credibility would
require further investigations.

The simplest alternative is to describe in words the
design to the process planning system. The relationships
between parts are obtained from answers to a series of
questions about the mating of part pairs. However this
could be very cumbersome and tedious from the user's
point of view.

5. Recommendations and Conclusions

The main aim of integration is to provide generative
systems so that these systems could still cope when
changes in design and manufacturing management occur.

The logic of obtaining the sequence of operations in
process planning is believed to be analogous to AI
planning which provide a robust foundation for developing
planning systems in general. This formal approach of
applying AI techniques is preferred to simply adding
manufacturing rules which could lead to very large and
unmanageable databases.

Future design systems should contain manufacturing
knowledge which could be defined explicitly with the

design data so that it could be used in the "ready form" further downstream in the manufacturing system such as in the process planning system. Consideration of other factors such as resources (e.g. machines, workers, budgets) and timing which have not been included are also important aspects of a manufacturing system.

When the difficulties of developing generative systems were realised, a semi-generative (combination of both variant and generative) approach was suggested [14].

However in order to achieve true integration of the manufacturing system, the generative approach should be adopted. This could be a costly and long term exercise which is why generative systems are still in their infancy. However as the generative approach and AI techniques become better understood, working generative systems are not an impossibility.

References

[1] Nilsson, N.J.: Principles of Artificial Intelligence, Palo Alto, California, Tioga Pub. Co., 1980, pp 275 - 319

[2] Cheung, Y.P.; Dowd, A.L.;: Artificial Intelligence in Process Planning, Computer Aided engineering Journal, August 1988, pp 153 - 156.

[3] Green, C.C.: Theorem Proving by resolution as a basis for question-answering systems, Machine Intelligence, Vol 4 (eds. Meltzer, B.; Michie, D.), New York, Elsevier Pub. Co. 1969.

[4] Genesereth, M.R.; Nilsson, N.: Logical Foundations of AI, Morgan Kaufmann, 1987 pp 263 - 283.

[5] Fikes, R.E.; Nilsson, N.J.: STRIPS: A new approach to the application of theorem proving to problem solving, AI, Vol 2, 1971 ,pp 189 - 208.

[6] Sarcedoti, E.A.: A structure for plans and behaviour, American Elsevier 1977.

[7] Chapman, D.: Planning for Conjunctive Goals, AI 32 (1987) pp 333 - 377.

[8] Descotte, Y.; Latombe, J.C.: GARI: A problem solver that plans how to machine mechanical parts, Proceedings of 7th International Joint Conference of AI, Vancouver, Canada, August 1981.

[9] Matsushita, K., et al: The integration of CAD and CAM by application of Artificial Intelligence techniques, 1982, Annals of the CIRP, 31.

[10] Wang, H.P.; Wysk, R.A.: A knowledge-based computer-aided process planning system, 19th CIRP International Seminar on Manufacturing Systems, Penn. State, USA, 1-2 June 1987.

[11] Houtzeel A, Schilperoort B, A chain-structured part classification system (MICLASS) and Group Technology, Proc. 13th Annual meeting and Technicla conference, Cincinnati, Ohio, March 1976, pp 383 - 400

[12] Opitz, A.: A Classification System to describe workpieces, Optiz H. (ed. MacConnell W R), Pergoamon Press, Elmsford, N. Y. 1970.

[13] Wang, H.P.: Intelligent Reasoning For Process Planning, Ph. D. thesis, Pennsylvania State University, Dept of Industrial & Management Systems Engineering, Dec 1986, pp 46 - 73.

[14] Emerson, C.; Ham, I.: An automated coding and process planning system using a DEC PDP-10, Computers and Industrial Engineering, 6(2), 1982.

A Knowledge-Based Approach for Small Parts Manual Assembly in the Space Environment

R. Luis Roman, S.N. Dwivedi

Department of Mechanical
and Aerospace Engineering,
West Virginia University,
Morgantown WV, 26506

ABSTRACT

The deployment of robots and their use for assembly in the space environment will provide for an increase in safety in the U.S. Manned Space Program. Robotic assembly, on the other hand, will be constrained by the capabilities of the particular robot gripper. When the part is of such size or complexity that it is inefficient for the assembly to be performed by a robot, it might be necessary for the labor to be performed manually, inside the space station or shuttle where the technician will not be obstructed in his labor by the space suit nor will he be completely exposed to the dangers of the space environment. This paper describes the development and formulation of rules and guidelines for manual assembly of small parts in the micro-gravity environment within a Knowledge Base methodology. The application of a Rule Based Sensitivity Analysis framework for the Artificial Intelligence decision stage of the design process is presented. The use of the DICETALK object oriented paradigm for the development of the Knowledge base is thoroughly explained. An outlook about further R&D work on this special subject closes the presentation.

INTRODUCTION

In recent years, because of the many advantages associated with automation and robotics, much work has been done in the development of systems to assist the designer and influence his thinking towards a design that is easier to assemble, be it by manual, automated or robotic means. Considerable knowledge is available as Design for Assembly rules. While the current rules and guidelines for assembly available in works such as BOOTHROYD/DEWHURST [1] are effective in the evaluation of parts for earth-based assembly, they were not developed for the implementation of assembly in space and because of this lack of consideration for the micro-gravity conditions some of the guidelines have no validity when space based considerations take into effect. For example, on earth environmental factors such as the weight of a part are important in the determination of part manipulation index, while in space this is less of a factor.

EVALUATION OF ASSEMBLABILITY

Efficiency of design in terms of assembly has traditionally been measured with the use of general guidelines and examples to aid the designer. Tipping [2] provides some insight to the formal procedure to be followed when designing a product with ease of assemblage in mind. At the time of publication of his book two methods of assembly were considered, manual and automated. His methodology, which is currently still valid was divided in three stages. In the first stage manual assembly is considered, on the second, automated and in the third a compromise is reached between manual and automated in order to achieve the optimum economical assembly method.

Product design for ease of assembly has four major goals[2]:
- Improvement of the effectiveness of assembly, i.e. increased productivity in relation to manpower and investment resources
- Improvement of Product Quality, i.e. improved product value from the buyer's standpoint in relation to product price
- Improvement of the Assembly System Profitability, i.e. increased utilization of equipment
- Improvement of working environment within the assembly system

The most important general design rule for ease of assembly is to minimize the number of parts in an assembly[3]. In determining the theoretical minimum number of parts, Salford University developed four criteria to decide whether or not a part may be combined with another. If a part meets any one of the criteria it must remain as a separate part. The four criteria are[1]:
- The part or subassembly moves relative to any other part in an assembly during the normal function of the assembly.
- The part must be made of a different material from its mating part.
- The part needs to be disassembled for servicing.
- The part, combined with its mating component, would prevent assembly of other parts that meet the first three criteria.

Tipping [2] provides us with a group of general guidelines for assembly, as presented in figure 1.

KNOWLEDGE BASE DEVELOPMENT

Before attempting to develop the knowledge base, the procedure leading to a solution must be found. To aid this, the work of a human expert should be analyzed and his approach to finding a solution should be examined. This knowledge acquisition is the first and the most important step in developing the knowledge base. This knowledge about the domain, can be found from experts in the field and the knowledge compiled in books, articles and other references.

After discovering the method for finding a solution, the specification for the knowledge base must be compiled. The domain specific knowledge must be represented in an efficient way. This representation must store the rules and facts about the system. Once the knowledge is organized, an efficient mechanism to deal with the rules and facts has to be developed.

Various knowledge sources are needed to build the knowledge base, one of the major information sources can be experts in the field, and lacking these, textbooks, reports, data bases, case studies, and empirical data.

Rule Based Sensitivity Analysis
Ishii [4],[5] introduces us the idea of rule based sensitivity analysis (RSA), a tool which we propose to use in order to apply Artificial Intelligence to the decision stage of the design process. RSA combines qualitative and quantitative AI tool within a single framework.

Rule 1. Components should be symmetrical if possible.
Rule 2. If components are not symmetrical, they should have marked polar properties by geometry and/or weight.
Rule 3. Components should have the least possible number of important directions.
Rule 4. Components which can tangle when in a mass should be avoided.
Rule 5. Consistency in the dimensions used in feeding, orienting and locating the component is essential.
Rule 6. Components should be designed for easy manual assembly. It follows that if the product is easy to assemble by hand, it will be easier to mechanize the assembly.
Rule 7. The product should be so designed that it has a datum surface or datum point on which to build the assembly.
Rule 8. The product should have location points.
Rule 9. Design the product so that one component can be placed on top of another.
Rule 10. Never turn the assembly over if it can be avoided.
Rule 11. Never bury important components.
Rule 12. Standardize.
Rule 13. Eliminate as many components as possible from the assembly.
Rule 14. Eliminate separate fastening whenever possible.

Figure 1. Rules and Guidelines for Mechanical Assembly

The use of RSA in KB Development
Designers must, upon initial design, evaluate particular proposed design solutions in order to optimize them. They do this by analyzing the proposed design against a set of performance constraints such as weight, volume, etc., in order to remedy a ny inconsistencies and improve the design. At this stage of redesign, designers use their knowledge of relationships between the design variables and performance. Discrete variables such as part characteristics can also be

handled with the introduction of a criteria order list, a list of possible choices ordered with respect to a certain criterion. Figure 2 shows an example of how such knowledge can be expressed. In it we show the four criteria for remaining a separate part, as developed by Salford University, in the rule based format.

```
!KnowledgeBase rules !
IF part movement is movementRelativeToOtherPartDuringNormal
FunctionOfTheAssembly
then part elimination is notPossible!
IF part movement is noMovement
  and part material is differentFromMatingPart
then part elimination is notPossible!
IF part movement is noMovement
  and part material is sameAsMatingPart
  and part disassemblyForServicing is needed
then part elimination is notPossible!
IF part movement is noMovement
  and part material is sameAsMatingPart
  and part disassemblyForServicing is notNeeded
  and part oombinationWithPartMate is preventingAssemblyOfOtherParts
then part elimination is notPossible!
IF part movement is noMovement
 and part material is sameAsMatingPart
 and part disassemblyForServicing is notNeeded
 and part oombinationWithPartMate is notPreventingAssemblyOfOtherParts
then part elimination is recommended! !
```

Figure 2. Requirements for part elimination, IF-THEN format.

The Rule based sensitivity table (RST) is a table of the above rules at a particular design point. Rule based sensitivity analysis is the use of inference techniques on the RST and is an interactive procedure that applies knowledge engineering to the iterative design process. The RST is composed of relationships between design variables and performance levels expressed in a production rule format.

		Part Movement							
		Moves				Fixed			
		Part Material				Part Material			
		Different		Same		Different		Same	
		Disassembly		Disassembly		Disassembly		Disassembly	
		Needed	Not Needed	Needed	Not Needed	Needed	Not Needed	Needed	Not Needed
Part Combination with Mate	Prevents Assemblage	Elimination not Possible	Elimination not Possible	Elimination not Possible	Elimination not Possible	Elimination not Possible	Elimination not Possible	Elimination not Possible	Elimination not Possible
	Does not Prevents Assemblage	Elimination not Possible	Elimination not Possible	Elimination not Possible	Elimination not Possible	Elimination not Possible	Elimination not Possible	Elimination not Possible	Recommend Elimination

Figure 3. RST for Part Elimination

BOOTHROYD/DEWHURST tables [4] for DFA represent a good RST representation of empirical knowledge on assembly. A sample table for the case of part elimination is shown in figure 3. From the table it can be deduced that the objective function is part elimination. This table can then be reduced to a tree structure form, figure 4, translated to the SMALLTALK object oriented language form and embedded into the DICETALK environment.

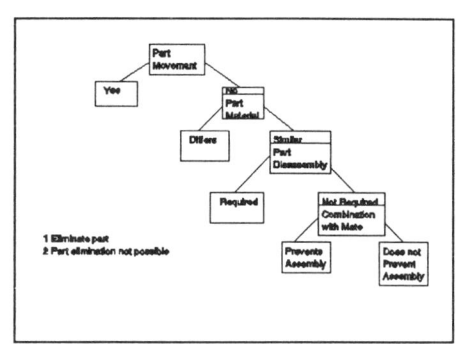

Figure 4. Binary Tree

Use of the Manual Assembly Knowledge Base in an Iterative Design Process

Figure 5 shows a typical use of the Knowledge base to obtain the part elimination decision. Based on the designer inputs, the objective function is reached in the RST.

During the iterative design process, the designer modifies the current design to improve it. Modifications may be required for a variety of reasons. The proposed design may be unacceptable with respect to a particular specification. Even if the designer has a working design, an additional specification or constraint could easily make the design unsatisfactory. Given a goal, such as an increase in performance, inference techniques applied to the RST can help find candidate modifications that satisfy the goal while satisfying all constraints.

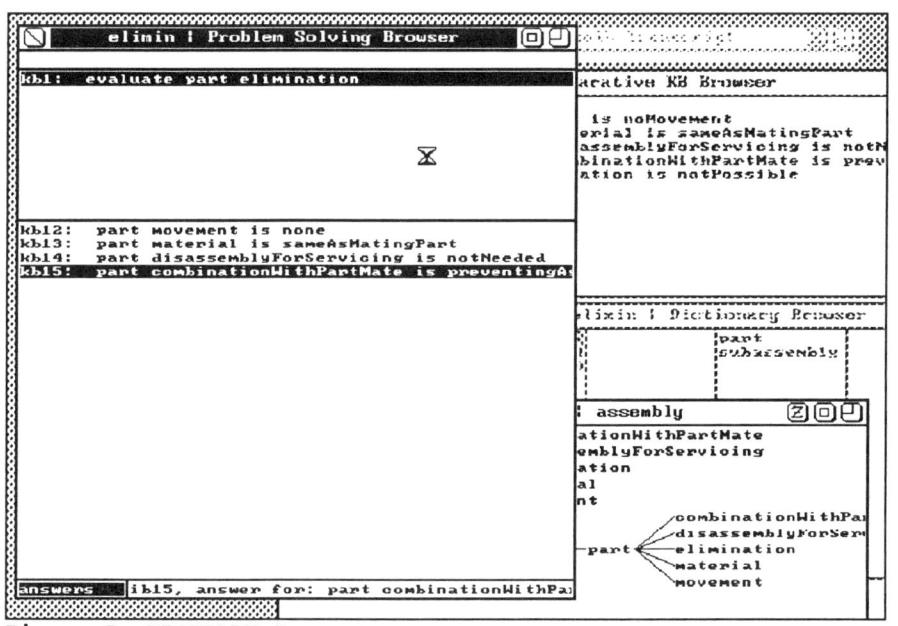

Figure 5. DICETALK Screen

434

As mentioned before, RSA involves the use of inference techniques on the RST in conjunction with conventional computational tools. Two major functions of RSA are:
    1) Advise on meeting design constraints during the iterative design process.
    2) Aid in design optimization

DISCUSSION OF RESEARCH AND ITS SIGNIFICANCE
This study sustains that it is indeed feasible to design and develop a Knowledge base for Small parts manual assembly in the space environment. The current research work is concentrated in the integration of various declarative knowledge bases with the procedural knowledge bases and the development of a metaknowledge base to improve performance of the system.

ACKNOWLEDGEMENT
Part of this work was carried out at the Goddard Space Center under a Grant by the National Aeronautics and Space Administration.

REFERENCES
[1] Geoffrey Boothroyd and Peter Dewhurst. Product Design for Assembly. Boothroyd Dewhurst, Inc. 212 Main St. Suite C-3, Wakefield, Rhode Island, 1987.

[2] Tipping, William Victor An Introduction to Mechanical Assembly. Business Books Limited: Mercury House, Waterloo Road, London, S.E.1. Great Britain 1969.

[3] Lane, Jack D. Automated Assembly. Dearborn, Michigan: Society of Manufacturing Engineers, 1986.

[4] Ishii, Kosuke. Knowledge Based design of complex mechanical systems PH. D. Dissertation. Mechanical Engineering Department, Stanford University, 1988.

[5] Ishii, K. and Barkan, P. Rule-based Sensitivity Analysis--a framework for expert systems in mechanical system design, in Gero, J (ed.), Expert Systems in Computer aided Design, North-Holland, Amsterdam, pp.179-198.

Building Industrial Expert Systems with Flex

V. Devedzic[*] and D. Velasevic[**]

[*] Mihailo Pupin Institute, Belgrade, Yugoslavia

[**] School of Electrical Engineering, University of Belgrade, Belgrade, Yugoslavia

Summary

FLEX, a specialized expert system building tool, is described in the paper. FLEX is capable of coupling numerical and symbolic computing, heuristic and model-based reasoning, and knowledge-based and sensor-based operation. It integrates several knowledge representation techniques, both standard, like rules, frames, parameters and variables, and nonstandard, which are not usually found in other expert system building tools. An outstanding FLEX feature is the possibility of building and testing real-time expert systems. FLEX can be qualified as a second generation expert system building tool. It is specialized for building expert systems in the field of flexible manufacturing. An example of applying FLEX for building an industrial expert system for controlling robot operation in a flexible manufacturing cell is also presented.

Introduction

FLEX (FLexible EXpert) is a special-purpose Expert System (ES) building tool. Its primary purpose is development of ESs in the field of Flexible Manufacturing Systems (FMS), which has a great impact on current industrial engineering.

There are four large domains in the field of FMS: design, planning, scheduling and control (Stecke, 1985). *Design* refers to specification and selection among the manufacturing requirements, FMS types, the range of part types or components to be produced, the numbers and types of machine tools and robots that are required, the types and amounts of flexibilities desired, the type and the capacity of the material handling system and the buffers, computer control requirements, the layout of the FMS, the number and type of fixtures and pallets, the strategies for running the FMS, etc. *Planning* includes part type selection, grouping the machines of each type, production ratio determination, pallets, fixtures and tools allocation among selected part types, etc. In the domain of *scheduling*, it is important to determine the optimal sequence at which the parts of the selected part types are to be input into the system, the priorities among the part types when there are several parts waiting to be processed, the sequence of operations that should be performed on each machine tool, etc. Finally, *control* means determining policies for handling machine tool and other breakdowns, for maintenance, and for inspection of parts, procedures for tool life and process monitoring and data collection, as well as for updating the estimates of tool life, etc.

A lot of the above problems are suitable for ES approach. FLEX is designed to support development of ESs that are used to solve such problems. The focus of the design was on the possibility of building real-time ESs, capable of performing on-line tasks in the domains of FMS scheduling and FMS control. This paper describes the FLEX design objectives and current implementation. It also presents an example of building a real-time ES for industrial robot control.

Architecture of FLEX

Global architecture of FLEX is shown in Fig. 1. Particular blocks are described in details in the following sections.

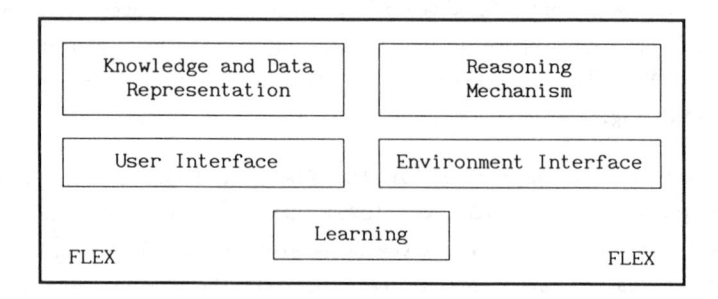

Fig 1. Architecture of FLEX

Knowledge and data representation

Due to the specific purpose of FLEX, the following knowledge and data can be represented when building an ES with this shell:

- basic concepts and heuristics
- application independent FMS knowledge and data
- time-varying data

To represent basic concepts and heuristics in a FLEX-based ES, three standard knowledge representation techniques are provided: rules, attributes and frames. Rules and attributes can be defined within a frame, or on top level, out of the frame tree (metarules and global attributes).

Application independent knowledge and data from the FMS domains described in the Introduction make a distinction between FLEX and more general ES shells. This kind of knowledge and data is built into FLEX to an extent, but it can be easily extended. There are three techniques for representing this kind of knowledge and data in a FLEX-based ES: models and algorithms, objects, and facts. By means of models and algorithms, qualitative and procedural knowledge in the four FMS domains is represented. For example, there are various models and algorithms for part type selection and for production ratio determination (Stecke, 1985). This provides the possibility to combine numeric and symbolic computation in FLEX-based ES, as well as heuristic and model-based reasoning. Such a possibility is typical for second generation expert systems (Devedžić and Velašević, 1990). Objects are used to represent structured data records (for example certain machine types, tool constructs, general application constraints, etc.). Facts are used to represent unstructured fixed data.

Another distinction between FLEX and more general shells are time-varying data that can be represented by FLEX. FLEX can be used to build consulting ES in certain domains of FMS (off-line ES), as well as real-time ES (on-line ES), for monitoring, scheduling and control tasks. In a real-time FMS-ES for industrial application, it is necessary to provide the information reflecting dynamics and changes in the manufacturing environment. FLEX provides two techniques for representing this information: sensor data, for establishing links with the external sensor system, and status data, for on-line feedback processing in cases when ES is applied as a control module in a more complex FMS control system (Devedžić, 1990).

Reasoning mechanism

Several reasoning and search techniques feature FLEXIE, the FLEX Inference Engine (backward and forward chaining, inheritance down the frame tree, reasoning with uncertain and multiple values, the possibility of combining conflict resolution strategies, reasoning with time varying data and variables, metarules, calling external procedures, etc.). It is up to the user to select the desired combination of these techniques for his application, or to use the default combination. This possibility features tools for building second generation expert systems (see Tong et al., 1987, and Wu et al., 1990).

FLEXIE has two modes of operation: off-line and on-line. In the off-line operation, reasoning is linked with user interface, in order to provide consultation, explanation, tracing and other facilities. In the on-line operation, these facilities are skipped for the sake of speed and real-time performance. In both modes of operation FLEXIE can communicate with the interface to the environment, in order to exchange information with other software modules, devices and systems.

User interface and environment interface

FLEX user interface supports communication with the user when building, testing and applying an ES, and provides means for simulating changes in the manufacturing environment of a real-time FLEX-based ES. It is based on windows, menus and dialog boxes, includes specialized editors for building knowledge and data bases, interface to external graphics packages, HELP messages, etc.

The FLEX environment interface provides means for embedding FLEX-based ESs into more complex systems, and for including asynchronous external information into the reasoning process (Sabharwal et al., 1988). It is essential for real-time operation of FLEX-based ESs, particularly in the domain of FMS control, when signals from external sensors and other devices, operating independently and asynchronously regarding the inference engine, must be included into the ES reasoning process.

Learning

The capability of acquiring new knowledge through machine learning is another essential feature of second generation expert systems (Devedžić and Velašević, 1990). Two kinds of learning can be achieved with FLEX: off-line learning, when building a knowledge base, and on-line learning, during the operation of a real-time ES. Off-line learning includes algorithms for learning rules from examples and for learning the order in which certain search and reasoning strategies are tried (Carbonell, 1989). Two on-line learning algorithms are built into FLEX, in order to improve ES task solutions and the performance of models and algorithms (Devedžić, 1990).

A FLEX-based real-time expert system for robot control

As an illustration of how FLEX can be applied for building ESs for industrial applications, the SCLES system is briefly described here. It was previously built by applying another shell (Devedžić, 1990), and it was rebuilt later using FLEX.

SCLES means Strategic Control Level Expert System. It performs the robot control task of selecting control algorithms for executing simple robot tasks (like streight-line moves of robot hand, gripping parts, compliant motion, etc.). SCLES is a real-time ES, capable of communicating with an external sensor system. It is a part of a more complex robot control system, which is intended for controlling robot operation in a flexible flexible manufacturing cell (Devedžić, 1990). Status information flow between SCLES and other robot control levels is provided.

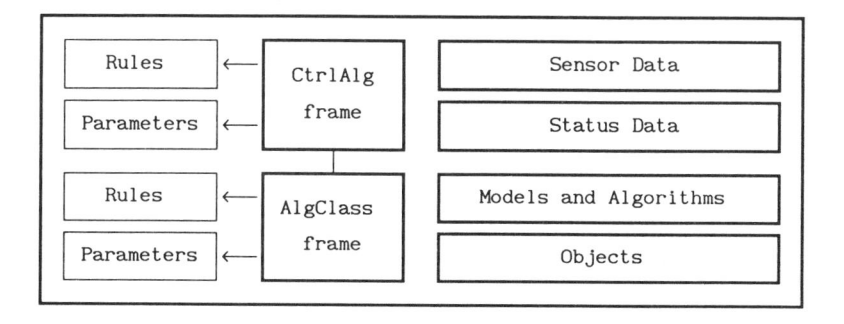

Fig. 2. Knowledge base of SCLES

Fig. 2 illustrates the SCLES knowledge base, which is developed with

FLEX. It contains a top-level frame (CtrlAlg) and a subframe (AlgClass), containing about 70 rules in total, and several attributes, objects, sensor and status data, models and algorithms.

Conclusions

FLEX is a tool for building ESs for design, planning, scheduling and control tasks in the field of FMS. Apart from traditional knowledge and data representation techniques, like rules and frames, it also includes some specific techniques, not typical for other ES building tools. Its multiple reasoning methods can be easily combined, in order to adapt the resulting ES to the application needs. FLEX can be used for building real-time ESs, and can be embedded into larger and more complex systems.

FLEX is capable of building second generation ESs, featured by heuristic and model-based reasoning, both numerical and symbolic computation, and the possibility of acquiring new knowledge thruogh machine learning.

References

1. Carbonell, J.G.: Introduction: Paradigms for Machine Learning. Artificial Intelligence 40 (1989), Special Volume on Machine Learning, 1-10.

2. Devedžić, V.: A Knowledge-Based System for the Strategic Control Level of Robots in Flexible Manufacturing Cells. The International Journal of Flexible Manufacturing Systems 2 (1990) 263-287.

3. Devedžić, V.; Velašević, D.: Features of Second-Generation Expert Systems - An Extended Overview. Engineering Applications of Artificial Intelligence 3 (1990) 255-270.

4. Sabharwal, A.S.; Iyengar, S.S.; de Saussure, G.; Weisbin, C.R.; Pin, F.P.: Asynchronous Production Systems for Real-Time Expert Systems. Proc. of 8th Int. Workshop on Expert Systems and Their Applications, Specialized Conferences, Avignon, France (May 30.-June 3. 1988) 113-146.

5. Stecke, K.E.: Design, Planning, Scheduling and Control Problems of Flexible Manufacturing Systems. Annals of Operations Research 3 (1985) 3-12.

6. Tong, X.; He, Z.; Yu, R.: A Tool for Building Second Generation Expert Systems. Proc. of 10th IJCAI Vol.1, Milan, Italy (August 1987) 91-96.

7. Wu, Z.; Yang, T.; Ni, F.; He, Z.; Yu, R.Z.: A C-Oriented Tool for Building Second Generation Expert Systems. Future Generation Computer Systems 6 (1990) 77-83.

A Distributed Knowledge-Based System for Total Manufacturing Support

Vevek Ram
Department of Computer Science
University of Natal
Pietermaritzburg
South Africa

Abstract

Companies worldwide have realized that the application of sophisticated computer technology in the manufacturing area is an effective means of coping with the demands of intensified global competition. Computer Integrated Manufacturing (CIM) is an example of such technology whose widespread use is an indication of the trend. Unfortunately, many view the use of technologies such as CIM as panaceas for manufacturing management problems. Although the use of computer-based technology in manufacturing does generate tangible benefits in scheduling, utilization of facilities and inventory management, its effectiveness in acquiring competitive advantage can only be optimized if it is viewed in a strategic context. In this context, the competitive use of CIM is dependent on the interrelationship between the organization's internal competencies and the external environment.

This paper describes a distributed knowledge-based, blackboard control architecture that is capable of integrating the various diverse aspects of manufacturing. Three separate knowledge modules interact in order to share knowledge and to cooperatively optimize the manufacturing function. The interaction and cooperation of these knowledge modules can ensure that a CIM strategy is aligned both with the organization's internal configuration as well as its external environment and that the strategic effectiveness of CIM extends well beyond the shop floor.

Introduction

The success of organizations worldwide that make extensive use of CIM has made managers realize that they have little choice in moving toward automated manufacturing. Unfortunately, many of them believe that this move will automatically enable their organizations to remain competitive. Researchers in the past few years have shown that the success of an organization arising out of the use of CIM does not come directly from the technology itself but rather from the interaction between the environmental forces and the organization's internal competencies and response mechanisms (Jaikumar [1], Kaplan [2], Knight [3], Wheelright [4], Williams and Novak [5]). As aptly quoted in [5],

".. CIM is not just a matter of setting up computer-controlled machines and turning a key..".

The task of identifying the relationships between environmental conditions and organizational capability in order to decide on an optimum CIM strategy is difficult because of its inherent complexity. This is due not only to the multiplicity and uncertainty of the input that needs to be considered but also because of its dynamic nature. Environmental forces are constantly changing and the relationship must therefore also be continually reassessed. Computer-Based support in the form of Expert Systems has been successfully used in complex management domains (Ernst [6], Feigenbaum, McCorduck and Nii [7], Ram [8], Ram [9]), and can also be used to support the decision making task in this case. The prototype system developed for this application makes use of a distributed architecture with a blackboard as the controlling and communication mechanism. The system configuration is shown in Figure 1.

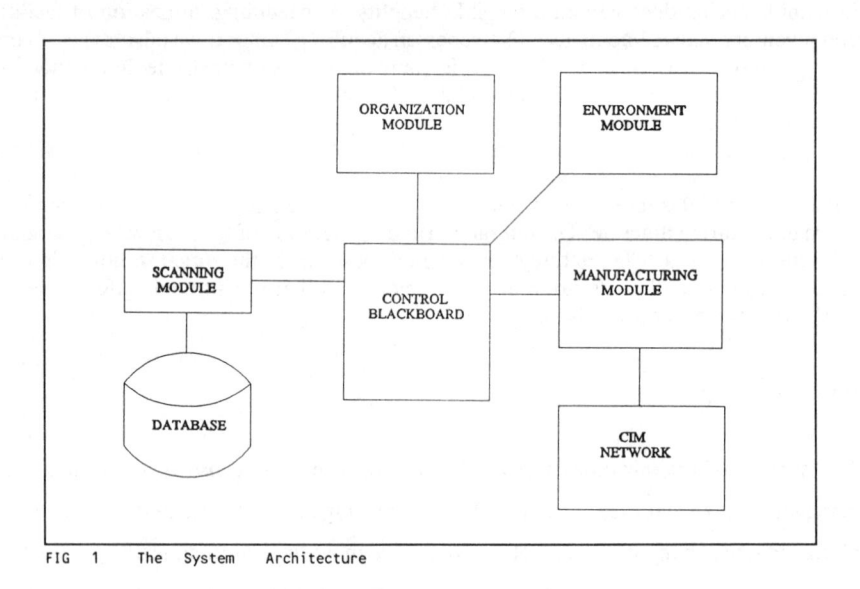

FIG 1   The System   Architecture

The system consists of a network of modules which interact through the blackboard in order to co-operatively solve a problem. Apart from the Scanning module, all the modules contain knowledge about their respective domains. This knowledge or domain expertise is represented as a combination of production rules and frames. A brief description of the knowledge sources or modules and a discussion of their major roles in the distributed network follows. As the scope of this paper limits the degree to which the constructional details of the system can be described, only the more important aspects of the system will be discussed.

The System Components

The Control Module. The control module acts as the general management expert and also as the manager of the network. As the general management expert, it controls the direction and format of the network problem solving process. It contains meta-knowledge, which is knowledge about how the rest of the system's knowledge is distributed throughout the network. This meta-knowledge allows the control module to decide that interest rates concern the financial expert, raw material costs concerns the manufacturing expert and so on. As the network controller, the control module controls the execution of individual modules as well as the management of the status of the blackboard.

The Scanning Module. The scanning sub-system acts as the machine interface between the network and the organization. The scanning module monitors a set of strategic factors and reports all variances to the control module via the blackboard. It performs a simple but nevertheless important role in the network. The module is non-intelligent in that it reports all variances. The control module decides on the severity of an occurred variance.

The Organizational Module. This module contains knowledge about the marketing, finance and human resource functions and policies of the organization. Typical marketing knowledge includes market relationships in terms of market coverage, nature of markets, market development, breadth of product line, degree of customization, nature of distribution, pricing and credit policies and promotional and advertising policies. Typical financial knowledge includes debt structure, investment structure and policies regarding inventories, receivables, expenditure and taxation. Typical human resources knowledge includes policies regarding recruitment, rewards, training, and promotion.

The Environmental Module. This module contains knowledge about the organization's environment in terms of Competitive forces in terms of direct or potential competition, substitute products, forward integration by suppliers or backward integration by customers; Legislative forces, Economic forces, Technological forces, Social forces and Pricing and Demand sensitivity.

The Manufacturing Module. This module acts as the production expert in the network. It uses information derived from the blackboard to perform production planning and scheduling functions. It is also connected to and is able to control various functions in the CIM network. Because the continuous communication between this module and the database via the blackboard, the module is able to make use of embedded knowledge to control the balance between push and pull systems in order to optimize the production output. Typical knowledge in this regard is the relationships between lead-time variability, production stage and production process (Karmarkar [10]). The manufacturing module also performs the important task of intelligent scheduling. This is accomplished by using the multiple worlds framework and assumption based truth maintenance (Filman [11]) which has several advantages over the conventional scheduling methods. A world is a set of facts and assumptions. Once a world is created, facts can be added or deleted from it. A world can thus represent a set of changes over time. Truth maintenance systems preserve the consistency of assumptions in changing worlds. In the scheduling application, the problem is broken down into subproblems and solutions are generated in a time unit. The time unit chosen (minutes or hours) together with the production assumptions made in that time unit is a world. When a world is consistent, that is, there are no constraints violated, the next set of worlds are generated. Constraints are propagated across worlds and invalid solutions are immediately discarded. The process of generating worlds ends when there are no more remaining time units: the end of a production day. The partial solutions represented by each world can be merged to form a complete solution or schedule.

Operational Overview.

Each knowledge source is responsible for maintaining a set of strategic variables in its own domain. Variables are categorised as either internal or external depending on whether the entity that a variable relates to is changed from within or outside of the organisation's boundary. As long as the values of these variables remain within predefined limits, there is a balance between the organisational ability, environmental pressure and a chosen manufacturing strategy. Values for the internal variables are held in the organisational database which is constantly updated through the organisation's information system. External variables are updated through manual input on a regular basis. All variables are monitored by the scanning subsystem. When the value of a variable changes, the scanning subsystem communicates this change to the control module. The control module decides on the degree of severity of the variance (and others which may occur simultaneously), assigns priorities and then decides on which modules need to be called in order to resolve the problem. It then posts a request with parameters describing the nature of the variance on the blackboard and activates the appropriate expert module or knowledge source. A major problem for the control module in the execution of this function is "knowing" which module to call for a given request. A simple and effective way to overcome this problem is to maintain a list that links all the relevant organizational data items with the modules responsible for them. Such a list represents Meta-level knowledge since it represents knowledge about the use of the distributed expertise in the most efficient way. When a request that can be resolved by a single module is received, the control module need only scan the list in order to identify the module best suited to resolving the request. A problem arises when a request is received that cannot be resolved by one module alone. Such a request has to be decomposed into subrequests that can be resolved by individual modules. This decomposition process can be implemented by organizing the decomposition relationships into a taxonomy of meta-knowledge frames. The individual knowledge source assesses the impact of the change in relation to the present strategic posture and communicates the result back to the control module via the blackboard. If the result concerns other knowledge sources, these are then activated by the control module. The process continues until a final result is obtained that is consistent with all the experts individual results. If two or more experts put forward recommendations that are conflicting, the control module can resolve the conflict by choosing the recommendation with the highest utility value or by modifying and reposting variables on the blackboard so that the individual experts reassess their respective recommendations and in so doing resolve the conflict themselves after a number of cycles.

Conclusion

As pointed out by Novak and Williams [5], successes brought on by the advances of CIM flow fundamentally, not from the technology per se, but from the responsiveness of the organization's internal configuration to external forces. This paper has described the design of a prototype distributed manufacturing support system that takes these factors into consideration and enables the CIM strategy not only to increases quality and lower costs but also to be an integral part of the overall corporate strategy.

References

1. Jaikumar, R., Postindustrial Manufacturing, Harvard Business Review, Nov/Dec 1986, 69-76.

2. Kaplan, R.S., Must CIM be justified by faith alone?, Harvard Business Review, Mar/Apr 1986, 87-93.

3. Knight, R., Plan first then use CIM, Software News, Dec 1987, 61-64.

4. Wheelwright, S.C., Japan: Where operations really are Strategic, Harvard Business Review, Jul/Aug 1981, 67-74.

5. Williams, J.R. and Novak, R.S., Aligning CIM Strategies to Different Markets, Long Range Planning, 23(1), 1990, 126-135.

6. Ernst, C.J., Management Expert Systems, Addison Wesley, UK, 1988.

7. Feigenbaum, E., McCorduck, P. and Nii, H., The Rise of the expert Company, Macmillan, London, 1988.

8. Ram, V., A System of Cooperating Experts for Strategic Planning Support in Business, in Mallouppas, A., (Ed), Management Technology: Control Tools for the 90's, London, Peter Perigrinus Ltd, 1990.

9. Ram, V., A Distributed Knowledge-Based Support System for Strategic Management, Unpublished PhD Thesis, University of Natal, Pietermaritzburg, 1990.

10. Karmarkar, U., Getting Control of Just-In-Time, Harvard Business Review, Sep/Oct 1989, 122-131.

11. Filman, R.E., Reasoning with Worlds and Truth Maintenance in a Knowledge-Based Programming Environment, CACM, 33(4), 1988, 382-401.

Intelligence Computer-Aided Industrial Ecology System

V.V. KUPRIYANOV, N.I. KUPREEV

Mechanical Engineering Research Institute
USSR Academy of Sciences

Summary

Knowledge presentation schema, inference engine for the research, sample of expert system surface used for consultancy and ecology system acquisition are described. A description fragment of domain model is examined. The research prototype of the expert system is oriented to the development of consultancy models designed for the solution of classification problems which have a limited set of facts. The expert system shell icludes the knowledge presentation system, the inference engine and the explanation component. The knowledge base is determined by object-oriented programming facilities. The employed inference rules of knoledge processing make possible the inference of new facts and rules out of a certain initial set. The employed user language is interactival, menu-type. The development of shell prototype version for the expert system was conducted at the knowledge presentation language layer in terms of expert concepts. Shell software support for the consultative model was implemented in an IBM PC/AT computer in C language within the MS-DOS environment.

Knowledge Presentation

The basis for the knowledge presentation model is the assumption that the described problem domain is characterized by a finite set of facts serving as inference during the interpretation of the finite problem data volume. The knowledge is described by facts, hypotheses and decision rules. The knoledge facts is determined during user dialogue. Every hypothesis possesses a certain amount of confidence [1].
The interpretation of facts and the acception/rejection of hypotheses is implemented by the decision rules in the form of the following propositions:

$$A_1 \& A_2 \& \ldots \ldots A_n \rightarrow I \qquad (1)$$

where A_i, i=1, n - the i-th antecedent of the decision rule to check whether the fact/hypothesis belongs to the confidence interval; I - the consequent, which produces the new knowledge of fact/hypothesis during the implementation of the left-hand part of the rule.

The following rule types are implemented in the system: fact ⟶ fact, fact ⟶ hypothesis, hypothesis ⟶ hypothesis. The first rule type designates the assumptions inferred from facts already established; the second rule type is employed for the description of facts used for confirming the validity of a certain hypothesis(designation of the hypothesis),whereas the third rule type is used to describe the correlation existing between different hypotheses.

The designation of facts and hypothesis is performed by system's infere engine. System of estimates containing the numerical value, the logical value (true, false, undef) and the confidence interval (a finite subset of integers, which the numerical values are brought down to belong to is employed for the interpretation of logical connectives between the knowledge components. The logical values are expressed via 1.0 and - 1.0.

Formalization of Knowledge Presentation

The process of identification of the RP problem was applied to describe the formal model of knowledge presentation, where the RP problem was expressed by the following:

$$RP = \langle F, H, I, V \rangle , \qquad (2)$$

where $F = \{f_n / n = \overline{1, i}\}$ - suppositions-facts;
$H = \{h_k / k = \overline{1, j}\}$ - hypotheses-facts; I - set of decision rules $A \rightarrow B$; $V = \{II\}$ - set of integers to determine the system of estimates; R - set of rules.

The problem domain contained:

$F = \{f_1, f_2, f_3, f_4\}$; $H = \{h_1, h_2, h_3, h_4, h_5, h_6\}$;
$I = \{R_1, R_2, R_3, R_4, R_5\}$; $V = \{II\}$

Every I element was defined as: (3)

$R_1 : f_1 \& f_2 \rightarrow h_1$; $R_2 : f_2 \& f_3 \& f_4 \rightarrow h_2$; $R_3 : f_1 \& f_4 \rightarrow h_3$;
$R_4 : h_1 \& h_2 \rightarrow h_5$; $R_5 : h_3 \& h_5 \rightarrow h_6$;

The semantic system (3) may be presented as a GI (F, H, R) inference graph. The RP model is inferred by means of I in case there exists at least one pathway to connect all F vortex. The RP model may be made more complicated by introducing additional rules like fact - fact or adding estimation weights. The algorithm of search for the minimal cost decision tree using the connectivity matrix for the estimate weights is supposed to be examined.

Inference Engine

The inference engine contains the interpreter of the decision rules and the designation mechanism. The interpreter controls the rule application chain to achieve the fact. The designation mechanism implements both the selection of the most reasonable queries and their designation, i.e., the assignment of a certain confidence estimate to the concent (fact/hypothesis). Inference management included the following phases: selection of an advisable query, designation of the fact and model revaluation. The phases were repeated in cycles until the complete list of facts was designated.

The following operations and predicate symbols were used to describe the inference algorithm: f(Vol (concent)) - concent designation; f(Recost (   )) - search for the designated hypothesis (h) with the maximum estimate Vh_{max} = true, if the hypothesis exists; $p(Fh_{max}(h, V))$ - search for hypothesis (h) in FH, which is estimated higher than V, Hh_{max} = true, if the hypothesis exists; "goto" - transfer to the specified step; p(Notvol (C)) - the predicate returns "true" if concent C is not designed; p(Not Empty (Ant)) - the predicate returns "true" if the rule antecedent list is not empty; FH is the set of all fact ⟶ hypothesis type rules; (kFH (i)) consequent is the hypothesis; Ant(j) rule antecedent is the fact; Ant(j)FH(i) is the j-th antecedent of the i-th FH rule; Ant FH (i) is the antecedent list of the i-th FH rule; HH is the set of all hypothesis ⟶ hypothesis rules; HH(i), kHH(i) and Ant(j)HH(i) are concepts similar to FH case.

Description of Language Objects

The expert system language was designed with regard to the semantics of knowledge presentation components, including facts, hypothesis and rules[2]. The expert system language syntax rules were developed using the Backus-Naur description forms for formal agreements. The fact descriptor includes the ⟨hdr fact⟩ :: = "+"/"-" symbol to indicate whether the fact is active or not; the fact ⟨name⟩ containing alpha-numeric characters and special symbols; the ⟨numeric⟩ fact body to determine the type of the answer and the method of its interpretation. It is expressed by an integral value (1 to 3). Example of fact description:

+ Fact ≠ 1-2 (2 [REGULATOR CUTS PROFILE COMPLETELY]);
Fact 1-2 is active and should be introduced first. The described fact will be displayed by user interface as:
regulator cuts profile completely
⟨ T ⟩

The ⟨hypotes⟩ hypothesis specifier starts with the symbol: ⟨hdr hypo⟩:: = "?", followed by the ⟨name⟩ and the body ⟨sent⟩ of the hypothesis. The hypothesis semantics are described by ⟨sent⟩ and will be shaped in accordance with the rules similar to the ⟨sent⟩ of the fact.
Example of hypothesis description:
? - Type - Accidents ([SECTION NVS2 BLOCK CHECKING DUE TO CABLE BREAK]).
Hypothesis are displayed by user interface only after their designation and the "Display current inference" instruction. The ⟨rule⟩ descriptor consists of the left-hand and the right-hand parts divided by an "⟶" arrow symbol. The ⟨antec⟩ left-hand part of the rule is inscribed as a list of bracketed structures. The right-hand part is inscribed as the fact/hypothesis ⟨name⟩, followed by one of the values ⟨numeric⟩= i; i=1,3, which determine the designated fact/hypothesis.
Example of rule description:

Fact ≠ 1-2 (true), set SN(2)⟶ TYPE - ACCIDENTS (5).
The rule says: if true fact is FACT ≠ 1-2 and the fact value SET SN is 2 then accept hypothesis TYPE - ACCIDENTS with "5" validity. The described implementation possesses all the basic functions of the problem - independent expert system shell. The system consists of 11 modules containing 1500 statements. The

The problem of operational consultancy of the operator of computer-aided industrial ecology monitoring system was employed as a possible domain for the expert system operability testing. Examine several rules which correspond to the given problem as an example of software support.

2 METHANE - 1 (1.2) & AFR (3) & PFV (1) ⟶ AIR 1

The hypothesis interpretation is as follows: if methane concentration = 1.2%, air flow regulator (AFR) is completely closed and primary flow ventilator (PFV) operates at minimal flow rate, then the section lacks air.

3 AIR (1) ⟶ REGULATOR (10)

The latter rule describes the possible connectives between the condition of the regulating devices and the air flow velocity. System's further development implies the creation of model database; the construction of a vertical inference engine and the perfection of inference algorithm; the creation of an interface with one of the industrial DBMS's to provide access to digital databases; the production of industrial shell version to architect consultancy and diagnostic expert systems; the automated formation of knowledge bases in the formal logic language. The user is supposed to be able to immediately apply the recommendations of the expert system, correct them in the conversation mode and receive recommendations in accordance with the changing situation. Exclusive orientation of expert systems to the class of management tasks as compared to the general purpose source systems makes them easily adaptable.

References

1. Ukolov I.S., Kuprianov V.V., Suslov M.L. Method to Solve the Problem of Knowledge Representation of the Complex Dynamic Objects // Proceed. of the 41st Inter. Symposium on Spase Systems. - Dresden, October 6-13, 1990.

2. Kupriyanov V.V., Ukolov I.S. Information-Theory Methods in Artificial Intelligence // Proceed. of the 11th Prague Conf. on Information Theory, Statis. Decision Functions. -Prague, August 23-31, 1990.

An Expert Systems Approach to Setup Reduction

Seung Lae Kim, Bay Arinze, and Snehamay Banerjee

Department of Management
Drexel University
Philadelphia, PA 19104

ABSTRACT

This paper describes a knowledge based decision support system (KBDSS) for supporting manufacturing managers involved in making investment decisions for setup reduction. It incorporates expert knowledge from a number of knowledge domains within its knowledge base, and employs user/system interaction to facilitate the diagnosis of different manufacturing contexts. The KBS further incorporates knowledge of alternative setup reduction methods and assumptions associated with the different setup cost functions e.g., straight-line, exponential, and logistic functions etc.

INTRODUCTION

There had been no serious analytical work on this topic before the work of Porteus (1985), as setup cost was generally assumed to be constant. Some authors, such as Porteus (1985, 1986), Chakravarty and Schtub (1985), Billington (1987), and Kim (1990) used mathematical functions to model the behavior of setup cost when investments are infused to reduce it. Other authors, such as Flynn (1987), Spence and Porteus (1987), Zangwill (1987), Hahn, Bragg, and Shin (1988), Fine and Porteus (1989), Moily (1990), have demonstrated that setup reduction has beneficial effects such as smaller lot size, improved quality, an increase in production capacity, and production flexibility.

Decisions relating to setup reduction made by a manufacturing manager are typically difficult and complex, involving multiple criteria. In addition, consequences of erroneous decisions in a particular manufacturing setting may be heavy, ranging from little or no observable benefits from (insufficient) investment, to overspending on setup reduction (Billington, 1987; Kim, 1990).
The decision further involves expert assessments of the decision maker and other experts pertaining to the level of existing technology, plant layout, throughput sought, and other relevant factors. A considerable amount of expertise is therefore required in the consideration of these factors, the determination of an adequate setup cost function for the existing context, and the derivation of the optimal investment amount (Billington, 1987; Kim, 1990).

THE KNOWLEDGE BASED MODEL

Any Knowledge Based Decision Support System or KBDSS for supporting investment decisions in setup reduction must possess both a procedural and declarative capability to deal with both the derivation of the optimal investment through algorithmic means, as well as the heuristics or rules utilized in the practice of setup cost reduction. Parsaye and Chignell (1988) provide a useful framework for knowledge representation that facilitates effective knowledge engineering (KE) for the KBDSS. They define KE as consisting of the following processes, namely: (a) Naming: the relevant entities of interest; (b) Describing: these entities i.e., entity purpose, required inputs, internal procedures, and outputs; (c) Organizing: the entities or objects based on their characteristics; (d) Relating: the entities to understand their interrelationships; and (e) Constraining: the entities within a range of expected behavior.

The primary entity of interest is the investment decision, that is, whether or not to invest in setup reduction, and if so, how much investment to make. There are correspondingly, four basic elements viewed as requiring determination by the KBDSS, forming the goals (or diagnosis and recommendations) of the system. These are (1) The form of the mathematical function that matches the description of the manufacturing setting. The three types of functions that analyzed by the KBDSS were straight-line, exponential, and logistic functions. (2) The second KBDSS function is the investment decision, namely, whether or not to invest in setup reduction. The KBDSS' recommendations are: an out right decision to invest (at, or as near to the optimal investment point as possible) when benefits are shown to clearly result; and not investing in setup reduction, when negative circumstances exist, or when no gains are indicated by the total inventory costs (TRC). (3)The focus of the investment is a further decision, specifically, whether investment should primarily be in automation, worker training/hiring, or divided between the two (i.e., a mixed strategy). (4) The final diagnosis involves a heuristic-based assessment of the risk involved in investing in setup reduction. The risk involved is that of not attaining the desired efficiencies following investment in setup reduction.

Determining a set of rules for investing in setup reduction suffers from the unavailability of empirically-derived guidelines for suitable strategy selection. However, the work of Hall (1983), Schonberger (1982a & b) and Fogarty, Blackstone and Hoffman (1991, ch.17) indicate the existence of factors which appear to be determinants of both the investment decision and accompanying recommendations of the KBDSS (i.e., focal point of investment, possible risks accompanying such an investment and the setup reduction function). These will subsequently be reflected in the form of rules by the KBS. Six factors identified by these and other authors as central determinants of the investment strategy are: (1) Automation: refers to whether the manufacturing processes are manual or automated. (2)-(3) Worker skills and motivation: these represent two determinants of the investment decision for evaluating possible risks in investing in setup reduction. (4) The existing level of technology in the manufacturing setting. (5) Layout: three layouts were considered for inclusion into the KBDSS knowledge base, i.e., the assembly line, group technology, and job shop alternatives. (6) The Variety of Products: was considered to be very important to the investment decision. A large variety of products mandated investment in setup reduction, while low product variety suggested investments in setup reduction only if benefits could be shown through derivation of the optimal investment.

The knowledge areas represented in the KBS consisted of six determinant factors and four areas of recommendations described above. Consideration of the goals and their corresponding conditions through a literature survey and analysis resulted in development of 24 rules for providing expertise-based decision support for the setup reduction investment decision. These rules (see Table 1) were transformed into knowledge-based representations in the 1st class expert system shell. The overall structure of the KBDSS for investment decisions in setup reduction is shown in Figure 1. This structure shows the chained knowledge bases linked by the main or calling knowledge base.

Table 1 : Rules For Investment In Setup Reduction

Rule#	Automation	Technology	Layout	No. of Products	Worker Skill	Worker Motivation	Setup Reduction Function	Investment Decision	Investment Focus	Investment Risk
1	Automated	High-Tech	Assem-Line	High Variety	*	*	S-shaped	Yes, if + ve	Automation	High
2	Automated	High-Tech	Group-Tech	High Variety	*	*	S-shaped	Yes, if + ve	Mixed	Moderate
3	Automated	High-Tech	Assem-Line	Low Variety	*	*	S-shaped	Yes, if + ve	Automation	Moderate
4	Automated	High-Tech	Group-Tech	Low Variety	*	*	S-shaped	Yes, if + ve	Mixed	Moderate
5	Automated	Medium-Tech	Assem-Line	High Variety	*	*	Linear	Yes, if + ve	Automation	Moderate
6	Automated	Medium-Tech	Group-Tech	High Variety	*	*	Linear	Yes, if + ve	Mixed	Moderate
7	Automated	Medium-Tech	Assem-Line	Low Variety	*	*	Linear	Yes, if + ve	Automation	Moderate
8	Automated	Medium-Tech	Group-Tech	Low Variety	*	*	Linear	Yes, if + ve	Mixed	Low
9	Manual	Medium-Tech	*	High Variety	High	High	Linear	Yes, if + ve	Mixed	Low
10	Manual	Medium-Tech	*	Low Variety	High	High	Linear	Yes, if + ve	Mixed	Low
11	Manual	Low-Tech	*	High Variety	High	High	Convex	Yes, if + ve	Worker	Low
12	Manual	Low-Tech	*	Low Variety	High	High	Convex	No	Worker	Low
13	Manual	Medium-Tech	*	High Variety	High	Low	Linear	Yes, if + ve	Mixed	Moderate
14	Manual	Medium-Tech	*	Low Variety	High	Low	Linear	Yes, if + ve	Mixed	Moderate
15	Manual	Low-Tech	*	High Variety	High	Low	Convex	Yes, if + ve	Worker	Moderate
16	Manual	Low-Tech	*	Low Variety	High	Low	Convex	No	Worker	Moderate
17	Manual	Medium-Tech	*	High Variety	Low	High	Linear	Yes, if + ve	Mixed	Moderate
18	Manual	Medium-Tech	*	Low Variety	Low	High	Linear	Yes, if + ve	Mixed	Moderate
19	Manual	Low-Tech	*	High Variety	Low	High	Convex	Yes, if + ve	Worker	Moderate
20	Manual	Low-Tech	*	Low Variety	Low	High	Convex	No	Worker	Moderate
21	Manual	Medium-Tech	*	High Variety	Low	Low	Linear	Yes, if + ve	Mixed	High
22	Manual	Medium-Tech	*	Low Variety	Low	Low	Linear	Yes, if + ve	Mixed	High
23	Manual	Low-Tech	*	High Variety	Low	Low	Convex	Yes, if + ve	Worker	High
24	Manual	Low-Tech	*	Low Variety	Low	Low	Convex	No	Worker	High

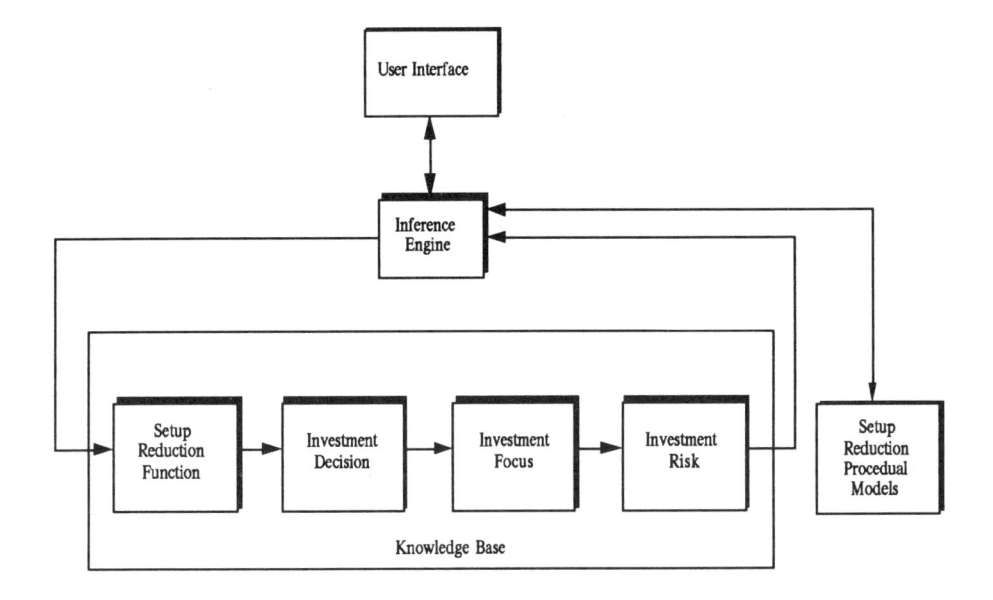

Figure 1 : KBDSS Structure for Investment in Setup Reduction

CONCLUSIONS

This paper has outlined the structure of a knowledge based decision support system or KBDSS for supporting investment decisions for setup reduction, and has demonstrated the use of a developed prototype system. Setup reduction is increasingly important to manufacturing entities as JIT techniques become more widespread, and many firms seek to both increase the variety of manufactured products, and to create further efficiencies in the manufacturing process. The KBDSS proposed in this paper utilizes acquired knowledge of setup reduction, and user-entered facts about the manufacturing context to propose recommendations relating to investments in setup reduction.

REFERENCES

1. Billington, P.J., "The Classic Economic Production Quantity Model with Setup Cost as a Function of Capital Expenditure," *Decision Sciences,* Vol.18, No. 1, pp.25-40, 1987.

2. Chakravarty, A. K., and Shtub, A., "New Technology Investments in Multistage Production Systems," *Decision Sciences,* Vol.16, No.3, pp.248-264, 1985.

3. Fine, C.H. and Porteus, E.L., "Dynamic Process Improvement," *Operations Research,* Vol.37, No.4, pp.580-591, 1989.

4. Flynn, B.B., "The Effects of Setup Time on Output Capacity in Cellular Manufacturing," *Int.J.Prod.Res.,* Vol.25, No.12, pp.1761-1772, 1987.

5. Fogarty, W.D., Blackstone, H.J., and Hoffmann, R.T., *Production and Inventory Management,* South-Western Publishing Co., Cincinnati, OH, 1991.

6. Hahn, C.K., Bragg, D.J., and Shin, D., "Impact of the Setup Variable on Capacity and Inventory Decisions," *Academy of Management Review,* Vol.13, No.1, pp.91-103, 1988.

7. Hall, R.W., *Zero Inventories,* Homewood, Illinois: Dow Jones-Irwin, 1983.

8. Kim, S.L., *Setup Cost Reduction Models and Synergistic Effects,* unpublished doctoral dissertation, Penn State University, 1990.

9. Moily, J.P., "The Economic Manufacturing Quantity Model and Its Implications," *Working Paper, Merrick School of Business,* University of Baltimore, 1989.

10. Parsaye, K., and Chignell, M., *Expert Systems for Experts,* Wiley, New York, NY, (1988).

11. Porteus, E.L., "Investing in Reduced Setups in the EOQ Model," *Management Science,* Vol.31, No.8, pp.998-1010, 1985.

12. _____, "Investing in New Parameter Values in the Discounted EOQ Model," *Naval Research Logistics Quarterly,* Vol.33, No.1, pp.137-144, 1986.

13. Schonberger, R.J., *Japanese Manufacturing Techniques,* New York, The Free Press, 1982a.

14. _____, Some Observations on the Advantages and Implementation Issues of JIT Production Systems," *Journal of Operations Management,* Vol.3, No.1, pp.1-11, 1982b.

15. Spence, A.M. and Porteus, E.L., "Setup Reduction and Increased Effective Capacity," *Management Science,* Vol.33, No.10, pp.1291-1301, 1987.

16. Turban, E., "Expert Systems - Another Frontier for Industrial Engineering," *Computers and Industrial Engineering,* 10, No. 3, pp. 227-235, 1987.

17. Zangwill, W.I., "From EOQ Toward ZI," *Management Science,* Vol.33, No.10, pp.1209-1223, 1987.

Application of an Expert System to Robot Gripper Selection

Chien-Te Ho, Michael DeVore, and A. Sherif El-Gizawy
Industrial and Technological Development Center
University of Missouri-Columbia
Columbia, MO 65211

Summary

An expert system is presented in this paper for the characterization of part handling tasks in robotic assembly systems. The knowledge of part geometry, surface condition, and working environment which are particularly critical in the selection of robot grippers are organized in the system. This system is designed to replace some experienced personnel in the process of selecting a set of suitable robot grippers for handling tasks. In addition, the group technology coding system for both parts and grippers should prove valuable in the integration of products, processes, and assembly systems design.

1.0 Introduction

Assembly is often the most expensive and labor intensive process in a manufacturing system. Consequently, the use of the industrial robot for automatic assembly has become widespread. Since a robot is a reprogrammable manipulator and is able to carry out many different tasks, the end effectors of a given robot can vary greatly. Among these end effectors, the most common is known as a gripper, the grasp mechanism of a robot. In general, the capability of a single gripper is very limited in that it can handle only certain types of objects. The ability of a gripper usually is confined by the object's shape, size, weight, surface condition and temperature. Besides object limitations, the working environment can also affect the performance of a gripper. Grippers may encounter unexpected conditions during their operation. These conditions can be easily overlooked by the operator until the gripper (or even the robot) is damaged. It is therefore very important to be able to take into consideration the many factors involved in selecting a gripper for a given job.

Several attempts have been reported to develop classification systems for grippers [1,2,3]. Early on, Lundstrom, et. al. [1] and Chen [2] proposed very simple classification systems for grippers. Very little, however, was mentioned about rational gripper selection in their work. Wilson and Tan [3] have developed a seven digit group technology coding system for robot gripper selection, which is useful in developing a short list of suitable grippers for particular application. Nevertheless, the program is not user friendly and does not help a user with the coding and selection process. Most recently, Pham and Yeo [4,5] proposed a strategy for robot gripper design and selection utilizing knowledge-based technologies. Their system uses a ranking technique to select the most cost effective gripper. The system, however, is limited to only three types of grippers: jaw, vacuum, and magnetic types. Therefore, there is a need for a computer aided method that would indicate in a rational fashion how a component should be gripped and with what type of gripper.

2.0 Knowledge-Based Gripper Selector

The present knowledge-based gripper selector consists of three major components: part families, gripper families, and the domain knowledge. The block diagram of this system is shown in Figure 1.

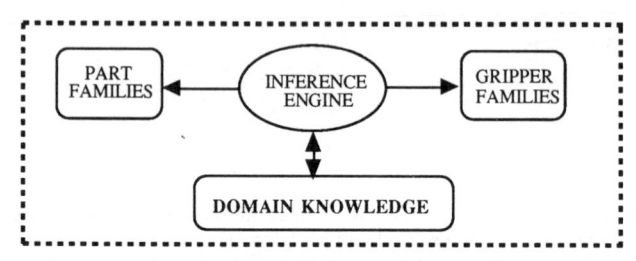

Figure 1: Components of the Expert System

Both input parts description and output grippers specifications are classified into families using group technology coding systems. The inference engine applies the domain knowledge to the classified input part to find feasible grippers. Gripper families are stored in a built-in library. If the system finds that one of the grippers is well suited to the input part, it will be retrieved from the library and displayed to the user. The domain knowledge consists of expertise in the field of grippers and other relevant areas. This knowledge was collected from literature in this specific area and arranged into appropriate rules.

2.1 Parts Classification System

The code structure used for describing different part families is shown in Figure 2. The first digit of the classification system is the part class code. The four major families represented within this code are rotational, triangular or square prismatic, rectangular, and odd shaped parts. Each family is subdivided into classes depending on its geometric aspect ratio. For example, the ratio between L (part thickness) and D (diameter of rotational parts or the edge length of prismatic parts) or the ratio between A (length), B (width), and C (thickness) of rectangular parts, will determine the class of the part.

The second digit describes part composite features. Since a part may have more than one feature, this second digit can have multiple values. A list is used to keep these values together. The third digit describes the symmetry conditions of the part. The fourth digit represents part surface conditions which may affect the operation of the grippers. Since part surface conditions could include more than one item of interest, this digit can also have multiple values. Consideration is also given to the environmental conditions in which the gripper will be operating. The fifth digit is used to code important environmental factors.

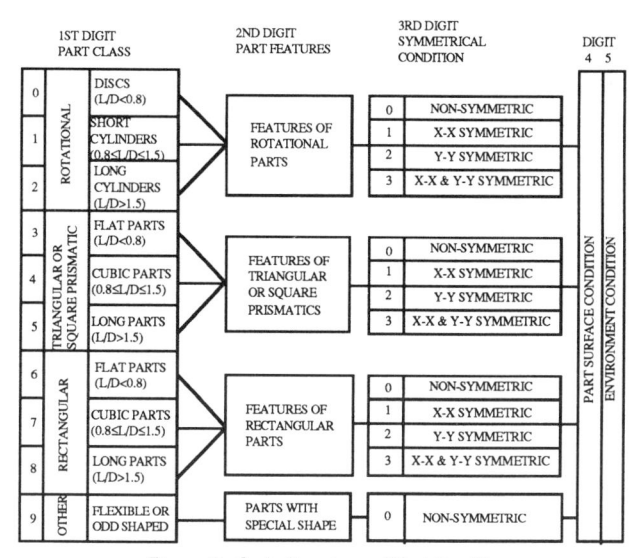

Figure 2: Code Structure of Part Families

2.2 Gripper Classification System

A proposed classification system for gripper families is illustrated in Figure 3. This classification is based on a grippers capabilities and characteristics.

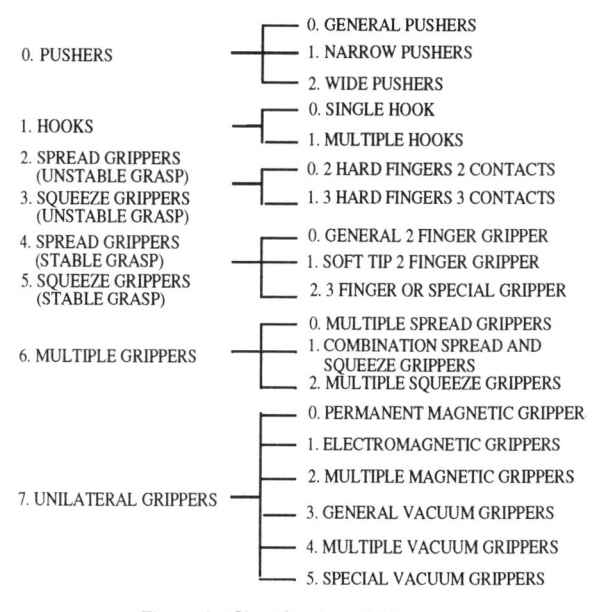

0. PUSHERS
- 0. GENERAL PUSHERS
- 1. NARROW PUSHERS
- 2. WIDE PUSHERS

1. HOOKS
- 0. SINGLE HOOK
- 1. MULTIPLE HOOKS

2. SPREAD GRIPPERS (UNSTABLE GRASP)
3. SQUEEZE GRIPPERS (UNSTABLE GRASP)
- 0. 2 HARD FINGERS 2 CONTACTS
- 1. 3 HARD FINGERS 3 CONTACTS

4. SPREAD GRIPPERS (STABLE GRASP)
5. SQUEEZE GRIPPERS (STABLE GRASP)
- 0. GENERAL 2 FINGER GRIPPER
- 1. SOFT TIP 2 FINGER GRIPPER
- 2. 3 FINGER OR SPECIAL GRIPPER

6. MULTIPLE GRIPPERS
- 0. MULTIPLE SPREAD GRIPPERS
- 1. COMBINATION SPREAD AND SQUEEZE GRIPPERS
- 2. MULTIPLE SQUEEZE GRIPPERS

7. UNILATERAL GRIPPERS
- 0. PERMANENT MAGNETIC GRIPPER
- 1. ELECTROMAGNETIC GRIPPERS
- 2. MULTIPLE MAGNETIC GRIPPERS
- 3. GENERAL VACUUM GRIPPERS
- 4. MULTIPLE VACUUM GRIPPERS
- 5. SPECIAL VACUUM GRIPPERS

Figure 3: Classification of Grippers

2.3 Domain Knowledge

Once the description of the part to be manipulated is complete, Gripper Selector begins narrowing the field of possible grippers for the part. This is done by applying a filter-like set of rules to the part in a specified order and then rejecting the grippers that are found to be improper [6]. The order in which the rules are applied has been chosen to correspond to a priority assigned to each gripper on the basis of low cost, versatility, and availability. This is done so that the grippers that rate the highest in these three areas are reported by Gripper Selector before those that do not rate as high.

3.0 A Sample Consultation Session

The part to be manipulated is a cylindrical metal roller as specified in Figure 4 below.

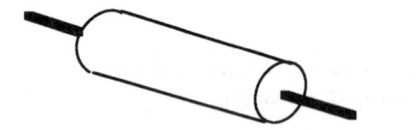

Length: 4"
Diameter: 1"
Material: Aluminum alloy
Surface: Normal

Figure 4: Specification of Cylindrical Part

We begin by entering information about this part into the program via the screen of Figure 5. The overall part shape is best described by item 2 in this screen which describes a long cylinder. The program will respond by prompting for main features of the cylinder as in Figure 6. Since this cylinder has no features of interest, select item 0 only. This part is symmetric in two directions, so at the prompt for symmetry information, select item 3. Gripper Selector will then request information pertaining to the surface conditions of the part as in Figure 7. Assuming that the surface conditions are normal, select item zero. After all of this information has been entered, Gripper Selector will respond with a recommendation for a two-fingered gripper with a squeeze grasp as shown in Figure 8. The V-shaped features inside the fingers will be used to grasp the rotational sides of the roller. After the program presents the grippers it has selected, an opportunity is given to allow for consideration of the working environment of the gripper. The screen of Figure 9 is presented to allow the user to describe a range of possibilities. The program will take this information and return points of interest that pertain when using the recommended grippers in the described environment.

Figure 5: Overall Part Shape

Figure 6: Main Features of the Cylinder

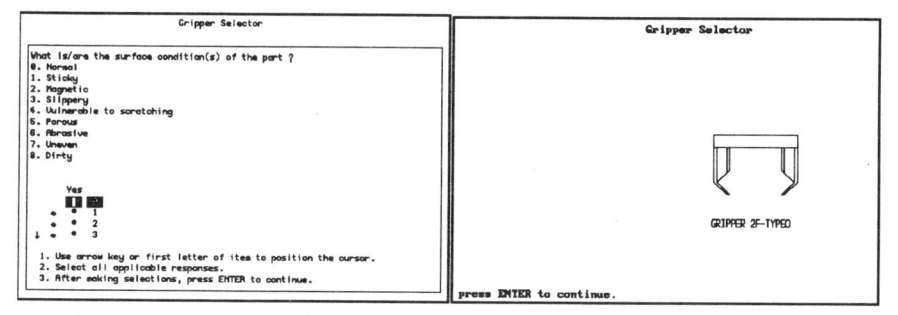

Figure 7: Possible Surface Conditions

Figure 8: Recommended Gripper

Figure 9: Possible Environmental Conditions

4.0 Conclusion

The proposed system is designed to be a relatively inexpensive means of facilitating the automation of process planning. It is designed to aid in this development by helping to select the most efficient and yet best suited machine grippers for the assembly process. Gripper Selector runs on the IBM PC line of personal computers, which are relatively inexpensive and readily available. It accepts information and provides results through a very simple and easily understood graphical interface.

This program accepts information from the user concerning the shape of the objects to be manipulated, the handling environment, and any relevant surface or other material conditions. When this information has been entered, the program searches through its list of possibilities and returns, in order of lowest cost, those grippers which will be able to perform the job requirements. Gripper Selector also analyses, from the surface and environmental conditions, potential problems which may arise and gives recommendations to help alleviate those problems.

References

1. Lundstrom, G., Glemme, B., and Rooks, B. W., "Industrial Robots-Gripper Review," International Fluidic Service Ltd., Bedford, England, 1977.

2. Chen, F. Y., "Gripping Mechanisms for Industrial Robots," J. of Mechanism and Machine Theory, Vol. 17, No 5, 1982, pp. 299-311.

3. Wilson, W. R. D. and Tan, K. T., "Use of Group Technology in Robot Gripper Selection," Proceedings of the 15th NAMRC, SME, 1987, pp. 649-655.

4. Pham, D. T., and Yeo, S. H., "A Knowledge-Based System for Robot Gripper Selection: Implementation Details," International Journal of Machine Tools and Manufacture, Vol. 28, No. 4, 1988, pp. 315-324.

5. Pham, D. T., and Yeo, S. H., "Strategies for Gripper Design and Selection in Robotic Assembly," Int. J. of Production Research, Vol. 29, No. 2, 1991, pp. 303-315.

6. Ho, Chien-Te, "Application of Expert System to Gripper Selection," A Master of Science Thesis, University of Missouri-Columbia, 1990.

An Object-Oriented Model for Manufacturing Database

P. Ji and R. S. Ahluwalia
Department of Industrial Engineering
West Virginia University
Morgantown, WV 26505

Abstract

The most important factor affecting the development and evolution of a computer integrated manufacturing system (CIMS) is a common database. An effective CIMS database is an invaluable tool that provides numerous design and manufacturing benefits. Traditional data models (hierarchic, network, and relational) have difficulty handling manufacturing data due to the complexity of manufacturing objects, data types, and relationships. This paper discusses the application of object-oriented data model to conceptual manufacturing database schema design. In a manufacturing database, workpieces, manufacturing features, process operations, machine tools, cutting tools, and fixtures can be treated as objects of an object-oriented model.

Introduction

An integrated database is essential for the development of a Computer Integrated Manufacturing System (CIMS) [2] and a Manufacturing Data Base (MDB) is one of the necessary components of an integrated database. The MDB needs to have information on part geometry features, operation processes, operating parameters, machine tools, cutting tools, jigs and fixtures. The MDB should also provide capabilities to model, store and manipulate manufacturing data in a manner suitable to the users. The traditional data models (hierarchical, network, and relational) are not widely used in manufacturing application primarily because the conceptual and internal modeling tools that they provide do not meet the requirements of manufacturing users. The object-oriented data model is a new and conceptually powerful alternatives to the traditional data models. This paper discusses the application of the object-oriented data model to the manufacturing database schema design.

Object-Oriented Model (OOM)

Object-oriented data models are derived from object-oriented programming [4]. In the traditional data models (hierarchical, network, relational), data is viewed as a collection of record types (or relations), each having a collection of records (or tuples) stored in a file. In an object-oriented approach, a database is considered as a collection of objects, or real-world objects are represented directly by database objects. Objects are described by their properties, or instance variables in object-oriented data models [3] [5].

An object-oriented data model has three essential features [1]: encapsulation, inheritance, and object identity. These specific features differentiate the object-oriented data model from traditional data models. In an object-oriented data model, classes are used to describe a set of objects having the same properties. Encapsulation, or abstract data typing, extends the notion of a class through 'hiding' the implementation of user-defined operations associated with the data type. That is, an abstract data type is a set of similar objects with an associated collection of operators. Classes may have a "tree" structure, or in the words of object-oriented data model, the subclass/superclass structure. A subclass can inherit all properties of all its superclass. Inheritance is a tool for organizing, building, and using reusable classes. Without inheritance, every class would be a free-standing unit, each developed from the ground up. In an object-oriented data model, a hidden identifier is used to refer to an object, which replaces the primary key used in relational data models to uniquely identify a record. However, object identifiers differ from identity keys in conventional data models in many ways.

Manufacturing Data Base (MDB)

A manufacturing system can be modeled as shown in Figure 1. The inputs of a manufacturing system are raw materials, equipment (machine tools), tooling and fixtures, energy, and labor. The outputs of the system are the desired products, scrap and waste. However, in most cases the equipment, tooling and fixtures are fixed at the outset. They can be regarded as the basis rather than inputs of the manufacturing system.

```
     Raw Materials  ────▶┌──────────────┐
       Equipment   ────▶│              │──▶ Product
    Tooling, Fixtures ──▶│ MANUFACTURING│
        Energy    ────▶ │    System    │──▶ Scrap and Waste
        Labor     ────▶ └──────────────┘
```

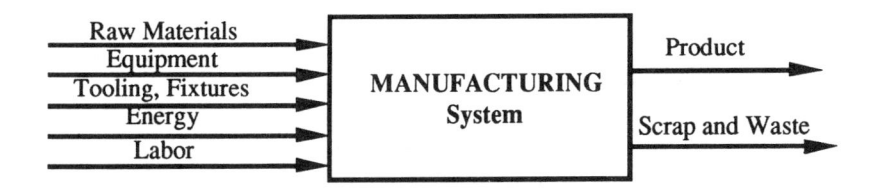

Figure 1 A Manufacturing System

Manufacturing processes are the operations that transfer the inputs into the outputs of the system. In other words, manufacturing processes add value to the raw materials by transforming them into more desirable state. A specific operation may require different operation parameters, such as feed, speed, and some auxiliaries, such as coolant, lubricant. The processes are usually carried out on manufacturing equipment that reflects a capital investment by the firm. The equipment is typically general-purpose in a traditional manufacturing system, or numerical controlled in a computer integrated manufacturing system. The equipment must be adapted to the particular product (workpiece) by the use of tools, fixtures, jigs, etc. This tooling must often be designed specifically for a given product, thus requiring dynamic schema. Energy is typically electrical in a computer integrated manufacturing system and is consumed to operate the machine tools, move the workpiece, load and unload the part. Labor is required to operate machine tools, check, maintain the machines, cutting tools, and so forth. Product refers to the desired output of the manufacturing system. Since the workpiece must be routed through several operations, only the last operation yields the finished product. The other operations produces work-in-process for succeeding processes. As a by-product of all manufacturing processes, some scrap materials and waste results. The scrap can be in the form of metal chips. The waste results from some unpredictable factors, such as inconsistent material, wrong operation of machine tool, out of tolerance, etc. Thus product (or workpiece), equipment (machine tools), tooling (cutting tools, machining parameters), fixtures, materials need to be considered as superclasses in a MDB.

Enhanced OOM for MDB

Manufacturing data has some special characteristics which are difficult to model by "common" object-oriented data models. In order to model the complex data of a manufacturing system, three aspects of a object-oriented data model need to be enhanced: dynamic schema, multiple inheritance and engineering data types. A schema describes a data structure and a database schema is typically static, i.e., not changed after it is defined. Since manufacturing systems have dynamic characteristics, it requires dynamic schema. Manufacturing systems typically have a long life and require huge initial investment. Often, new equipment is added to the existing system, such as a new Laser Beam Machining (LBM) machine. It is almost impossible to anticipate new kinds of machines and their properties during the database schema design. Physically, when a new machine is set up in the manufacturing system, it causes the production line to be changed and the problems for the information system. Since the function of the new machine are not known during the database design, it is impossible to define the schema for the new machines early. To resolve this problem, a new object types need to be added to database system, or in other words, a dynamic schema is needed. Traditionally, a database can have many instances, like a student administration database. So it is natural to define the data schema and instance when the new object comes in, not during the database system design. A best way to do this is to add this new object to the database system directly. For example, a dynamic object, LBM machine, shown in Figure 2 with a dashed line box can be added to the existing database. Dynamic objects are not limited to machine tools only, other examples of dynamic objects are specially-designed fixtures, cutting tools (like form tools in lathes) for specific products, new measurement devices, and material handling devices.

Inheritance from more than one superclass is called multiple inheritance. In the machine tools example, machine tools can be classified according to single process function, such as

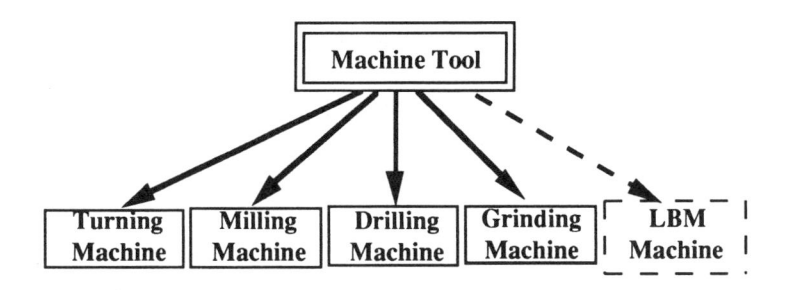

Figure 2 A Dynamic Schema

turning machine (lathe), milling machine, grinding machine, etc. However, with the development of numerical control (NC), machine tools have more than one function. An NC machining center may have several machining functions, such turning, drilling, milling and grinding. The NC machining center is the combination of these machining function, so it inherits all properties of these machine tools. However, such a structure is no longer hierarchical, or tree-like, as shown in Figure 3. As a matter of fact, it is a network, or a lattice.

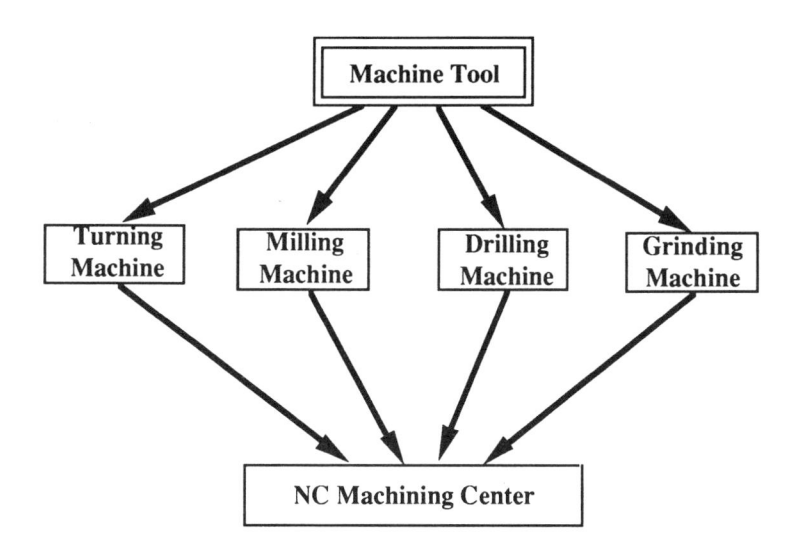

Figure 3 Structure of NC Machining Center

Database management systems have some built-in data types, such as, currency, date, time, besides the common data types, integer, floating point, string. However, manufacturing needs additional data types, such as unit, vector, matrix, range, tolerance, etc.

Conclusions

The object-oriented data model offers several advantages over the traditional data models. Several object-oriented data models have been proposed. None of the proposed models offer a capability to handle the complex data types and dynamic data requirements of a manufacturing system.

References

1. Bancilhon, F., "Object-Oriented Database systems", *Proceedings of the Symposium on Principles of Database Systems*, March, 1988.

2. Beeby, William; "The Heart of Integration: a Sound Data Base," IEEE Spectrum, May 1983, pp. 44-48.

3. Bertino, Elisa, Martino, Lorenzo, "Object-Oriented Database Management Systems: Concepts and Issues", *Computer*, Vol. 24, No. 4, pp. 33- 47, April, 1991.

4. Cox, Brad J., "Object-Oriented Programming: an Evolutionary Approach", Addison-Wesley Publishing Company, 1987.

5. Dittrich, K. R., (editor), "Advances in Object-Oriented Database Systems", *Proceedings of Second International Workshop on Object-Oriented Database Systems*, Bad Münster am Stein-Ebernburg, FRG, September 27-30, 1988.

Object-Oriented Knowledge Bases in Engineering Applications

M. SOBOLEWSKI

Concurrent Engineering Research Center
West Virginia University, Morgantown 26506
E-Mail: sobol@cerc.wvu.wvnet.edu

Summary

An object-oriented knowledge-based system has been developed, using a new knowledge representation paradigm. Its high-level architecture includes a problem-solving engine, a user interface, and tools. Its low-level architecture is described by more than 50 different kinds of objects that are implemented in the system as Smalltalk classes. A procedural attachment mechanism, bidirectional data exchange between the inference engine and procedure calls are basic mechanisms used for object-oriented integration of declarative and procedural knowledge bases.

Introduction

A number of problems encountered in engineering design are not amenable to algorithmic solutions, for example finite elements methods and circuit simulators. These problems are often ill structured (the term *ill-structured problems* is used here to denote problems that do not have an explicit, clearly defined algorithmic solution). Experienced engineers deal with them using judgement and experience. Knowledge-based programming technology offers a methodology to tackle these ill-structured design problems. We have developed such a system called DICEtalk.

The DICEtalk knowledge-based system [7, 8] has been designed for engineers. The low-level architecture of the DICEtalk system is described by special kinds of objects that are implemented as more than 50 Smalltalk classes. The DICEtalk high-level architecture includes a problem-solving engine, a high-level user interface, and high-level tools. It features object-orientation (high-level structured programming) which provides procedural knowledge-based programming integrated with simplified English as declarative knowledge-based programming. The integration of both declarative and procedural knowledge bases is supported by an extended graphical user interface with access to several knowledge bases. The Smalltalk environment is accessible from within a knowledge base and through Smalltalk; programs in other conventional languages, for example the C language, also can be executed. This integration has been verified in a series of applications [2,3,8,10].

---

*This work has been sponsored by Defense Advanced Research Projects Agency (DARPA), under contract No. MDA972-88-C-0047 for DARPA Initiative in Concurrent Engineering (DICE).

In the rest of this paper we describe the DICEtalk representation scheme, which involves several language levels. We also delve into the object-oriented problem solving aspects on DICEtalk. Finally a specific example is discussed.

DICEtalk Knowledge Representation Scheme

In DICEtalk, a knowledge description scheme is based on a surface language, an intermediate language, and a deep language [4,6]. The surface language sentences appear as simplified English sentences that allow engineers to create knowledge bases and metaknowledge bases in a natural way, without their having to learn specialized data description and manipulation languages. The expressions of intermediate language are logical formulas of the formalized percept language [6]. In this case, language primitives of surface sentences and percept formulas are the same, i.e., a subject of a sentence and its complements, which allow the conversion of natural sentences into percept formulas to be much easier and more natural. The deep language is Smalltalk-80 [1] for implementing structured objects that represent logical formulas at a computer level.

The knowledge description language *SPDL* (Surface Percept Description Language) is used to express declarative (factual) knowledge and metaknowledge bases, and goal knowledge bases, whereas the Smalltalk-80 is dedicated for representing a procedural knowledge base. SPDL is defined by 44 EBNF (Extended Backus-Naur Form) rules [9]. In order to give a general idea of what SPDL is, we list below 12 basic SPDL rules:

1   *sentence = assumption | rule | goal | initData | question.*
2   *assumption = [entry] clause.*
3   *goal = [entry] clause.*
4   *rule = [entry] "IF" clause "THEN" clause.*
7   *clause = [["not"] subject] complements.*
8   *subject = path term ":" | inputVariableConstant path ":".*
9   *path = attributeName {attributeName | variableName}.*
10   *complements = complement {("," | "and" | "or") complement}.*
11   *complement = path ("is" | "are" | "=") term [definiteness] |*
            path ["not"] decisionAttributeName [definiteness] |
            outputVariableConstant ["is" | "are" | "="] path |
            ("if" | "whether") decisionAttributeName inversePath |
            "no" attribute | "has" attribute | predicate.
14   *term = literal {("and" | "or") literal} | number | interval | point | vector |*
            date | time | multivalue | "[" SmalltalkExpression "]".
15   *literal = ["not"] (valueName | variableName).*
17   *predicate = "[" booleanExpression "]" |*
            "[" SmalltalkExpression "for" variableName "]".

These rules explain the declarative and procedural knowledge-base integration which is based on expressions included in brackets in rules 14 and 17. Generally speaking, a clause is a description of an entity in terms of subjects and complements, as in natural sentences. Subjects represent main qualities and quantities, whereas complements represent complementary qualities and quantities. Qualities are expressed by paths — sequences of attributes, and quantities by values, i.e., numbers, intervals, points, names, etc. The following example describes the boring machine HBM2:

> *boring machine HBM2:*
>   *table area is 1000,*
>   *boring surface finish is 63,*
>   *boring tolerance is 0.01*

where each quality is expressed by a sequence of two attributes, and where the main quantity is a name and complementary quantities are numbers. Logical connectives (*and, or,* and *not*) are allowed to build compound complements and values, including variables. When a subject is omitted in a clause, it means that one is understood.

SmalltalkExpression stands for any sequence of statements (expression series [1]) of the Smalltalk language, possibly with variables including the special variable "kb" which represents the current knowledge base of the DICEtalk system (a knowledge base is an instance of class **KnowledgeBase** or its subclass); *booleanExpression* denotes *SmalltalkExpression* evaluating to true or false (boolean objects). These two forms of Smalltalk expressions provide a procedural attachment mechanism. This mechanism is especially essential to any engineering design tasks, where many pieces are given in the form of formulas and calculations, most effectively carried out by appropriate procedures or external programs.

Problem-solving

The DICEtalk problem-solving model is object-oriented and based on a so-called dispatch-managing problem-solving model. We can view such a dispatch-managing model as a natural study of how a group of individual solvers can combine to solve a goal (problem). The presented approach is to split the goal into simpler tasks and to solve each of these tasks by a so-called *dispatch-managing module* (DM module). A dispatch-managing module consists of a task *dispatcher* and its *manager*. We suppose that tasks are not independent, i.e., they interrelate in some way.

The dispatch-managing model deals with problem-solving by separating a goal into a hierarchical structure of subordinate tasks solved by DM modules. A dispatch-managing architecture of the problem-solving engine consists of the following basic components:

1. The *knowledge*:
    The main repository of goals, facts, procedures, and control advices.

2. The *supervisor*:
   The master DM module deals with solving the user defined goal. It creates the top DM module and controls a DM network activity.
3. The *DM network*:
   The problem-solving tasks are organized into the hierarchical structure related to the current state of goal-solving. Each local DM module deals with local task-solving, according to its local control strategy and a local knowledge-base taken as a subtask perspective of the knowledge base. Managers of local DM modules are responsible to their dispatchers for local strategy. They decide what actions to take next for their dispatchers. Communication and interaction among parent DM modules and their child modules take place through their dispatchers.
4. The *working memory*:
   Subtasks are created by dispatchers and supplied for solving by their managers according to control advices. DM modules produce changes in the working memory that lead incrementally to a global solution as a unification of all local solutions. Parameters are user-defined characteristics evaluated by dispatchers and then used as arguments of procedure calls.
5. The *results*:
   Problem solving results (answers, findings, and conclusions) are created by the supervisor and dispatchers and can be transfered to other knowledge bases as the part of their descriptive knowledge for distributed problem solving.

These five components form a DICEtalk global knowledge base implemented by the highly structured class **GlobalKnowledgeBase**. An instance of this class can be considered as a kind of object-oriented distributed blackboard [11]. DICEtalk can contain many global knowledge bases during complex cooperative problem solving.

The DICEtalk object-oriented framework of the presented model is implemented by three types of dispatchers: atomic-dispatcher, and-dispatcher, or-dispatcher; and three types of managers: task-manger, rule-forward-manager, and rule-backward-manager. The problem-solving engine introduces an architecture that treats a declarative and a procedural knowledge base as one active object. Fundamental to this object-oriented integration is the notion of a DICEtalk knowledge base as an instance of a class created when new knowledge-based application is defined by a user. This new class, say **SubKnowledgeBase**, is a subclass of the predefined **KnowledgeBase** class. Facts and goals of a declarative knowledge base are stored in instance variables of **KnowledgeBase** class, whereas instance variables and user defined methods of **SubKnowledgeBase** class form a procedural knowledge base. Thus, this procedural knowledge base inherits declarative knowledge base from its superclass **KnowledgeBase**. Procedures defined as methods of the procedural knowledge base can be called by the inference engine when tasks to be solved contain messages sent to *kb* (*kb* refers to the current DICEtalk knowledge base).

Example of an object-oriented knowledge base

Let us consider the following rule in the Turbine Blade Fabrication Cost Advisor [10]:

IF
    cluster blade number is x1
    and pattern assembly time is x2
    and cluster dress time is x3
    and cluster inspection time is x4
    and [kb kpp for y]
THEN
    pattern wax cost is y

ast1: parameter is Ncb, ast2: parameter is Tpa, ast3: parameter is Tcd, ast4: parameter is Tci

where *Ncb, Tpa, Tcd, Tci* are user defined parameters (for antecedent subtasks denoted by *ast* with indices) for *cluster blade number is x1, pattern assembly time is x2, cluster dress time is x3*, and *cluster inspection time is x4*, respectively. When this rule is used in backward direction and the tasks *ast1, ast2, ast3*, and *ast4* are proved, then the inference engine assigns the values of variables *x1, x2, x3*, and *x4* to parameters *Ncb, Tpa, Tcd*, and *Tci*, respectively. Next, when the task *[kb kpp for y]* is executed, then *kpp* is sent to the current knowledge base. The result of that message is used as the value substituted for the variable *y*, and then the inference engine returns the task *cost wax pattern preparation is y* appropriately instantiated as a finding. The message *kpp* is implemented as the following Smalltalk method:

kpp
  | *kppCost* |
  *kppCost := 1 / Ncb value * (Rpl value + Tpa value + Tcd value + Tci value)*
    * kc1Cost + kpCost.
  tbfCost := tbfCost + kppCost.
  ^kppCost asFloat

where *tbfCost, kc1Cost* and *kpCost* are instance variables (for other costs) in the procedural knowledge base, and *Ncb, Rpl, Tpa, Tcd, Tci* are parameters defined above in the declarative knowledge base for the rule and treated as Smalltalk pool variables in the procedural knowledge base and so in the method *kpp*.

As the above example illustrates, the presented scheme of declarative and procedural knowledge representation is based on the following mechanisms:

1. Transfer of data from the inference engine to the procedural knowledge-base via user defined parameters associated with antecedent tasks in rules.
2. Procedure calls by the inference engine to execute tasks in the form *[booleanExpression]*. If the variable *kb* appears in *booleanExpression*, it means that methods of the current procedural knowledge base are executed (SPDL rule 17).

3. Transfer of data from the procedural knowledge base to the inference engine by executing tasks in the form [*SmalltalkExpression for variableName*] (SPDL rule 17).

Conclusions

This paper has presented a knowledge-based system with integration of declarative and procedural knowledge bases based on object-oriented paradigm. The use of a percept knowledge representation scheme is utilized to create the rules, and the procedural knowledge base is developed in Smalltalk-80 and other languages. A knowledge base is treated as one active object consisting of both declarative and procedural knowledge bases. During problem solving there exists bidirectional data exchange between declarative and procedural knowledge. The feasibility of using an object-oriented paradigm in developing knowledge-based systems for engineering applications has been demonstrated in this paper. In the DICE (DARPA Initiative in Concurrent Engineering) program the DICEtalk system is considered as a knowledge-based integration shell for the integration of tools used by a single designer and communicating between different designers by using the DICE generic services called from DICEtalk.

References

1. Goldberg, A., and Robson, D. 1989. *Smalltalk-80: The Language,* Addison-Wesley.
2. Padhy, S.K., and Dwivedi S.N. 1990. A Knowledge Based Approach for Manufacturability of Printed Wiring Boards. Proc. of the Fifth Int. Conference on CAD/CAM, Robotics and Factories of the Future.
3. Padhy, S.K. 1990. A Knowledge-Based System for PWB Manufacturability in Concurrent Engineering Environment, Master Thesis, College of Engineering WVU.
4. Sobolewski, M. 1987. Percept Knowledge-base Systems. In I. Plander (Ed.), *Artificial Intelligence and Information - Control Systems of Robots.* North-Holland.
5. Sobolewski, M. 1989. EXPERTALK: An Object-Oriented Knowledge-based System. In I. Plander (Ed.), Artificial Intelligence and Information-Control Systems of Robots. North-Holland.
6. Sobolewski, M. 1989. Percept Knowledge Description and Representation, *ICS PAS Reports* No. 663. Institute of Computer Science of Polish Academy of Sciences.
7. Sobolewski, M. 1990. Percept Knowledge and Concurrency. *Proc. The Second National Symposium on Concurrent Engineering,* February 1990, Morgantown, WV.
8. Sobolewski, M. 1990. DICEtalk: An Object-Oriented Knowledge-Based Engineering Environment. Proc. of the Fifth Int. Conference on CAD/CAM, Robotics and Factories of the Future, Norfolk, VA.
9. Sobolewski, M. 1991. DICEtalk Knowledge-Based System, Tutorial and Manual. Concurrent Engineering Center, WVU, Morgantown, WV.
10. Saidi, M. 1991. Turbine Blade Investment Casting Cost Advisor Model. Proc. of AACE's 1991 Annual Meeting, Seattle, WA.
11. Jagannathan, V., Dodhiawala, R., and Baum, L.S. (Eds.) 1989. Blackboard Architectures and Applications. Academic Press, Inc., Boston.